Birkhäuser

Birkhäuser Advanced Texts

For further volumes:
http://www.springer.com/series/4842

Jaume Llibre • Antonio E. Teruel

Introduction to the Qualitative Theory of Differential Systems

Planar, Symmetric and Continuous Piecewise Linear Systems

Birkhäuser

Jaume Llibre
Departament de Matemàtiques
Universitat Autònoma de Barcelona
Bellaterra, Barcelona
Spain

Antonio E. Teruel
Departament de Matemàtiques i Informàtica
Universitat de les Illes Balears
Palma-Illes Balears
Spain

ISSN 1019-6242 ISSN 2296-4894 (electronic)
ISBN 978-3-0348-0656-5 ISBN 978-3-0348-0657-2 (eBook)
DOI 10.1007/978-3-0348-0657-2
Springer Basel Heidelberg New York Dordrecht London

Mathematics Subject Classification (2010): 35R35, 35K40, 35Q30, 35Q79, 76T10

Printed on acid-free paper

Springer Basel is part of Springer Science+Business Media (www.birkhauser-science.com)

To Alba, Montse and Sara.

Contents

Preface

Ordinary differential equations (ODEs) are the preferred language for the investigation and understanding of various natural phenomena. Employed extensively in natural sciences, engineering, and technology, ODEs are nowadays integrated in any standard undergraduate science curriculum, while continuing to be the subject of intensive research.

Although ODEs model a large number of natural phenomena, it is well known that not many admit explicit solution. For this reason, the qualitative theory and associated methods are often employed as an alternative investigative tool. When successful, the qualitative approach leads to a broader picture of important open subsets of solutions (sometimes the entire set), providing information about the ODEs' flow, parametric stability and bifurcations.

However, few families of ODEs allow a full treatment from the qualitative theory standpoint. The family of systems of linear differential equations is one of them. In the context of the qualitative theory, the importance of this family is evident when much of the local analysis of nonlinear ODEs is reduced to the study of their linear part. Nevertheless, this family exhibits limited richness from a dynamical systems standpoint.

In this book we consider planar systems of piecewise linear differential equations (PWLS), to which we apply the full program of the qualitative theory. PWLS may be considered as some of the most tractable nonlinear ODEs and they display a rich and interesting dynamical behaviour, comparable to that of general nonlinear ODEs.

Beyond the academic-theoretical significance, the study of PWLS has practical relevance. The interest in these class of systems is driven by concrete applications in engineering, in particular in control theory and the design of electric circuits.

This book is addressed to mathematicians, engineers, and scientists in general, who are interested in the qualitative theory of ODEs, PWLS in particular. It is also a reference book for anyone interested in the global phase portraits and the bifurcation sets of all the symmetric three-piece linear differential systems (here called *fundamental systems*), since their full characterization is presented here for the first time.

The book is divided into five chapters. Chapter 1 introduces fundamental systems, describes their global phase portraits (including behaviour at infinity) and the bifurcations occurring when parameters vary. To emphasize the importance of fundamental systems in applications, we discuss two well-known examples: the motor position control and the Wien bridge circuit. For the later and for specific values of the parameters, we describe the evolution of the phase portrait.

In Chapter 2 we collect the basic results of the qualitative theory of planar ODEs which are used in the rest of the book. To simplify the exposition of some concepts we have confined ourselves to ODEs having a complete flow. For this reason some of the results presented here are more restrictive than those that normally appear in the literature. In Section 2.5 we treat planar linear differential systems. We refer frequently to this section throughout the book. In Section 2.9 we formalize some aspects of the compactification of flows in order to apply this technique to the fundamental systems. As known, the Poincaré compactification is widely used in polynomial differential systems to study the behaviour of the flow near the infinity. However, although some differential equations can be compactified satisfactorily, we have not found a systematization of its use outside the class of polynomial differential systems.

Chapter 3 begins with the study of the fundamental systems. We show that within this class the existence and uniqueness theorem and the theorem on continuous dependence on initial conditions and parameters are valid. We further prove that the behaviour of these systems is determined by a pair of matrices, called fundamental matrices. This justifies that, except in very singular cases, we use the trace and the determinant of the two matrices as fundamental parameters to describe the dynamics of these systems. Additionally, we study the local phase portrait at the singular points, both finite and infinite, and we give some results about the existence and configuration of periodic orbits.

Poincaré maps of PWLS are determined by the linear differential systems which act in each of the pieces. For fundamental systems, one of these linear differential systems is homogeneous, while the other two are non-homogeneous. Consequently, in Chapter 4 we study all the Poincaré maps of linear differential systems associated to two cross sections. These cross sections are parameterized in such a way that the Poincaré maps become invariant under linear transformations. We note that the parametrization introduced here has important implications. First, it allows the study of the Poincaré maps by choosing, in each case, the simplest expression for the fundamental matrices. Usually we will assume that the matrices are expressed in their real Jordan normal form. Second, we can characterize the region in the parameter space where we can guarantee the existence of the Poincaré maps. Thus the bifurcation set associated to the non-existence of the Poincaré maps in the parameter space is an algebraic manifold homeomorphic to the Whitney umbrella. Finally, this parametrization establishes a link between Poincaré maps of PWLS and the class of differential systems which are called observable in control theory.

By collecting the results obtained in the previous chapters, in Chapter 5 we are able to describe and classify all the phase portraits of fundamental systems. The description of the phase portraits is carried out via the characterization of all separatrices and canonical regions. This allows us to use in a rigorous way the Marcus–Newmann–Peixoto Theorem on the topological classification of planar flows and to describe explicitly the bifurcation manifolds. Each of the sections of the chapter is devoted to fundamental systems having fixed the sign of two fundamental parameters. All sections of this chapter are structured similarly. First, we collect the results about singular points (both finite and infinite) and limit cycles. Second, we locate the rest of the separatrices of the system and we describe the behaviour of the canonical regions. Finally, we organize all the information in propositions which describe and classify fundamental systems when we vary the two parameters. At the end of each section we describe the bifurcations set and provide a picture of the parameter space representing the bifurcation manifolds and the corresponding phase portraits.

Readers interested only in such results can read the introductory Chapter 1 and then skip directly to Chapter 5, where they may find at the end of each section a complete list of phase portraits and their bifurcations.

The book has been organized in such a way so that the full classification of the global dynamics of the fundamental systems is obtained by using the qualitative theory of ODEs. Since there are many cases that must be considered, some propositions are very similar to each other and following all of them at the first reading becomes a little tedious. It may be recommended that at first reading only some of the proofs presented in Sections 3.11, 4.4 and 4.5 be followed in detail, so that the main arguments are understood. For instance, in Chapter 5, it may be useful to focus on one class of fundamental systems given by fixing the sign of the two fundamental parameters, and then follow the rest of the results in more detailed subsequent readings.

We thank Christina Stoica for her careful reading of the text of this book and her improvements to our poor English.

Jaume Llibre
Antonio E. Teruel
Barcelona, 2013.

Chapter 1

Introduction and statement of the main results

Nowadays most scientific research is written in the language of ordinary differential equations (ODEs). Since the times these equations appeared first in the works of G.W. Leibnitz (1646–1716) and I. Newton (1642–1727), more and more fields of knowledge found and continue to find in them an accurate language to determine and to develop their knowledge. Astronomy, and in particular Celestial Mechanics, Physics and Chemistry found in differential equations the most natural way of expressing their laws. Engineering, Economics, Ecology, Epidemiology, Neuroscience, etc., use this language in order to model natural phenomena and to simulate their behaviour in theoretical and numerical experiments that hardly could be carried out in a laboratory. As a result, the study of ODEs became one of the areas of mathematics with a very large number of applications.

The determination of explicit expressions of the solutions of ODEs has been the objective of the first mathematicians who studied ODEs, even though it soon became clear that not all equations admit solutions that can be expressed terms of elementary functions, see J. Liouville's work (1809–1882). In fact, in spite of the multiple attempts to progress along this line, the number of differential equations that can be solved explicitly is insignificantly small compared with the totality of equations. Moreover, even when it is possible to find an expression for the solution, this could be so complicated that its analysis would encounter significant difficulties.

At the end of 19th century, H. Poincaré (1854–1912) [in his "*Mémoire sur les courbes définies par une équation différentielle* (1881–1886)] inaugurates a new direction in the study and understanding of ODEs. Thanks to Poincaré's perspective, solutions started to be considered geometric elements (orbits). This new point of view did lead to the qualitative theory of differential equations. Research of A. Lyapunov (1857–1918) about the stability of the motion, of I.O. Bendixson

(1861–1935) and G.D. Birkhoff (1884–1944), among others, joined the direction set forward by Poincaré's ideas.

The new approach tries to understand the dynamics of a system modeled by a family of (ordinary) differential equations, $\dot{\mathbf{x}} = \mathbf{f}(t, \mathbf{x}; \lambda)$ without needing to find an explicit expression of their solutions. From the point of view of the qualitative theory of differential equations, this understanding involves:

(1) the description of the *phase portrait* of every differential equation in the family;

(2) the introduction of an equivalence relation between the different phase portraits and their classification according to this relation;

(3) the description of the changes (bifurcations) in the phase portrait which occur when the equations change from one class of equivalence to another.

The phase portrait of a differential equation describes the domain where the differential equation is defined (*phase space*) as the union of all its orbits. Since orbits are manifolds of dimension less than or equal to 1, these could be: points, and in this case we call them *singular points*; curves homeomorphic to the circle $\mathbb{S}^1$ (*periodic orbits*); or curves homeomorphic to the straight line $\mathbb{R}$. Usually only a finite number of orbits determine the phase portrait. The set S formed by these special orbits is closed and $\mathbb{R}^2 \setminus S$ is formed by open connected components, each of them called *canonical regions.* The union of the separatrices and an orbit of each canonical region is called the *separatrix configuration.* A graphical representation homeomorphic to the separatrix configuration is called a description of the phase portrait.

The qualitative theory provides results and tools for the local analysis of phase portraits. For instance, the Hartman–Grobman Theorem [30] (1963) describes, under general hypotheses, the behaviour of orbits in a neigbourhood of singular points. Nevertheless, the results in the description of the global phase portraits are mainly significative when we work with equations in dimension 1 or 2. A specific example is the Poincaré–Bendixson Theorem, which guarantees, under compactness assumptions, that the *limit sets* of the orbits are: singular points, periodic orbits or *separatrix cycles.*

In fact, we do not have complete knowledge of global phase portraits of differential equations, not even in the plane. Important questions, such as the number of *limit cycles* (isolated periodic orbits inside the set of all periodic orbits) and their distribution in the plane, are still to be answered beyond the field of *linear differential equations* [61]. This question, focused on planar polynomial equations, is known as the *second part of 16th Hilbert's problem*, which was formulated by D. Hilbert (1862–1943) in 1900.

Two differential equations can be equivalent from the point of view of qualitative theory, even if they are different in some other aspects. The most used equivalence relation, which preserves the topological structure of the phase portrait, is the so-called *topological equivalence.* Two systems are said to be *topolog-*

ically equivalent if there exists a homeomorphism between their respective phase portraits, transforming the orbits of one system into the orbits of the other and preserving their orientation. For some authors, as for instance M.M. Peixoto [51], topological equivalence is not required to preserve the orbit orientation.

Generalizing L. Markus' works [48] and referring to vector fields on 2-dimensional manifolds with isolated singular points, D.A. Neumann [50] established that the separatrix configuration determines the topological class of equivalence of phase portraits. Another characterization of the topological equivalence classes for differential equations on 2-dimensional manifolds is due to M.M. Peixoto [51].

Since 1937, when the physicist A.A. Andronov (1901–1952) and the mathematician L.S. Portryagin (1908–1988) introduced the concept of *structural stability*, the analysis of changes in separatrix configurations acquired great importance in the qualitative theory of differential equations. Without going more deeply into the subject, a differential equation is said to be structurally stable if its separatrix configuration is equivalent to the separatrix configuration of any vector field "close" to it. A characterization of structurally stable 2-dimensional vector fields was obtained by Peixoto [52].

On the other hand, separatrix configurations of phase portraits can change when the parameters change. These changes are called *bifurcations* and the value of the parameter where they take place are called *bifurcation values*. Both the graphical representation of bifurcation values and the description of the changes of the separatrix configurations are called *bifurcation set*.

To sum up, we can assert that a phase portrait grasps the essence of the dynamical behaviour of a differential equation. In a similar way, a bifurcation set grasps the essence of the dynamical behaviour of a family of differential equations.

In this book we apply the whole program of the qualitative theory of differential equations to the symmetric (with respect to the origin) family of three-piece piecewise linear differential systems in the plane. The richness of the dynamic behaviours observed in this family is, in general, comparable to that of general nonlinear differential systems in the plane.

1.1 Piecewise linear differential systems

After presenting the book's purpose, in this section we introduce the family of systems under study, that is the family of piecewise linear differential systems, elsewhere called piecewise affine systems. In particular we deal with planar continuous and symmetric ones. We also consider two examples of these systems that show their relevance in applications.

A differential system defined on an open region $S \subseteq \mathbb{R}^n$ is said to be a *piecewise linear differential system* (PWLS) on S if there exists a set of 3-tuples $\{(A_i, \mathbf{b}_i, S_i)\}_{i\in I}$ such that: A_i is a $n \times n$ real matrix; $\mathbf{b}_i \in \mathbb{R}^n$; $S_i \subseteq S$ is an open set in $\mathbb{R}^n$ satisfying that $S_i \cap S_j = \emptyset$ if $i \neq j$ and $\bigcup_{i\in I} \mathrm{Cl}(S_i) = S$; and $A_i\mathbf{x} + \mathbf{b}_i$ is the vector field defined by the system when $\mathbf{x} \in S_i$. As usual $\mathrm{Cl}(S_i)$ denotes the closure of S_i. Thus the vector field defined by a PWLS is a linear map on each of the disjoint regions S_i, but is not globally linear on the whole S.

Example 1. From a given planar differential system $\dot{\mathbf{x}} = \mathbf{f}(\mathbf{x})$ with a differentiable vector field $\mathbf{f}$ one can construct a set of different PWLS. For instance, let us suppose that $\mathbf{p}_1$ and $\mathbf{p}_2$ are two zeros of $\mathbf{f}$, and let $\mathbf{k}$ be a vector in $\mathbb{R}^2$ such that $\mathbf{k}^T\mathbf{p}_1 < 0$ and $\mathbf{k}^T\mathbf{p}_2 > 0$. The straight line $\Gamma = \{\mathbf{x} \in \mathbb{R}^2 : \mathbf{k}^T\mathbf{x} = 0\}$ divides $\mathbb{R}^2$ into the two open regions $S_1 = \{\mathbf{x} \in \mathbb{R}^2 : \mathbf{k}^T\mathbf{x} < 0\}$ and $S_2 = \{\mathbf{x} \in \mathbb{R}^2 : \mathbf{k}^T\mathbf{x} > 0\}$. Denoting by $D\mathbf{f}(\mathbf{p}_i)$ the Jacobian matrix of the vector field $\mathbf{f}$ at the point $\mathbf{p}_i$, it follows that $\{(D\mathbf{f}(\mathbf{p}_i), -D\mathbf{f}(\mathbf{p}_i)\mathbf{p}_i, S_i)\}_{i=1,2}$ is a piecewise differential system on the whole $\mathbb{R}^2$.

In this example it can be observed that the piecewise linear vector field coincides on each region S_i with the Taylor expansion up to order one of the map $\mathbf{f}$ around the point $\mathbf{p}_i$. In this sense the PWLS is a kind of global linearization of the differential system $\dot{\mathbf{x}} = \mathbf{f}(\mathbf{x})$. Just as linear differential systems arise by local linearization of differential systems, PWLS can be thought of as a global linearization of differential systems. Unfortunately, there are no results about the relationship between the dynamics of the two systems in the global case (as they are available in the local case, for instance the Hartman–Grobman Theorem). However, the intuition says that important features of the global dynamical behaviour will persist when we change from the differential system to the piecewise linear one [10, 11, 12, 43, 54, 56].

We note that the definition of a PWLS does not contain information about the behavior of the flow at the boundaries ∂S_i of the regions S_i. PWLS can be classified depending on how we can extend the vector field to ∂S_i. Let $\Gamma_{ij} = \partial S_i \cap \partial S_j$ be the common boundary of the regions S_i and S_j. If $A_i\mathbf{p}+\mathbf{b}_i = A_j\mathbf{p}+\mathbf{b}_j$ for every $\mathbf{p} \in \Gamma_{ij}$, then the PWLS is said to be continuous, otherwise the PWLS is said to be discontinuous.

Discontinuous systems (not necessarily piecewise linear ones) are very important, see the recent excellent book by Di Bernardo et al. [19] and references therein. The use of discontinuous models for mechanical systems in which impacts occur, or for electronic systems employing electronic switches, allows to faithfully represent the real dynamics of these types of systems.

From now on we restrict ourselves to continuous PWLS. In this case we have that $A_i\mathbf{p} + \mathbf{b}_i = A_j\mathbf{p} + \mathbf{b}_j$ for every $\mathbf{p} \in \Gamma_{ij}$. Hence, the boundary Γ_{ij} is contained in the linear manifold defined by the solutions of the linear system $(A_i - A_j)\mathbf{x} = \mathbf{b}_j - \mathbf{b}_i$. Therefore, the boundary of the region S_i is formed by pieces of hyperplanes in $\mathbb{R}^n$.

Planar vector fields defined by continuous PWLS are globally Lipschitz, but are not differentiable at the boundaries Γ_{ij}. A great part of the qualitative theory of differential equations, for instance bifurcation theory, is developed under the assumption of differentiability of the vector field. This explains why the results obtained in that framework cannot be directly applied to the study of PWLS. Nevertheless, the piecewise linear behaviour of these systems allows, in some cases, to carry out completely the program of the qualitative theory, from the description of phase portraits to the study of the bifurcation set. See, for instance, the pioneering work of Andronov [3]. A complete description of the phase portraits and bifurcation sets of two-piece piecewise linear systems can be found in [26]. Some aspects of the phase portraits of non-symmetric three-piece piecewise linear systems appear in [55]. Contributions on the phase portraits of fundamental systems in dimension three can be found in [41].

In this book we deal with the special family of planar and continuous PWLS given by $\{(A, \mathbf{b}, S_+), (B, \mathbf{0}, S_0), (A, \mathbf{b}, S_-)\}$, where A is a 2×2 real matrix, $\mathbf{b} \in \mathbb{R}^2 \setminus \{\mathbf{0}\}$, $B = A + \mathbf{b}\mathbf{k}^T$, the regions S_+ and S_- are the half-planes $\{\mathbf{x} \in \mathbb{R}^2 : \mathbf{k}^T\mathbf{x} > 1\}$ and $\{\mathbf{x} \in \mathbb{R}^2 : \mathbf{k}^T\mathbf{x} < -1\}$, respectively, and the region S_0 is the central strip $\{\mathbf{x} \in \mathbb{R}^2 : |\mathbf{k}^T\mathbf{x}| < 1\}$. The boundary of S_0 is formed by two symmetric straight-lines $\Gamma_+ := \{\mathbf{x} \in \mathbb{R}^2 : \mathbf{k}^T\mathbf{x} = 1\}$ and $\Gamma_- := \{\mathbf{x} \in \mathbb{R}^2 : \mathbf{k}^T\mathbf{x} = -1\}$. Following J. Llibre and J. Sotomayor [44], we call these systems *fundamental systems.*

Fundamental systems can also be written in the *piecewise linear form*

$$\dot{\mathbf{x}} = \begin{cases} A\mathbf{x} + \mathbf{b}, & \text{if } \mathbf{k}^T\mathbf{x} > 1, \\ B\mathbf{x}, & \text{if } |\mathbf{k}^T\mathbf{x}| \leq 1, \\ A\mathbf{x} - \mathbf{b}, & \text{if } \mathbf{k}^T\mathbf{x} < -1, \end{cases} \tag{1.1}$$

or in the *Lur'e form*

$$\dot{\mathbf{x}} = A\mathbf{x} + \varphi(\mathbf{k}^T\mathbf{x})\mathbf{b}, \tag{1.2}$$

where the function $\varphi : \mathbb{R} \to \mathbb{R}$ is the odd three-piece linear function

$$\varphi(\sigma) = \begin{cases} -1, & \text{if } \sigma < -1, \\ \sigma, & \text{if } |\sigma| \leq 1, \\ 1, & \text{if } \sigma > 1. \end{cases}$$

Restricted to each of the half-planes S_+ and S_-, the fundamental system (1.1) is a non-homogeneous linear system. Then the dynamical behavior of the fundamental system in these regions is determined by the trace t and the determinant d of the matrix A. On the other hand, when restricted to the central strip S_0, the fundamental system (1.1) is a homogeneous linear system. Therefore, the dynamical behaviour of the fundamental system in S_0 is determined by the trace T and the determinant D of the matrix B.

The values of D, T, d and t will be called the *fundamental parameters* of the family, and we will describe all the bifurcations of the fundamental family in dependence on them.

We emphasize that fundamental systems are the canonical representatives of a wide class of PWLS. So for given $m_0, m_1, u \in \mathbb{R}$ with $m_0 \neq m_1$ and $u > 0$, the change of variables $\widetilde{\mathbf{k}} = u\mathbf{k}$, $\widetilde{\mathbf{b}} = u^{-1}(m_0 - m_1)^{-1}\mathbf{b}$ and $\widetilde{A} = A - m_1\widetilde{\mathbf{b}}\widetilde{\mathbf{k}}^T$ transforms the fundamental system (1.2) into the PWLS

$$\dot{\mathbf{x}} = \widetilde{A}\mathbf{x} + \widetilde{\varphi}(\widetilde{\mathbf{k}}^T\mathbf{x})\widetilde{\mathbf{b}}, \tag{1.3}$$

where

$$\widetilde{\varphi}(\sigma) = \begin{cases} m_1\sigma - (m_0 - m_1)\,u, & \text{if} \quad \sigma < -u, \\ m_0\sigma, & \text{if} \quad -u \le \sigma \le u, \\ m_1\sigma + (m_0 - m_1)\,u, & \text{if} \quad u < \sigma. \end{cases}$$

Therefore, the two systems have the same orbits and thus the same dynamic behaviour.

1.1.1 Examples

Many publications on piecewise linear differential systems come from applications, as for instance control theory and electric circuits design. The list of published papers devoted to these systems gives an idea of their increasing importance. We refer the reader to the following works and the references therein: M. Komuro [37], L.O. Chua and A.C. Deng [15], L.O. Chua and R. Lum [47] and [46], Chai Wah and L.O. Chua [59], and J. Álvarez, R. Suárez and J. Álvarez [1].

Non-linearities that appear in real dynamical systems are very often modeled by smooth functions. Hence, results and tools from smooth dynamics and local bifurcation theory can be fruitfully applied. But, in some cases, considering piecewise linear functions is an alternative that fits better, qualitatively and quantitatively, the experiments [3], [24]. Standard piecewise linear functions are: saturation, to model amplifiers and motors, see Figure 1.1(a); dead zone, to model valves and motors, see Figure 1.1(b); friction, to model the static friction of motors, see Figure 1.1(c); and sign, to model relays, see Figure 1.1(d).

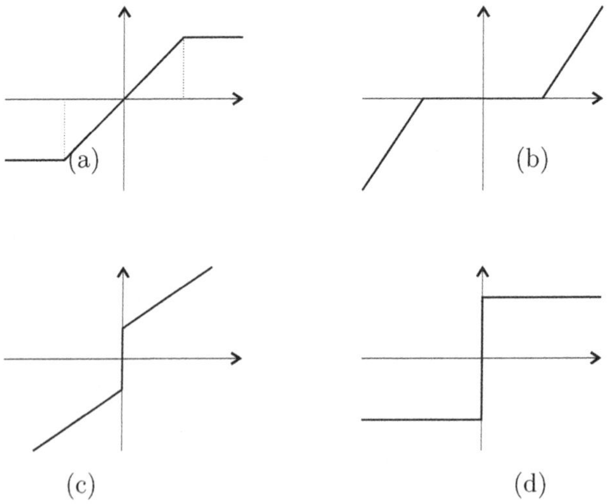

Figure 1.1: Piecewise linear functions: (a) saturation; (b) dead zone; (c) friction; (d) sign.

In the following examples, we show usual applications where PWLS arise in a natural way. In our opinion these applications justify the interest in these systems.

Motor position control

A classical problem in control theory is the motor position control. This problem consists in designing a device fed by a motor, which is able to place the motor very precisely in a position θ_i called the reference position. This problem appears very often in industrial automatization and in robotics control when we try to control the position of a mechanical arm. An introduction to control theory can be found in the books of S. Lefschetz [39], D.P. Atherton [9], and K.S. Narendra and J.M. Taylor [49].

A direct current motor position control device is sketched in Figure 1.2. The system is formed by four elements: an operational amplifier with characteristic function $f_a(v)$; a DC motor with characteristic function $T_M(v)$ and with a tachometer included; a set of gears with a velocity relation from n to 1; and a proportional control device with constant K. This design corresponds to a full-state feedback control design.

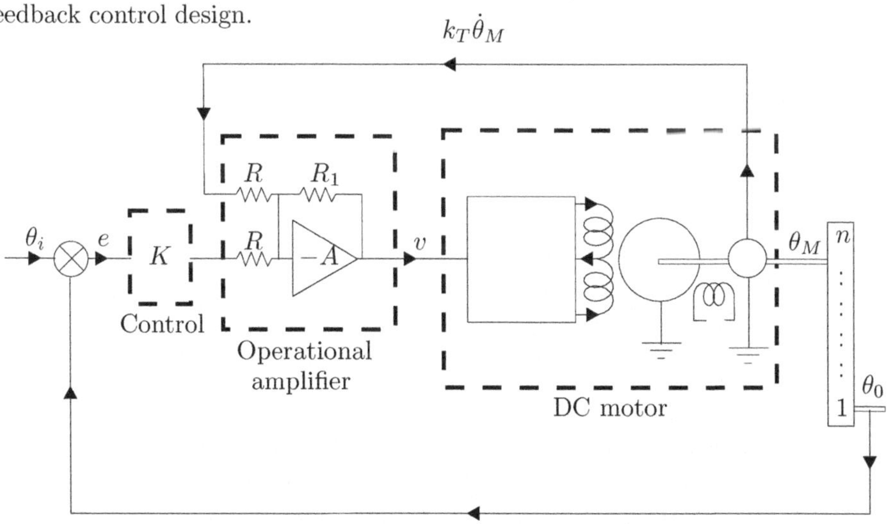

Figure 1.2: Sketch of a DC motor position control device.

Let θ_o, I_o and $F_o(\dot{\theta}_o)$ denote the position, the inertia momentum and the friction force of the outer axis, and let θ_M, I_M and $F_M(\dot{\theta}_M)$ denote the position, the inertia momentum and the friction force of the motor axis. The equation of motion of the device is

$$\left(n^2 I_M + I_o\right)\ddot{\theta}_o = nT_M\left(f_a\left(K\left(\theta_i - \theta_o\right) - K\,n\dot{\theta}_o\right)\right) - n^2 F_M \dot{\theta}_o - F_o\dot{\theta}_o, \quad (1.4)$$

see [9]. In this equation we assume that friction forces are proportional to angular velocities.

THE LINEAR CASE: FIRST APPROACH. Some electronic components are designed in such a way that their outputs are proportional to the inputs. Consequently, we can suppose that the characteristic functions of both the amplifier and the motor are linear, that is $f_a(v) = Gv$ and $T_M(v) = K_T v$. Considering $\theta_i = 0$, equation (1.4) can be written as the linear homogeneous differential equation

$$\left(n^2 I_M + I_0\right) \ddot{\theta}_0 + \left[n^2 \left(K_T G K + F_M\right) + F_o\right] \dot{\theta}_o + \left(n K_T G K\right) \theta_o = 0.$$

COMPONENTS SATURATION. It is not a restriction to assume that the characteristic function of the amplifier is linear. However, this behaviour cannot be kept for every input voltage. An operational amplifier has a finite output range which cannot be exceeded, even for high input voltages. Therefore, a more realistic model has to consider a characteristic function for the amplifier of saturation type

$$f_a(v) = \begin{cases} G\,\delta_a, & \text{if } v > \delta_a, \\ G\,v, & \text{if } |v| \leq \delta_a, \\ -G\,\delta_a, & \text{if } v < -\delta_a, \end{cases}$$

see Figure 1.1(a). Hence, by considering

$$A = \begin{pmatrix} 0 & 1 \\ 0 & -\left(n^2 F_M + F_0\right) / \left(n^2 I_M + I_0\right) \end{pmatrix},$$

$x_1 = \theta_o$, $x_2 = \dot{\theta}_o$, $\mathbf{k} = (K, nK)^T$, $\mathbf{b} = \left(0, -nK_T/(n^2 I_M + I_0)\right)^T$ and $\theta_i = 0$, equation (1.4) can be written as $\dot{\mathbf{x}} = A\mathbf{x} + f_a\left(\mathbf{k}^T\mathbf{x}\right)\mathbf{b}$, which is a fundamental system, see (1.3).

In the same way that the operational amplifier does, the DC motor works in a range of set voltages. If we exceed this range, the output of the motor will remain constant. Consequently, a more realistic model will take it into account a piecewise linear characteristic function of saturation type for the motor T_M. Taking $\mathbf{k}^T = (-GK, -GKn)^T$, $\mathbf{b} = \left(0, n/\left(n^2 I_M + I_o\right)\right)^T$ and the values for x_1, x_2, θ_i and A, as in the previous case, equation (1.4) can be written as $\dot{\mathbf{x}} = A\mathbf{x} + T_M\left(\mathbf{k}^T\mathbf{x}\right)\mathbf{b}$, which is a fundamental system.

COULOMB FRICTION IN THE MOTOR. We can suppose that the saturation problems in the operational amplifier and motor can be avoided by choosing components whose features are higher than those which are usually required. Nevertheless, the motor needs a minimum tension to overcome inner frictions and to start turning. In order to take this into account, we will consider a piecewise linear characteristic function for the motor of dead zone type, see Figure 1.1(b). Setting

$$T_M(v) = \begin{cases} K_T\left(v - \delta_M\right), & \text{if } v > \delta_M, \\ 0, & \text{if } |v| \leq \delta_M, \\ K_T\left(v + \delta_M\right), & \text{if } v < -\delta_M, \end{cases}$$

and A, $\mathbf{k}$, $\mathbf{b}$, $\mathbf{x}$ and θ_i as in the previous case, expression (1.4) transforms into expression $\dot{\mathbf{x}} = A\mathbf{x} + T_M\left(\mathbf{k}^T\mathbf{x}\right)\mathbf{b}$, which is a fundamental system.

Wien bridge

In electronic circuits design also arises a large family of examples modeled by fundamental systems [3], [16]. In the following example we introduce a well-known circuit, the Wien bridge oscillator formed by two resistors, two capacitances and one operational amplifier (op-amp) with negative feedback see Figure 1.3.

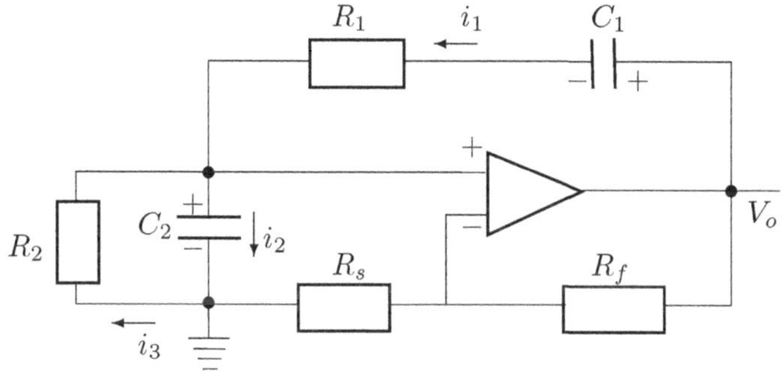

Figure 1.3: Wien bridge circuit.

The circuit is formed by two loops. The first one contains the resistor R_1 and the capacitors C_1 and C_2. The second loop is formed by the resistor R_2 and the capacitor C_2. For the sake of simplicity, we consider that the circuit is clockwise oriented in the first loop and anticlockwise oriented in the second one.

Kirchhoff's laws can be used to describe the evolution of the voltages V_{C_1} and V_{C_2} across the capacitors C_1 and C_2, respectively, leading to the differential equations

$$\begin{cases} R_1C_1\dot{V}_{C_1} = -V_{C_1} - V_{C_2} - V_0, \\ R_1C_2\dot{V}_{C_2} = -V_{C_1} - \left(1 + \dfrac{R_1}{R_2}\right)V_{C_2} - V_0, \end{cases} \tag{1.5}$$

where V_0 is the output voltage of the op-amp.

The characteristic function of an op-amp depends only on the difference between the voltage at the non-inverting terminal and the voltage at the inverting terminal (V_{C_2} and 0, respectively, in the Wien bridge). In an ideal framework, this function is considered to be linear and the slope of the function is called the open-loop gain of the amplifier. In practice, the op-amp has a limited response range $(-E, E)$, beyond which the amplifier is saturated. Taking this into account, a more realistic characteristic function for the op-amp is given by

$$V_0 = \begin{cases} E \operatorname{sign}(-\alpha V_{C_2} + E), & \text{if } |\alpha V_{C_2}| > E, \\ -\alpha V_{C_2}, & \text{if } |\alpha V_{C_2}| \leq E, \end{cases} \tag{1.6}$$

where $\alpha = 1 + R_F/R_S$ is the gain of the op-amp.

Using the above expression of V_0 and making the change of variables

$$x_1 = \frac{\alpha V_{C_2}}{E}, \quad x_2 = \frac{\alpha V_{C_1}}{E},$$

the system of differential equations (1.5) can be rewritten as the fundamental system

$$\dot{\mathbf{x}} = \begin{cases} A\mathbf{x} + \mathbf{b}, & \text{if } x_1 > 1, \\ B\mathbf{x}, & \text{if } |x_1| \leq 1, \\ A\mathbf{x} - \mathbf{b}, & \text{if } x_1 < 1, \end{cases} \tag{1.7}$$

where

$$\mathbf{b} = \begin{pmatrix} \dfrac{\alpha}{R_1 C_2} \\ \\ \dfrac{\alpha}{R_1 C_1} \end{pmatrix}, \quad A = \begin{pmatrix} -\left(\dfrac{1}{R_1 C_2} + \dfrac{1}{R_2 C_2}\right) & -\dfrac{1}{R_1 C_2} \\ \\ -\dfrac{1}{R_1 C_1} & -\dfrac{1}{R_1 C_1} \end{pmatrix},$$

and $B = A + \mathbf{b}^T \mathbf{e}_1$.

1.2 Main results

What follows is the presentation of the main new results obtained in this book on the classification of the fundamental systems. The readers who are not familiar with the qualitative theory of differential equations are referred to the next chapters, where they will find the definitions of the notions which appear here.

Since Andronov and his colleagues began the study of the piecewise linear differential equations in [3], and in particular the study of fundamental systems, part of their phase portraits have been described by different authors. Andronov also established the existence of limit cycles in the family of fundamental systems and used the Poincaré map between the lines Γ_+ and Γ_- as a tool for the search for limit cycles and in the analysis of their stability.

Some questions about the local phase portrait in a neighbourhood of the singular points of the fundamental systems with parameter $D > 0$ can be found in [1]. However, the study of fundamental systems from the point of view of the qualitative theory of differential equations starts with the work of Llibre and Sotomayor [44]. In that paper the authors describe the phase portraits and the bifurcation set of all the fundamental systems with parameters $D > 0$ and $T < 0$. We note that in [44] the authors do not study the behaviour of the system in a neighbourhood of

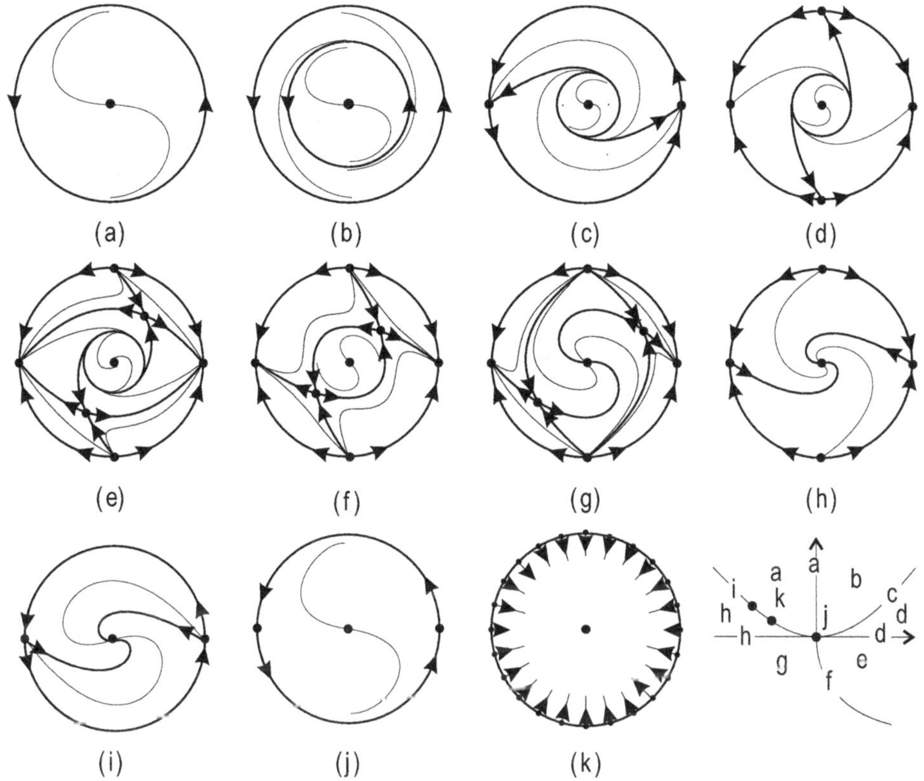

Figure 1.4: Phase portrait of fundamental systems with $D > 0$ and $T < 0$.

infinity. Therefore, from the eleven equivalence classes depicted in Figure 1.4, the authors only identified five. Moreover, the techniques used there for the study of the limit cycles differ from those introduced by Andronov. A review of this work appears in Section 5.2.

Based on the study of the Poincaré maps, in [46] and [47] R. Lum and L.O. Chua studied the configuration of the limit cycles appearing in two-piece and in three-piece linear differential systems, respectively. The two studies are based on a conjecture which is true in the first case, as it has been proved by E. Freire, E. Ponce and F. Torres [24], but it is erroneous in the second, as we will show in Section 5.5.

The bifurcation set of fundamental systems has also been subject to analysis by other authors. For example, Llibre and Ponce [42] characterize the values of the parameters in which the system exhibits a Hopf bifurcation at infinity.

Following the point of view of the qualitative theory of differential equations,

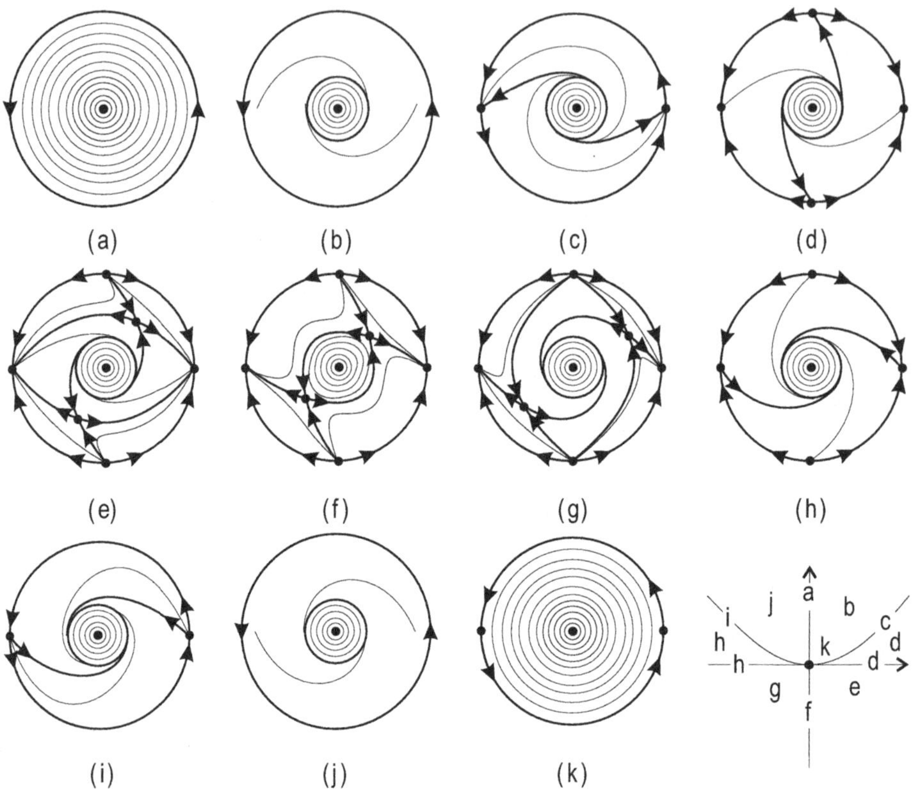

Figure 1.5: Phase portrait of fundamental systems with $D > 0$ and $T = 0$.

in this book we make a topological classification of the family of the fundamental systems with parameter $D \neq 0$; we provide the global phase portrait for each of the 56 topological equivalence classes; and we describe the bifurcation set in the fundamental parameter space (D, T, d, t).

The main results we provide in this book can be summarized in the following four theorems.

Theorem 1.2.1. *The phase portrait of a fundamental system with fundamental parameter $D > 0$ and given (t, d) is topologically equivalent to the corresponding one shown in Figure* 1.4 *when $T < 0$; or in Figure* 1.5 *when $T = 0$; or in Figure* 1.6 *when $T > 0$.*

Theorem 1.2.2. *Figure* 1.7 *shows the bifurcation set of the fundamental systems for which the fundamental parameter D is positive and constant.*

Theorem 1.2.3. *The phase portrait of a fundamental system with fundamental parameter $D < 0$ is topologically equivalent to the corresponding one shown in*

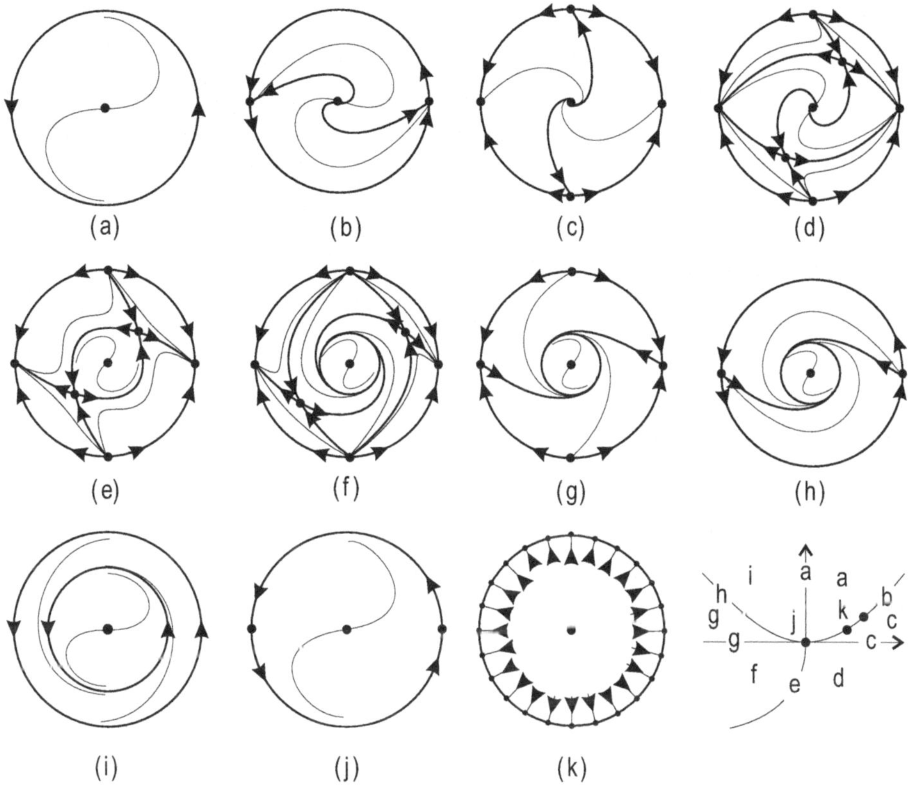

Figure 1.6: Phase portrait of fundamental systems with $D > 0$ and $T > 0$.

Figure 1.8 *when* $T \leq 0$ *or* $T > 0$ *and* $t^2 - 4d \geq 0$*; or in Figure* 1.9 *when* $T > 0$ *and* $t^2 - 4d < 0$.

Theorem 1.2.4. *Figure* 1.10 *corresponds to the bifurcation set of the fundamental systems for which the fundamental parameter* D *is negative and constant.*

The last picture in Figure 1.4, 1.5, 1.6, 1.8 and 1.9 corresponds to the bifurcation set of the fundamental systems where the fundamental parameters D and T are constant. To easily follow the evolution of the phase portraits when the parameters (t, d) vary and for a better understanding of the nature of the bifurcations, we have ordered the phase portraits clockwise.

When $D > 0$ the bifurcation set is formed by the three-dimensional manifolds $\mathcal{H}_\infty$, $\mathcal{SN}_\infty$, $\mathcal{N}$, $\mathcal{H}_e\mathcal{L}$, $\{T = 0\}$, and the surfaces $\mathcal{O}$, $\mathcal{VB}_1$ and $\mathcal{VB}_2$, see Figure 1.7. We remark that in Figure 1.7 we are considering a positive fixed value for the parameter D. This allows us to represent the three-dimensional manifolds by

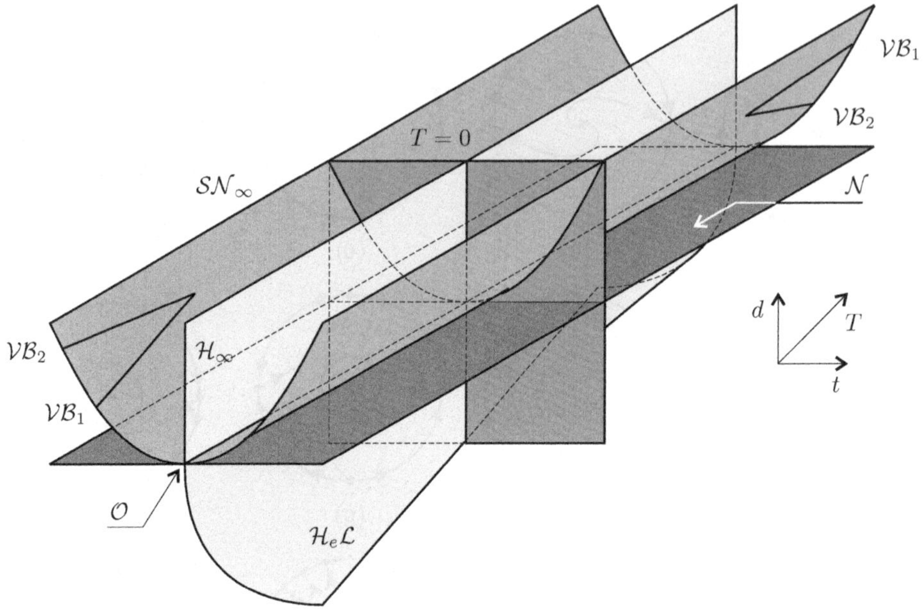

Figure 1.7: Bifurcation set when D is a positive constant.

surfaces and the surfaces by curves.

The manifold $\mathcal{H}_\infty$ corresponds to a Hopf bifurcation at infinity, see Figure 1.4(a) and (b). The manifold $\mathcal{SN}_\infty$ corresponds to a saddle-node bifurcation of two singular points at infinity, see Figure 1.4(b), (c) and (d) in the supercritical case and Figure 1.4(h), (i) and (a) in the subcritical case. The manifold $\mathcal{N}$ corresponds to a pitchfork bifurcation at infinity, see Figure 1.4(d) and (e) in the supercritical case and Figure 1.4(g) and (h) in the subcritical case. Finally, the manifold $\mathcal{H}_e\mathcal{L}$ corresponds to a *heteroclinic bifurcation*, see Figure 1.4(f). We remark that in the surface $\mathcal{O}$, where the above manifolds intersect, we have the four bifurcations simultaneously.

The bifurcation manifolds $\mathcal{VB}_1$ and $\mathcal{VB}_2$ do not correspond to any dynamical bifurcation. These manifolds appear when the real Jordan normal forms of the fundamental matrices of the system are not uniquely determined. In such a case two different phase portraits are possible for the same parameter value.

The manifold $\{T = 0\}$ corresponds to a *vertical-Hopf bifurcation*. This bifurcation occurs when the periodic orbit at the boundary of a bounded center persists as a limit cycle. This phenomena has been widely studied by Freire, Ponce and Torres [24].

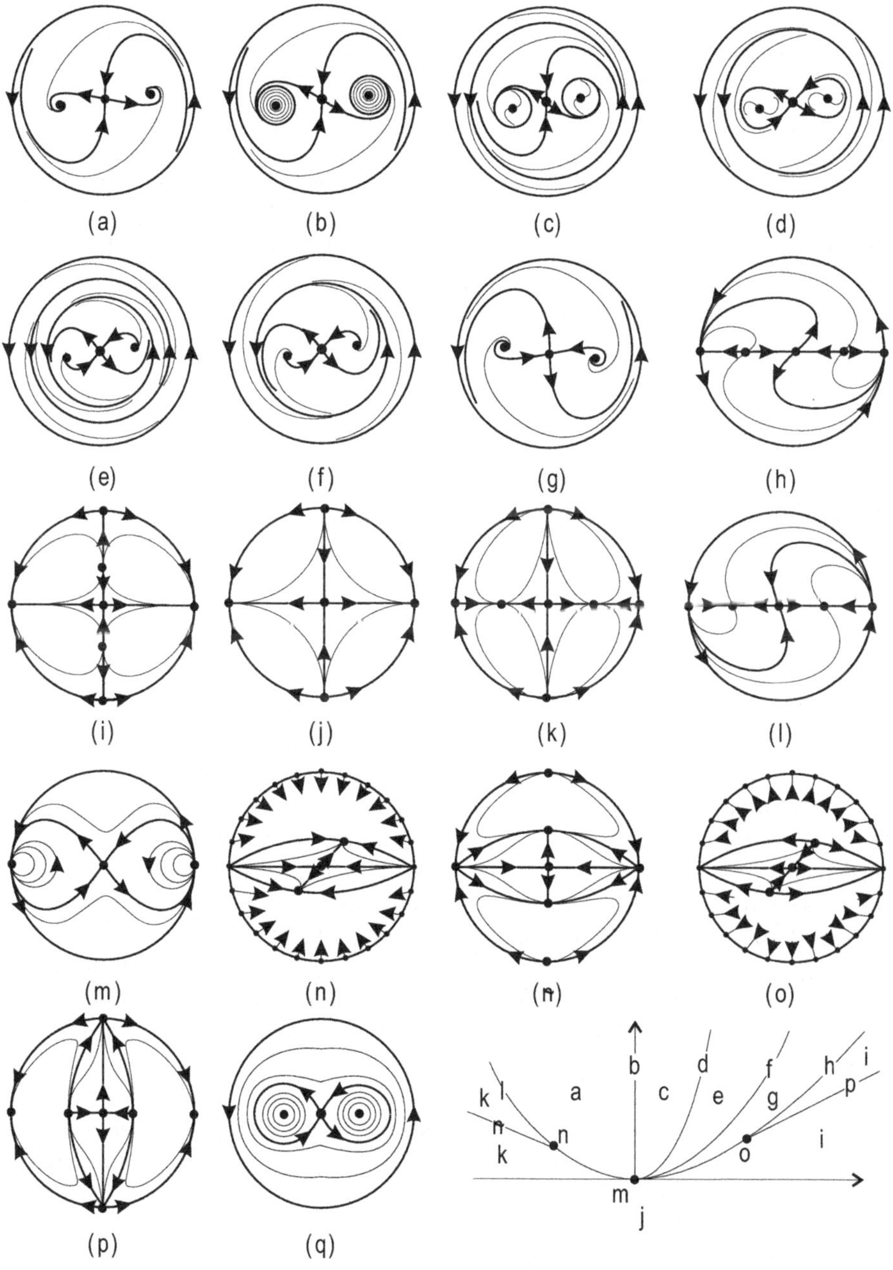

Figure 1.8: Phase portrait of fundamental systems with $D < 0$ and $T \leq 0$.

The bifurcation set in the case when $D < 0$ is formed by the three-dimensional manifolds $\mathcal{H}_\infty$, $\mathcal{H}_o\mathcal{L}$, $\mathcal{NH}_{lc}$, $\mathcal{SN}_\infty$, W_1^*, W_2^*, $\mathcal{N}$, and the surfaces $\mathcal{O}$, $\mathcal{VB}_1$, $\mathcal{VB}_2$, and $\{T = 0\}$, see Figure 1.10. In Figure 1.10 we consider the case when the fundamental parameter D takes a fixed negative value. For this reason the three-dimensional manifolds are represented by surfaces and the surfaces by curves.

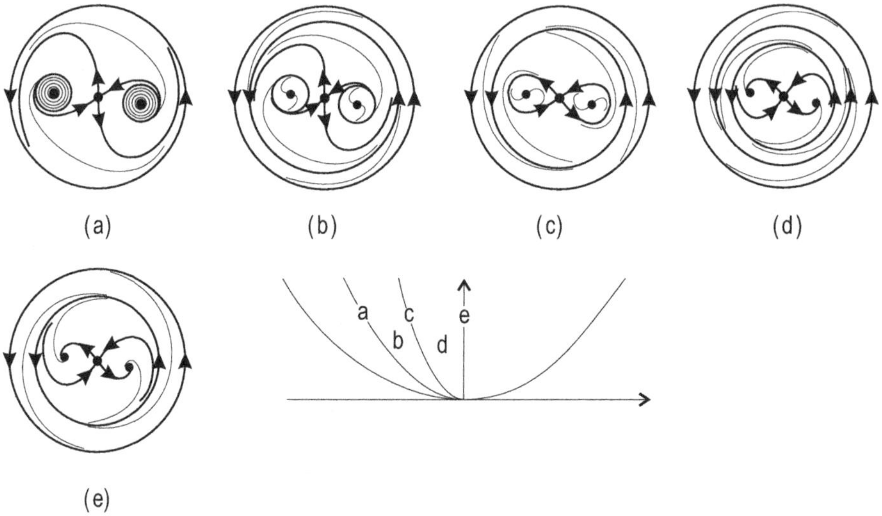

Figure 1.9: Phase portrait of fundamental systems with $D < 0, T > 0$ and $t^2 - 4d < 0$.

The manifold $\mathcal{H}_\infty$ corresponds to a bifurcation in which a Hopf bifurcation at infinity and a vertical-Hopf bifurcation at two finite singular points occur simultaneously, see Figure 1.8(a), (b) and (c). The manifold $\mathcal{H}_o\mathcal{L}$ corresponds to homoclinic bifurcation, see Figure 1.8(d). The manifold $\mathcal{NH}_{lc}$ corresponds to a saddle-node bifurcation of limit cycles, see Figure 1.8(f).

Like in the case $D > 0$, the manifolds $\mathcal{SN}_\infty$ and $\mathcal{N}$ correspond to a saddle-node bifurcation and a pitchfork bifurcation of singular points at infinity. Also, the surfaces $\mathcal{VB}_1$ and $\mathcal{VB}_2$ describe the same type of bifurcations as in the case $D > 0$.

The bifurcations associated with surfaces W_1^* and W_2^* cannot be described locally; rather, they correspond to global bifurcations which arise when an eigenvector of the fundamental matrix A is parallel to the straight lines Γ_+ and Γ_-, see Figure 1.8(i) and (p), or Figure 1.8(k) and (ñ).

Just as in the case $D > 0$ on the bifurcation surface $\mathcal{O}$, which is the intersection of the bifurcation manifolds $\mathcal{H}_\infty$, $\mathcal{SN}_\infty$, $\mathcal{H}_o\mathcal{L}$ and $\mathcal{N}$, the four bifurcations occur simultaneously. The bifurcation surface $\{T = 0, t = 0\}$, where the manifolds $\mathcal{H}_\infty$ and $\mathcal{H}_o\mathcal{L}$ intersect, also involves a combination of the two bifurcations.

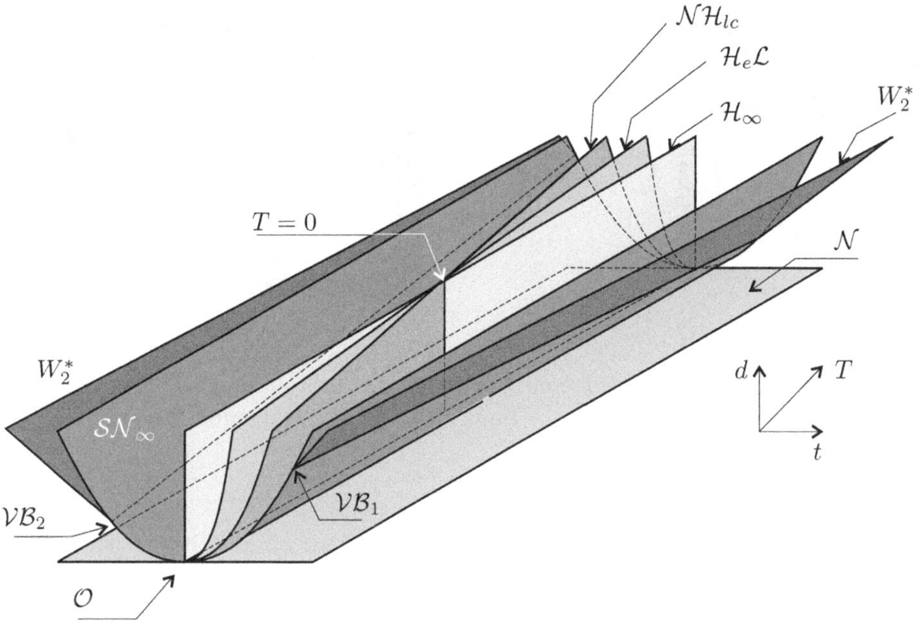

Figure 1.10: Bifurcation set when D is a negative constant.

Applications

We finish this chapter by describing the evolution of the phase portrait and the bifurcations occurring in the example of the Wien bridge when we fix the values of the components R_1, R_2, R_S, C_1 and C_2, and vary only the value of R_f.

Straightforward computations show that the trace t and the determinant d of the matrix A are expressed in terms of the values of the components by

$$t = -\left(\frac{1}{R_1C_2} + \frac{1}{R_2C_2} + \frac{1}{R_1C_1}\right), \quad d = \frac{1}{R_1C_1R_2C_2}.$$

From this we obtain that $t < 0$ and $d > 0$. Similarly, the trace T and the determinant D of the matrix B are given by

$$T = t + \frac{1}{R_1C_2}\left(1 + \frac{R_F}{R_S}\right), \quad D = d.$$

Therefore, $D > 0$ and the sign of T depends on the value of the resistor R_F.

Set $R_1 = 2.188\,K\Omega$, $R_2 = 2.167\,K\Omega$, $R_S = 2.192\,K\Omega$, $C_1 = 646\,nF$ and $C_2 = 328\,nF$. For these values one has that $t^2 - 4d > 0$. This circuit was actually built in laboratory, see Figure 1.11(a) and the experiences described below have been confirmed on the oscilloscope.

(a) (b)

Figure 1.11: (a) A Wien bridge circuit built on a protoboard. (b) Experimental observation on a oscilloscope of the limit cycle, for the real value $R_F = 10.63\,K\Omega$.

Setting the value of R_F less than $3.326\,K\Omega$ we obtain that $T < 0$. The corresponding phase portrait of the PWLS (1.7) is depicted in Figure 1.4(h). We see that the origin, which is a focus type singular point, is a global attractor. Also, two separatrices born at the two saddles at infinity are observed. The remaining singular points at infinity are two nodes.

When $R_F = 3.326\,K\Omega$ one has $T = 0$. The corresponding phase portrait is depicted in Figure 1.6(h). Now the global attractor, namely, the singular point at the origin, is replaced by a central region foliated by periodic orbits (bounded period annulus). The remainder of the phase portrait persists without changes.

For values of R_F greater than $3.326\,K\Omega$ we have that $T > 0$. The corresponding phase portrait is depicted in Figure 1.9(g). As it can be observed, the period annulus disappears and only one periodic solution persists. This limit cycle is a global attractor and bifurcates from the boundary of the period annulus. This bifurcation is called a focus-center-limit cycle bifurcation. The stability of the singular point at the origin changes, the point becoming unstable, see Figure 1.11(b). Qualitative and quantitative aspects of the focus-center-limit cycle bifurcation are studied in [23].

Chapter 2

Basic elements of the qualitative theory of ordinary differential equations

In this chapter we collect some basic ideas and results from the qualitative theory of ordinary differential equations. We present only the tools needed in our later analysis and the theoretical context where they appear. Most of these results have extensions to more general contexts. To not make our presentation too long we will restrict ourselves to the most relevant facts.

A deeper and more detailed introduction can be found in the following books: A.A. Andronov, E.A. Leontovich, I.I. Gordon and A.G. Maier [4], [5], M.W. Hirsch and S. Smale [33], V.I. Arnold [7], J. Sotomayor [57], [58], P. Hartman [30], S. Lefschetz [40], L. Perko [53], C. Chicone [14], and recently the book of F. Dumortier, J. Llibre and J.C. Artés [21].

2.1 Differential equations and solutions

2.1.1 Existence and uniqueness of solutions

Let U be a subset of $\mathbb{R}^n$ and W an open subset of U. We say that the function $\mathbf{f} : U \to \mathbb{R}^n$ is *Lipschitz on* W, if there exists a constant $L \in \mathbb{R}$, such that for every $\mathbf{x}$, $\mathbf{y} \in W$

$$||\mathbf{f}(\mathbf{x}) - \mathbf{f}(\mathbf{y})|| \leq L\,||\mathbf{x} - \mathbf{y}||\,.$$

The constant L is called a *Lipschitz constant for* $\mathbf{f}$ on W. Here and in the sequel $||\cdot||$ denotes the Euclidean norm of $\mathbb{R}^n$. Since $\mathbb{R}^n$ is a finite-dimensional vector space, if $\mathbf{f}$ is Lipschitz with respect to a norm of $\mathbb{R}^n$, then $\mathbf{f}$ is Lipschitz with respect to any other norm of $\mathbb{R}^n$. Hence, the definition of Lipschitz functions does not depend on the chosen norm. However, this is not true for the Lipschitz constants. For instance, if $\mathbf{f}$ is Lipschitz on W, with Lipschitz constant L with respect to the Euclidean norm of $\mathbb{R}^n$, then $\sqrt{n}L$ is a Lipschitz constant of $\mathbf{f}$ with respect to the *maximum norm* of $\mathbb{R}^n$

$$\|\mathbf{x}\|_\infty = \max_{1\le k\le n} \{|x_k|\},$$

where $\mathbf{x} = (x_1, x_2, \ldots, x_n)^T$, and $(\cdot)^T$ denotes the transposed vector.

In particular when $\mathbf{f}$ is Lipschitz on the whole domain U, we call $\mathbf{f}$ *globally Lipschitz*. On the other hand if for every $\mathbf{x}_0 \in U$ there exists a neighbourhood W of $\mathbf{x}_0$ in U such that $\mathbf{f}$ is Lipschitz on W, then we call $\mathbf{f}$ *locally Lipschitz* on U.

Example 2 (Linear function). Consider the function $\mathbf{f}(\mathbf{x}) = A\mathbf{x}$, where A is a $n \times n$ matrix. Since $\|\mathbf{f}(\mathbf{x}) - \mathbf{f}(\mathbf{y})\| = \|A\mathbf{x} - A\mathbf{y}\| \le \|A\|\|\mathbf{x} - \mathbf{y}\|$, $\mathbf{f}$ is both locally and globally Lipschitz in $\mathbb{R}^n$, with $L = \|A\|$ as a Lipschitz constant.

Example 3. Consider the quadratic function $f(x) = x^2$. Since

$$|f(x) - f(y)| = |x + y||x - y|, \tag{2.1}$$

for any $x_0 \in \mathbb{R}$ one has $|f(x) - f(y)| < 2(|x_0| + \varepsilon)|x - y|$ in $W = (x_0 - \varepsilon, x_0 + \varepsilon)$. Therefore, f is a locally Lipschitz function in $\mathbb{R}$. However, f is not globally Lipschitz in $\mathbb{R}$. Indeed, assuming that there exists a constant L such that $|f(x) - f(y)| < L|x - y|$ for every $x, y \in \mathbb{R}$, we contradict (2.1).

Example 4 (Piecewise linear function). Consider the piecewise linear function $f(x) = |x|$. From the triangle inequality we have $|f(x) - f(y)| = ||x| - |y|| \le |x - y|$, which implies that f is both locally and globally Lipschitz, with Lipschitz constant equal to 1.

For the purposes of this book it is enough to consider a *differential equation* or a *system of ordinary differential equations* as

$$\dot{\mathbf{x}} = \mathbf{f}(\mathbf{x}), \tag{2.2}$$

where $\mathbf{x} = \mathbf{x}(s) \in U$, U is an open subset of $\mathbb{R}^n$ and $\mathbf{f} : U \to \mathbb{R}^n$ is a locally Lipschitz function on U. From now on $\dot{\mathbf{x}}$ denotes the derivative of $\mathbf{x}(s)$ with respect to s. As usual, the domain of $\mathbf{f}$ (the set U) is called the *phase space*, the variable $\mathbf{x}$ is called the *dependent variable*, and s is called the *independent variable* or *time*. We use the variable s instead of the standard variable t because t will denote the trace of some matrices which will appear later on.

In a more general context equation (2.2) is known as an *autonomous ordinary differential equation* (as opposed to *non-autonomous differential equations*), because the function $\mathbf{f}$ does not depend explicitly on the independent variable s.

A smooth function $\phi : I \to U$, where I is an open interval of $\mathbb{R}$, is said to be a *solution* of the differential equation (2.2) if $\dot{\phi}(s) = \mathbf{f}(\phi(s))$ for every $s \in I$.

Geometrically, a differential equation (2.2) assigns to every point $\mathbf{x}$ in the phase space U a vector $\mathbf{f}(\mathbf{x})$ in the tangent space at $\mathbf{x}$. Then a solution of the differential equation is a curve $\phi : I \to U$ whose tangent vector at $\dot{\phi}(s)$ coincides with the vector $\mathbf{f}(\phi(s))$ for any s, see Figure 2.1. From this reason we call the function $\mathbf{f}$ a *vector field*.

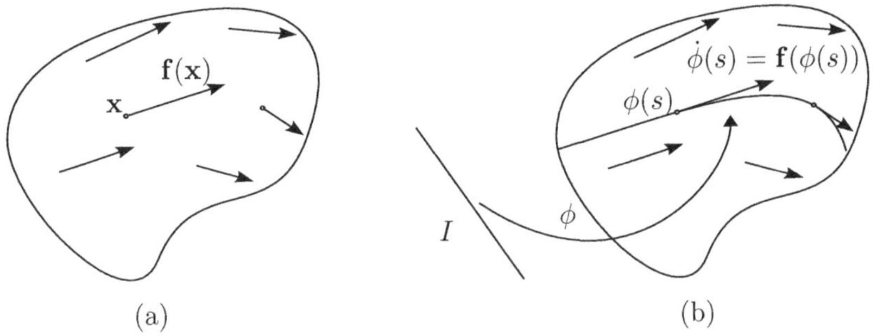

Figure 2.1: (a) Vector field $\mathbf{f}$ defined in the phase space U. (b) A solution $\phi(s)$ of the differential equation $\dot{\mathbf{x}} = \mathbf{f}(\mathbf{x})$.

The existence of solutions of differential equations (2.2) is not obvious and it depends on some properties of the vector field $\mathbf{f}$. The same is true for the uniqueness of the solution which satisfies the *initial conditions* $(s_0, \mathbf{x}_0)$, i.e. $\phi(s_0) = \mathbf{x}_0$. The following theorem states the basic result in this direction.

Theorem 2.1.1 (Existence and uniqueness). *Let U be an open subset of $\mathbb{R}^n$, $\mathbf{f} : U \to \mathbb{R}^n$ be a locally Lipschitz function on U, $s_0 \in \mathbb{R}$ and $\mathbf{x}_0 \in U$. There exist a constant $c > 0$ and a unique solution $\phi : (s_0 - c, s_0 + c) \to U$ of the differential equation $\dot{\mathbf{x}} = \mathbf{f}(\mathbf{x})$ such that $\phi(s_0) = \mathbf{x}_0$.*

For a proof of this theorem we refer the reader to [33].

To emphasize the dependence of the solutions on the *initial conditions* $(s_0, \mathbf{x}_0)$, we denote the solution of the differential equation (2.2) passing through $\mathbf{x}_0$ at time $s = s_0$ by $\phi(s; s_0, \mathbf{x}_0)$.

2.1.2 Prolongability of solutions

From the existence and uniqueness theorem we obtain conditions on the vector field $\mathbf{f}(\mathbf{x})$ so that it has exactly one solution passing through an a-priori fixed point. This solution is defined at least on a sufficiently small open interval. In the next result we find the maximal interval of existence. First, we need to introduce the following definitions.

We say that $\phi : I \to U$, with $\phi = \phi(s; s_0, \mathbf{x}_0)$, is a *maximal solution* of equation (2.2), if for every solution $\psi : J \to U$, with $\psi = \psi(s; s_0, \mathbf{x}_0)$, we have $J \subseteq I$. We call *maximal interval of definition* the interval of definition of the maximal solution $\phi(s; s_0, \mathbf{x}_0)$, and we denote it by $I_{(s_0, \mathbf{x}_0)}$. *From now on we will only consider maximal solutions.* The differential systems (vector fields) such that all their solutions have the maximal interval of definition equal to $\mathbb{R}$ are called *complete*. In the following proposition we present sufficient conditions on the vector field of a differential equation for it to be complete.

Proposition 2.1.2. *Consider the differential equation* $\dot{\mathbf{x}} = \mathbf{f}(\mathbf{x})$*, where* $\mathbf{f} : \mathbb{R}^n \to \mathbb{R}^n$ *is a globally Lipschitz function. Then for every initial conditions* $(s_0, \mathbf{x}_0) \in \mathbb{R} \times \mathbb{R}^n$ *it holds that* $I_{(s_0,\mathbf{x}_0)} = \mathbb{R}$.

A proof of Proposition 2.1.2 can be found in [53, Section 3.1, Theorem 3] or in [57, Proposition 4, p. 15]. We emphasize that the differentiability condition imposed on the vector field in the first reference is not essential for the proof and can be removed. Note that the hypothesis in Proposition 2.1.2 is very restrictive. As we will see in Section 3.3, fundamental systems satisfy it.

Example 5. As we saw in Example 2 linear differential systems $\dot{\mathbf{x}} = A\mathbf{x}$ are globally Lipschitz, and hence complete.

2.1.3 Dependence on initial conditions and parameters

Consider the family of differential equations

$$\dot{\mathbf{x}} = \mathbf{f}(\mathbf{x}, \lambda),$$

where $\mathbf{f} : U \times V \to \mathbb{R}^n$, U is an open subset of $\mathbb{R}^n$, and V is an open subset of $\mathbb{R}^p$. The set V is called the *parameter space* of the differential equation.

Assuming that $\lambda_0 \in V$, $s_0 \in \mathbb{R}$ and $\mathbf{x}_0 \in \mathbb{R}^n$, there exists exactly one solution of the differential equation $\dot{\mathbf{x}} = \mathbf{f}(\mathbf{x}, \lambda_0)$ passing through $\mathbf{x}_0$ at time s_0. We denote this solution by $\phi(s; s_0, \mathbf{x}_0, \lambda_0)$. In the next theorem we summarize the behaviour of the solution $\phi(s; s_0, \mathbf{x}_0, \lambda_0)$ when we vary s_0, $\mathbf{x}_0$ or λ_0. First we introduce some additional definitions.

Let W be an open subset of U. The function $\mathbf{f}(\mathbf{x}, \lambda)$ is said to be *Lipschitz with respect to the first variable in* W, if there exists a positive constant $L \in \mathbb{R}$, such that for every $\mathbf{x}, \mathbf{y} \in W$ and $\lambda \in V$

$$||\mathbf{f}(\mathbf{x}, \lambda) - \mathbf{f}(\mathbf{y}, \lambda)|| \leq L\, ||\mathbf{x} - \mathbf{y}||.$$

In particular, if $\mathbf{f}$ is Lipschitz with respect to the first variable in U, then we say that $\mathbf{f}$ is *globally Lipschitz with respect to the first variable*. The function $\mathbf{f}$ is said to be *locally Lipschitz with respect to the first variable* if for every $\mathbf{x}_0 \in U$ there exists a neighbourhood W of $\mathbf{x}_0$ in U such that $\mathbf{f}$ is Lipschitz with respect to the first variable in W. For simplicity we will call $\mathbf{f}$ globally or locally Lipschitz without a reference to the first variable when no confusion can arise.

Theorem 2.1.3 (Dependence on initial conditions and parameters). *Let* U *and* V *be open subsets of* $\mathbb{R}^n$ *and* $\mathbb{R}^p$*, respectively. Let* $\mathbf{f} : U \times V \to \mathbb{R}^n$ *be a locally Lipschitz function with respect to the first variable in* U *and* $\mathbf{f} \in C^r(U \times V)$ *for some* $r \geq 0$*. Then for every* $(s_0, \mathbf{x}_0, \lambda_0) \in \mathbb{R} \times U \times V$ *the solution* $\phi(s; s_0, \mathbf{x}_0, \lambda_0)$

of the differential equation $\dot{\mathbf{x}} = \mathbf{f}(\mathbf{x}, \lambda_0)$ *is* r *times continuously differentiable with respect to* $\mathbf{x}_0$ *and* λ_0 *and* $r+1$ *times continuously differentiable with respect to* s.

A proof of this theorem can be found in Hartman [30, pp. 93–96] or Lefschetz [40, pp. 36–43].

Example 6 (A family of piecewise linear differential equations). Consider the family of differential equations $\dot{x} = |x| + \lambda$, with $\lambda > 0$, which is defined on whole $\mathbb{R}$. With respect to the vector field, the phase space splits into two regions, $\{x < 0\}$ and $\{x > 0\}$, and in both the field is given by a linear function. Moreover, it is continuously differentiable with respect to the parameter λ, but is only globally Lipschitz with respect to the variable x.

Straightforward computations show that the solution $\phi(s; 0, x_0, \lambda)$ of the differential equation passing through $x_0 < 0$ at time $s = 0$ is given by

$$\phi(s; 0, x_0, \lambda) = \begin{cases} \lambda + e^{-s}(x_0 - \lambda), & \text{if } s \leq s^*, \\ \lambda\left(\dfrac{\lambda}{\lambda - x_0}e^s - 1\right), & \text{if } s > s^* \end{cases}$$

where $s^* = \ln(1 - x_0/\lambda)$ is the time required for the solution to reach the origin, see Figure 2.2. Note that the maximal interval of definition of the solution is $\mathbb{R}$.

Taking the first and the second derivative with respect to s one has that

$$\frac{d^k\phi}{ds^k}(s; 0, x_0, \lambda) = \begin{cases} (-1)^k e^{-s}(x_0 - \lambda), & \text{if } s \leq s^*, \\ \dfrac{\lambda^2}{\lambda - x_0}e^s, & \text{if } s > s^* \end{cases}$$

for $k = 1, 2$. Thus the solution $\phi(s; 0, x_0, \lambda)$ is an analytical function of s in $\mathbb{R}\setminus\{s^*\}$ and once continuously differentiable at $s = s^*$, but it is not twice continuously differentiable at $s = s^*$.

Taking derivatives with respect to λ it is easy to conclude that $\phi(s; 0, x_0, \lambda)$ is once continuously differentiable in $\mathbb{R}$ but is not twice continuously differentiable at $s = s^*$.

This example shows that solutions of piecewise linear differential equations lose regularity at the boundary between the regions where the vector field is linear.

2.1.4 Other properties

We recall now some other properties of the solutions of differential equations. We say that $\phi : \mathbb{R} \to \mathbb{R}^n$ is a *periodic function*, if there exists a positive constant T such that $\phi(s + T) = \phi(s)$ for every $s \in \mathbb{R}$. The smallest value of T satisfying this property is called the *period* of the function ϕ.

Proposition 2.1.4. *Consider the differential equation* $\dot{\mathbf{x}} = \mathbf{f}(\mathbf{x})$ *with* $\mathbf{f} : \mathbb{R}^n \to \mathbb{R}^n$ *a globally Lipschitz function.*

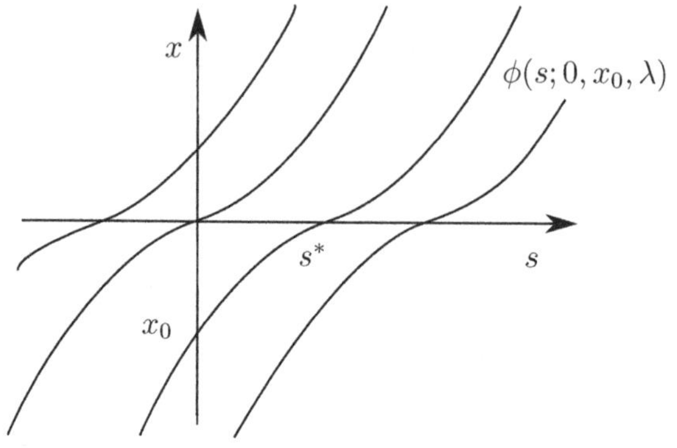

Figure 2.2: Solutions $\phi(s;0,x_0,\lambda)$ of the differential equation $\dot{x} = |x| + \lambda$.

(a) *Let $\phi(s;s_0,\mathbf{x}_0)$ be a solution. Then for every $\tau \in \mathbb{R}$, $\phi(s+\tau,s_0,\mathbf{x}_0)$ is also a solution.*

(b) *Let $\phi(s;s_1,\mathbf{x}_1)$ and $\phi(s;s_2,\mathbf{x}_2)$ be two solutions satisfying $\phi(\tau_1;s_1,\mathbf{x}_1) = \phi(\tau_2;s_2,\mathbf{x}_2)$ for fixed $\tau_1,\tau_2 \in \mathbb{R}$. Then $\phi(s-(\tau_2-\tau_1);s_1,\mathbf{x}_1) = \phi(s;s_2,\mathbf{x}_2)$ for every $s \in \mathbb{R}$.*

(c) *Let $\phi(s;s_0,\mathbf{x}_0)$ be a solution and suppose that there exist $\tau_1,\tau_2 \in \mathbb{R}$, $\tau_1 < \tau_2$, such that $\phi(\tau_1;s_0,\mathbf{x}_0) = \phi(\tau_2;s_0,\mathbf{x}_0)$. Then, $\phi(s;s_0,\mathbf{x}_0)$ is a periodic function whose period is a multiple of $\tau = \tau_2 - \tau_1$.*

For a proof of this result we refer the reader to [60, pp. 8–9]. Note that in this reference the author assumes that the vector field is differentiable, but it is easy to check that this hypothesis can be substituted by requiring the uniqueness of the solutions.

2.2 Orbits

In this section we present some dynamical features of solutions to differential equations. Take $s_0 \in \mathbb{R}$ and $\mathbf{x}_0 \in U$, and let $\phi(s;s_0,\mathbf{x}_0)$ be a maximal solution of the differential equation (2.2). We call the set

$$\gamma(s_0,\mathbf{x}_0) := \left\{\mathbf{x} \in U : \mathbf{x} = \phi(s;s_0,\mathbf{x}_0) \text{ and } s \in I_{(s_0,\mathbf{x}_0)}\right\}$$

the orbit of the solution $\phi(s;s_0,\mathbf{x}_0)$.

When the phase space is the whole $\mathbb{R}^n$ and the vector field $\mathbf{f}$ is globally Lipschitz in $\mathbb{R}^n$, the maximal interval of definition of all solutions is $\mathbb{R}$, see Proposition 2.1.2. Then $\gamma(t_0,\mathbf{x}_0) = \gamma(t_0+\tau,\mathbf{x}_0)$ for every $\tau \in \mathbb{R}$, see Proposition 2.1.4(a).

Hence we will simply use $\gamma(\mathbf{x}_0)$ to denote the orbit through $\mathbf{x}_0$. Moreover, if $\mathbf{x}_1 \in \gamma(\mathbf{x}_0)$, then there exists $s_1 \in \mathbb{R}$ such that $\mathbf{x}_1 = \phi(s_1; s_0, \mathbf{x}_0)$. Applying Proposition 2.1.4(b) to the solutions $\phi(s; s_0, \mathbf{x}_0)$ and $\phi(s; s_1, \mathbf{x}_1)$ one obtains that $\gamma(\mathbf{x}_1) = \gamma(\mathbf{x}_0)$. Therefore orbits are independent on the point of reference, and we can avoid the reference to such point when no confusion can arise.

Suppose that $\mathbf{x}_2 \in \gamma(\mathbf{x}_1) \cap \gamma(\mathbf{x}_0) \neq \emptyset$. Since orbits do not depend on the point of reference, $\gamma(\mathbf{x}_0) = \gamma(\mathbf{x}_1) = \gamma(\mathbf{x}_2)$. Therefore, *if two orbits intersect at a point, then they coincide.*

Example 7. Consider the planar piecewise linear differential system $\dot{x} = x, \dot{y} = |y|$. Since the two variables are decoupled, the corresponding differential equation can be easily solved. Indeed, the solution with initial condition (x_0, y_0) is given by $\phi(s; 0, (x_0, y_0)) = (x(s), y(s))$, where

$$x(s) = e^s x_0, \quad y(s) = \begin{cases} e^s y_0, & \text{if } y_0 \geq 0, \\ e^{-s} y_0, & \text{if } y_0 < 0, \end{cases}$$

see Figure 2.3(a) and (b).

Set $x_0 \in \mathbb{R}$ and $y_0 < 0$. The orbit through the point $\mathbf{p} = (x_0, y_0)$ is defined by $\gamma(\mathbf{p}) = \{(e^s x_0, e^{-s} y_0) : s \in \mathbb{R}\}$, and so $y(s) = x_0 y_0 / x(s)$, which is the branch of an hyperbola passing through $\mathbf{p}$, see Figure 2.3(c).

On the other hand, if $y_0 > 0$, then the orbit through $\mathbf{p}$ is defined by $\gamma(\mathbf{p}) = \{(e^s x_0, e^s y_0) : s \in \mathbb{R}\}$, and so $\gamma(\mathbf{p})$ is a half-line, see Figure 2.3(c).

2.3 The flow of a differential equation

Consider the differential equation

$$\dot{\mathbf{x}} = \mathbf{f}(\mathbf{x}), \tag{2.3}$$

where $\mathbf{f} : U \to \mathbb{R}^n$ is locally Lipschitz in an open subset U of $\mathbb{R}^n$. Suppose that for every $\mathbf{x} \in U$, the solution $\phi(s; 0, \mathbf{x})$ is defined on whole $\mathbb{R}$, i.e., $I_{(0,\mathbf{x})} = \mathbb{R}$. The *flow of the differential equation* (2.3) is defined to be the function

$$\Phi : \mathbb{R} \times U \to \mathbb{R}^n$$

given by $\Phi(s, \mathbf{x}) = \phi(s; 0, \mathbf{x})$. The notion of flow introduced here is sometimes referred as *completed flow*. That is because the maximal interval of definition of the solutions is the whole $\mathbb{R}$. Since the differential systems considered in this work are complete, we can use both terms. In particular, if $\mathbf{f} : \mathbb{R}^n \to \mathbb{R}^n$ is globally Lipschitz, then the flow of the differential equation $\dot{\mathbf{x}} = \mathbf{f}(\mathbf{x})$ is complete, see Proposition 2.1.2.

Other authors denote the flow of a differential equation by the pair consisting of the function Φ and the phase space U. It is also usual to denote by $\Phi_s(\mathbf{x})$ the function $\Phi(s, \mathbf{x})$ (see [29] or [53]). Some properties of flows are collected in the following result.

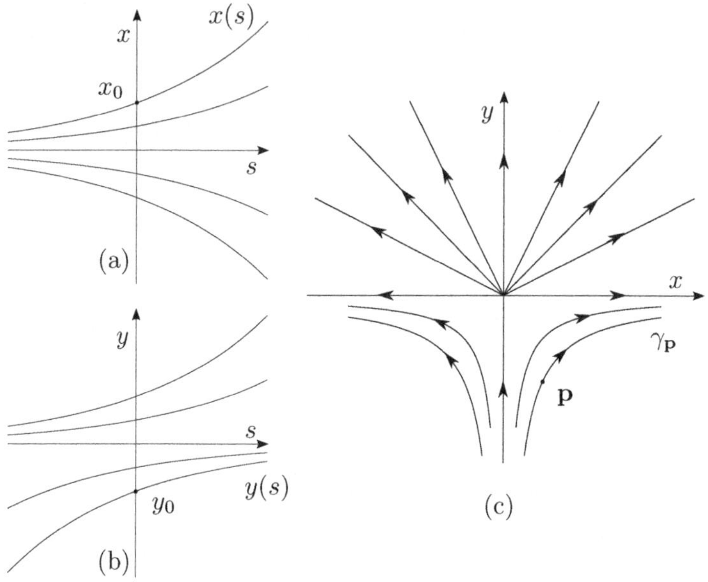

Figure 2.3: Solutions $\phi(s;0,(x_0,y_0)) = (x(s),y(s))$ and orbits of the differential equation $\dot{x} = x$, $\dot{y} = |y|$. (a) Dependence of the first coordinate $x(s)$ of the solution $\phi(s;0,(x_0,y_0))$ on s. (b) Dependence of the second coordinate $y(s)$ of the solution $\phi(s;0,(x_0,y_0))$ on s. (c) Orbit $\gamma_{\mathbf{p}}$ with $\mathbf{p} = (x_0,y_0)$ depicted in the phase space (x,y).

Proposition 2.3.1. *Let $\Phi(s,\mathbf{x})$ be the flow defined by the differential equation* (2.3)*.*

(a) *For every $\mathbf{x} \in U$, $\Phi(0,\mathbf{x}) = \mathbf{x}$.*

(b) *For every $s,t \in \mathbb{R}$ and $\mathbf{x} \in U$, $\Phi(s+t,\mathbf{x}) = \Phi(s,\Phi(t,\mathbf{x}))$.*

(c) *Φ is a continuous function.*

Proof. Statement (a) follows from the definition of Φ. Statement (b) follows by taking $\mathbf{x}_1 = \mathbf{x}$, $\mathbf{x}_2 = \phi(t;0,\mathbf{x})$, $\tau_1 = t$, $\tau_2 = 0$ and $s_1 = s_2 = 0$ and applying Proposition 2.1.4(b). Statement (c) is a consequence of the continuous dependence of the solutions on the initial conditions and parameters, see Theorem 2.1.3. □

In the classical point of view, the objective of the theory of differential equations is to find explicit expressions for the flow $\Phi(s,\mathbf{x})$. In the qualitative theory it is more important to describe the topological properties of the flow and the asymptotic behaviour of its orbits, i.e., the behaviour of the orbits when s tends to $\pm\infty$. The *phase portrait* of a differential equation (2.3) is defined as the union of all the orbits of (2.3).

Let $\Phi(s, \mathbf{x})$ be the flow of the differential equation (2.3) and take $\mathbf{p} \in U$. By the continuous dependence of the solutions on the initial conditions and parameters, the function $\Phi_{\mathbf{p}} : \mathbb{R} \to U$ given by $\Phi_{\mathbf{p}}(s) := \Phi(s, \mathbf{p})$ is continuously differentiable. Furthermore, since $\dot{\Phi}_{\mathbf{p}}(s) = \mathbf{f}(\Phi_{\mathbf{p}}(s))$, if there exists s_0 such that $\dot{\Phi}_{\mathbf{p}}(s_0) = \mathbf{0}$, then (by the uniqueness of the solutions) we have $\Phi_{\mathbf{p}}(s) = \mathbf{p}$ for every $s \in \mathbb{R}$. In this case, the orbit $\gamma(\mathbf{p}) = \{\mathbf{p}\}$ is called a *singular point*. To simplify the notation, if $\gamma(\mathbf{p})$ is a singular point, we denote it by $\mathbf{p}$. Therefore $\mathbb{R}^n \setminus \gamma(\mathbf{p})$, $\mathbb{R}^n \setminus \{\mathbf{p}\}$ and $\mathbb{R}^n \setminus \mathbf{p}$ are identical notations. If $\dot{\Phi}_{\mathbf{p}}(s_0) \neq \mathbf{0}$ for some $s_0 \in \mathbb{R}$, then $\Phi_{\mathbf{p}}(\mathbb{R}) = \gamma(\mathbf{p})$ is a one-dimensional manifold and we call $\mathbf{p}$ a *regular point*. The flow in a sufficiently small neighbourhood of a regular point is said to be *parallel*. For the definition of a parallel flow in a neighbourhood of a singular point see Subsection 2.6.3. By the classification of one-dimensional manifolds (see [38]), $\gamma(\mathbf{p})$ is diffeomorphic either to $\mathbb{R}$, or to $\mathbb{S}^1$. When $\gamma(\mathbf{p})$ is diffeomorphic to $\mathbb{S}^1$ the orbit $\gamma(\mathbf{p})$ is called a *periodic orbit*.

Theorem 2.3.2. *Every orbit of a differential equation* (2.3) *is diffeomorphic either to a point, or to a circle* $\mathbb{S}^1$, *or to a straight line* $\mathbb{R}$.

Example 8. By Example 7, the flow of the piecewise linear differential system $\dot{x} = x$, $\dot{y} = |y|$ is given by $\Phi(s, (x_0, y_0)) = (e^s x_0, e^s y_0)$ when $y_0 \geq 0$ and by $\Phi(s, (x_0, y_0)) = (e^s x_0, e^{-s} y_0)$ when $y_0 < 0$. The corresponding phase portrait is shown in Figure 2.3(c). In this example, each orbit, except the one that passes through the origin, is diffeomorphic to the line $\mathbb{R}$. The orbit through the origin is diffeomorphic to a point. Therefore, it is a singular point.

2.4 Basic ideas in qualitative theory

After analysing the topology of the orbits we present some basic definitions for studying their asymptotic behaviour. Consider the differential equation (2.3) and let E be a subset of U. The set E is said to be *positively invariant (under the flow)* if for every $\mathbf{q} \in E$ we have $\Phi(s, \mathbf{q}) \in E$ for all $s \geq 0$. The set E is said to be *negatively invariant (under the flow)* if for every $\mathbf{q} \in E$ we have $\Phi(s, \mathbf{q}) \in E$ for all $s \leq 0$. A set E is said to be *invariant (under the flow)* when it is both positively and negatively invariant (under the flow).

An invariant set E is *stable*, if for any neighbourhood W of E, there exists a neighbourhood V of E such that, for every $\mathbf{p} \in V$ and $s > 0$ it holds that $\Phi(s, \mathbf{p}) \in W$. An invariant set E is *unstable* when it is not stable.

Given $\mathbf{p}, \mathbf{q} \in U$, the point $\mathbf{q}$ is called an *α-limit point of* $\mathbf{p}$ if there exists a sequence $\{s_n\}_{n=0}^{+\infty}$ satisfying $\lim_{n \nearrow +\infty} s_n = -\infty$ and such that $\lim_{n \nearrow +\infty} \Phi(s_n, \mathbf{p}) = \mathbf{q}$. The point $\mathbf{q}$ is called an *ω-limit point of* $\mathbf{p}$ if there exists a sequence $\{s_n\}_{n=0}^{+\infty}$ satisfying $\lim_{n \nearrow +\infty} s_n = +\infty$ and such that $\lim_{n \nearrow +\infty} \Phi(s_n, \mathbf{p}) = \mathbf{q}$.

The *α-limit set* of a point $\mathbf{p} \in U$, denoted by $\alpha(\mathbf{p})$, is defined as the union of the α-limit points of $\mathbf{p}$. Analogously the *ω-limit set* of a point $\mathbf{p} \in U$, denoted by $\omega(\mathbf{p})$, is defined as the union of the ω-limit points of $\mathbf{p}$.

Let $\gamma(\mathbf{p})$, or simply γ, be the orbit passing through the point $\mathbf{p} \in U$. The *α-limit set of the orbit* γ is the α-limit set of the point $\mathbf{p}$, the *ω-limit set of the orbit* γ is the ω-limit set of $\mathbf{p}$. As it is easy to check, these definitions do not depend on the chosen point $\mathbf{p}$ of the orbit. Therefore, we denote the α- and the ω-limit set of an orbit by $\alpha(\gamma)$ and $\omega(\gamma)$, respectively.

Given an invariant set E, the *stable manifold* of E, denoted by $W^s(E)$, is the set of points in the phase space U whose ω-limit set is contained in E. The *unstable manifold* of E, denoted by $W^u(E)$, is the set of points in U whose α-limit set is contained in E.

A set E is called *asymptotically stable* if its stable manifold $W^s(E)$ is a neighbourhood of E. A set E is called *asymptotically unstable* if its unstable manifold $W^u(E)$ is a neighbourhood of E. In particular, every asymptotically stable (respectively, unstable) set is stable (respectively, unstable).

A *limit cycle* of the differential equation (2.3) is a periodic orbit isolated in the set of all the periodic orbits of (2.3). A limit cycle is called *stable* (respectively, *unstable*) if it is asymptotically stable (respectively, unstable). Another kind of limit cycle, called semistable limit cycle, can be also defined and we will introduce it in Section 2.8.

Example 9. In this example we consider a fundamental system

$$\dot{\mathbf{x}} = \begin{cases} A\mathbf{x} + \mathbf{b}, & \text{if } \mathbf{k}^T\mathbf{x} > 1, \\ B\mathbf{x}, & \text{if } |\mathbf{k}^T\mathbf{x}| \leq 1, \\ A\mathbf{x} - \mathbf{b}, & \text{if } \mathbf{k}^T\mathbf{x} < -1, \end{cases}$$

with parameters $d = \det(A) < 0$, $t = \operatorname{trace}(A) < 0$, $D = \det(B) > 0$ and $T = \operatorname{trace}(B) = 0$. In Section 5.3 we prove that its phase portrait in a neighbourhood of the origin is which is the one shown in Figure 2.4.

Different invariant sets can be easily identified. For instance, invariant sets are present in both the grey and the central white region formed by periodic orbits. This is because every orbit contained in one of these regions does not leave the region, neither in positive time, nor in negative time. Of course, sets formed by singular orbits are also invariant. Hence the singular points $\mathbf{e}_+$, $\mathbf{0}$ and $\mathbf{e}_-$, and the periodic orbit Γ are invariant.

Note that Γ is a stable invariant set. In fact, its stable manifold $W^s(\Gamma)$ is the whole grey region. However, it is not asymptotically stable, because $W^s(\Gamma)$ is not a neighbourhood of Γ. The origin $\mathbf{0}$ is also a stable invariant set which is not asymptotically stable.

On the other hand, the singular point $\mathbf{e}_-$ is the ω-limit set of the orbits γ_1^- and γ_2^-, see Figure 2.4. It is also the α-limit set of the orbits γ_3^- and γ_4^-. The periodic orbit Γ is the ω-limit set of the orbit γ_4^-.

Let γ be an orbit of the flow $\Phi(s, \mathbf{x})$ and $\mathbf{p}$ be a point on γ. We define the positive and negative semiorbit of γ as the sets $\gamma^+(\mathbf{p}) := \{\Phi(s, \mathbf{p}) : s \geq 0\}$ and

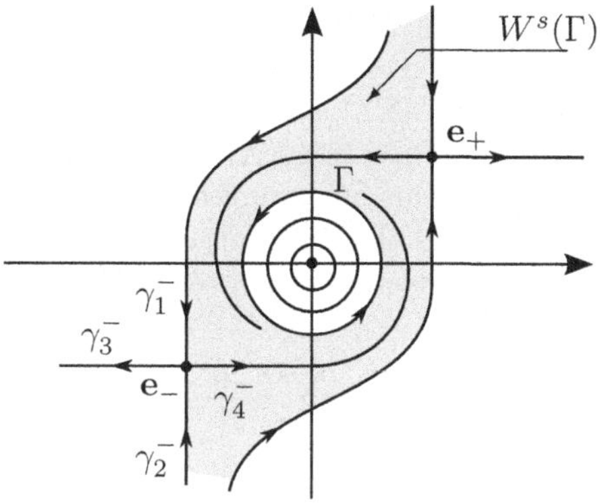

Figure 2.4: Phase portrait of the fundamental system with $D > 0$ and $T = 0$ in a neighbourhood of the origin $\mathbf{0}$. Invariant regions: the singular points $\mathbf{e}_+$, $\mathbf{0}$, $\mathbf{e}_-$; the periodic orbit Γ; and the open region $W^s(\Gamma)$ (in grey) and the open region in the interior of Γ (in white and foliated by periodic orbits).

$\gamma^-(\mathbf{p}) := \{\Phi(s, \mathbf{p}) : s \leq 0\}$, respectively. The orbit γ is called *positively bounded* if there exist a point $\mathbf{p} \in \gamma$ and a compact subset K of U such that $\gamma^+(\mathbf{p}) \subset K$. The orbit γ is called *negatively bounded* if there exist a point $\mathbf{p} \in \gamma$ and a compact subset K of U such that $\gamma^-(\mathbf{p}) \subset K$. Finally, γ is said to be *bounded* if it is positively and negatively bounded.

Proposition 2.4.1. *Let γ be an orbit of the differential system* (2.3). *If γ is positively bounded (respectively, negatively bounded), then $\omega(\gamma)$ (respectively, $\alpha(\gamma)$) is a non-empty set.*

For a proof of this result we refer the reader to [53, p. 191] or [57, p. 245]. Note that in references above, authors require the differentiability of the vector field. It is easy to check that instead of this hypothesis we can require the uniqueness of the solutions and the completeness of the flow.

2.5 Linear systems

Linear systems of differential equations, or briefly, linear systems, are one of the families of differential equations for which there exists a complete theory. We review some of the standard facts on linear systems because, as we will see later, there exists a close relationship between linear and general non-linear differential systems. The nature of this relationship is such that linear systems can be considered as a first natural step in the study of the differential systems.

As usual, $L(\mathbb{R}^n)$ denotes the vector space of the linear maps from $\mathbb{R}^n$ to $\mathbb{R}^n$, and $GL(\mathbb{R}^n)$ the group of the invertible linear maps. Consider $T \in L(\mathbb{R}^n)$ and let

A be the matrix representation of T. In the sequel we will identify the linear map T with its matricial representation A, and write $A \in L(\mathbb{R}^n)$. If T is invertible; i.e., $\det(A) \neq 0$, we will write $A \in GL(\mathbb{R}^n)$.

If $A \in L(\mathbb{R}^n)$ we denote by t or trace(A) the trace of A, and by d or $\det(A)$ the determinant of A. This explains our use of the variable s, instead of the more usual one t, to denote the time in the differential equation. Let $A \in L(\mathbb{R}^n)$. Then for every $s \in \mathbb{R}$ we define the *exponential matrix* of the matrix sA as the formal power series

$$e^{sA} := \sum_{k=0}^{\infty} \frac{s^k A^k}{k!},$$

where A^0 denotes the identity matrix Id and $A^k = A^{k-1}A$ for $k \geq 1$. Two matrices $A, B \in L(\mathbb{R}^n)$ are said to be *equivalent* if there exists $P \in GL(\mathbb{R}^n)$ such that $B = PAP^{-1}$. We summarize some properties of the exponential matrix in the following proposition.

Proposition 2.5.1. *Let $A \in L(\mathbb{R}^n)$.*

(a) *For every $s \in \mathbb{R}$, the series*

$$\sum_{k=0}^{\infty} \frac{s^k A^k}{k!}$$

is absolutely convergent. Moreover, if $s_0 > 0$, the series is uniformly convergent in $(-s_0, s_0)$.

(b) *If $A, B \in L(\mathbb{R}^n)$ are equivalent matrices with $B = PAP^{-1}$ for a $P \in \mathrm{GL}(\mathbb{R}^n)$, then $e^{sB} = Pe^{sA}P^{-1}$ for every $s \in \mathbb{R}$.*

(c) *If $B \in L(\mathbb{R}^n)$ is such that $AB = BA$, then $e^{s(A+B)} = e^{sA}e^{sB}$ for every $s \in \mathbb{R}$.*

(d) *For every $s \in \mathbb{R}$, $\left(e^{sA}\right)^{-1} = e^{-sA}$.*

(e) *For every $s \in \mathbb{R}$, $de^{sA}/ds = Ae^{sA}$.*

(f) *Let $\mathbf{v} \in \mathbb{R}^n$ be an eigenvector of A with eigenvalue $\lambda \in \mathbb{R}$. Then $\mathbf{v}$ is an eigenvector of e^{sA} with eigenvalue $e^{s\lambda}$.*

A proof of these results can be found in [7, Chapter 3] or [53, pp. 10–13].

In this section we consider the linear system (more precisely, *the homogeneous linear system*)

$$\dot{\mathbf{x}} = A\mathbf{x}, \tag{2.4}$$

where $A \in L(\mathbb{R}^n)$, and denote $d = \det(A)$ and $t = \mathrm{trace}(A)$.

The linear vector field $\mathbf{f}(\mathbf{x}) = A\mathbf{x}$ is a globally Lipschitz function with Lipschitz constant $L = ||A||$. From the existence and uniqueness theorem it follows that for every $\mathbf{x}_0 \in \mathbb{R}^n$ there exists a unique solution of system (2.4) passing through $\mathbf{x}_0$ at $s = 0$. Moreover, this solution is defined for all $s \in \mathbb{R}$ (see Proposition 2.1.2). The following result provides an explicit expression for the linear flows.

Theorem 2.5.2 (Linear flow). *The linear differential equation* $\dot{\mathbf{x}} = A\mathbf{x}$, *with* $A \in L(\mathbb{R}^n)$, *defines a flow* $\Phi : \mathbb{R} \times \mathbb{R}^n \to \mathbb{R}^n$ *given by* $\Phi(s, \mathbf{x}) = e^{sA}\mathbf{x}$.

A proof of this theorem can be obtained as a corollary of Proposition 2.5.1(e).

We denote by $\ker(A)$ the vector subspace formed by the singular points of the linear system (2.4). This subspace is called the *kernel of the linear map A*. Notice that the origin always belong to $\ker(A)$. Moreover, when $A \in GL(\mathbb{R}^n)$, the origin is the unique singular point.

Let $\mathbf{v}_1, \mathbf{v}_2, \ldots, \mathbf{v}_{n_s}$ be the generalized eigenvectors corresponding to the eigenvalues of the matrix A with negative real part. The *stable subspace* is the vector subspace generated by the vectors $\mathbf{v}_1, \mathbf{v}_2, \ldots, \mathbf{v}_{n_s}$, i.e.,

$$E^s := \langle \mathbf{v}_1, \mathbf{v}_2, \ldots, \mathbf{v}_{n_s} \rangle .$$

Let $\mathbf{u}_1, \mathbf{u}_2, \ldots, \mathbf{u}_{n_u}$ be the generalized eigenvectors corresponding to the eigenvalues of the matrix A with positive real part. The *unstable subspace* is the vector subspace

$$E^u := \langle \mathbf{u}_1, \mathbf{u}_2 \ldots, \mathbf{u}_{n_u} \rangle .$$

Let $\mathbf{w}_1, \ldots, \mathbf{w}_{n_c}$ be the generalized eigenvectors corresponding to the eigenvalues of the matrix A with zero real part. The *center subspace* is the vector subspace

$$E^c := \langle \mathbf{w}_1, \mathbf{w}_2 \ldots, \mathbf{w}_{n_c} \rangle .$$

Theorem 2.5.3 (Dynamical behaviour of linear systems). *Consider the linear differential system* $\dot{\mathbf{x}} = A\mathbf{x}$ *with* $A \in GL(\mathbb{R}^n)$. *Then:*

(a) $\mathbb{R}^n = E^s \oplus E^u \oplus E^c$.

(b) $W^s(\mathbf{0}) = E^s$.

(c) $W^u(\mathbf{0}) = E^u$.

For a proof of this result, see [53, Section 1.9].

2.5.1 Non-homogeneous linear systems

Differential systems of the form

$$\dot{\mathbf{x}} = A\mathbf{x} + \mathbf{b}, \tag{2.5}$$

with $A \in L(\mathbb{R}^n)$ and $\mathbf{b} \in \mathbb{R}^n \setminus \{\mathbf{0}\}$ are called *non-homogeneous linear (differential) systems*. By Proposition 2.5.1(e), the flow of systems (2.5) is given by

$$\Phi(s, \mathbf{x}) = e^{sA}\mathbf{x} + \int_0^s e^{(s-r)A}\mathbf{b}\, dr.$$

If the non-homogeneous linear system (2.5) has a singular point $\mathbf{x}^*$, the change of coordinates $\mathbf{z} = \mathbf{x} - \mathbf{x}^*$ transforms it into the homogeneous linear system $\dot{\mathbf{z}} = A\mathbf{z}$. Thus the flow of the non-homogeneous linear system (2.5) is a translation of the flow of a homogeneous linear system, namely $\Phi(s, \mathbf{x}) = e^{sA}(\mathbf{x} - \mathbf{x}^*) + \mathbf{x}^*$. Finally, note that if the non-homogeneous linear system has no singular points, then $\det(A) = 0$.

2.5.2 Planar linear systems

In the following two subsections we restrict our attention to planar linear systems. We begin by showing the following version of the real Jordan normal form theorem [33].

Theorem 2.5.4 (Real Jordan normal form). *Consider a matrix $A \in L(\mathbb{R}^2)$ with $d = \det(A)$ and $t = \text{trace}(A)$. A is equivalent to one of the following matrices J:*

(a) *If $d = 0$ and $t = 0$, then $J = \begin{pmatrix} 0 & 0 \\ 0 & 0 \end{pmatrix}$ or $J = \begin{pmatrix} 0 & 1 \\ 0 & 0 \end{pmatrix}$.*

(b) *If $d = 0$ and $t \neq 0$, then $J = \begin{pmatrix} t & 0 \\ 0 & 0 \end{pmatrix}$.*

(c) *If $d > 0$ and $t = 0$, the eigenvalues of A are complex numbers with zero real part and imaginary part $\beta > 0$, and $J = \begin{pmatrix} 0 & -\beta \\ \beta & 0 \end{pmatrix}$.*

(d) *If $d > 0$ and $t^2 - 4d = 0$, there exists exactly one real eigenvalue of A with multiplicity two, λ_1, and $J = \begin{pmatrix} \lambda_1 & 0 \\ 0 & \lambda_1 \end{pmatrix}$ or $J = \begin{pmatrix} \lambda_1 & 1 \\ 0 & \lambda_1 \end{pmatrix}$.*

(e) *If $d > 0$ and $t^2 - 4d > 0$, there exist two real eigenvalues of A, $\lambda_1 > \lambda_2$, and $J = \begin{pmatrix} \lambda_1 & 0 \\ 0 & \lambda_2 \end{pmatrix}$.*

(f) *If $d > 0$, $t \neq 0$ and $t^2 - 4d < 0$, the eigenvalues of A are complex numbers with real part $\alpha \neq 0$ and imaginary part $\beta > 0$, and $J = \begin{pmatrix} \alpha & -\beta \\ \beta & \alpha \end{pmatrix}$.*

(g) *If $d < 0$, there exist two real eigenvalues of A, $\lambda_1 > 0 > \lambda_2$, and $J = \begin{pmatrix} \lambda_1 & 0 \\ 0 & \lambda_2 \end{pmatrix}$.*

The matrix J defined in the preceding theorem is called the *real Jordan normal form* of A. Note that, except when $t^2 - 4d = 0$, the real Jordan normal form of A is determined by the parameters t and d. If $t^2 - 4d = 0$, then there exist two possibilities, one diagonal and the other non-diagonal, depending on the coefficients of A.

Consider the linear system

$$\dot{\mathbf{x}} = A\mathbf{x}, \tag{2.6}$$

with $A \in L(\mathbb{R}^2)$, and let $P \in GL(\mathbb{R}^2)$ be the matrix which transforms A into its real Jordan normal form J, i.e., $J = PAP^{-1}$. The linear change of coordinates $\mathbf{y} = P\mathbf{x}$ transforms the linear system (2.6) into the system

$$\dot{\mathbf{y}} = J\mathbf{y}. \tag{2.7}$$

To obtain the expression of the linear flow of (2.7) it is enough to consider the following cases:

$$J = \begin{pmatrix} \lambda_1 & 0 \\ 0 & \lambda_2 \end{pmatrix}, \quad J = \begin{pmatrix} \lambda & 1 \\ 0 & \lambda \end{pmatrix} \text{ and } J = \begin{pmatrix} \alpha & -\beta \\ \beta & \alpha \end{pmatrix},$$

see Propositon 2.5.1(b) and Theorem 2.5.4

Proposition 2.5.5. *Consider $J \in L(\mathbb{R}^2)$ and $s \in \mathbb{R}$.*

(a) *If* $J = \begin{pmatrix} \lambda_1 & 0 \\ 0 & \lambda_2 \end{pmatrix}$, *then* $e^{sJ} = \begin{pmatrix} e^{s\lambda_1} & 0 \\ 0 & e^{s\lambda_2} \end{pmatrix}$.

(b) *If* $J = \begin{pmatrix} \lambda & 1 \\ 0 & \lambda \end{pmatrix}$, *then* $e^{sJ} = e^{s\lambda} \begin{pmatrix} 1 & s \\ 0 & 1 \end{pmatrix}$.

(c) *If* $J = \begin{pmatrix} \alpha & -\beta \\ \beta & \alpha \end{pmatrix}$, *then* $e^{sJ} = e^{s\alpha} \begin{pmatrix} \cos(\beta s) & -\sin(\beta s) \\ \sin(\beta s) & \cos(\beta s) \end{pmatrix}$.

For a proof of this proposition see [7], [53], or [57].

Let $\Phi(s, \mathbf{x})$ and $\Psi(s, \mathbf{y})$ be the flows of systems (2.6) and (2.7), respectively. If $\mathbf{x}_0 \in \mathbb{R}^2$, then $\Phi(s, \mathbf{x}_0) = e^{sA}\mathbf{x}_0 = P^{-1}e^{sJ}P\mathbf{x}_0 = P^{-1}\Psi(s, P\mathbf{x}_0)$. Therefore,

$$\Phi(s, \mathbf{x}) = P^{-1}\Psi(s, P\mathbf{x}). \tag{2.8}$$

From this we obtain the expressions of the flow of any planar linear system.

Theorem 2.5.6. *Consider the flow $\Phi(t, \mathbf{x})$ of the linear system $\dot{\mathbf{x}} = A\mathbf{x}$, with $A \in L(\mathbb{R}^2)$, $d = \det(A)$ and $t = \operatorname{trace}(A)$. Let J be the real Jordan normal form of A and P be the matrix such that $J = PAP^{-1}$.*

(a) *If $t^2 - 4d > 0$, then*

$$\Phi(s, \mathbf{x}) = P^{-1} \begin{pmatrix} e^{s\lambda_1} & 0 \\ 0 & e^{s\lambda_2} \end{pmatrix} P\mathbf{x}.$$

(b) *If $t^2 - 4d = 0$, then either $\Phi(s, \mathbf{x}) = e^{s\lambda}\mathbf{x}$ or*

$$\Phi(s, \mathbf{x}) = P^{-1} \begin{pmatrix} e^{s\lambda} & s \\ 0 & e^{s\lambda} \end{pmatrix} P\mathbf{x},$$

depending on whether J is diagonal or not.

(c) *If $t^2 - 4d < 0$, then*

$$\Phi(s, \mathbf{x}) = e^{s\alpha} P^{-1} \begin{pmatrix} \cos(\beta s) & -\sin(\beta s) \\ \sin(\beta s) & \cos(\beta s) \end{pmatrix} P\mathbf{x}.$$

2.5.3 Planar phase portraits

In this subsection we describe the phase portrait of planar linear systems. We also present the notation of singular points of such systems. For a general classification of these singular points, see Subsection 2.7.1.

From relation (2.8) it follows that given $\mathbf{x}_0 \in \mathbb{R}^2$, the orbit of system (2.6) through $\mathbf{x}_0$ and the orbit of system (2.7) through $P\mathbf{x}_0$, $\gamma(\mathbf{x}_0)$ and $\gamma(P\mathbf{x}_0)$, respectively, satisfy $\gamma(\mathbf{x}_0) = P^{-1}\gamma(P\mathbf{x}_0)$. Therefore, the phase portrait of system (2.6) is a linear transformation of the phase portrait of system (2.7). Hence, it is enough to describe the phase portrait of a linear system (2.7), where J is the real Jordan normal form of the matrix A.

Case $d < 0$

If the determinant of the matrix A is strictly negative, then A has two real eigenvalues $\lambda_1 > 0 > \lambda_2$. Hence, the stable and unstable subspaces (E^s and E^u) are each generated by an eigenvector, and the central subspace is the origin, $E^c = \{\mathbf{0}\}$. The real Jordan normal form of A is

$$J = \begin{pmatrix} \lambda_1 & 0 \\ 0 & \lambda_2 \end{pmatrix}.$$

The phase portrait of the system $\dot{\mathbf{y}} = J\mathbf{y}$ is represented in Figure 2.5, the phase portrait of system $\dot{\mathbf{x}} = A\mathbf{x}$ is a linear transformation of it.

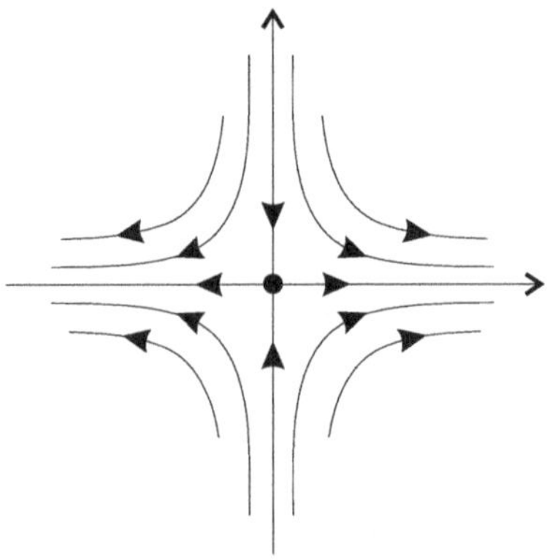

Figure 2.5: A saddle point and its stable and unstable separatrices.

In this case the singular point at the origin is called a *saddle point*. The two orbits in the stable subspace are called the *stable separatrices of the saddle* and the orbits in the unstable subspace are called the *unstable separatrices of the saddle*.

Case $d = 0$

Suppose that A is the zero matrix, i.e., the dimension of $\ker(A)$ is 2. In this case, any point in the phase plane is a singular point, so the case is of no interest. Assume now that $\ker(A)$ has dimension equal to 1, i.e., $\ker(A)$ is a straight line through the origin formed by all the singular points of the system. Hence, the singular points are not isolated. The real Jordan normal form of A changes according to whether $t = \text{trace}(A)$ is equal to zero or not. Thus, when $t = 0$ the matrix J is not diagonal and the straight line $\ker(A)$ is called a *non-isolated nilpotent manifold*, see Figure 2.6(b). When $t < 0$ (respectively $t > 0$) the matrix J is diagonal and the straight line $\ker(A)$ is called a *stable (respectively unstable) normally hyperbolic manifold*, see Figure 2.6(a) (respectively, (c)). The term "normally hyperbolic manifold" is motivated by [34].

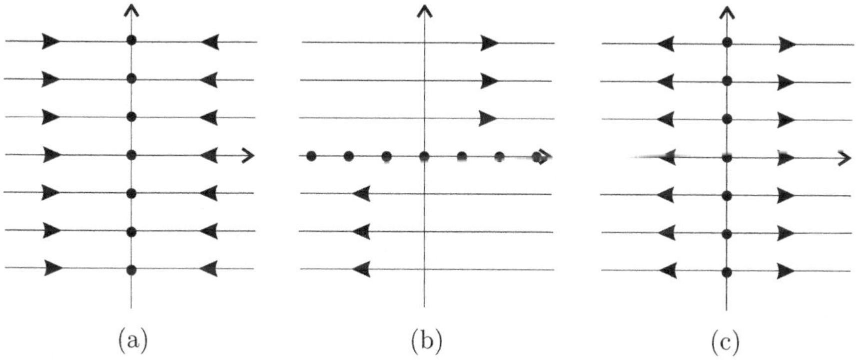

Figure 2.6: Non-isolated singular points: (a) Stable normally hyperbolic manifold for $t < 0$; (b) Non-isolated nilpotent manifold for $t = 0$; and (c) unstable normally hyperbolic manifold for $t > 0$.

Case $d > 0$

We distinguish three cases, depending on the sign of $t^2 - 4d$. When $t^2 - 4d > 0$, the matrix A has two real eigenvalues with the same sign, $\lambda_1 > \lambda_2$. Therefore if $t < 0$, then $E^s = \mathbb{R}^2$ and $E^u = E^c = \{\mathbf{0}\}$; and if $t > 0$, then $E^u = \mathbb{R}^2$ and $E^s = E^c = \{\mathbf{0}\}$. The phase portrait of the system $\dot{\mathbf{y}} = J\mathbf{y}$ is shown in Figure 2.7, depending on t. The corresponding phase portrait of the system $\dot{\mathbf{x}} = A\mathbf{x}$ is obtained by a linear transformation. The origin is called an *asymptotically stable node* if $t < 0$, and an *asymptotically unstable node* if $t > 0$.

When $t^2 - 4d = 0$, there exists a unique eigenvalue λ, which is real, and the real Jordan normal form of A is

$$J = \begin{pmatrix} \lambda & 0 \\ 0 & \lambda \end{pmatrix} \quad \text{or} \quad J = \begin{pmatrix} \lambda & 1 \\ 0 & \lambda \end{pmatrix}.$$

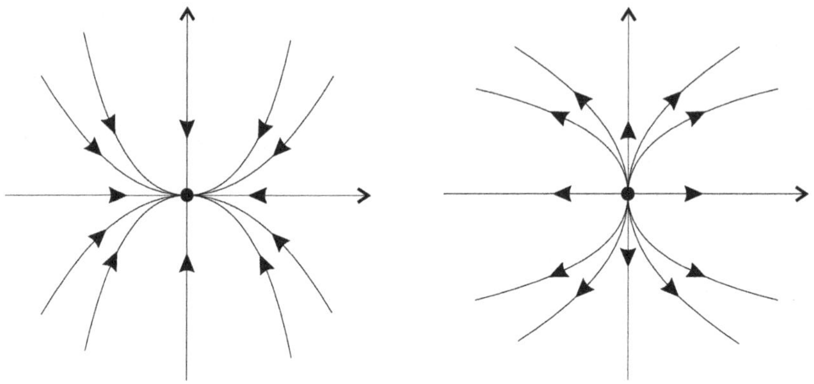

(a) (b)

Figure 2.7: (a) Asymptotically stable node. (b) Asymptotically unstable node.

For each of these matrices we have to consider the cases $t < 0$ and $t > 0$. The phase portrait of the system $\dot{\mathbf{y}} = J\mathbf{y}$ is shown in Figure 2.8, depending on t and J. The corresponding phase portrait of the system $\dot{\mathbf{x}} = A\mathbf{x}$ is a linear transformation of it. The origin is called a *degenerated diagonal node* in the first case, and a *degenerated node* in the second one.

When $t^2 - 4d < 0$, the eigenvalues of A are a pair of conjugate complex numbers and

$$J = \begin{pmatrix} \alpha & -\beta \\ \beta & \alpha \end{pmatrix}.$$

The phase portrait of the system $\dot{\mathbf{y}} = J\mathbf{y}$ is shown in Figure 2.9, depending on the sign of $t = 2\alpha$. The corresponding phase portrait of the system $\dot{\mathbf{x}} = A\mathbf{x}$ is a linear transformation of it. When $t = 0$, the origin is called a *center*. When $t < 0$, the origin is called an *asymptotically stable focus*. When $t > 0$, the origin is called an *asymptotically unstable focus*.

2.6 Classification of flows

Every classification criterion involves appropriate definitions for invariant sets, as specialized to different classes. If the list of the selected invariant sets is large, then the number of elements in each class is small and the classification is not effective. If the list of invariant sets is small, then we can collect systems with different behaviours and assign them to the same class. Thus the first step is to find an optimal classification criterion. In the theory of flows the criterion chosen is the preservation of the "orbit structure", a notion that will be defined in the following subsection.

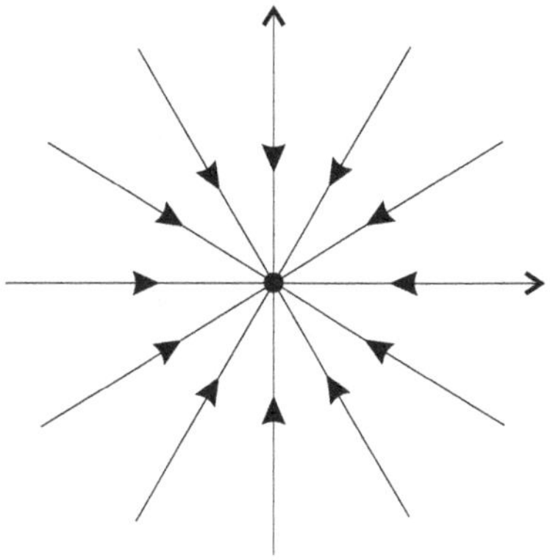
(a)

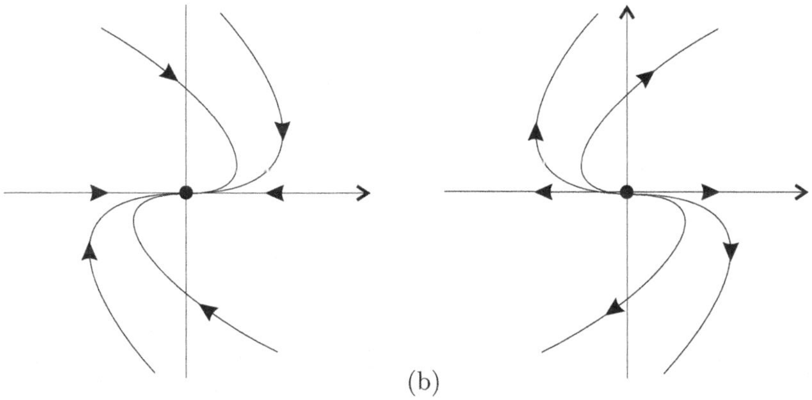
(b)

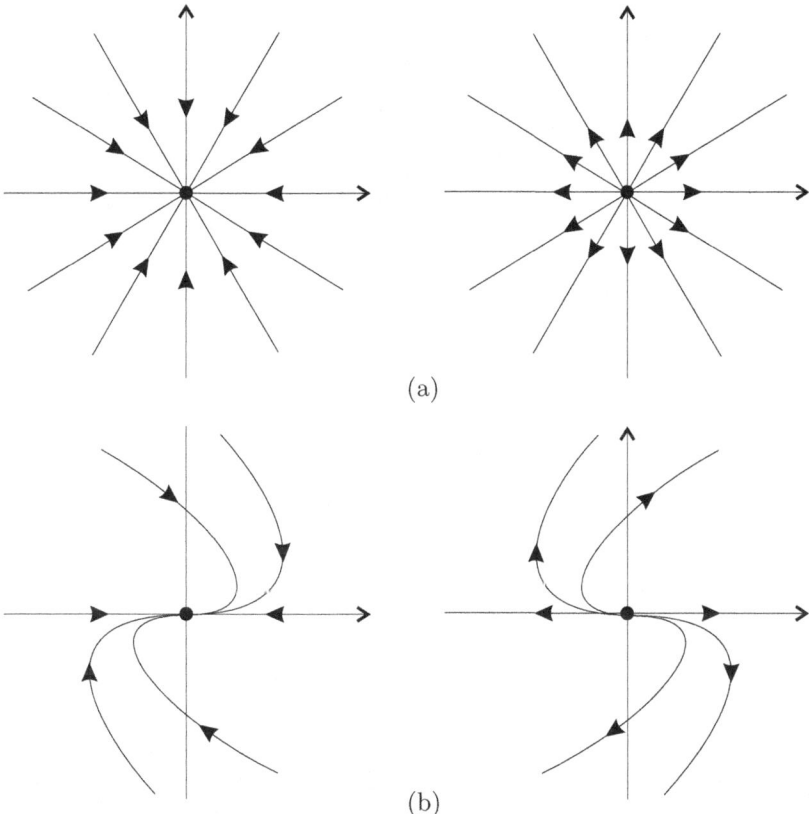

Figure 2.8: (a) Degenerated diagonal nodes. (b) Degenerated nodes.

2.6.1 Classification criteria

We begin by defining equivalence relations for flows, in correspondence to the algebraic, the differentiable and the topological points of view.

Consider the differentiable systems $\dot{\mathbf{x}} = \mathbf{f}(\mathbf{x})$ and $\dot{\mathbf{y}} = \mathbf{g}(\mathbf{y})$, with $\mathbf{f} : U \to \mathbb{R}^n$ a locally Lipschitz function defined on $U \subset \mathbb{R}^n$ and $\mathbf{g} : V \to \mathbb{R}^n$ a locally Lipschitz function defined on $V \subset \mathbb{R}^n$. Let $\Phi(s, \mathbf{x})$ and $\Phi^*(s, \mathbf{y})$ be the respective flows. We recall that in this work we consider only complete flows, i.e., the interval of definition of all the solutions is the entire $\mathbb{R}$.

Two flows are said to be *conjugate* if there exists a bijection $\mathbf{h} : U \to V$ (called *conjugacy*), such that $\Phi^*(s, \mathbf{h}(\mathbf{x})) = \mathbf{h}(\Phi(s, \mathbf{x}))$ for every $s \in \mathbb{R}$ and $\mathbf{x} \in U$. The flows are said to be *equivalent* if there exists a bijection $\mathbf{h} : U \to V$ (called *equivalence*), such that γ is an orbit of the first system if and only if $\mathbf{h}(\gamma)$ is an

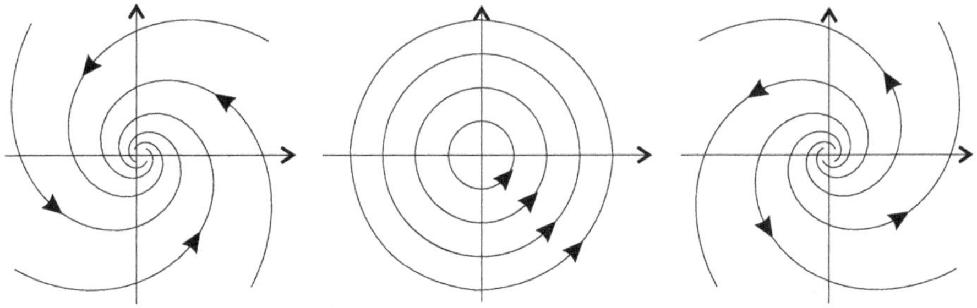

Figure 2.9: (a) Asymptotically stable focus. (b) Center. (c) Asymptotically unstable focus.

orbit of the second one and in addition $\mathbf{h}$ preserves the orientation of the orbit. It is easy to check that if two flows are conjugate, then they are equivalent. The converse is not always true.

An equivalence $\mathbf{h}$ transforms singular points into singular points and periodic orbits into periodic orbits. When $\mathbf{h}$ is a conjugacy, the period of the periodic orbits is also preserved.

Consider two conjugate (respectively equivalent) flows. The flows are said to be *linearly conjugate* (respectively, *linearly equivalent*) if $\mathbf{h}$ is a linear isomorphism. The flows are said to be C^r*-conjugate* (respectively, C^r*-equivalent*), with $r \in \{1, 2, \ldots, \infty, \omega\}$, if $\mathbf{h}$ is a diffeomorphism such that $\mathbf{h}$, $\mathbf{h}^{-1} \in C^r$ (recall here that C^ω denotes the class of analytic functions). The flows are said to be *topologically conjugate* (respectively *topologically equivalent*) if $\mathbf{h}$ is a homeomorphism.

Two differential equations are said to be linearly, C^r, or topologically equivalent (respectively, conjugate) if their flows are linearly, C^r, or topologically equivalent (respectively, conjugate). Further, they are said to present the same *qualitative behaviour* or the same *dynamical behaviour* if they are topologically equivalent.

In the next result we relate the different classification criteria.

Proposition 2.6.1. *Consider two differential equations.*

(a) *If they are linearly conjugate (respectively, equivalent), then they are* C^r*-conjugate (respectively,* C^r*-equivalent) for every* $r \in \{1, 2, \ldots, \infty, \omega\}$.

(b) *If they are* C^r*-conjugate (respectively,* C^r*-equivalent) with* $r \in \{1, \ldots \infty, \omega\}$, *then they are topologically conjugate (respectively, equivalent).*

(c) *If they are linearly,* C^r*, or topological conjugate, then they are linearly,* C^r*, or topologically equivalent.*

The conjugacy of flows is also a conjugacy of vector fields. In the next lemma we characterize the C^r-conjugacy via the conjugacy of vector fields. As usual, given a diffeomorphism $\mathbf{h} : U \to V$, $D\mathbf{h}(\mathbf{x})$ denotes the *Jacobian matrix* of $\mathbf{h}$ evaluated at the point $\mathbf{x}$.

Lemma 2.6.2. *Consider two differential equations $\dot{\mathbf{x}} = \mathbf{f}(\mathbf{x})$ and $\dot{\mathbf{y}} = \mathbf{g}(\mathbf{y})$, with $\mathbf{f} : U \to \mathbb{R}^n$ and $\mathbf{g} : V \to \mathbb{R}^n$ locally Lipschitz functions on U and V, respectively. Their flows are C^r-conjugate if and only if there exists a diffeomorphism $\mathbf{h} : U \to V$ in C^r such that $D\mathbf{h}(\mathbf{x})\mathbf{f}(\mathbf{x}) = \mathbf{g}(\mathbf{h}(\mathbf{x}))$ for every $\mathbf{x} \in U$.*

A proof of this result can be found in [58, p. 19, Lemma 3.4]

2.6.2 Classification of linear flows

Given a linear isomorphism $\mathbf{h} : U \to V$, with U and V open subsets of $\mathbb{R}^n$, there exists a matrix $M \in GL(\mathbb{R}^n)$ such that $\mathbf{h}(\mathbf{x}) = M\mathbf{x}$ for any $\mathbf{x} \in U$.

Lemma 2.6.3. *If the linear map $\mathbf{h}(\mathbf{x}) = M\mathbf{x}$ is constant on an open subset $U \subset \mathbb{R}^n$, then M is the zero matrix.*

Proof. Suppose that M is not the zero matrix. Then there exists a vector $\mathbf{e} \in U$ such that $M\mathbf{e} \neq \mathbf{0}$. Take $\mathbf{x}_0 \in U$. Since U is open, $\mathbf{x}_1 = \mathbf{x}_0 + \delta\mathbf{e} \in U$ for $\delta > 0$ small enough. Therefore, $\delta M\mathbf{e} = M\mathbf{x}_1 - M\mathbf{x}_0 = \mathbf{0}$, a contradiction. □

Proposition 2.6.4 (Linear conjugacy of linear flows). *Consider two linear systems $\dot{\mathbf{x}} = A\mathbf{x}$ and $\dot{\mathbf{y}} = A^*\mathbf{y}$, with $A, A^* \in L(\mathbb{R}^2)$, and denote $d = \det(A)$, $t = \operatorname{trace}(A)$, $d^* = \det(A^*)$ and $t^* = \operatorname{trace}(A^*)$.*

(a) *The systems are linearly conjugate if and only if there exists $M \in GL(\mathbb{R}^2)$ such that $A^* = MAM^{-1}$, i.e., the matrices of the systems are equivalent.*

(b) *If the systems are linearly conjugate, then $d = d^*$ and $t = t^*$.*

(c) *If $d = d^*$, $t = t^*$ and $t^2 - 4d \neq 0$, then the systems are linearly conjugate.*

Proof. (a) Suppose that the given systems are linearly conjugate. By definition there exists a linear map $M \in GL(\mathbb{R}^2)$ such that, for any given solution of the first system, $\mathbf{x}(s) = \phi(s; 0, \mathbf{x}_0)$, the function $\mathbf{y}(s) = M\mathbf{x}(s)$ is a solution of the second one. Moreover, $\dot{\mathbf{y}} = MAM^{-1}\mathbf{y}$. Applying Lemma 2.6.3 to the linear map $\mathbf{h}(\mathbf{y}) = \left(A^* - MAM^{-1}\right)\mathbf{y}$, we conclude that $A^* = MAM^{-1}$.

Conversely, suppose that $A^* = MAM^{-1}$ with $M \in GL(\mathbb{R}^2)$. By Proposition 2.5.1.(b), $e^{sA^*} = Me^{sA}M^{-1}$ for all $s \in \mathbb{R}$. The flows of the linear systems are $\Phi(s, \mathbf{x}) = e^{sA}\mathbf{x}$ and $\Phi^*(s, \mathbf{y}) = e^{sA^*}\mathbf{y}$, respectively, see Theorem 2.5.2. Hence, $\Phi^*(s, M\mathbf{x}) = e^{sA^*}M\mathbf{x} = Me^{sA}\mathbf{x} = M\Phi(s, \mathbf{x})$. Therefore, the systems are linearly conjugate.

Statement (b) follows from statement (a). For a proof of statement (c) see Arnold [7, p. 169]. □

Proposition 2.6.5 (C^r-conjugacy of linear flows). *Two linear flows are C^r-conjugated for $r \in \{1, 2, \ldots, \infty, \omega\}$ if and only if they are linearly conjugate.*

For a proof of the previous proposition see Arnold [7, p. 170].

Corollary 2.6.6. *Consider two linear systems* $\dot{\mathbf{x}} = A\mathbf{x}$ *and* $\dot{\mathbf{y}} = A^*\mathbf{y}$ *and denote* $d = \det(A)$, $t = \operatorname{trace}(A)$, $d^* = \det(A^*)$ *and* $t^* = \operatorname{trace}(A^*)$. *If the flows are* C^r*-conjugate with* $r \in \{1, 2, \ldots, \infty, \omega\}$, *then* $d = d^*$ *and* $t = t^*$.

Proof. The proof follows from Propositions 2.6.5 and 2.6.4 (b). □

In the next result we present a characterization of the topological conjugacy of linear flows.

Proposition 2.6.7 (Topological conjugacy of linear flows). *The flows of two linear systems whose eigenvalues have no zero real part are topologically conjugate if and only if they have the same number of eigenvalues with positive and the same number of eigenvalues with negative real part.*

For a proof of this result see Arnold [7, pp. 172–182].

2.6.3 Topological equivalence of non-linear flows

As we have seen, in the case of linear flows there exists a characterization of the three different classification criteria. To our knowledge a complete characterization of topological equivalence exists only for planar non-linear flows. To introduce it we need some new notations and results analogous to the ones in the previous subsection. Essentially all these definitions and results can be found in [48, pp. 127–148] and [50, pp. 73–81], where they are applied in a more general context. Similar results are due to Peixoto [52].

Consider a differential equation $\dot{\mathbf{x}} = \mathbf{f}(\mathbf{x})$ with $\mathbf{f}$ a Lipschitz function defined in $\mathbb{R}^2$, and let $\Phi(s, \mathbf{x})$ be its flow. Following Markus and Neumann, we denote this flow by $(\mathbb{R}^2, \Phi)$. By the continuous dependence of solutions on the initial conditions and parameters, the flow $(\mathbb{R}^2, \Phi)$ is continuous in both variables. The flow $(\mathbb{R}^2, \Phi)$ is said to be *parallel* if it is topologically equivalent to one of the following flows:

(a) The flow defined in $\mathbb{R}^2$ by the differential system $\dot{x} = 1$, $\dot{y} = 0$, called *strip flow*.

(b) The flow defined in $\mathbb{R}^2 \setminus \{\mathbf{0}\}$ by the differential system in polar coordinates $\dot{r} = 0$, $\dot{\theta} = 1$, called *annular flow*.

(c) The flow defined in $\mathbb{R}^2 \setminus \{\mathbf{0}\}$ by the differential system in polar coordinates $\dot{r} = r$, $\dot{\theta} = 0$, called *spiral* or *radial flow*.

An orbit $\gamma(\mathbf{p})$ of the flow $(\mathbb{R}^2, \Phi)$ is called a *separatrix* if

(a) is a singular point, or

(b) is a limit cycle, or

(c) $\gamma(\mathbf{p})$ is homeomorphic to $\mathbb{R}$ and there is no tubular neighbourhood N of $\gamma(\mathbf{p})$ with the following properties:

  (c.1) Every point $\mathbf{q}$ in N has the same α-limit and ω-limit sets of $\mathbf{p}$, i.e., $\alpha(\mathbf{q}) = \alpha(\mathbf{p})$ and $\omega(\mathbf{q}) = \omega(\mathbf{p})$.

(c.2) The boundary of N, i.e., $\mathrm{Cl}(N) \setminus N$, is formed by $\alpha(\mathbf{p})$, $\omega(\mathbf{p})$ and two orbits $\gamma(\mathbf{q}_1)$ and $\gamma(\mathbf{q}_2)$ such that $\alpha(\mathbf{p}) = \alpha(\mathbf{q}_1) = \alpha(\mathbf{q}_2)$ and $\omega(\mathbf{p}) = \omega(\mathbf{q}_1) = \omega(\mathbf{q}_2)$, see Figure 2.10. As usual $\mathrm{Cl}(N)$ denotes the closure of N, i.e., the smallest closed set containing N.

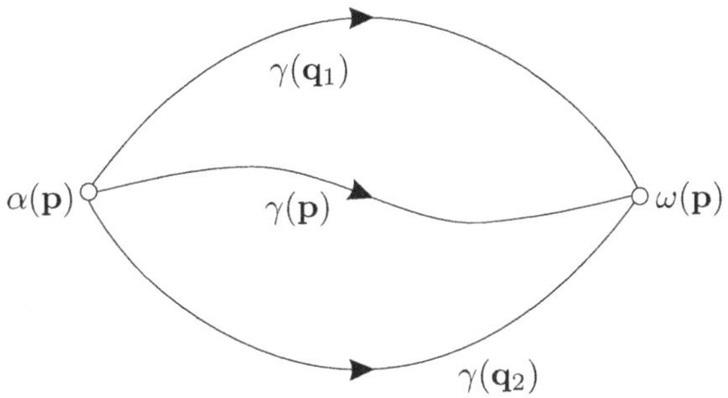

Figure 2.10: The boundary of N.

Let S be the union of the separatrices of the flow $(\mathbb{R}^2, \Phi)$. It is easy to check that S is an invariant closed set. If N is a connected component of $\mathbb{R}^2 \setminus S$, then N is also an invariant set, and the flow $(N, \Phi|_N)$ is called a *canonical region* of the flow $(\mathbb{R}^2, \Phi)$.

Proposition 2.6.8. *Every canonical region of the flow* $(\mathbb{R}^2, \Phi)$ *is parallel.*

For a proof of this proposition see [50].

The *separatrix configuration* of a flow $(\mathbb{R}^2, \Phi)$ is the union of all separatrices of the flow together with an orbit belonging to each canonical region. Given two flows $(\mathbb{R}^2, \Phi)$ and $(\mathbb{R}^2, \Phi^*)$, let S and S^* be the union of their separatrices, respectively. The separatrix configuration C of the flow $(\mathbb{R}^2, \Phi)$ is said to be topologically equivalent to the separatrix configuration C^* of the flow $(\mathbb{R}^2, \Phi^*)$ if there exists an orientation preserving homeomorphism from $\mathbb{R}^2$ to $\mathbb{R}^2$ which transforms orbits of C into orbits of C^*, and orbits of S into orbits of S^*.

Theorem 2.6.9 (Markus–Neumann–Peixoto). *Let* $(\mathbb{R}^2, \Phi)$ *and* $(\mathbb{R}^2, \Phi^*)$ *be two continuous flows with only isolated singular points. Then they are topologically equivalent if and only if their separatrix configurations are topologically equivalent.*

For a proof of this result we refer the reader to [50].

It follows from the previous theorem that in order to classify the flows of planar differential equations, it is enough to describe their separatrix configuration.

Example 10. Consider the local phase portrait depicted in Figure 2.11(a). The set S of all separatrices is formed by the singular points $\mathbf{e}_+$, $\mathbf{e}_-$ and $\mathbf{0}$, the periodic orbits Γ_+ and Γ_-, and the homoclinic orbits γ_+ and γ_-. Therefore, S is an invariant closed set. In Figure 2.11(b) we represent the set of all canonical regions.

Note that Figure 2.11(a) presents clearly the set of all separatrices together with an orbit for each canonical region which shows the asymptotic behaviour of the orbits contained in its interior. Thus Figure 2.11(a) also represents the separatrix configuration of the phase portrait. From this it is easy to understand that the separatrix configuration is the skeleton of the phase portrait.

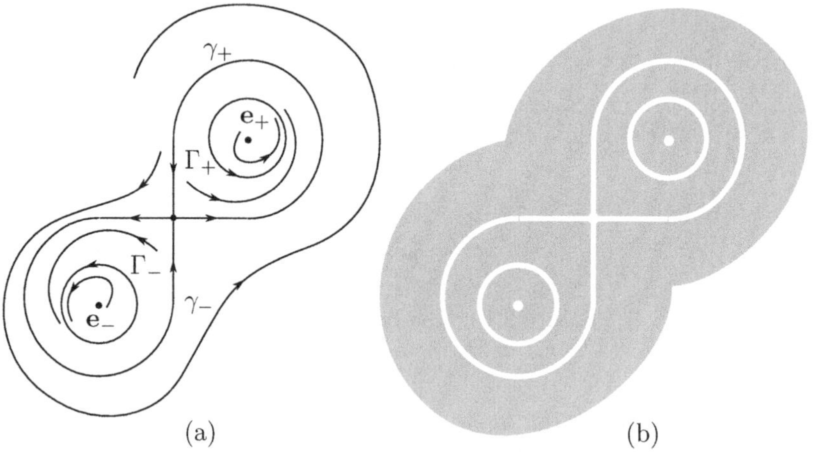

Figure 2.11: (a) Separatrix configuration correspondig to a fundamental system with parameters $D < 0$, $T < 0$ and $t = w_1(d)$, see Section 5.5. (b) Canonical regions associated to the phase portrait.

2.7 Non-linear systems

In this section we return to non-linear flows. Let $U \subseteq \mathbb{R}^n$ be an open subset, $\mathbf{f} : U \to \mathbb{R}^n$ be a locally Lipschitz function in U and $\Phi(s, \mathbf{x})$ be the flow defined by the differential equation $\dot{\mathbf{x}} = \mathbf{f}(\mathbf{x})$. Recall that we consider only complete flows, i.e., solutions are defined for every value of time $s \in \mathbb{R}$.

2.7.1 Local phase portraits of singular points

We begin by studying the local behaviour of flows in a neighbourhood of singular points, i.e., points $\mathbf{x} \in U$ such that $\mathbf{f}(\mathbf{x}) = \mathbf{0}$.

Theorem 2.7.1 (Lyapunov function). *Consider the differential equation $\dot{\mathbf{x}} = \mathbf{f}(\mathbf{x})$, with $\mathbf{f} : U \to \mathbb{R}^n$ a locally Lipschitz function in U. Let $\mathbf{x}_0$ be a singular point. If there exist a neighbourhood W of $\mathbf{x}_0$ in U and a function $V : W \to \mathbb{R}$ satisfying*

(a) $V(\mathbf{x}_0) = 0$ *and* $V(\mathbf{x}) > 0$ *when* $\mathbf{x} \neq \mathbf{x}_0$,

(b) $\frac{dV(\mathbf{x}(s))}{ds} \leq 0$ *in* $W \setminus \{\mathbf{x}_0\}$, *where* $\mathbf{x}(s)$ *is a solution of the differential equation,*

then $\mathbf{x}_0$ *is stable. Moreover,*

(c) *if* $\frac{dV(\mathbf{x}(s))}{ds} < 0$ *in* $W \setminus \{\mathbf{x}_0\}$, *then* $\mathbf{x}_0$ *is asymptotically stable.*

The function V figuring in this theorem is called a *Lyapunov function.* For a proof of the Lyapunov function theorem we refer the reader to [33, p. 192].

Now we classify the singular points according to the linear part of the vector field. Let $\mathbf{x}_0$ be a singular point of the differential system $\dot{\mathbf{x}} = \mathbf{f}(\mathbf{x})$, where $\mathbf{f}$ is a C^1 function in a neighbourhood of $\mathbf{x}_0$. Let $D\mathbf{f}(\mathbf{x}_0)$ be the Jacobian matrix of $\mathbf{f}$ evaluated at $\mathbf{x}_0$. The point $\mathbf{x}_0$ is said to be a *hyperbolic singular point* if all the eigenvalues of $D\mathbf{f}(\mathbf{x}_0)$ have non-zero real part.

For a planar differential system we say that a singular point $\mathbf{x}_0$ is an *elementary non-degenerate* singular point if the determinant of $D\mathbf{f}(\mathbf{x}_0)$ is not zero. In particular, every hyperbolic singular point is an elementary non-degenerate one. The converse is not true. Since elementary non-degenerate singular points with determinant of $D\mathbf{f}(\mathbf{x}_0)$ less than zero are saddle points, we call *antisaddle* any non-degenerate singular point at which the Jacobian matrix has positive determinant. The singular point $\mathbf{x}_0$ is said to be an *elementary degenerate* singular point if the determinant of $D\mathbf{f}(\mathbf{x}_0)$ is zero and the trace of $D\mathbf{f}(\mathbf{x}_0)$ is non-zero. The singular point $\mathbf{x}_0$ is said to be *nilpotent* if the determinant and the trace of the matrix $D\mathbf{f}(\mathbf{x}_0)$ are both zero and $D\mathbf{f}(\mathbf{x}_0)$ is not the zero matrix.

Since the concept of a flow introduced in our textbook corresponds to the concept of a complete flow used by other authors (see Subsection 2.3), in the following version of the Hartman–Grobman theorem we impose the condition that the maximal interval of definition of all solutions is$\mathbb{R}$.

Theorem 2.7.2 (Hartman–Grobman). *Let* U *be an open subset of* $\mathbb{R}^n$, $\mathbf{f} : U \to \mathbb{R}^n$ *be a* $C^1(U)$ *function,* $\Phi(s, \mathbf{x})$ *be the flow of the differential equation* $\dot{\mathbf{x}} = \mathbf{f}(\mathbf{x})$, *and* $\mathbf{x}_0$ *be a hyperbolic singular point. Then there exist a neighbourhood* W *of* $\mathbf{x}_0$, *a neighbourhood* V *of the origin, a homeomorphism* $\mathbf{h} : W \to V$ *with* $\mathbf{h}(\mathbf{x}_0) = \mathbf{0}$, *and an interval* $I \subseteq \mathbb{R}$ *containing the origin, such that*

$$\mathbf{h} \circ \Phi(s, \mathbf{x}) = e^{sD\mathbf{f}(\mathbf{x}_0)}\mathbf{h}(\mathbf{x})$$

for every $s \in I$ *and* $\mathbf{x} \in U$.

For a proof of the previous theorem see Section 4.3 in [14] or [51, p. 294].

The Hartman–Grobman theorem asserts that the differential systems $\dot{\mathbf{x}} = \mathbf{f}(\mathbf{x})$ and $\dot{\mathbf{x}} = D\mathbf{f}(\mathbf{x}_0)$ are topologically equivalent in a neighbourhood W of a hyperbolic singular point $\mathbf{x}_0$ and V of the origin. This is why we use the same names for non-linear hyperbolic singular points and for the linear hyperbolic ones. Even for non-hyperbolic singular points, when the system is topologically equivalent to a linear system, we use the same terminology for both singular points. Accordingly, the singular point $\mathbf{x}_0$ of a non-linear differential system $\dot{\mathbf{x}} = \mathbf{f}(\mathbf{x})$ is said to be a *stable normally hyperbolic singular point* if $\mathbf{f}$ is topologically equivalent to the differential system $\dot{x} = 0$, $\dot{y} = -y$ in a neighbourhood of $\mathbf{x}_0$ and $\mathbf{0}$. The singular point $\mathbf{x}_0$ is said to be an *unstable normally hyperbolic singular point* if $\mathbf{f}$ is topologically equivalent to the differential system $\dot{x} = 0$, $\dot{y} = y$ in a neighbourhood of $\mathbf{x}_0$ and $\mathbf{0}$. The singular point $\mathbf{x}_0$ is said to be a *non-isolated nilpotent singular point* if $\mathbf{f}$ is topologically equivalent to the differential system $\dot{x} = y$, $\dot{y} = 0$.

The standard tool for studying the flow in a neighbourhood of a planar non-hyperbolic singular point is a change of variables called *blow-up*, see [8], [20] and [21] for more details. Here, we summarize a description of this change of variables in the case of planar vector fields $\mathbf{f}(x, y) = (P(x, y), Q(x, y))$, where P and Q are analytic functions. Without loss of generality we can assume that the origin is a singular point of the system (otherwise we can translate the singular point to the origin by a convenient change of variables).

Consider the differentiable function $\mathbf{h}_x : \mathbb{R}^2 \to \mathbb{R}^2$ defined by $\mathbf{h}_x(\bar{x}, \bar{y}) = (\bar{x}, \bar{x}\,\bar{y})$. Using the Jacobian matrix of $\mathbf{h}_x$ and the vector field $\mathbf{f}$ we can define a vector field $\mathbf{f}_x$ on $\mathbb{R}^2$ satisfying the equality

$$D\mathbf{h}_x(\mathbf{f}_x(\bar{x}, \bar{y})) = \mathbf{f}(\mathbf{h}_x(\bar{x}, \bar{y})) = \mathbf{f}(\bar{x}, \bar{x}\,\bar{y}).$$

From here, one obtains the following expression for $\mathbf{f}_x$ when $\bar{x} \neq 0$

$$\mathbf{f}_x(\bar{x}, \bar{y}) = \left(P(\bar{x}, \bar{x}\,\bar{y}), \frac{Q(\bar{x}, \bar{x}\bar{y}) - \bar{y}P(\bar{x}, \bar{x}\bar{y})}{\bar{x}} \right). \tag{2.9}$$

Since the origin is a singular point, i.e., $P(0, 0) = Q(0, 0) = 0$, expression (2.9) can be extended to $\bar{x} = 0$ to yield an analytic vector field on $\mathbb{R}^2$. Such a vector field is called a *blow-up in the x-direction*.

The vector fields $\mathbf{f}$ and $\mathbf{f}_x$ are topologically equivalent in $\mathbb{R}^2 \setminus \{\mathbf{0}\}$ and $\mathbb{R}^2 \setminus \{\bar{x} = 0\}$, respectively. Moreover, since $\mathbf{h}_x$ maps the straight line $\bar{x} = 0$ into the origin, the behaviour of the flow of $\mathbf{f}$ in a neighbourhood of the origin can be obtained from the behaviour of the flow of $\mathbf{f}_x$ in a neighbourhood of $\bar{x} = 0$ in the following sense. Let γ be an orbit of the differential system $\dot{\mathbf{x}} = \mathbf{f}(\mathbf{x})$ such that the origin is contained in its α- or ω-limit set. If $m = \tan\theta$, with $\theta \in (-\pi/2, \pi/2)$, is the slope of γ at the origin, then the angle θ is called a *characteristic direction* of the origin and the point $(0, m)$ is a singular point of the blow-up system $\dot{\mathbf{u}} = \mathbf{f}_x(\mathbf{u})$. The study of the local phase portrait at the point $(0, m)$ is easier than the one of the origin, because such singular points are less degenerate.

If $m = \pm\infty$, then another change of variables applies. Specifically, consider the function $\mathbf{h}_y : \mathbb{R}^2 \to \mathbb{R}^2$ given by $\mathbf{h}_y(\bar{x}, \bar{y}) = (\bar{x}\,\bar{y}, \bar{y})$, and the vector field $\mathbf{f}_y$ satisfying $D\mathbf{h}_y(\mathbf{f}_y(\bar{x}, \bar{y})) = \mathbf{f}(\bar{x}\,\bar{y}, \bar{y})$. It follows that

$$\mathbf{f}_y(\bar{x}, \bar{y}) = \left(\frac{P(\bar{x}\,\bar{y}, \bar{y}) - \bar{x}Q(\bar{x}\,\bar{y}, \bar{y})}{\bar{y}}, Q(\bar{x}\,\bar{y}, \bar{y}) \right). \tag{2.10}$$

Thus $(0, 0)$ is a singular point of the blow-up system $\dot{\mathbf{u}} = \mathbf{f}_y(\mathbf{u})$. In general, if $m = \tan\theta$ with $\theta \in (0, \pi)$, then $(1/m, 0)$ is a singular point of the blow-up system

$\dot{\mathbf{u}} = \mathbf{f}_y(\mathbf{u})$. Hence, going back to the original variables a finite number of curves are present, splitting any neighbourhood of the origin into *hyperbolic, elliptic and parabolic sectors*, see Figure 2.12.

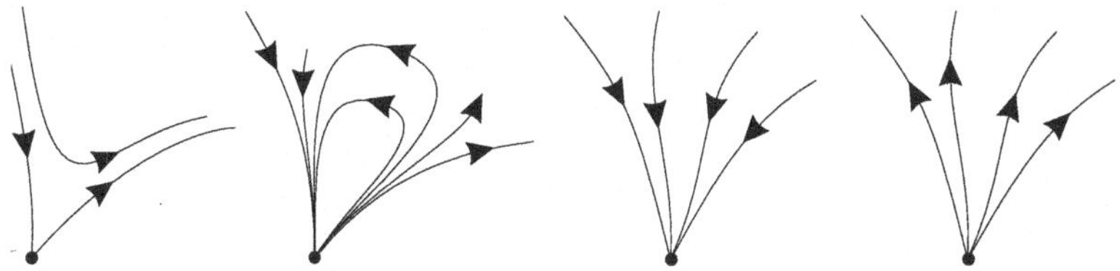

Figure 2.12: Sectors in the neighbourhood of a singular point.

A singular point $\mathbf{x}_0$ is called a *saddle-node* if a neighbourhood of $\mathbf{x}_0$ is the union of a unique parabolic sector and two hyperbolic sectors. Thus a saddle-node has three separatrices: two of them, called the *hyperbolic manifolds* or *separatrices of the saddle-node*, are related to the boundary of the parabolic sector; and the remainder, called the *central manifold* or *separatrice of the saddle-node*, is related to the boundary between the two hyperbolic sectors. Note that this terminology is appropriate only when the singular point is elementary and degenerate. To simplify notation we continue using this terminology not only for nilpotent saddle-nodes, but also for more degenerated saddle points.

Theorem 2.7.3 (Elementary non-degenerate singular points). *Let* $(0,0)$ *be an isolated singular point of the differential system*

$$\dot{x} = X(x,y), \quad \dot{y} = y + Y(x,y),$$

where X *and* Y *are analytic functions in a neighbourhood of the origin and their series expansion involve only terms of second order and higher. Let* $f(x)$ *be a solution of the equation* $y + Y(x,y) = 0$ *in a neighbourhood of the origin and suppose that the function* $g(x) = X(x, f(x))$ *can be written in the form* $g(x) = a_m x^m + O(x^{m+1})$ *where* $O(x^k)$ *stands for an analytic function with terms of order greater or equal than* k *in its series expansion,* $m \geq 2$*, and* $a_m \neq 0$*.*

(a) *If* m *is odd and* $a_m > 0$*, then the origin is topologically equivalent to a stable node.*

(b) *If* m *is odd and* $a_m < 0$*, then the origin is topologically equivalent to a saddle with the stable manifold tangent to the* x*-axis and the unstable manifold tangent to the* y*-axis.*

(c) *If* m *is even, then the origin is a saddle-node. Its hyperbolic manifold is unstable and tangent to the* y*-axis. Its central manifold is tangent to the*

x-axis and when $a_m > 0$ (respectively, $a_m < 0$) it is unstable (respectively, stable) in the 0 direction and stable (respectively, unstable) in the π direction.

For a proof of Theorem 2.7.3 we refer the reader to [4, p. 340] or [21, p. 74].

Theorem 2.7.4 (Nilpotent singular points). *Let $(0,0)$ be an isolated singular point of the system*

$$\dot{x} = y + X(x,y), \quad \dot{y} = Y(x,y),$$

where X and Y are analytic functions in a neighbourhood of the origin and their series expansions involves only terms of second order and higher. Let $y = f(x) = a_2x^2 + a_3x^3 + O(x^4)$ be a solution of the equation $y + X(x,y) = 0$ in a neighbourhood of the origin, and suppose that $F(x) = Y(x, f(x)) = Ax^\alpha(1 + O(x))$ and $\Phi(x) = (\partial X/\partial x + \partial Y/\partial y)(x, f(x)) = Bx^\beta(1 + O(x))$, with $A \neq 0$, $\alpha \geq 2$ and $\beta \geq 1$.

(a) *If α is even, then*

(a.1) *if $\alpha > 2\beta + 1$, the origin is a saddle-node with the three separatrices tangent to the x-axis;*

(a.2) *if $\alpha < 2\beta + 1$ or $\Phi \equiv 0$, then a neighourhood of the origin is the union of two hyperbolic sectors.*

(b) *If α is odd and $A > 0$, then the origin is a saddle whose stable and unstable separatrices are tangent to the x-axis.*

(c) *If α is odd and $A < 0$, then*

(c.1) *if $\alpha > 2\beta+1$ and β even; or $\alpha = 2\beta+1$, β even and $B^2+4A(\beta+1) \geq 0$, then the origin is a node, stable when $B < 0$ and unstable when $B > 0$;*

(c.2) *if $\alpha > 2\beta+1$ and β odd; or $\alpha = 2\beta+1$, β odd and $B^2+4A(\beta+1) \geq 0$, then the origin is the union of a hyperbolic sector and an elliptic sector;*

(c.3) *if $\alpha = 2\beta + 1$ and $B^2 + 4A(\beta + 1) < 0$, then the origin is a focus;*

(c.4) *if $\alpha < 2\beta + 1$; or $\Phi \equiv 0$, then the origin is a center.*

A proof of the previous theorem can be found in [4, pp. 357–362], in [2], or in [21, p. 116].

2.7.2 Periodic orbits: Poincaré map

One of the most important tools in the study of flows in the neighbourhood of periodic orbits is the so called Poincaré map. Consider a locally Lipschitz vector field $\mathbf{f} : U \to \mathbb{R}^n$ and let $\Phi(s, \mathbf{x})$ be the flow defined by the differential equation $\dot{\mathbf{x}} = \mathbf{f}(\mathbf{x})$. Let Σ be a hypersurface in $\mathbb{R}^n$ and take a point $\mathbf{p}$ in $\Sigma \cap U$. The flow Φ is said to be *transverse* to Σ at the point $\mathbf{p}$ if $\mathbf{f}(\mathbf{p})$ is not contained in $T_\mathbf{p}\Sigma$ (the tangent space to Σ at point $\mathbf{p}$). If $\mathbf{f}(\mathbf{p}) \in T_\mathbf{p}\Sigma$, then $\mathbf{p}$ is called a *contact point* of the flow with Σ.

Let V be an open subset of Σ. We say that the flow is *transverse to* Σ at V if the flow is transverse to Σ at every point in V.

Consider now two open hypersurfaces Σ_1, Σ_2 and two points $\mathbf{p}_1 \in \Sigma_1 \cap U$, $\mathbf{p}_2 \in \Sigma_2 \cap U$ such that $\mathbf{p}_2 = \Phi(s_1, \mathbf{p}_1)$. There exist a neighbourhood V_1 of $\mathbf{p}_1$ in $\Sigma_1 \cap U$, a neighbourhood V_2 of $\mathbf{p}_2$ in $\Sigma_2 \cap U$, and a function $\tau : V_1 \to \mathbb{R}$ satisfying $\tau(\mathbf{p}_1) = s_1$ and $\Phi(\tau(\mathbf{q}), \mathbf{q}) \in V_2$ for every $\mathbf{q} \in V_1$. Moreover, if the vector field $\mathbf{f}$ is globally Lipschitz, C^r with $r \geq 1$, or analytic, then the function τ is also continuous, C^r with $r \geq 1$, or analytic, respectively. For more details see [53, pp. 193–194] or [57, pp. 226–227]. In this situation we define the *Poincaré map* as the map $\pi : V_1 \to V_2$ given by

$$\pi(\mathbf{q}) = \Phi(\tau(\mathbf{q}), \mathbf{q}),$$

for every $\mathbf{q} \in V_1$, see Figure 2.13.

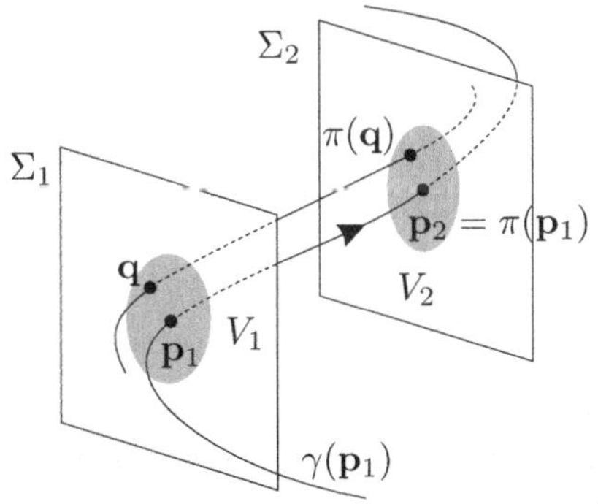

Figure 2.13: Poincaré map π.

When the vector field is globally Lipschitz, C^r with $r \geq 1$, or analytic, the Poincaré map π is also continuous, C^r with $r \geq 1$, or analytic, respectively.

By reversing the sense of the flow it is easy to conclude that the Poincaré map is invertible and the inverse map π^{-1} is continuous, C^r with $r \geq 1$, or analytic, respectively. In the particular case when $\Sigma_1 = \Sigma_2$ the Poincaré map π is called a *return map*.

Consider $\mathbf{p} \in \Sigma_1$ and let $\gamma(\mathbf{p})$ be a periodic orbit. From the continuous dependence of the flow on the initial conditions, it follows that a return map π can be defined in a neighbourhood of $\mathbf{p}$, and $\mathbf{p}$ is a fixed point of π. Conversely, if $\mathbf{p} \in \Sigma_1$ is a fixed point of a return map π, then $\gamma(\mathbf{p})$ is a periodic orbit. Hence, limit cycles are associated to isolated fixed points of return maps. A limit cycle $\gamma(\mathbf{p})$ is called a *hyperbolic limit cycle* if the absolute value of all the eigenvalues of the Jacobian matrix $D\pi(\mathbf{p})$ is different from 1; otherwise $\gamma(\mathbf{p})$ is called a *non-hyperbolic limit cycle*. Note that this definition does not depend on the chosen point $\mathbf{p}$ or on the chosen cross section Σ_1.

Theorem 2.7.5. *Let* $\mathbf{f} : U \subset \mathbb{R}^n \to \mathbb{R}^n$ *be a Lipschitz function in* U, $\gamma(\mathbf{p})$ *be a hyperbolic limit cycle of the differential equation* $\dot{\mathbf{x}} = \mathbf{f}(\mathbf{x})$ *and* π *be a return map defined in a neighbourhood of* $\gamma(\mathbf{p})$. *Suppose that* π *is differentiable in a neighbourhood of* $\mathbf{p}$.

(a) *If the absolute value of every eigenvalue of* $D\pi(\mathbf{p})$ *is less than* 1, *then* $\gamma(\mathbf{p})$ *is a stable limit cycle.*

(b) *If the absolute value of at least one eigenvalue of* $D\pi(\mathbf{p})$ *is greater than* 1, *then* $\gamma(\mathbf{p})$ *is an unstable limit cycle.*

A proof of this result can be found in [21] or in [57, Chapter IX].

2.8 α- and ω-limit sets in the plane

In this section we deal with the asymptotic behaviour of the remainder orbits. These orbits are diffeomorphic to straight lines, see Theorem 2.3.2. In this section we restrict ourselves to planar flows. In this context the following version of the Jordan curve theorem will be useful later on.

A curve in the plane is said to be a *Jordan curve* if it is homeomorphic to $\mathbb{S}^1$, i.e., if it is a closed curve without autointersections.

Theorem 2.8.1 (Jordan curve). *The complementary set of a Jordan curve* γ *in the plane is the union of two open, disjoint and connected sets. Furthermore, one of these sets is bounded and its boundary is the curve* γ.

Since orbits of a flow are disjoint, from the Jordan curve theorem it follows that a periodic orbit γ splits the phase plane into two invariant regions, one of which is bounded. This bounded region will be called the *interior* of γ and be denoted by Σ_γ.

Periodic orbits are not the unique Jordan curves formed by solutions. We define a *separatrix cycle* to be a finite union of n singular points $\mathbf{p}_1, \mathbf{p}_2, \ldots, \mathbf{p}_n$ (some of these points may coincide) and n orbits $\gamma_1, \gamma_2, \ldots, \gamma_n$, with the property that $\alpha(\gamma_k) = \{\mathbf{p}_k\}$ for $k = 1, 2, \ldots, n$, $\omega(\gamma_k) = \{\mathbf{p}_{k+1}\}$ if $k = 1, 2, \ldots, n-1$, and $\omega(\gamma_n) = \{\mathbf{p}_1\}$, see Figure 2.14. The singular points $\mathbf{p}_1, \mathbf{p}_2, \ldots, \mathbf{p}_n$ will be called the *vertices of the cycle*.

We define a *homoclinic cycle* to be a separatrix cycle formed by one singular point (*homoclinic point*) and one orbit (*homoclinic orbit*), see Figure 2.14(a). A *double homoclinic cycle* is a separatrix cycle formed by one singular point (in this case $\mathbf{p}_1$ and $\mathbf{p}_2$ are identified) and two orbits, see Figure 2.14(b). Finally, a *heteroclinic cycle* is a separatrix cycle formed by two singular points and two orbits, see Figure 2.14(c).

A periodic orbit γ is said to be *inside asymptotically stable* (respectively, *inside asymptotically unstable*) if there exists a neighbourhood V of γ such that

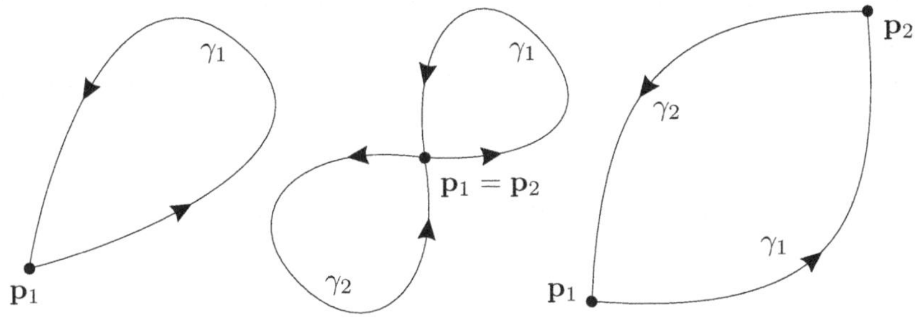

Figure 2.14: Separatrix cycles: (a) homoclinic cycle; (b) double homoclinic cycle; (c) heteroclinic cycle.

$V \cap \Sigma_\gamma \subset W^s(\gamma)$ (respectively, $V \cap \Sigma_\gamma \subset W^u(\gamma)$). A periodic orbit γ is said to be *outside asymptotically stable* (respectively, *outside asymptotically unstable*) if there exists a neighbourhood V of γ such that $V \cap (\mathbb{R}^2 \setminus \mathrm{Cl}(\Sigma_\gamma)) \subset W^s(\gamma)$ (respectively, $V \cap (\mathbb{R}^2 \setminus \mathrm{Cl}(\Sigma_\gamma)) \subset W^u(\gamma)$).

A limit cycle γ is said to be *semistable* if γ is either inside asymptotically stable and outside asymptotically unstable, or inside asymptotically unstable and outside asymptotically stable.

The following result asserts that the α- and ω-limit set of orbits of planar differential systems are simple sets: singular points, periodic orbits, or separatrix cycles.

Theorem 2.8.2 (Poincaré–Bendixson). *Let $\mathbf{f} : U \subset \mathbb{R}^2 \to \mathbb{R}^2$ be a locally Lipschitz function in the open subset U, and let γ be an orbit of the differential system $\dot{\mathbf{x}} = \mathbf{f}(\mathbf{x})$. Suppose that γ is positively bounded (respectively, negatively bounded) and the number of singular points in $\omega(\gamma)$ (respectively, in $\alpha(\gamma)$) is finite.*

(a) *If $\omega(\gamma)$ (respectively, $\alpha(\gamma)$) has no singular points, then $\omega(\gamma)$ (respectively, $\alpha(\gamma)$) is a periodic orbit.*

(b) *If $\omega(\gamma)$ (respectively, $\alpha(\gamma)$) has singular points and regular points, then $\omega(\gamma)$ (respectively, $\alpha(\gamma)$) is a separatrix cycle.*

(c) *If $\omega(\gamma)$ (respectively $\alpha(\gamma)$) has no regular points, then $\omega(\gamma)$ (respectively, $\alpha(\gamma)$) is a singular point.*

A proof of this result can be found in the book of Hartman [30, Chapter 7] or in [21]. The following results are corollaries of the Poincaré–Bendixson Theorem, see [21].

Corollary 2.8.3. *Let $\mathbf{f} : U \subset \mathbb{R}^2 \to \mathbb{R}^2$ be a Lipschitz function in an open set U and let γ be a periodic orbit of the differential system $\dot{\mathbf{x}} = \mathbf{f}(\mathbf{x})$. If $\eta, \varrho \subset \Sigma_\gamma$ are orbits* *and* $\omega(\eta)=\gamma$ *(respectively,* $\alpha(\eta)=\gamma$*), then* $\alpha(\varrho)\neq\gamma$ *(respectively,* $\omega(\varrho)\neq\gamma$*).*

Corollary 2.8.4. *Let* $\mathbf{f}: U\subset\mathbb{R}^2\to\mathbb{R}^2$ *be a Lipschitz function in an open and simply connected set* U *and let* $\gamma\subset U$ *be a periodic orbit of the differential system* $\dot{\mathbf{x}}=\mathbf{f}(\mathbf{x})$*. Then there exists a singular point in* Σ_γ*.*

2.9 Compactified flows

The aim of this section is to describe the asymptotic behaviour of unbounded orbits, i.e. the behaviour of flows near the infinity.

To do this, we use the so called Poincaré compactification. The French mathematician H. Poincaré was the first to use this technique, in the study of polynomial vector fields. We will only consider some aspects of this technique. More information can be found in [58], [4] and [21].

2.9.1 Poincaré compactification

We define the following sets in $\mathbb{R}^3$

$$\begin{aligned}\mathbb{S}^2 &:= \left\{(x,y,z)\in\mathbb{R}^3 : x^2+y^2+z^2=1\right\},\\ H_+ &:= \left\{(x,y,z)\in\mathbb{S}^2 : z>0\right\},\\ \mathbb{S}^1 &:= \left\{(x,y,z)\in\mathbb{S}^2 : z=0\right\},\\ H_- &:= \left\{(x,y,z)\in\mathbb{S}^2 : z<0\right\}.\end{aligned}$$

$\mathbb{S}^2$ is called the *unit sphere* of $\mathbb{R}^3$, and H_+, $\mathbb{S}^1$ and H_- are called the north hemisphere, the *equator* and the *south hemisphere* of $\mathbb{S}^2$, respectively.

We say that a function $\mathbf{f}:\mathbb{R}^2\to\mathbb{R}^2$ satisfies the *Łojasiewicz property at infinity* if there exists a positive integer n such that the function $\mathbf{f}_n$ defined by

$$\mathbf{f}_n(x,y,z) := z^n\,\mathbf{f}\left(\frac{x}{z},\frac{y}{z}\right) \tag{2.11}$$

can be extended to $z=0$ and this extension is locally Lipschitz in the whole $\mathbb{S}^2$. Since $\mathbb{S}^2$ is a compact set, if $\mathbf{f}_n$ is locally Lipschitz in $\mathbb{S}^2$, then $\mathbf{f}_n$ is also globally Lipschitz in $\mathbb{S}^2$.

Given a function $\mathbf{f}$, if there exists a non-negative integer n_0 such that the function $\mathbf{f}_{n_0}$ is globally Lipschitz in $\mathbb{S}^2$, then for every $n\geq n_0$ the function $\mathbf{f}_n$ is also globally Lipschitz in $\mathbb{S}^2$. We call the *degree of* $\mathbf{f}$ *at infinity*, and denote it by $n=n(\mathbf{f})$, the least positive integer m such that $\mathbf{f}_m$ is well defined and Lipschitz in $\mathbb{S}^2$.

Lemma 2.9.1. *If the function* $\mathbf{f}:\mathbb{R}^2\to\mathbb{R}^2$ *satisfies the Łojasiewicz property at infinity with degree at infinity equal to* n*, then there exist positive constants* R *and* M*, such that*

$$||\mathbf{f}(\mathbf{x})||\leq M\,||\mathbf{x}||^n,$$

for every $||\mathbf{x}||>R$*.*

Proof. Given a point (x,y) in $\mathbb{R}^2$ we consider the point $(\bar{x},\bar{y},\bar{z})$ in the north hemisphere H_+, where $\bar{z}=(1+x^2+y^2)^{-\frac{1}{2}}>0$, $\bar{x}=x\bar{z}$, and $\bar{y}=y\bar{z}$. Conversely, for every point $(\bar{x},\bar{y},\bar{z})\in H_+$ the point $(\bar{x}/\bar{z},\bar{y}/\bar{z})$ belongs to $\mathbb{R}^2$.

By the hypothesis, there exists a positive integer n such that the function $\mathbf{f}_n$ is (globally) Lipschitz in $\mathbb{S}^2$, and consequently $\mathbf{f}_n$ is continuous in $\mathbb{S}^2$. Since the unit sphere is a compact manifold, there exists a positive constant N for which $||\mathbf{f}_n(x,y,z)||<N$ for every $(x,y,z)\in\mathbb{S}^2$, or, equivalently $||\mathbf{f}_n(x,y,z)||<N\,||(x,y,z)||^n$. Therefore,

$$|\bar{z}|^n\left|\left|\mathbf{f}\left(\frac{\bar{x}}{\bar{z}},\frac{\bar{y}}{\bar{z}}\right)\right|\right|<N\,||(\bar{x},\bar{y},\bar{z})||^n.$$

Here $||\cdot||$ denotes the Euclídean norm in $\mathbb{R}^2$ or in $\mathbb{R}^3$, depending on the context.

Dividing by $|\bar{z}|^n$ and returning to the original variables, we obtain $||\mathbf{f}(x,y)||<N\,||(x,y,1)||^n$. Taking a positive constant R such that

$$\frac{N}{N+1}<\left(\frac{R}{\sqrt{1+R^2}}\right)^n,$$

we have $(N+1)\,||(x,y)||^n>N\,||(x,y,1)||^n$ for every $||(x,y)||>R$. The lemma follows by taking $M=N+1$. □

The inequality in Lemma 2.9.1 justifies the name of the Łojasiewicz property at infinity (see [20] for more information). From this inequality it is also easy to understand the degree of a function at infinity.

The rest of this section is devoted to the compactification of vector fields satisfying the Łojasiewicz property at infinity. We also provide an explicit expression of a flow near infinity and a technique for studying this flow in a neighbourhood of a singular point at infinity.

Let $\mathbf{f}:\mathbb{R}^2\to\mathbb{R}^2$ be a local Lipschitz function satisfying the Łojasiewicz property at infinity and let n be the degree of $\mathbf{f}$ at infinity. Consider the diffeomorphisms $\mathbf{h}_+:\mathbb{R}^2\to H_+$ and $\mathbf{h}_-:\mathbb{R}^2\to H_-$ defined by

$$\mathbf{h}_+(x,y):=\frac{1}{\sqrt{1+x^2+y^2}}(x,y,1)\quad\text{and}\quad\mathbf{h}_-(x,y):=-\mathbf{h}_+(x,y).\tag{2.12}$$

The functions $\mathbf{h}_+$ and $\mathbf{h}_-$ are the central projections (with center at the origin) of the tangent plane to $\mathbb{S}^2$ at the point $(0,0,1)$ onto H_+ and H_-, respectively, see Figure 2.15.

The diffeomorphisms $\mathbf{h}_+$ and $\mathbf{h}_-$ and the vector field $\mathbf{f}$ define two vector fields $\mathbf{f}_+$ and $\mathbf{f}_-$ on the hemispheres H_+ and H_-, respectively, given by

$$\begin{aligned}\mathbf{f}_+(x,y,z)&:=D\mathbf{h}_+\left(\mathbf{h}_+^{-1}(x,y,z)\right)\mathbf{f}\left(\mathbf{h}_+^{-1}(x,y,z)\right),\\ \mathbf{f}_-(x,y,z)&:=D\mathbf{h}_-\left(\mathbf{h}_-^{-1}(x,y,z)\right)\mathbf{f}\left(\mathbf{h}_-^{-1}(x,y,z)\right).\end{aligned}\tag{2.13}$$

Therefore, the rule

$$\widetilde{\mathbf{f}}(x,y,z) := \begin{cases} \mathbf{f}_+(x,y,z), & \text{if } (x,y,z) \in H_+, \\ \mathbf{f}_-(x,y,z), & \text{if } (x,y,z) \in H_-, \end{cases}$$

defines a vector field over $H_+ \cup H_- = \mathbb{S}^2 \setminus \mathbb{S}^1$ which, by (2.12) and (2.13), can be written as

$$\widetilde{\mathbf{f}}(x,y,z) = z \begin{pmatrix} 1-x^2 & -xy \\ -xy & 1-y^2 \\ -xz & -yz \end{pmatrix} \mathbf{f}\left(\frac{x}{z},\frac{y}{z}\right).$$

In general the vector field $\widetilde{\mathbf{f}}$ cannot be extended to the equator of the sphere. However since $\mathbf{f}$ has degree n at infinity, the vector field $\mathbf{f}_{\mathbb{S}^2}(x,y,z) := z^{n-1}\widetilde{\mathbf{f}}(x,y,z)$ obtained by multiplying by z^{n-1} satisfies

$$\mathbf{f}_{\mathbb{S}^2}(x,y,z) = \begin{pmatrix} 1-x^2 & -xy \\ -xy & 1-y^2 \\ -xz & -yz \end{pmatrix} \mathbf{f}_n(x,y,z). \tag{2.14}$$

Therefore, $\mathbf{f}_{\mathbb{S}^2}$ is defined and Lipschitz on whole $\mathbb{S}^2$. Since $\mathbf{f}_{\mathbb{S}^2}|_{H_+} = z^{n-1}\mathbf{f}_+$ and $\mathbf{f}_{\mathbb{S}^2}|_{H_-} = z^{n-1}\mathbf{f}_-$, the vector field $\mathbf{f}_{\mathbb{S}^2}$ can be understood as an extension to $\mathbb{S}^2$ of the vector field $\widetilde{\mathbf{f}}$ multiplied by the analytic function z^{n-1}. This multiplicative factor is not important in the analysis of the asymptotic behaviour of the flow because it only represents a change in the scale of time. In particular, if we change the variable s to the variable τ by $ds = z^{n-1}d\tau$, the vector field $\mathbf{f}_{\mathbb{S}^2}$ over $\mathbb{S}^2$ can be understood as two copies (each defined on a hemisphere) of the vector field $\mathbf{f}$ defined on $\mathbb{R}^2$. Therefore, the behaviour of $\mathbf{f}$ near infinity follows from the behaviour of $\mathbf{f}_{\mathbb{S}^2}$ in a neighbourhood of the equator. Note that the equator, $z = 0$, is invariant under the flow of $\mathbf{f}_{\mathbb{S}^2}$.

For polynomial planar vector fields $\mathbf{f}(x,y) = (P(x,y), Q(x,y))$, with P and Q polynomials, it is easy to prove that $\mathbf{f}$ satisfies Łojasiewicz's property at infinity and $n = \max\{\deg P, \deg Q\}$ is the degree of $\mathbf{f}$ at infinity. Furthermore the vector field $\mathbf{f}_{\mathbb{S}^2}$ is analytic on $\mathbb{S}^2$, see [21] or [58, pp. 57–60] for details.

Consider the *Poincaré disc*, $\mathbb{D} := \{(x,y) \in \mathbb{R}^2 : x^2 + y^2 \le 1\}$, and the so called *gnomonic projection* $\mathbf{p}_+ : H_+ \cup \mathbb{S}^1 \to \mathbb{D}$, given by

$$\mathbf{p}_+(x,y,z) := \frac{1}{1+z}(x,y).$$

The vector field $\mathbf{f}_{\mathbb{S}^2}|_{H_+\cup\mathbb{S}^1}$ and the diffeomorphism $\mathbf{p}_+$ define a vector field $\mathbf{f}_{\mathbb{D}}$ on $\mathbb{D}$ given by

$$\mathbf{f}_{\mathbb{D}}(x,y) := D\mathbf{p}_+\left(\mathbf{p}_+^{-1}(x,y)\right)\mathbf{f}_{\mathbb{S}^2}\left(\mathbf{p}_+^{-1}(x,y)\right).$$

For a differential system $\dot{\mathbf{x}} = \mathbf{f}(\mathbf{x})$, where $\mathbf{f}$ is a locally Lipschitz function in $\mathbb{R}^2$ and satisfies the Łojasiewicz property at infinity with degree n at infinity, we call the differential system $\dot{\mathbf{x}} = \mathbf{f}_{\mathbb{D}}(\mathbf{x})$ *Poincaré's compactification.* The vector fields $\mathbf{f}$ and $\mathbf{f}_{\mathbb{D}}|_{\mathrm{Int}(\mathbb{D})}$ are C^r-equivalent and $\mathbf{h}_{\mathbb{D}} := \mathbf{p}_+ \circ \mathbf{h}_+$ is the equivalence map. Here $\mathrm{Int}(\mathbb{D})$ denotes the interior of $\mathbb{D}$; that is the biggest open subset contained in $\mathbb{D}$. In this sense we identify the behaviour of $\mathbf{f}_{\mathbb{D}}$ at the boundary $\partial\mathbb{D}$ with the behaviour of $\mathbf{f}$ at infinity.

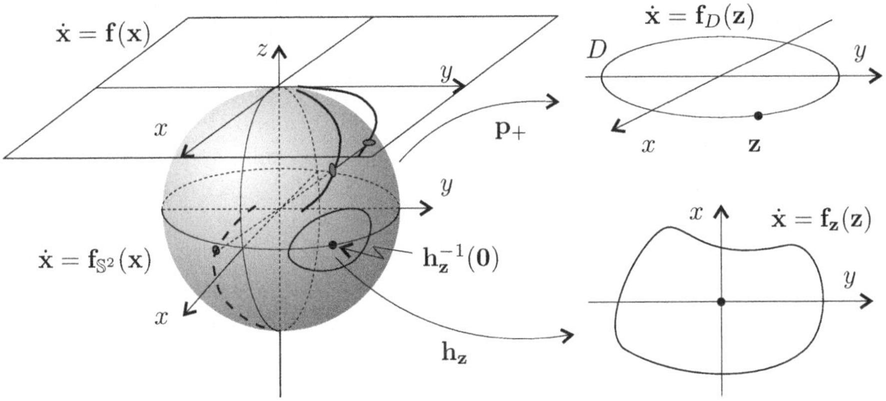

Figure 2.15: Poincaré's compactification.

Finally, for every $\mathbf{x} \in \mathbb{R}^2$ it is easy to prove that

$$\mathbf{h}_{\mathbb{D}}(\mathbf{x}) = \frac{1}{1 + \sqrt{1 + ||\mathbf{x}||^2}}\mathbf{x} \tag{2.15}$$

and

$$\mathbf{f}_{\mathbb{D}}(x, y) = \begin{pmatrix} \frac{1-x^2+y^2}{2} & -xy \\ -xy & \frac{1+x^2-y^2}{2} \end{pmatrix} \mathbf{f}_n \left(\frac{2x}{1+x^2+y^2}, \frac{2y}{1+x^2+y^2}, \frac{1-x^2-y^2}{1+x^2+y^2} \right). \tag{2.16}$$

2.9.2 The behaviour of a flow at infinity

Since the equator of $\mathbb{S}^2$ is invariant under the flow defined by $\mathbf{f}_{\mathbb{S}^2}$, the boundary of the Poincaré disc $\partial\mathbb{D}$ is invariant under the flow defined by $\mathbf{f}_{\mathbb{D}}$. Then $\partial\mathbb{D}$ is a circle formed by solutions called the *infinity manifold.* A point $\mathbf{p} \in \partial\mathbb{D}$ is said to be a *singular point at infinity* if $\mathbf{f}_{\mathbb{D}}(\mathbf{p}) = \mathbf{0}$. If there are no singular points at infinity, we say that there exists a *periodic orbit at infinity,* or that *the infinity is a periodic orbit.*

Let $\mathbf{p} \in \partial\mathbb{D}$ be a singular point at infinity. As we know, the stable manifold of $\mathbf{p}$, $W^s(\mathbf{p}) \subset \mathbb{D}$, is formed by the orbits γ of the Poincaré compactification satisfying $\mathbf{p} \in \omega(\gamma)$. Consider the subset $\mathbf{h}_{\mathbb{D}}^{-1}(W^s(\mathbf{p}))$ of $\mathbb{R}^2$. For simplicity we call this set the *stable manifold of the singular point at infinity* $\mathbf{p}$ and we also denote it by $W^s(\mathbf{p})$. Note that orbits in $\mathbb{R}^2$ belonging to the stable manifold of a singular point at infinity escape to infinity in forward time. In a similar way we define in $\mathbb{R}^2$ *the unstable manifold of a singular point at infinity* and denote it by $W^u(\mathbf{p})$. Note that orbits in $\mathbb{R}^2$ belonging to the unstable manifold of a singular point at infinity escape to infinity in backward time. In general, we will use the same name for a subset E of $\mathbb{R}^2$ and for the subset $\mathbf{h}_{\mathbb{D}}(E)$ of $\mathrm{Int}(\mathbb{D})$.

When there are no singular points at $\partial\mathbb{D}$, we denote the stable and the unstable manifold of the periodic orbit at infinity by $W^s(\infty)$ and $W^u(\infty)$, respectively.

Let $\mathbf{z} = (x_0, y_0)^T \in \partial\mathbb{D}$ be a singular point at infinity of the differential system $\dot{\mathbf{x}} = \mathbf{f}(\mathbf{x})$, that is, a solution of the equation $\mathbf{f}_{\mathbb{D}}|_{\partial\mathbb{D}}(\mathbf{x}) = \mathbf{0}$. To determine the behaviour of the flow in a neighbourhood of $\mathbf{z}$ we use the local chart $(H_{\mathbf{z}}, \mathbf{h}_{\mathbf{z}})$ of $\mathbb{S}^2$, where $H_{\mathbf{z}} = \left\{(x, y, z) \in \mathbb{S}^2 : xx_0 + yy_0 > 0\right\}$ is the hemisphere centered at the point $\mathbf{p}_+^{-1}(\mathbf{z}) = (x_0, y_0, 0)^T$ and

$$\mathbf{h}_{\mathbf{z}}(x, y, z) := \frac{1}{xx_0 + yy_0}(yx_0 - xy_0, z)$$

is the inverse of the central projection (with center at the origin) of the tangent plane to $\mathbb{S}^2$ at the point $(x_0, y_0, 0)^T$. Thus, the vector field $\mathbf{f}_{\mathbb{S}^2}$ and the diffeomorphism $\mathbf{h}_{\mathbf{z}}$ define a locally Lipschitz vector field $\mathbf{f}_{\mathbf{z}} : \mathbb{R}^2 \to \mathbb{R}^2$, given by

$$\mathbf{f}_{\mathbf{z}}(x, y) := D\mathbf{h}_{\mathbf{z}}\left(\mathbf{h}_{\mathbf{z}}^{-1}(x, y)\right) \mathbf{f}_{\mathbb{S}^2}\left(\mathbf{h}_{\mathbf{z}}^{-1}(x, y)\right).$$

Since $\mathbf{h}_{\mathbf{z}}(x_0, y_0, 0) = \mathbf{0}$, the origin is a singular point of the flow defined by $\mathbf{f}_{\mathbf{z}}$, see Figure 2.15.

The vector fields $\mathbf{f}_{\mathbb{S}^2}$ and $\mathbf{f}_{\mathbf{z}}$ are differentiably conjugate in a neighbourhood of the singular points $\mathbf{p}_+^{-1}(\mathbf{z})$ and $\mathbf{0}$. Therefore, to describe the behaviour of the flow generated by $\mathbf{f}_{\mathbb{D}}$ in a neighbourhood of $\mathbf{z}$ it is sufficient to describe the behaviour of the flow generated by $\mathbf{f}_{\mathbf{z}}$ in a neighbourhood of $\mathbf{0}$ with $y \geq 0$.

We end the section by giving explicit expressions of the vector field $\mathbf{f}_{\mathbf{z}}(x, y)$. From

$$\mathbf{h}_{\mathbf{z}}^{-1}(x, y) = \frac{1}{\sqrt{1 + x^2 + y^2}}(x_0 - xy_0, y_0 + xx_0, y) \tag{2.17}$$

and

$$D\mathbf{h}_{\mathbf{z}}(x, y, z) = \frac{1}{(xx_0 + yy_0)^2}\begin{pmatrix} -y & x & 0 \\ -zx_0 & -zy_0 & xx_0 + yy_0 \end{pmatrix}$$

it follows that

$$\mathbf{f}_{\mathbf{z}}(x, y) = z(x, y)\begin{pmatrix} -y_0 - xx_0 & -xy_0 + x_0 \\ -yx_0 & -yy_0 \end{pmatrix} \mathbf{f}_n\left(\frac{x_0 - xy_0}{z(x,y)}, \frac{y_0 + xx_0}{z(x,y)}, \frac{y}{z(x,y)}\right), \tag{2.18}$$

where $z(x, y) = \sqrt{1 + x^2 + y^2}$.

In particular, for polynomial vector fields $\mathbf{f}(x,y) = (P(x,y), Q(x,y))$, if we take charts centered at the points $\mathbf{z}_x = (1,0)$ and $\mathbf{z}_y = (0,1)$, we obtain

$$\mathbf{f}_{\mathbf{z}_x}(x,y) = y^n m(x,y) \begin{pmatrix} Q\left(\frac{1}{y}, \frac{x}{y}\right) - xP\left(\frac{1}{y}, \frac{x}{y}\right) \\ -yP\left(\frac{1}{y}, \frac{x}{y}\right) \end{pmatrix},$$

$$\mathbf{f}_{\mathbf{z}_y}(x,y) = y^n m(x,y) \begin{pmatrix} -P\left(\frac{-x}{y}, \frac{1}{y}\right) - xQ\left(\frac{-x}{y}, \frac{1}{y}\right) \\ -yQ\left(\frac{-x}{y}, \frac{1}{y}\right) \end{pmatrix},$$

where $m(x,y) = z(x,y)^{1-n}$. These expressions are the usual ones found in the literature, see for instance [58] or [4] or [21]. To obtain the expression of $\mathbf{f}_{\mathbf{z}_y}$ which appears in [58, p. 59] we have to perform the change of variables $(x,y) \to (-x,y)$ which only change the orientation of the base. If we remove $m(x,y)$ from the expressions of $\mathbf{f}_{\mathbf{z}_x}$ and $\mathbf{f}_{\mathbf{z}_y}$ by rescaling the time, these vector fields are polynomial. Note that, in general, $\mathbf{f}_{\mathbb{D}}$ is not C^1.

2.10 Local bifurcations

The qualitative behaviour of a parametric family of differential equations, $\dot{\mathbf{x}} = \mathbf{f}(\mathbf{x}, \lambda)$, can change by the value of the parameter λ; that is, the qualitative behaviour can change from one topological equivalence class to another. From Theorem 2.6.9, a change of the topological equivalence class implies a change of the separatrix configuration. This change in the separatrix configuration is called a *bifurcation* and the value of the parameter λ in which it takes place is called a *bifurcation value*. In a more general context, the word bifurcation refers not only to other changes in the behaviour of the flow, but also to changes in the topological equivalence class. For details about bifurcation theory see the books of J. Guckenheimer and P. Holmes [29], J. Hale and H. Koçak [31], and S. Chow and J. Hale [17].

In this section we introduce the basic notions of the theory and offer a brief summary of the most usual bifurcations, at least in the context of this book. It is not our purpose to study analytical aspects of bifurcation theory. Here we consider only its geometrical aspects. Some bifurcations described below take place in a neighbourhood of a singular point, hence they are refered as *local bifurcations*.

The set of all bifurcation values in the parameter space is called the *bifurcation set* of the parametric family. When the bifurcation values form a manifold in the parameters space we refer to it as *bifurcation manifold.* The representation in the product space $V \times U$ (where V is the parameter space and U is the phase space) of the invariant sets (singular points, periodic orbits, separatrix cycles, etc. ...) is called *bifurcation diagram.* When the invariant set represented in a bifurcation diagram is a periodic orbit, it is customary to use in the representation the amplitude or the period of the periodic orbit instead of the orbit itself.

2.10.1 Bifurcations from a singular point

Now we describe some of the bifurcations which take place in a neighbourhood of a singular point. We distinguish between uniparametric bifurcations or bifurcations of codimension 1 (saddle-node bifurcation, transcritical bifurcation, pitchfork bifurcation and Hopf bifurcation), and biparametric bifurcations or bifurcations of codimension 2 (cusp bifurcation).

For a bifurcation value λ_0 we say that the differential system $\dot{\mathbf{x}} = \mathbf{f}(\mathbf{x}, \lambda)$ has a *supercritical saddle-node bifurcation* at the singular point $\mathbf{x}_0$ if

(i) for $\lambda < \lambda_0$, the differential system has no singular points in a neighbourhood U of $\mathbf{x}_0$;

(ii) when $\lambda = \lambda_0$, $\mathbf{x}_0$ is the unique singular point in U and it is a saddle-node;

(iii) for $\lambda > \lambda_0$, the differential system has exactly two singular points at U, one of which is a saddle and the other a node.

In Figure 2.16(a) we represent the bifurcation diagram of the supercritical saddle-node bifurcation. When this bifurcation occurs to the left of the bifurcation value, it is called a *subcritical saddle-node bifurcation*, see Figure 2.16(b).

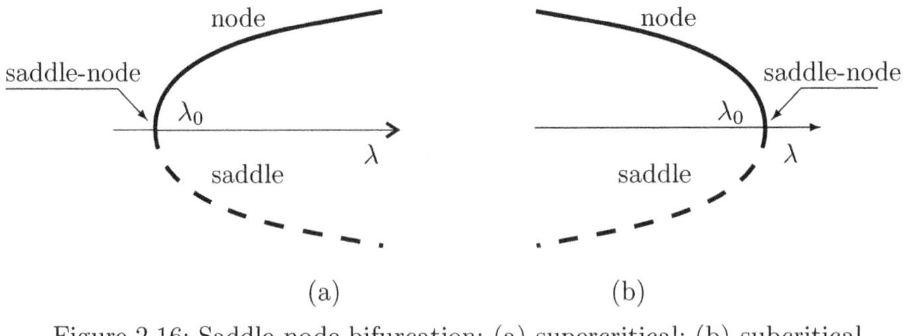

Figure 2.16: Saddle-node bifurcation: (a) supercritical; (b) subcritical.

The differential system $\dot{\mathbf{x}} = \mathbf{f}(\mathbf{x}, \lambda)$ is said to have a *transcritical bifurcation* at $\mathbf{x}_0$ for the bifurcation value λ_0 if

(i) for $\lambda < \lambda_0$, there exist exactly two singular points (one stable and one unstable) in a neighbourhood U of $\mathbf{x}_0$;

(ii) for $\lambda = \lambda_0$, the two singular points collapse into one at $\mathbf{x}_0$, which in general is a non-hyperbolic singular point;

(iii) for $\lambda > \lambda_0$, there exist exactly two singular points in U (one stable and one unstable).

In Figure 2.17 we represent the bifurcation diagram of a transcritical bifurcation.

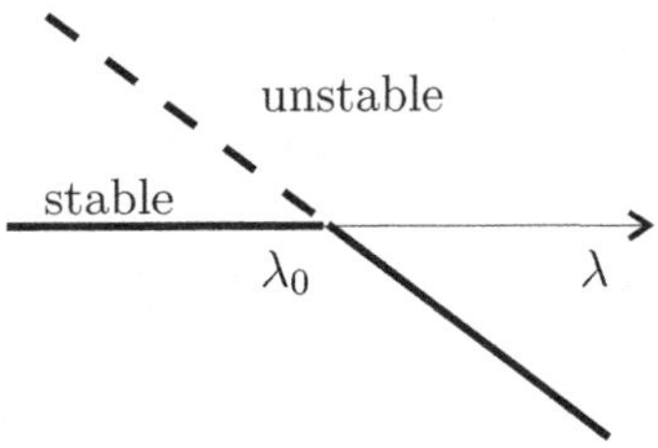

Figure 2.17: Transcritical bifurcation diagram

The differential system $\dot{\mathbf{x}} = \mathbf{f}(\mathbf{x}, \lambda)$ is said to have a *supercritical pitchfork bifurcation* at the bifurcation value λ_0 for the singular point $\mathbf{x}_0$ if

(i) for $\lambda < \lambda_0$, there exists exactly one singular point in a neighbourhood U of $\mathbf{x}_0$ and it is a node (respectively, a saddle);

(ii) for $\lambda = \lambda_0$, $\mathbf{x}_0$ is the unique singular point in U;

(iii) for $\lambda > \lambda_0$, there exist exactly three singular points in U. Two of them are nodes (respectively, saddles) and have the same stability as the singular point which exists for $\lambda < \lambda_0$. The other singular point is a saddle (respectively, a node).

When the bifurcation occurs for values of $\lambda < \lambda_0$, it is called a *subcritical pitchfork bifurcation*. In Figure 2.18 we represent the bifurcation diagram of the pitchfork bifurcation. Note that in this bifurcation we can choose different behaviours for the singular points.

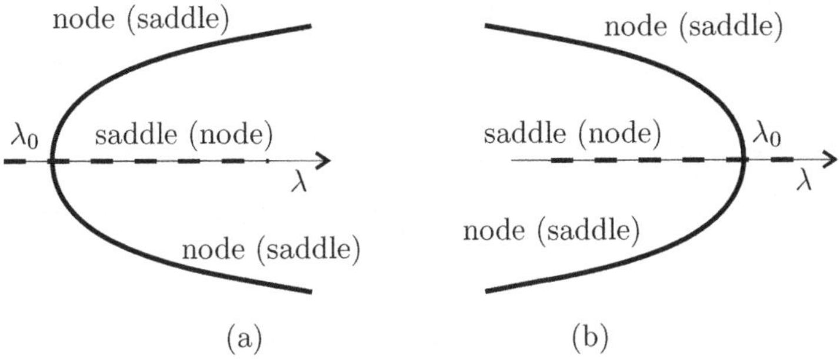

Figure 2.18: Pitchfork bifurcation: (a) supercritical; (b) subcritical. The names in parentheses correspond to the other choice of the singular points.

2.10.2 Bifurcations from orbits

This subsection is devoted to the local bifurcations that involve singular points, periodic orbits and separatrix cycles.

We say that the differential equation $\dot{\mathbf{x}} = \mathbf{f}(\mathbf{x},\lambda)$ has a *vertical bifurcation* at the singular point $\mathbf{x}_0$ for the bifurcation value λ_0, if

(i) for $\lambda < \lambda_0$, there exists exactly one singular point in a neighbourhood U of $\mathbf{x}_0$;

(ii) for $\lambda = \lambda_0$, $\mathbf{x}_0$ is the unique singular point in U and U is foliated by periodic orbits;

(iii) for $\lambda > \lambda_0$, there exists exactly one singular point in U and it has opposite stability compared with the singular point which appears in (i).

In Figure 2.19(a) we represent the bifurcation diagram of the vertical bifurcation. There the vertical variable corresponds to the amplitude of the periodic orbit.

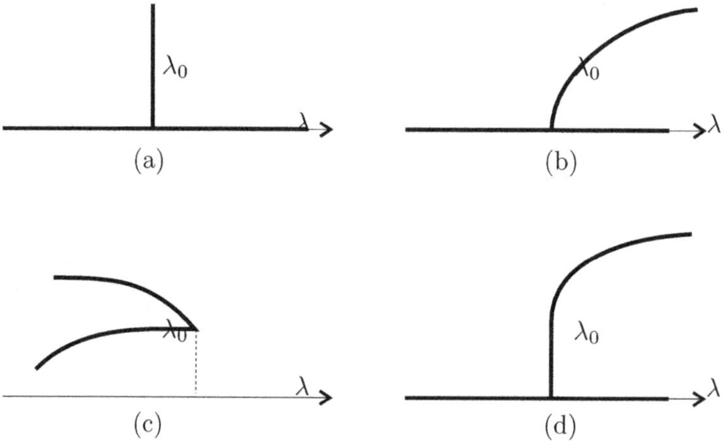

Figure 2.19: Bifurcation diagram involving periodic orbits. The y-axis represent the amplitude of the periodic orbits: (a) vertical bifurcation; (b) Hopf bifurcation; (c) saddle-node bifurcation of limit cycles; and (d) focus-center-limit cycle bifurcation.

The differential equation $\dot{\mathbf{x}} = \mathbf{f}(\mathbf{x}, \lambda)$ has a *supercritical Hopf bifurcation* at the singular point $\mathbf{x}_0$ for the bifurcation value λ_0, if

(i) for $\lambda < \lambda_0$, there exists exactly one singular point and it is stable (respectively, unstable) in a neighbourhood U of $\mathbf{x}_0$;

(ii) for $\lambda = \lambda_0$, $\mathbf{x}_0$ is the unique singular point in U;

(iii) for $\lambda > \lambda_0$, the system has exactly one singular point $\mathbf{x}_0$ and one limit cycle γ in U. Moreover, the singular point is unstable (respectively stable) the limit cycle is stable (respectively unstable) and the amplitude of γ tends to 0 as λ tends to λ_0.

In Figure 2.19(b) we represent the supercritical Hopf bifurcation diagram. The variable in the vertical axis is the amplitude of the limit cycle γ. When the limit cycle γ appears for $\lambda < \lambda_0$ and disappears for $\lambda > \lambda_0$ we say that it is a *subcritical Hopf bifurcation.*

We say that the differential equation $\dot{\mathbf{x}} = \mathbf{f}(\mathbf{x}, \lambda)$ has a *supercritical saddle-node bifurcation of limit cycles* at λ_0 for the limit cycle γ if

(i) for $\lambda < \lambda_0$, the system has no limit cycles in a neighbourhood U of γ;

(ii) for $\lambda = \lambda_0$, γ is the unique limit cycle in U and it is semistable;

(iii) for $\lambda > \lambda_0$, the system has exactly two limit cycles in U, one stable and the other unstable. Moreover, both limits cycles tend to γ as λ tends to λ_0.

In Figure 2.19(c) we show the supercritical saddle-node bifurcation of limit cycles. When the limit cycles appear for $\lambda < \lambda_0$, we say that a *subcritical saddle-node bifurcation of limit cycles* occurs.

The differential equation $\dot{\mathbf{x}} = \mathbf{f}(\mathbf{x}, \lambda)$ is said to have a *supercritical focus-center-limit cycle bifurcation* in the periodic orbit γ if

(i) for $\lambda < \lambda_0$, there exists a convex neighbourhood U of γ with exactly one singular point $\mathbf{x}_0$, which is stable (respectively, unstable);

(ii) for $\lambda = \lambda_0$, the singular point $\mathbf{x}_0$ is a local center, with γ in the boundary;

(iii) for $\lambda > \lambda_0$, there exists a unique limit cycle borning at γ and it is stable (respectively, unstable), and there exists exactly one singular point, which is unstable (respectively, stable).

In Figure 2.19(d) we represent the bifurcation diagram of a supercritical Hopf-vertical bifurcation. When the bifurcation takes place for $\lambda < \lambda_0$, it is called *subcritical focus-center-limit cycle bifurcation.*

The differential equation $\dot{\mathbf{x}} = \mathbf{f}(\mathbf{x}, \lambda)$ is said to have a *homoclinic cycle bifurcation* at point $\mathbf{x}_0$ if

(i) for every $\lambda \neq \lambda_0$, the system has exactly one singular point in a neighbourhood U of $\mathbf{x}_0$ and that point is a saddle;

ii) for $\lambda = \lambda_0$, the system has a saddle point at $\mathbf{x}_0$ and the stable and unstable separatrices of $\mathbf{x}_0$ meet, forming a homoclinic cycle.

Chapter 3

Fundamental Systems

In this chapter we introduce the class of differential systems which is the focus of this book, namely, the fundamental systems. Vector fields associated to fundamental systems are continuous piecewise linear functions. This ensures the existence and uniqueness of solutions of fundamental systems, and the continuous dependence of solutions on the initial conditions and on parameters. From a geometric point of view we can think of a fundamental system as the union of three linear systems, each of them defined on a different region in the phase space. This enables us to use a matricial approach for studying this class.

We also present some results about the existence and localization of singular points (either finite or infinite), and periodic orbits.

3.1 Definition of fundamental systems

Consider a 2×2 real matrix $\widetilde{A}$, i.e., $\widetilde{A} \in L\left(\mathbb{R}^2\right)$, and two non-zero vectors $\widetilde{\mathbf{b}}$, $\widetilde{\mathbf{k}} \in \mathbb{R}^2$. Let $\widetilde{\varphi} : \mathbb{R} \to \mathbb{R}$ be a function given by

$$\widetilde{\varphi}(\sigma) = \begin{cases} m_1\sigma - (m_0 - m_1)\,u, & \text{if } \sigma < -u, \\ m_0\sigma, & \text{if } |\sigma| \leq u, \\ m_1\sigma + (m_0 - m_1)\,u, & \text{if } \sigma > u, \end{cases} \tag{3.1}$$

where $m_0 \neq m_1$ and $u > 0$, see Figure 3.1(a). We define a *fundamental system* as the following family of planar differential equations

$$\dot{\mathbf{y}} = \widetilde{A}\mathbf{y} + \widetilde{\varphi}\left(\widetilde{\mathbf{k}}^T\mathbf{y}\right)\widetilde{\mathbf{b}}. \tag{3.2}$$

The function $\widetilde{\varphi}$ is usually referred as the *characteristic function* of the fundamental system.

Since the characteristic function $\widetilde{\varphi}$ is a continuous nonlinear ($m_0 \neq m_1$ and $u > 0$) function, the vector field defined by system (3.2) is also continuous and nonlinear.

3.2 Normal forms

From equation (3.2) it follows that 11 parameters are used in order to define a fundamental system: 4 coefficients of the matrix $\widetilde{A}$, 2 components of the vector $\widetilde{\mathbf{k}}$, 2 components of the vector $\widetilde{\mathbf{b}}$, and 3 parameters of the characteristic function $\widetilde{\varphi}$. We now normalize the characteristic function to reduce the number of necessary parameters.

By substituting expression (3.1) in expression (3.2) we have

$$\dot{\mathbf{y}} = \begin{cases} \widetilde{A}\mathbf{y} + \left(m_1\widetilde{\mathbf{k}}^T\mathbf{y} - (m_0 - m_1)\,u\right)\widetilde{\mathbf{b}}, & \text{if } \mathbf{k}^T\mathbf{y} < -u, \\ \widetilde{A}\mathbf{y} + \left(m_0\widetilde{\mathbf{k}}^T\mathbf{y}\right)\widetilde{\mathbf{b}}, & \text{if } \left|\mathbf{k}^T\mathbf{y}\right| \le u, \\ \widetilde{A}\mathbf{y} + \left(m_1\widetilde{\mathbf{k}}^T\mathbf{y} + (m_0 - m_1)\,u\right)\widetilde{\mathbf{b}}, & \text{if } \mathbf{k}^T\mathbf{y} > u. \end{cases}$$

Using the fact that $(\widetilde{\mathbf{k}}^T\mathbf{y})\widetilde{\mathbf{b}} = (\widetilde{\mathbf{b}}\,\widetilde{\mathbf{k}}^T)\mathbf{y}$ and the notations $A = \widetilde{A} + m_1\widetilde{\mathbf{b}}\widetilde{\mathbf{k}}^T$, $\mathbf{b} = (m_0 - m_1)\widetilde{\mathbf{b}}$ and $\mathbf{k} = \widetilde{\mathbf{k}}$, the previous system can be written as

$$\dot{\mathbf{y}} = \begin{cases} A\mathbf{y} - u\mathbf{b}, & \text{if } \mathbf{k}^T\mathbf{y} < -u, \\ \left(A + \mathbf{b}\mathbf{k}^T\right)\mathbf{y}, & \text{if } \left|\mathbf{k}^T\mathbf{y}\right| \le u, \\ A\mathbf{y} + u\mathbf{b}, & \text{if } \mathbf{k}^T\mathbf{y} > u. \end{cases}$$

Finally, the change of variables $\mathbf{x} = (1/u)\mathbf{y}$ transforms system (3.2) into the system

$$\dot{\mathbf{x}} = A\mathbf{x} + \varphi\left(\mathbf{k}^T\mathbf{x}\right)\mathbf{b}, \tag{3.3}$$

with $A \in L(\mathbb{R}^2)$, $\mathbf{b}, \mathbf{k} \in \mathbb{R}^2 \setminus \{\mathbf{0}\}$ and

$$\varphi(\sigma) = \begin{cases} -1, & \text{if } \sigma < -1, \\ \sigma, & \text{if } |\sigma| \le 1, \\ 1, & \text{if } \sigma > 1, \end{cases} \tag{3.4}$$

see Figure 3.1(b).

The system (3.3) with the characteristic function (3.4) will be called the *normal form of the fundamental system* (3.2).

Normal forms of fundamental systems involve only 8 parameters $(A, \mathbf{b}, \mathbf{k})$ (we eliminated the 3 parameters of the characteristic function). Moreover, systems (3.2) and (3.3) are linearly conjugate, i.e. have the same qualitative behaviour. Hence *in the sequel we will restrict our attention to fundamental systems expressed in normal form.*

3.3 Existence and uniqueness of solutions

The continuity of the characteristic function φ implies the continuity of the vector field $\mathbf{f}(\mathbf{x}) = A\mathbf{x} + \varphi(\mathbf{k}^T\mathbf{x})\mathbf{b}$ defined by (3.3). In the following we prove that φ and

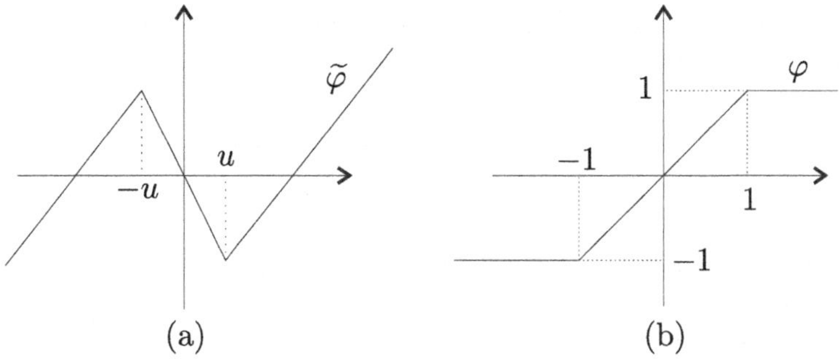

Figure 3.1: Characteristic functions $\widetilde{\varphi}$ and φ.

$\mathbf{f}$ are globally Lipschitz. This will allow us to conclude not only the existence and uniqueness of solutions of the fundamental systems, but also the completeness of their flow.

Lemma 3.3.1. *The characteristic function $\widetilde{\varphi}(\sigma)$ defined in* (3.1) *is globally Lipschitz in $\mathbb{R}$, with Lipschitz constant $L = \max\{|m_0|, |m_1|\}$.*

Proof. To show that $|\widetilde{\varphi}(\sigma_1) - \widetilde{\varphi}(\sigma_2)| \leq L|\sigma_1 - \sigma_2|$ for all $\sigma_1, \sigma_2 \in \mathbb{R}$, we divide the proof into 8 cases, depending on the intervals $I_- = (-\infty, -u)$, $I_0 = [-u, u]$ or $I_+ = (u, +\infty)$, where σ_1 and σ_2 lie.

If σ_1 and σ_2 are in the same interval, then

$$|\widetilde{\varphi}(\sigma_1) - \widetilde{\varphi}(\sigma_2)| = \begin{cases} |m_1|\,|\sigma_1 - \sigma_2|, & \text{if } \sigma_1, \sigma_2 \in I_-, \\ |m_0|\,|\sigma_1 - \sigma_2|, & \text{if } \sigma_1, \sigma_2 \in I_0, \\ |m_1|\,|\sigma_1 - \sigma_2|, & \text{if } \sigma_1, \sigma_2 \in I_+. \end{cases}$$

Therefore, $|\widetilde{\varphi}(\sigma_1) - \widetilde{\varphi}(\sigma_2)| \leq L\,|\sigma_1 - \sigma_2|$.

Suppose now that $\sigma_1 \in I_-$ and $\sigma_2 \in I_0$. Then

$$\begin{aligned} |\widetilde{\varphi}(\sigma_1) - \widetilde{\varphi}(\sigma_2)| &= |m_1\sigma_1 - (m_0 - m_1)\,u - m_0\sigma_2| \\ &= |m_1(\sigma_1 + u) - m_0(\sigma_2 + u)| \\ &\leq L\,(|\sigma_1 + u| + |\sigma_2 + u|)\,. \end{aligned}$$

Since $\sigma_1 \in I_-$ and $\sigma_2 \in I_0$, we have $\sigma_1 + u < 0$, $\sigma_2 + u \geq 0$ and $\sigma_1 < \sigma_2$. Therefore,

$$|\widetilde{\varphi}(\sigma_1) - \widetilde{\varphi}(\sigma_2)| \leq L\,|\sigma_2 - \sigma_1|\,.$$

Similar arguments apply to the remaining cases. □

Proposition 3.3.2. (a) *The vector field defined by the fundamental system* (3.3) *is Lipschitz in $\mathbb{R}^2$ with Lipschitz constant $L = \|A\| + \|\mathbf{b}\mathbf{k}^T\|$.*

(b) *The theorem on the existence and uniqueness theorem of solutions of ordinary differential equations holds for fundamental systems.*

(c) *Let* $\mathbf{x}(s)$ *be the solution of fundamental system* (3.3) *with the initial condition* $\mathbf{x}(0) = \mathbf{x}_0$*, then*

$$\mathbf{x}(s) = e^{As}\mathbf{x}_0 + \int_0^s e^{A(s-r)}\varphi\left(\mathbf{k}^T\mathbf{x}\right)\mathbf{b}\,dr. \tag{3.5}$$

(d) *Flows of fundamental systems are complete. Moreover, flows of fundamental systems depend in differential manner on time, and continuously on the initial conditions and on parameters.*

Proof. (a) By taking $m_0 = 1$, $m_1 = 0$ and $u = 1$ in Lemma 3.3.1, it follows that φ is a Lipschitz function in $\mathbb{R}$ with Lipschitz constant $L = 1$, i.e.,

$$\left|\varphi\left(\mathbf{k}^T\mathbf{x}_1\right) - \varphi\left(\mathbf{k}^T\mathbf{x}_2\right)\right| \leq \left|\mathbf{k}^T(\mathbf{x}_1 - \mathbf{x}_2)\right|.$$

Hence we have $\|(\varphi(\mathbf{k}^T\mathbf{x}_1) - \varphi(\mathbf{k}^T\mathbf{x}_2))\mathbf{b}\| \leq \|\mathbf{b}\mathbf{k}^T(\mathbf{x}_1 - \mathbf{x}_2)\|$. From this we conclude that the vector field $\mathbf{f}(\mathbf{x}) = A\mathbf{x} + \varphi(\mathbf{k}^T\mathbf{x})\mathbf{b}$ defined by (3.3) satisfies

$$\left\|\mathbf{f}(\mathbf{x}_1) - \mathbf{f}(\mathbf{x}_2)\right\| \leq \left(\|A\| + \left\|\mathbf{b}\mathbf{k}^T\right\|\right)\left\|\mathbf{x}_1 - \mathbf{x}_2\right\|,$$

which proves our statement.

(b) follows from Theorem 2.1.1 and statement (a).

(c) It is easy to check that the function $\mathbf{x}(s)$ defined by (3.5) satisfies equation (3.3) as well as the initial condition $\mathbf{x}(0) = \mathbf{x}_0$. The assertion follows by applying the statement (b).

(d) follows from Proposition 2.1.2 and Theorem 2.1.3. □

3.4 Symmetric orbits

As the function φ is odd (i.e., $\varphi(-\sigma) = -\varphi(\sigma)$), the vector field $\mathbf{f}(\mathbf{x}) = A\mathbf{x} + \varphi(\mathbf{k}^T\mathbf{x})\mathbf{b}$ defined by the fundamental system (3.3) is also an odd function. Therefore, if $\mathbf{x}(s)$ is a solution of (3.3), then $\mathbf{y}(s) = -\mathbf{x}(s)$ is also a solution of (3.3). Note that the orbits associated to the solutions $\mathbf{x}(s)$ and $\mathbf{y}(s)$ are symmetric to one another with respect to the origin. In particular, if $\mathbf{x}(s)$ and $\mathbf{y}(s)$ are associated to the same orbit, then this orbit is a symmetric periodic orbit.

3.5 Piecewise linear form

Since characteristic functions are nonlinear functions, fundamental systems are also nonlinear. In this section we show that fundamental systems can be understood as three linear systems, each defined on a different region in the plane.

Substituting φ in equation (3.3) and using the notations

$$\begin{aligned} S_- &:= \{\mathbf{x} \in \mathbb{R}^2 : \mathbf{k}^T\mathbf{x} < -1\}, \\ L_- &:= \{\mathbf{x} \in \mathbb{R}^2 : \mathbf{k}^T\mathbf{x} = -1\}, \\ S_0 &:= \{\mathbf{x} \in \mathbb{R}^2 : |\mathbf{k}^T\mathbf{x}| < 1\}, \\ L_+ &:= \{\mathbf{x} \in \mathbb{R}^2 : \mathbf{k}^T\mathbf{x} = 1\}, \\ S_+ &:= \{\mathbf{x} \in \mathbb{R}^2 : \mathbf{k}^T\mathbf{x} > 1\}, \end{aligned} \tag{3.6}$$

one can recast system (3.3) as

$$\dot{\mathbf{x}} = \begin{cases} A\mathbf{x} - \mathbf{b}, & \text{if } \mathbf{x} \in S_- \cup L_-, \\ B\mathbf{x}\,, & \text{if } \mathbf{x} \in L_- \cup S_0 \cup L_+, \\ A\mathbf{x} + \mathbf{b}, & \text{if } \mathbf{x} \in L_+ \cup S_+, \end{cases} \tag{3.7}$$

where, if $A = \begin{pmatrix} a_{11} & a_{12} \\ a_{21} & a_{22} \end{pmatrix}$, $\mathbf{b}^T = (b_1, b_2)$ and $\mathbf{k}^T = (k_1, k_2)$, then

$$B = A + \mathbf{b}\mathbf{k}^T = \begin{pmatrix} a_{11} + k_1 b_1 & a_{12} + k_2 b_1 \\ a_{21} + k_1 b_2 & a_{22} + k_2 b_2 \end{pmatrix}. \tag{3.8}$$

Expression (3.7) will be called *the piecewise linear form of the fundamental system* (3.3).

The straight lines L_+ and L_- are symmetric with respect to the origin. These lines split the phase plane into the half-planes S_+, S_- and the open strip S_0 (Figure 3.2). On each of these regions the system is linear, with matrices A and B, respectively; moreover, the system changes continuously on the straight lines L_+ and L_-. In fact, as we proved in Proposition 3.3.2(a), the vector field on L_+ and L_- is not only continuous, but also globally Lipschitz in $\mathbb{R}^2$.

3.6 Fundamental matrices

The matrices (A, B) in (3.7) are called the *fundamental matrices* of the fundamental system (3.3).

From (3.8) associated to a fundamental system with parameters $(A, \mathbf{b}, \mathbf{k})$, there exists a pair of fundamental matrices (A, B). However, not for all pairs of matrices A and B there exists a fundamental system with fundamental matrices (A, B). In the following we give sufficient and necessary conditions on the pair (A, B) so that they are fundamental matrices.

Proposition 3.6.1. *Let $A, B \in L(\mathbb{R}^2)$. Then (A, B) is a pair of fundamental matrices if and only if $A \neq B$ and* $\det(B - A) = 0$.

Proof. Suppose that (A, B) is a pair of fundamental matrices. From (3.8) it follows that there exist vectors $\mathbf{b}, \mathbf{k} \in \mathbb{R}^2 \setminus \{\mathbf{0}\}$ such that $B - A = \mathbf{b}\mathbf{k}^T$. Consequently, $\det(B - A) = k_1 b_1 k_2 b_2 - k_1 b_2 k_2 b_1 = 0$.

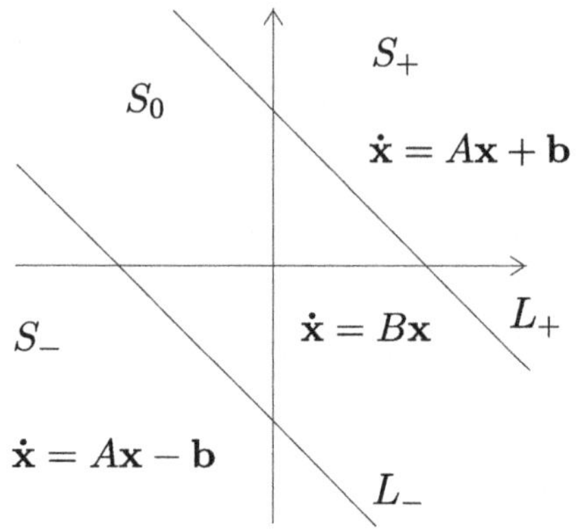

Figure 3.2: Piecewise linear phase plane of a fundamental system.

Suppose now $A = B$. Then we have $k_1b_1 = 0$, $k_1b_2 = 0$, $k_2b_1 = 0$ and $k_2b_2 = 0$. This clearly implies that $\mathbf{k} = \mathbf{0}$ or $\mathbf{b} = \mathbf{0}$, which contradicts our assumptions. So $A \neq B$. This proves the necessary condition.

Conversely, if $A \neq B$, the matrix

$$B - A = \begin{pmatrix} m_{11} & m_{12} \\ m_{21} & m_{22} \end{pmatrix},$$

is not the zero matrix. We assume that $m_{11} \neq 0$; the other three cases follow by using similar arguments.

Take $\mathbf{b}^T = (m_{11}, m_{21})$ and $\mathbf{k}^T = (1, m_{12}/m_{11})$, thus $\mathbf{b}, \mathbf{k} \in \mathbb{R}^2 \setminus \{\mathbf{0}\}$. Since $\det(B - A) = 0$ it is easy to check that $B = A + \mathbf{b}\mathbf{k}^T$. Therefore, (A, B) is the pair of fundamental matrices of the fundamental system with parameters $(A, \mathbf{b}, \mathbf{k})$. □

Given the pair of fundamental matrices (A, B) one can choose different vectors $\mathbf{b}$ and $\mathbf{k}$ satisfying (3.8). That is, there exist more than one fundamental system with (A, B) as fundamental matrices. The next proposition shows an important relationship between all these fundamental systems.

Proposition 3.6.2. *Two fundamental systems having the same pair of fundamental matrices are linearly conjugate.*

Proof. Let (A, B) be the fundamental matrices of the two fundamental systems

$$\dot{\mathbf{x}} = A\mathbf{x} + \varphi\left(\mathbf{k}^T\mathbf{x}\right)\mathbf{b} \quad \text{and} \quad \dot{\mathbf{y}} = A\mathbf{y} + \varphi\left(\mathbf{k}^{*T}\mathbf{y}\right)\mathbf{b}^*,$$

where $\mathbf{b}^T = (b_1, b_2)$, $\mathbf{k}^T = (k_1, k_2)$, $\mathbf{b}^{*T} = (b_1^*, b_2^*)$ and $\mathbf{k}^{*T} = (k_1^*, k_2^*)$.

By (3.8), the matrices $\mathbf{b}\mathbf{k}^T$ and $\mathbf{b}^*\mathbf{k}^{*T}$ are equal. Thus $k_1b_1 = k_1^*b_1^*$, $k_1b_2 = k_1^*b_2^*$, $k_2b_1 = k_2^*b_1^*$ and $k_2b_2 = k_2^*b_2^*$.

Since $\mathbf{b}^* \neq \mathbf{0}$, suppose that $b_1^* \neq 0$ (the case $b_1^* = 0$ and $b_2^* \neq 0$ follows using similar arguments). From the above equalities we have $b_1 \neq 0$, $\mathbf{k}^* = (b_1/b_1^*)\mathbf{k}$ and $\mathbf{b}^* = (b_1^*/b_1)\mathbf{b}$.

Consider $M = (b_1^*/b_1)\mathrm{Id} \in L(\mathbb{R}^2)$, where Id denotes the identity matrix. If $\mathbf{x}(s)$ is a solution of the first fundamental system and $\mathbf{y}(s) = M\mathbf{x}(s)$, then

$$\dot{\mathbf{y}} = \frac{b_1^*}{b_1}\dot{\mathbf{x}} = A\left(\frac{b_1^*}{b_1}\mathbf{x}\right) + \varphi\left(\mathbf{k}^T\frac{b_1}{b_1^*}\mathbf{y}\right)\left(\frac{b_1^*}{b_1}\right)\mathbf{b} = A\mathbf{y} + \varphi\left(\mathbf{k}^{*T}\mathbf{y}\right)\mathbf{b}^*,$$

i.e., $\mathbf{y}(s)$ is a solution of the second fundamental system. Hence the systems are linearly conjugate and M is the linear conjugation. □

Associated to a pair of fundamental matrices (A, B) there exists a one-parameter family of fundamental systems which have (A, B) as fundamental matrices. In fact, from the above proof, this family has parameters $(A, \alpha\mathbf{b}, (1/\alpha)\mathbf{k})$ with $\alpha \neq 0$. But all these systems have the same phase portrait up to a linear transformation; i.e., they are linearly conjugate, see Section 2.6 for more details. This allows us to use fundamental matrices to classify the behavior of fundamental systems.

Up to now we have not reduced the number of parameters used for describing the behavior of the systems. We have only reduced the number of fundamental systems to study by choosing a unique representant of the above family; i.e., for instance by choosing $\alpha = 1$.

3.7 Fundamental parameters

Consider a fundamental system (3.3) with fundamental matrices (A, B). The vector (D, T, d, t) where

$$D = \det(B),\ T = \mathrm{trace}(B),\ d = \det(A) \text{ and } t = \mathrm{trace}(A). \tag{3.9}$$

will be called the *fundamental parameters* of the fundamental system. Since fundamental matrices are related, fundamental parameters are also related. From (3.8) it follows that

$$\begin{aligned} D &= d + (a_{11}k_2b_2 + a_{22}k_1b_1 - a_{12}k_1b_2 - a_{21}k_2b_1) \\ &= d + \mathbf{k}^T\begin{pmatrix} a_{22} & -a_{12} \\ -a_{21} & a_{11} \end{pmatrix}\mathbf{b}. \end{aligned} \tag{3.10}$$

Therefore, when $d \neq 0$, we obtain

$$D = d\left(1 + \mathbf{k}^T A^{-1}\mathbf{b}\right). \tag{3.11}$$

In a similar way (3.8) yields

$$T = t + (k_1b_1 + k_2b_2) = t + \mathbf{k}^T\mathbf{b}. \tag{3.12}$$

Proposition 3.7.1. *If we change the direction of the flow (by the change of time $s \to -s$) of a fundamental system with fundamental parameters (D, T, d, t), then we obtain a fundamental system with fundamental parameters $(D, -T, d, -t)$.*

Proof. Let $\dot{\mathbf{x}} = A\mathbf{x} + \varphi(\mathbf{k}^T\mathbf{x})\mathbf{b}$ be a fundamental system with fundamental parameters (D, T, d, t). The change of variables $(\mathbf{x}, s) \to (\mathbf{y}, -s)$ transforms the system into the fundamental system $\dot{\mathbf{y}} = -A\mathbf{y} - \varphi(\mathbf{k}^T\mathbf{y})\mathbf{b}$ with parameters (D^*, T^*, d^*, t^*), where $d^* = d$, $t^* = -t$, $D^* = D$ and $T^* = -T$. □

3.8 Linear conjugacy

In Proposition 2.6.4(a) we have shown that two linear systems with equivalent matrices are linearly conjugate. This fact makes possible the classification of linear flows using the trace t and the determinant d of the matrix when $t^2 - 4d \neq 0$, see Proposition 2.6.4(c). As we shall prove in Theorem 3.8.2, a similar result holds for fundamental systems.

Two fundamental systems are said to be *equal* in an open subset $U \subset \mathbb{R}^2$ if the vector fields defined by them are equal in U.

Lemma 3.8.1. *Fundamental systems $\dot{\mathbf{x}} = A\mathbf{x} + \varphi(\mathbf{k}^T\mathbf{x})\mathbf{b}$ and $\dot{\mathbf{x}} = A^*\mathbf{x} + \varphi(\mathbf{k}^{*T}\mathbf{x})\mathbf{b}^*$ are equal in $\mathbb{R}^2$ if and only if $A = A^*$, $\mathbf{b} = n\mathbf{b}^*$ and $\mathbf{k} = n\mathbf{k}^*$, where $n \in \{1, -1\}$.*

Proof. Let

$$\dot{\mathbf{x}} = \begin{cases} A\mathbf{x} - \mathbf{b}, & \text{if } \mathbf{x} \in S_- \cup L_-, \\ B\mathbf{x}, & \text{if } \mathbf{x} \in L_- \cup S_0 \cup L_+, \\ A\mathbf{x} + \mathbf{b}\ , & \text{if } \mathbf{x} \in L_+ \cup S_+, \end{cases}$$

and

$$\dot{\mathbf{x}} = \begin{cases} A^*\mathbf{x} - \mathbf{b}^*, & \text{if } \mathbf{x} \in S_-^* \cup L_-^*, \\ B^*\mathbf{x}\ , & \text{if } \mathbf{x} \in L_-^* \cup S_0^* \cup L_+^*, \\ A^*\mathbf{x} + \mathbf{b}^*, & \text{if } \mathbf{x} \in L_+^* \cup S_+^*, \end{cases}$$

be the piecewise linear forms of the two fundamental systems, respectively.

Suppose that the two systems are equal in $\mathbb{R}^2$. If $S_- \cap S_-^* \neq \varnothing$, then $A\mathbf{x} - \mathbf{b} = A^*\mathbf{x} - \mathbf{b}^*$ for all $\mathbf{x}$ in the open set $S_- \cap S_-^*$, i.e., $(A - A^*)\mathbf{x} = \mathbf{b} - \mathbf{b}^*$ for $\mathbf{x} \in S_- \cap S_-^*$. From Lemma 2.6.3 it follows that $A = A^*$ and $\mathbf{b} = \mathbf{b}^*$. On the contrary, if $S_- \cap S_-^* = \varnothing$, then $S_- \cap S_+^* \neq \varnothing$. In this case applying similar arguments as before we obtain $A = A^*$, $\mathbf{b} = -\mathbf{b}^*$.

On the other hand, since $S_0 \cap S_0^*$ is a non-empty open set and $B\mathbf{x} = B^*\mathbf{x}$ when $\mathbf{x} \in S_0 \cap S_0^*$, applying Lemma 2.6.3 we obtain $B = B^*$. Now from equation (3.8) it follows that $\mathbf{b}\mathbf{k}^T = \mathbf{b}\mathbf{k}^{T*}$. Taking coordinates and noting that $\mathbf{b}, \mathbf{k}, \mathbf{k}^* \in \mathbb{R}^2 \setminus \{\mathbf{0}\}$, it is easy to check that $\mathbf{k} = \mathbf{k}^*$ or $\mathbf{k} = -\mathbf{k}^*$, depending on whether $\mathbf{b} = \mathbf{b}^*$ or $\mathbf{b} = -\mathbf{b}^*$.

Suppose now that $A = A^*$, $\mathbf{b} = n\mathbf{b}^*$ and $\mathbf{k} = n\mathbf{k}^*$. It is clear that $\varphi(\mathbf{k}^T\mathbf{x})\mathbf{b} = n^2\varphi(\mathbf{k}^{*T}\mathbf{x})\mathbf{b}^*$, which completes the proof. □

In the following theorem we give a characterization of the linear conjugacy classes of fundamental systems. In this characterization we use only fundamental matrices.

Theorem 3.8.2. (a) *Two fundamental systems with respective fundamental matrices* (A, B) *and* (A^*, B^*) *are linearly conjugate if and only if there exists a matrix* $M \in GL(\mathbb{R}^2)$ *such that* $A^* = MAM^{-1}$ *and* $B^* = MBM^{-1}$.

(b) *If two fundamental systems are linearly conjugate, then they have the same fundamental parameters.*

Proof. Let $\dot{\mathbf{x}} = A\mathbf{x} + \varphi(\mathbf{k}^T\mathbf{x})\mathbf{b}$ and $\dot{\mathbf{y}} = A^*\mathbf{y} + \varphi(\mathbf{k}^{*T}\mathbf{y})\mathbf{b}^*$ be two linearly conjugate fundamental systems, and let $M \in GL(\mathbb{R}^2)$ be the conjugacy matrix.

Take $\mathbf{q}$ in $\mathbb{R}^2$ and let $\mathbf{y}(s)$ be the solution of the second system with initial condition $\mathbf{q} = \mathbf{y}(0)$. Since the systems are conjugate with conjugacy M, it follows that $\mathbf{x}(s) = M^{-1}\mathbf{y}(s)$ is a solution of the first system. Multiplying by M and taking its derivative at $s = 0$, we have $M\dot{\mathbf{x}}(0) = \dot{\mathbf{y}}(0)$, which is equivalent to

$$MAM^{-1}\mathbf{q} + \varphi(\mathbf{k}^T M^{-1}\mathbf{q})M\mathbf{b} = A^*\mathbf{q} + \varphi(\mathbf{k}^{*T}\mathbf{q})\mathbf{b}^*.$$

Since $\mathbf{q}$ is an arbitrary point in $\mathbb{R}^2$ we obtain $A^* = MAM^{-1}$, $\mathbf{b}^* = nM\mathbf{b}$ and $\mathbf{k}^{*T} = n\mathbf{k}^T M^{-1}$ with $n \in \{1, -1\}$, see Lemma 3.8.1. Substituting these expressions into $B^* = A^* + \mathbf{b}^*\mathbf{k}^{*T}$, see (3.8), and taking into account that $B = A + \mathbf{b}\mathbf{k}^T$, we have

$$B^* = MAM^{-1} + n^2 M\mathbf{b}\mathbf{k}^T M^{-1} = M(A + \mathbf{b}\mathbf{k}^T)M^{-1} = MBM^{-1}.$$

Now let us prove the other implication. Let (A, B) and (A^*, B^*) be the fundamental matrices of the fundamental systems

$$\dot{\mathbf{x}} = A\mathbf{x} + \varphi(\mathbf{k}^T\mathbf{x})\mathbf{b}, \tag{3.13}$$

$$\dot{\mathbf{y}} = A^*\mathbf{y} + \varphi(\mathbf{k}^{*T}\mathbf{y})\mathbf{b}^*, \tag{3.14}$$

respectively. By hypothesis, there exists a regular matrix M such that $A^* = MAM^{-1}$ and $B^* = MBM^{-1}$.

The change of coordinates $\mathbf{z} = M\mathbf{x}$ transforms system (3.13) into the system

$$\dot{\mathbf{z}} = MAM^{-1}\mathbf{z} + \varphi\left(\mathbf{k}^T M^{-1}\mathbf{z}\right) M\mathbf{b}, \tag{3.15}$$

with fundamental matrices (MAM^{-1}, MBM^{-1}). Thus systems (3.13) and (3.15) are linearly conjugate with conjugacy matrix M.

Since systems (3.14) and (3.15) have the same pair of fundamental matrices, Proposition 3.6.2 implies that they are linearly conjugate. Thus systems (3.13) and (3.14) are also linearly conjugate. This completes the proof of statement (a).

Statement (b) is a straightforward consequence of statement (a). □

There are some important consequences of Theorem 3.8.2.

Remark 3.8.3. All fundamental systems in the same linear conjugacy class have topologically equivalent phase portraits. Thus, *there is not loss of generality in assuming that either matrix A or matrix B is given in its real Jordan normal form.* This usually simplifies the computations. Note that solutions of a linear system appear in their easiest form when we use their real Jordan normal form.

Remark 3.8.4. To each linear conjugacy class there is associated a unique point in the space $\mathbb{R}^4$ given by the fundamental parameters (D, T, d, t), see Theorem 3.8.2(b). However, two different classes can be associated to the same point. We call such points *virtual bifurcation points.* As we shall see in Chapter 5, the set of all virtual bifurcation points has zero Lebesgue measure in the set of all bifurcation points. Since almost always fundamental parameters characterize the linear conjugacy classes, *we choose* $\mathbb{R}^4 = \{(D, T, d, t) : D, T, d, t \in \mathbb{R}\}$ *as the parameter space of the fundamental systems.* Note that we have reduced the initial 8-dimensional parameter space to dimension 4.

Remark 3.8.5. The behaviour of fundamental systems with respect to the linear conjugacy relationship is very similar to that of linear systems, see Proposition 2.6.4.

3.9 Finite singular points

We are now interested in the study of phase portraits of fundamental systems. Since the separatrix configuration is the skeleton of a phase portrait, see Theorem 2.6.9, we start by studying the simplest separatrices of fundamental systems, i.e., singular points. In this section we give the number, the localization and the local phase portrait of finite singular points. Singular points at infinity will be studied in the following section.

Consider the fundamental system

$$\dot{\mathbf{x}} = A\mathbf{x} + \varphi\left(\mathbf{k}^T\mathbf{x}\right)\mathbf{b}. \tag{3.16}$$

Its singular points are determined by the zeros of the linear function $A\mathbf{x} + \mathbf{b}$ in the half-plane S_+, by the zeros of $B\mathbf{x}$ in the strip $L_+ \cup S_0 \cup L_-$, and by the zeros of $A\mathbf{x} - \mathbf{b}$ in S_-.

Since the origin $\mathbf{0} \in S_0$, $\mathbf{0}$ is always a singular point. Furthermore, if the fundamental parameter $D \neq 0$, then the origin is the unique singular point in $L_+ \cup S_0 \cup L_-$.

Suppose that the fundamental parameter $d \neq 0$. We define the points $\mathbf{e}_+$ and $\mathbf{e}_-$ as

$$\mathbf{e}_+ = -A^{-1}\mathbf{b} \quad \text{and} \quad \mathbf{e}_- = A^{-1}\mathbf{b}. \tag{3.17}$$

By (3.7), if $\mathbf{e}_+ \in S_+$, then $\mathbf{e}_+$ is the unique singular point in S_+. By the symmetry of the orbits with respect to the origin, $\mathbf{e}_-$ is the unique singular point in S_-. On the other hand, when $\mathbf{e}_+ \notin S_+$ it follows that $\mathbf{e}_- \notin S_-$. Hence $\mathbf{e}_+, \mathbf{e}_-$ are not singular points of (3.16). These points are usually called *virtual singular points* (see, [24] and [25]). In the next result we give necessary and sufficient conditions on the fundamental parameters so that $\mathbf{e}_+$ and $\mathbf{e}_-$ are genuine singular points.

Lemma 3.9.1. *Consider a fundamental system with fundamental parameter $d \neq 0$.*

(a) *The points $\mathbf{e}_+$ and $\mathbf{e}_-$ satisfy*

$$\mathbf{k}^T\mathbf{e}_+ = 1 - \frac{D}{d} \quad \text{and} \quad \mathbf{k}^T\mathbf{e}_- = \frac{D}{d} - 1. \tag{3.18}$$

(b) *$\mathbf{e}_+$ belongs to S_+ (respectively, $\mathbf{e}_- \in S_-$) if and only if $Dd < 0$.*

(c) *$\mathbf{e}_+$ belongs to L_+ (respectively, $\mathbf{e}_- \in L_-$) if and only if $D = 0$.*

Proof. (a) Since $d \neq 0$, from (3.11) we have $D = d(1 - \mathbf{k}^T\mathbf{e}_+)$, which establishes the formula. Statements (b) and (c) follow easily from the definition of S_+ and L_+ and (3.18). □

Now we classify the finite singular points depending on the fundamental parameters. Before doing this we prove that the number of singular points and the regions $S_+ \cup S_-$ and S_0 to which they belong are invariant under linear transformations.

Lemma 3.9.2. *Let $\dot{\mathbf{x}} = A\mathbf{x} + \varphi(\mathbf{k}^T\mathbf{x})\mathbf{b}$ and $\dot{\mathbf{x}} = A^*\mathbf{x} + \varphi(\mathbf{k}^{*T}\mathbf{x})\mathbf{b}^*$ be two linear conjugate fundamental systems, with conjugacy matrix $M \in GL(\mathbb{R}^2)$. If $\mathbf{q}$ is a singular point of the first system, then $\mathbf{q}^* = M\mathbf{q}$ is a singular point of the second one, and $|\mathbf{k}^T\mathbf{q}| \leq 1$ if and only if $|\mathbf{k}^{*T}\mathbf{q}^*| \leq 1$.*

Proof. Since M is a conjugacy matrix, it maps orbits into orbits. Thus $\mathbf{q}^*$ is a singular point for the second system.

From the proof of Theorem 3.8.2 we have $A^* = MAM^{-1}$, $\mathbf{b}^* = nM\mathbf{b}$, and $\mathbf{k}^{*T} = n\mathbf{k}^T M^{-1}$, with $n \in \{-1, 1\}$, which proves our assertion. □

According to Lemma 3.9.2, if $\mathbf{q} \in S_0$, then $\mathbf{q}^*$ belongs to the central open strip, written S_0^*, of the transformed system. But if $\mathbf{q} \in S_+$, we cannot be sure if $\mathbf{q}^*$ belongs to S_+^* or to S_-^*. In any case, if the singular point $\mathbf{q} \in S_+$, there is always a singular point of the transformed system in S_+^*. This point will be either $\mathbf{q}^*$ or $-\mathbf{q}^*$.

Theorem 3.9.3. *Consider a fundamental system with fundamental parameter $D \neq 0$. Under this assumption the origin $\mathbf{0}$ is the unique singular point in $L_- \cup S_0 \cup L_+$ and we have:*

(a) *Suppose $D > 0$. If $T^2 - 4D < 0$, then the origin is a stable focus when $T < 0$, a center when $T = 0$, and an unstable focus when $T > 0$. If $T^2 - 4D \geq 0$, then the origin is a stable node when $T < 0$, and an unstable node when $T > 0$.*

  (a.1) *If $d \geq 0$, then there are no singular points in $S_+ \cup S_-$.*

  (a.2) *If $d < 0$, then $\mathbf{e}_+$ (respectively $\mathbf{e}_-$) is the unique singular point in S_+ (respectively S_-). Moreover, $\mathbf{e}_+$ and $\mathbf{e}_-$ are saddle points.*

(b) *Suppose $D < 0$. Then the origin is a saddle point.*

(b.1) *If $d \leq 0$, then there are no singular points in $S_+ \cup S_-$.*

(b.2) *If $d > 0$, then $\mathbf{e}_+$ (respectively $\mathbf{e}_-$) is the unique singular point in S_+ (respectively S_-). Moreover, if $t^2 - 4d < 0$, then $\mathbf{e}_+$ and $\mathbf{e}_-$ are stable foci when $t < 0$, centers when $t = 0$, and unstable foci when $t > 0$. If $t^2 - 4d \geq 0$, then $\mathbf{e}_+$ and $\mathbf{e}_-$ are stable nodes when $t < 0$, and unstable nodes when $t > 0$.*

Proof. Let $\dot{\mathbf{x}} = A\mathbf{x} + \varphi(\mathbf{k}^T\mathbf{x})\mathbf{b}$ be the given fundamental system. In the central region $L_- \cup S_0 \cup L_+$ the system is $\dot{\mathbf{x}} = B\mathbf{x}$, with $B = A + \mathbf{b}\mathbf{k}^T$ and $D = \det(B) \neq 0$. Therefore the origin is the unique singular point in $L_- \cup S_0 \cup L_+$.

(a) Since the system is linear in a neighbourhood of the origin, its local phase portrait follows from Subsection 2.5.3.

(a.1) When $d > 0$, it follows that $Dd > 0$. By Lemma 3.9.1(b), there are no singular points in $S_+ \cup S_-$.

Suppose now that $d = 0$. By Theorem 3.8.2(a) and Lemma 3.9.2, we can assume without loss of generality that the matrix A is one of the following ones:

$$A = \begin{pmatrix} t & 0 \\ 0 & 0 \end{pmatrix}, \quad A = \begin{pmatrix} 0 & 1 \\ 0 & 0 \end{pmatrix} \quad \text{or} \quad A = \begin{pmatrix} 0 & 0 \\ 0 & 0 \end{pmatrix}.$$

In the three cases we arrive at a contradiction with the hypothesis. First suppose A is the zero matrix. Then $B = \mathbf{b}\mathbf{k}^T$ and $D = 0$, which contradicts the hypothesis. Suppose now that the matrix A is not the zero matrix and $\operatorname{trace}(A) = 0$. If there exists a singular point $\mathbf{q}$ in S_+, i.e., $A\mathbf{q} + \mathbf{b} = \mathbf{0}$, then $\mathbf{b}^T = (b_1, 0)$. In this case

$$B = \begin{pmatrix} b_1k_1 & 1 + b_1k_2 \\ 0 & 0 \end{pmatrix}.$$

Therefore $D = 0$, which contradicts the hypothesis. The remaining case follows in a similar way. So there are no singular points in S_+. By the symmetry of the orbits with respect to the origin, there are no singular points in S_-.

(a.2) When $d < 0$ we have $Dd < 0$. By Lemma 3.9.1(b), $\mathbf{e}_+$ is the unique singular point in S_+ and $\mathbf{e}_-$ is the unique singular point in S_-. Since the system is linear in S_+ and S_-, the local phase portraits of $\mathbf{e}_+$ and $\mathbf{e}_-$ follow from Subsection 2.5.3.

Statements (b), (b.1) and (b.2) follow using arguments similar to those of the proofs of statements (a), (a.1) and (a.2). □

Remark 3.9.4. All singular points of a fundamental system with fundamental parameter $D \neq 0$ are isolated, see Theorem 3.9.3. Furthermore, the existence in a fundamental system of more that one singular point is equivalent to $Dd < 0$.

In the following result we study the degenerate case $D = 0$. In this case singular points are not isolated. In particular, they are located on curves in the phase space. To compute the quantitative aspects of these curves one can assume with no loss of generality that the matrix B is given in its real Jordan normal form, see Lemma 3.9.2.

From now on, given two vectors $\mathbf{u}^T = (u_1, u_2)$, $\mathbf{v}^T = (v_1, v_2)$ in $\mathbb{R}^2$, we denote by $(\mathbf{u}, \mathbf{v})$ the matrix whose columns are the vectors $\mathbf{u}$ and $\mathbf{v}$. We recall that $\mathbf{v}^{\perp}$ denotes the vector $(-v_2, v_1)^T$ orthogonal to the vector $\mathbf{v}$.

Theorem 3.9.5. *Consider a fundamental system with fundamental parameter $D = 0$.*

(a) *If $d \neq 0$, the set formed by all the singular points of the system is the closed segment $s_0 := \{\lambda \mathbf{e}_+ : \lambda \in [-1, 1]\}$ contained in $L_- \cup S_0 \cup L_+$. If $T < 0$ (respectively, $T > 0$), then the segment without the endpoints is a stable (respectively, an unstable) normally hyperbolic manifold. If $T = 0$, then the segment without the endpoints is a non-isolated nilpotent manifold.*

(b) *Suppose that $d = 0$ and B is given in its real Jordan normal form.*

(b.1) *If B is the zero matrix, the set of singular points is the closed strip $L_- \cup S_0 \cup L_+$.*

(b.2) *Suppose that B is not the zero matrix and $B\mathbf{k}^{\perp} = \mathbf{0}$. Then the set of all the singular points in $L_- \cup S_0 \cup L_+$ is the straight line $r_0 := \{\lambda \mathbf{k}^{\perp} : \lambda \in \mathbb{R}\}$. Moreover, if $T < 0$ (respectively $T > 0$), then the straight line r_0 is a stable (respectively an unstable) normally hyperbolic manifold; if $T = 0$, then r_0 is a non-isolated nilpotent manifold.*

(b.2.1) *Suppose that either $Tt < 0$ and $\mathbf{k}^T\mathbf{b}^{\perp} = 0$, or $T = 0$, $t = 0$ and $\mathbf{k}^T\mathbf{b}^{\perp} > 1$. Then the set of all the singular points in S_+ is the straight line $r_+ := \{\mathbf{q} + \lambda \mathbf{k}^{\perp} : \lambda \in \mathbb{R}\}$ and the set of all the singular points in S_- is the straight line $r_- := \{-\mathbf{q} + \lambda \mathbf{k}^{\perp} : \lambda \in \mathbb{R}\}$, where $\mathbf{q}^T$ is equal to $(k_1^{-1}(1 - T/t), 0)$ or to $(0, b_1(k_2 b_1 - 1)^{-1})$ depending on whether $T \neq 0$ or $T = 0$.*
Moreover, if $T < 0$ (respectively, $T > 0$), the straight lines r_+, r_- are unstable (respectively, stable) normally hyperbolic manifolds; if $T = 0$, then r_+ and r_- are non-isolated nilpotent manifolds.

(b.2.2) *Otherwise, there are no singular points in $S_+ \cup S_-$.*

(b.3) *Suppose that B is not the zero matrix and $B\mathbf{k}^{\perp} \neq \mathbf{0}$. Then the set of all the singular points is the piecewise linear curve*

$$\begin{cases} s_- := \{-\mathbf{q} + \lambda \mathbf{w} : \lambda < 0\} \subset S_-, \\ s_0 := \{\eta \mathbf{q} : \eta \in [-1, 1]\} \subset L_- \cup S_0 \cup L_+, \\ s_+ := \{\mathbf{q} + \lambda \mathbf{w} : \lambda > 0\} \subset S_+, \end{cases}$$

where $\mathbf{q}^T$ is equal to $(0, k_2^{-1})$ or $(k_1^{-1}, 0)$ depending on whether $T \neq 0$ or $T = 0$; and $\mathbf{w}^T$ is equal to $(k_2 b_1, T - k_1 b_1)$ or $(1 - k_2 b_1, k_1 b_1)$ depending on whether $T \neq 0$ or $T = 0$.

Moreover, if either $T > 0$, or $T = 0$, or $T < 0$, then the segment s_0 is either an unstable normally hyperbolic manifold, a non-isolated nilpotent manifold, or a stable normally hyperbolic manifold, respectively. The segments s_+ and s_- are either unstable normally hyperbolic manifolds, non-isolated nilpotent manifolds, or stable normally hyperbolic manifolds depending on whether $t > 0$, $t = 0$, or $t < 0$, respectively.

Proof. Let $\dot{\mathbf{x}} = A\mathbf{x} + \varphi(\mathbf{k}^T\mathbf{x})\mathbf{b}$ be the given fundamental system. Therefore, $B = A + \mathbf{b}\mathbf{k}^T$ and $D = \det(B)$.

(a) Since $D = 0$ and $d \neq 0$, Lemma 3.9.1(c) implies that there are no singular points in S_+ and S_-. Moreover, the singular points $\mathbf{e}_+$ and $\mathbf{e}_-$ belong to L_+ and L_-, respectively. So all the singular points of the system are located in the closed strip $L_- \cup S_0 \cup L_+$.

By continuity, the vector fields $A\mathbf{x} + \mathbf{b}$ and $B\mathbf{x}$ are equal on the straight line L_+. Thus $B\mathbf{e}_+ = \mathbf{0}$ and $B\mathbf{e}_- = \mathbf{0}$. It is clear that if $\mathbf{q}$ belongs to the segment s_0 defined in the statement, then $B\mathbf{q} = 0$. Moreover, every point in s_0 is a singular point. Now we prove that there are no other singular points different from these.

Suppose that $\mathbf{q}$ is a singular point in $L_- \cup S_0 \cup L_+$ such that $\mathbf{q} \notin s_0$. Therefore $\{\mathbf{q}, \mathbf{e}_+\}$ is a basis of $\mathbb{R}^2$. Since the matrix B vanishes on each element of the basis, it is the zero matrix. Consequently, $A = -\mathbf{b}\mathbf{k}^T$ and $d = \det(-\mathbf{b}\mathbf{k}^T) = 0$, which contradicts our assumptions.

The local phase portrait of the singular point in s_0 follows from Subsection 2.5.3.

(b.1) Suppose that B is the zero matrix. Then every point in the closed strip $L_- \cup S_0 \cup L_+$ is a singular point.

Moreover, when B is the zero matrix, then $A = -\mathbf{b}\mathbf{k}^T$. Thus any point $\mathbf{q}$ in the half-plane $L_+ \cup S_+$ is a singular point if $\mathbf{b}(-\mathbf{k}^T\mathbf{q} + 1) = \mathbf{0}$. Since $\mathbf{b} \neq \mathbf{0}$, this is equivalent to $\mathbf{q} \in L_+$. By similar arguments we have that every singular point in the half-plane $S_- \cup L_-$ belongs to the straight line L_-.

(b.2) Since $B\mathbf{k}^\perp = 0$, every point in the segment r_0, defined in the statement, is a singular point. Moreover, there are no other singular points in $L_- \cup S_0 \cup L_+$. Otherwise, arguments similar to those in the proof of statement (a) show that B is the zero matrix, which contradicts our assumptions. The local phase portrait of the singular points in r_0 follows from Subsection 2.5.3.

Now we look for the singular points in S_+ and S_-. Since $D = 0$, either

$$B = \begin{pmatrix} T & 0 \\ 0 & 0 \end{pmatrix}, \quad \text{or} \quad B = \begin{pmatrix} 0 & 1 \\ 0 & 0 \end{pmatrix},$$

depending on whether $T \neq 0$, or $T = 0$. Therefore, if $T \neq 0$, then $k_1 \neq 0$ and $k_2 = 0$; and if $T = 0$, then $k_1 = 0$ and $k_2 \neq 0$.

Suppose $T \neq 0$. From expression (3.8) we have

$$A = \begin{pmatrix} T - k_1 b_1 & 0 \\ -k_1 b_2 & 0 \end{pmatrix}.$$

Therefore, the singular points in S_+ have to satisfy the equations $x_1(T - k_1 b_1) = -b_1$, $-x_1 k_1 b_2 = -b_2$ and $k_1 x_1 > 1$.

If $b_2 \neq 0$, then upon dividing by $-b_2$, isolating x_1 in the second equation, and substituting x_1 in the first one, one obtains that $T = 0$, contrary to our assumption. Thus, if $b_2 \neq 0$, there are no singular points in S_+ (by the symmetry of the orbits with respect to the origin, there are no singular points in S_-).

If $b_2 = 0$, then

$$A = \begin{pmatrix} T - k_1 b_1 & 0 \\ 0 & 0 \end{pmatrix}, \text{ where } t = T - k_1 b_1.$$

Suppose $t = 0$. In this case A is the zero matrix and $A\mathbf{x} + \mathbf{b} = \mathbf{b} \neq 0$. Thus, there are no singular points in $S_+ \cup S_-$.

Suppose now that $t \neq 0$. Since singular points in S_+ have to satisfy the equations $x_1 = -b_1/t$ and $k_1 x_1 = 1 - T/t > 1$, we have the following two possibilities. If $Tt > 0$, there are no singular points in $S_+ \cup S_-$. Otherwise, if $Tt < 0$, the set of singular points in S_+ is the straight line $r_+ = \{\mathbf{q} + \lambda \mathbf{k}^\perp : \lambda \in \mathbb{R}\}$, where $\mathbf{q}^T = \left(\frac{1}{k_1}\left(1 - \frac{T}{t}\right), 0\right)$. By the symmetry of the orbits, the set of all the singular points in S_- is the straight line $r_- = \{-\mathbf{q} + \lambda \mathbf{k}^\perp : \lambda \in \mathbb{R}\}$.

The translation $\mathbf{y} = \mathbf{x} - \mathbf{q}$ transforms the non-homogeneous system $\dot{\mathbf{x}} = A\mathbf{x} + \mathbf{b}$ into a homogeneous one. The local phase portraits of the singular points in r_+ and r_- follow from Subsection 2.5.3. This finishes the proof of statements (b.2.1) and (b.2.2) when $T \neq 0$.

Suppose now $T = 0$. From (3.8) we have

$$A = \begin{pmatrix} 0 & 1 - k_2 b_1 \\ 0 & -k_2 b_2 \end{pmatrix}.$$

Therefore the singular points in S_+ must satisfy the equations $x_2(1 - k_2 b_1) + b_1 = 0$, $b_2(1 - k_2 x_2) = 0$, and $k_2 x_2 > 1$. A direct computation shows that if either $b_2 \neq 0$; or $b_2 = 0$ and $1 - k_2 b_1 = 0$; or $b_2 = 0$, $1 - k_2 b_1 \neq 0$ and $-k_2 b_1/(1 - k_2 b_1) \leq 1$, there are no singular points in $S_+ \cup S_-$. Otherwise, we have $b_2 = 0$, $1 - k_2 b_1 \neq 0$, and $k_2 b_1/(k_2 b_1 - 1) > 1$. Thus the set of singular points in S_+ is the straight line $r_+ = \{\mathbf{q} + \lambda \mathbf{k}^\perp : \lambda \in \mathbb{R}\}$, where $\mathbf{q}^T = (0, b_1/(k_2 b_1 - 1))$. By the symmetry of the orbits with respect to the origin, the set of singular points in S_- is the straight line $r_- = \{-\mathbf{q} + \lambda \mathbf{k}^\perp : \lambda \in \mathbb{R}\}$. The local phase portrait of the singular points at r_+ and r_- follows from Subsection 2.5.3. This finishes the proof of statements (b.2.1) and (b.2.2) when $T = 0$.

(b.3) Consider $\mathbf{v} \in \mathbb{R}\backslash\{\mathbf{0}\}$ such that $B\mathbf{v} = \mathbf{0}$. Hence, the vectors $\mathbf{k}^\perp$ and $\mathbf{v}$ are linearly independent. From this we obtain that the straight lines $r = \{\lambda \mathbf{v} : \lambda \in \mathbb{R}\}$ and L_+ intersect. Let $\mathbf{q}$ be the intersection point. It is easy to check that the segment $s_0 = \{\lambda \mathbf{q} : \lambda \in [-1, 1]\}$ contains all the singular points in $L_- \cup S_0 \cup L_+$. The local phase portrait of the singular points follows from Subsection 2.5.3.

Now we shall look for the singular points in the half-plane S_+ and S_-. Suppose $T \neq 0$. From this it can be concluded that $k_2 \neq 0$, and the intersection point satisfies $\mathbf{q}^T = (0, k_2^{-1})$. Moreover, since $d = -Tk_2b_2 = 0$ it follows that $b_2 = 0$ and therefore

$$A = \begin{pmatrix} T - k_1b_1 & -k_2b_1 \\ 0 & 0 \end{pmatrix}.$$

Consider $\mathbf{w}^T := (k_2b_1, T - k_1b_1)$. Since $\det(\mathbf{w}, \mathbf{k}^\perp) = Tk_2 \neq 0$, the vectors $\mathbf{w}$ and $\mathbf{k}^\perp$ are linearly independent. Suppose that $s_+ = \{\mathbf{q} + \lambda\mathbf{w} : \lambda > 0\}$ is a half-line belonging to S_+. Otherwise, by taking $\mathbf{w}^T = (-k_2b_1, -T + k_1b_1)$, we get that $s_+ \subset S_+$. It is easy to check that $A\mathbf{w} = 0$. Therefore, all the points in s_+ are singular points. Now we will see that they are the only singular points in S_+.

If $\mathbf{p}$ is a singular point in S_+ such that $\mathbf{p} \notin S_+$, then $\{\mathbf{w}, \mathbf{p} - \mathbf{q}\}$ is a basis of $\mathbb{R}^2$ and $A\mathbf{w} = A(\mathbf{p} - \mathbf{q}) = \mathbf{0}$. Therefore, A is the zero matrix and $k_2b_1 = 0$, $k_2 \neq 0$, and $T = k_1b_1 = 0$, in contradiction with $T \neq 0$.

By the symmetry of the system, the set of singular points in S_- is the half-line s_-. Thus the set of singular points is the piecewise linear curve

$$\begin{cases} s_- & \text{in } S_-, \\ s_0 & \text{in } L_- \cup S_0 \cup L_+, \\ s_+ & \text{in } S_+. \end{cases}$$

The local phase portrait of the singular points follows from Subsection 2.5.3.

Suppose now that $T = 0$. Then $k_1 \neq 0$ and $\mathbf{q}^T = (k_1^{-1}, 0)$. Since $b_2 = 0$, we have

$$A = \begin{pmatrix} -k_1b_1 & 1 - k_2b_1 \\ 0 & 0 \end{pmatrix}.$$

A similar analysis to the one in the proof of the case $T \neq 0$ shows that, upon taking $\mathbf{w}^T = (1 - k_2b_1, k_1b_1)$, the set of all singular points is the piecewise linear curve

$$\begin{cases} s_- & \text{in } S_-, \\ s_0 & \text{in } L_- \cup S_0 \cup L_+, \\ s_+ & \text{in } S_+. \end{cases}$$

The local phase portrait of the singular points follows from Subsection 2.5.3. □

3.10 Compactification of the flow

In this section we calculate the number of singular points at infinity. We start by showing that the flows of fundamental systems can be extended to infinity via the Poincaré compactification. That is, for a fundamental vector field $\mathbf{f}$, we can define the Poincaré compactification $\mathbf{f}_{\mathbb{D}}$ of $\mathbf{f}$.

In general, when a vector field $\mathbf{f}$ defined in $\mathbb{R}^2$ can be compactified (i.e., can be extended to infinity), $\mathbf{f}$ and its Poincaré compactification $\mathbf{f}_{\mathbb{D}}$ are differentiably equivalent in their open domains, $\mathbb{R}^2$ and $\mathrm{Int}(\mathbb{D})$. This is because the behaviour of $\mathbf{f}_{\mathbb{D}}$ on the boundary $\partial\mathbb{D}$ corresponds to the behaviour of $\mathbf{f}$ at infinity.

In the particular case of fundamental systems it can be proved that the vector field and its compactification are differentiably conjugate in $\mathbb{R}^2$ and $\text{Int}(\mathbb{D})$, respectively. Moreover, if two fundamental systems are linearly conjugate in $\mathbb{R}^2$, then their Poincaré compactifications are differentiably conjugate in $\mathbb{D}$.

According to this and to Theorem 3.8.2(a), in order to study the compactified flow of a fundamental system, there is not loss of generality in assuming that either the matrix A, or the matrix B, is given in its real Jordan normal form.

We define on the unit sphere $\mathbb{S}^2 = \{\mathbf{x} \in \mathbb{R}^3 : \|\mathbf{x}\| = 1\}$ the following regions:

$$\begin{aligned}
\mathbb{S}_+ &:= \left\{\mathbf{z} \in \mathbb{S}^2 : z_3 \geq 0,\ (z_1, z_2)\,\mathbf{k} \geq z_3 \ \text{ or } \ z_3 \leq 0,\ (z_1, z_2)\,\mathbf{k} \leq z_3\right\},\\
\mathbb{S}_0 &:= \left\{\mathbf{z} \in \mathbb{S}^2 : \ |(z_1, z_2)\,\mathbf{k}| \leq |z_3|\right\},\\
\mathbb{S}_- &:= \left\{\mathbf{z} \in \mathbb{S}^2 : z_3 \geq 0,\ (z_1, z_2)\,\mathbf{k} \leq -z_3 \ \text{ or } \ z_3 \leq 0,\ (z_1, z_2)\,\mathbf{k} \geq -z_3\right\}.
\end{aligned}$$

Note that the closed regions $\mathbb{S}_+$, $\mathbb{S}_-$ and $\mathbb{S}_0$ are the images under the central projection on the unit sphere of the half-planes $L_+ \cup S_+$, $S_- \cup L_-$ and of the central strip $L_- \cup S_0 \cup L_+$, respectively.

Proposition 3.10.1. (a) *Fundamental vector fields* $\mathbf{f}(\mathbf{x}) = A\mathbf{x} + \varphi(\mathbf{k}^T\mathbf{x})\mathbf{b}$ *satisfy the Łojasiewicz property at infinity. Furthermore, the degree of any fundamental vector fields at infinity is equal to* 1, *i.e.,* $n(\mathbf{f}) = 1$.

(b) *The Poincaré compactification of a fundamental vector field is given by*

$$\mathbf{f}_{\mathbb{D}}(\mathbf{x}) = \begin{pmatrix} \frac{1+x_2^2-x_1^2}{1+\|\mathbf{x}\|^2} & \frac{-2x_1x_2}{1+\|\mathbf{x}\|^2} \\ \frac{-2x_1x_2}{1+\|\mathbf{x}\|^2} & \frac{1+x_1^2-x_2^2}{1+\|\mathbf{x}\|^2} \end{pmatrix} \left(A \begin{pmatrix} x_1 \\ x_2 \end{pmatrix} + \chi\left(x_1, x_2, \tfrac{1-\|\mathbf{x}\|^2}{2}\right)\mathbf{b}\right),$$

where $\chi : \mathbb{S}^2 \to \mathbb{R}$ *is the function given piecewise by*

$$\chi(\mathbf{z}) := \begin{cases} -z_3, & \textit{if } \mathbf{z} \in \mathbb{S}_-,\\ (z_1, z_2)\,\mathbf{k}, & \textit{if } \mathbf{z} \in \mathbb{S}_0,\\ z_3, & \textit{if } \mathbf{z} \in \mathbb{S}_+. \end{cases}$$

(c) *The Poincaré compactification* $\mathbf{f}_{\mathbb{D}}$ *is symmetric with respect to the origin; that is,* $\mathbf{f}_{\mathbb{D}}(-\mathbf{x}) = -\mathbf{f}_{\mathbb{D}}(\mathbf{x})$.

(d) *The fundamental vector field* $\mathbf{f}$ *and the Poincaré compactification* $\mathbf{f}_{\mathbb{D}}|_{\text{Int}(\mathbb{D})}$ *are differentiably conjugate.*

Proof. (a) To prove this statement it suffices to show that the vector field

$$\mathbf{f}_1(\mathbf{z}) = z_3\mathbf{f}\left(\tfrac{z_1}{z_3}, \tfrac{z_2}{z_3}\right) = A\begin{pmatrix} z_1 \\ z_2 \end{pmatrix} + z_3\varphi\left(k_1\tfrac{z_1}{z_3} + k_2\tfrac{z_2}{z_3}\right)\mathbf{b},$$

defined on $\mathbb{S}_+ \cup \mathbb{S}_-$, can be extended as a global Lipschitz function to whole $\mathbb{S}^2$, see Subsection 2.9.1.

Easy computations show that when $\mathbf{z} \in \mathbb{S}_+ \cup \mathbb{S}_-$, that is, $z_3 \neq 0$, the nonlinear term in the definition of $\mathbf{f}_1$ satisfies

$$z_3 \varphi\left(k_1 \frac{z_1}{z_3} + k_2 \frac{z_2}{z_3}\right) = \begin{cases} -z_3, & \text{if } \dfrac{k_1 z_1 + k_2 z_2}{z_3} < -1, \\ (z_1, z_2)\,\mathbf{k}, & \text{if } \left|\dfrac{k_1 z_1 + k_2 z_2}{z_3}\right| \leq 1, \\ z_3, & \text{if } \dfrac{k_1 z_1 + k_2 z_2}{z_3} > 1. \end{cases}$$

In new of the expression of $\chi(\mathbf{z})$ in the statement,

$$\mathbf{f}_1(\mathbf{z}) = A \begin{pmatrix} z_1 \\ z_2 \end{pmatrix} + \chi(\mathbf{z})\,\mathbf{b}$$

on $\mathbb{S}_+ \cup \mathbb{S}_-$. Therefore, $\mathbf{f}_1$ can be prolonged to $z_3 = 0$. For simplicity, we will use $\mathbf{f}_1$ to denote the extended vector field on the whole sphere $\mathbb{S}^2$.

Now we prove that $\mathbf{f}_1$ is a global Lipschitz function on $\mathbb{S}^2$. Since $\mathbf{f}_1$ is defined as the sum of two functions and one of them is linear, we only need to show that the other, χ, is a global Lipschitz function on $\mathbb{S}^2$, i.e., $|\chi(\mathbf{q}_1) - \chi(\mathbf{q}_2)| \leq L\|\mathbf{q}_1 - \mathbf{q}_2\|_\infty$ for all $\mathbf{q}_1, \mathbf{q}_2 \in \mathbb{S}^2$. Recall that $\|\cdot\|_\infty$ denotes the supremum norm in $\mathbb{R}^3$.

We divide the proof into cases according to the set $\mathbb{S}_+$, $\mathbb{S}_0$ or $\mathbb{S}_-$ where the points $\mathbf{q}_1$ and $\mathbf{q}_2$ belong. When $\mathbf{q}_1 = (x_1, y_1, z_1)^T$ and $\mathbf{q}_2 = (x_2, y_2, z_2)^T$ belong to the same set, then

$$|\chi(x_1, y_1, z_1) - \chi(x_2, y_2, z_2)| = \begin{cases} |z_1 - z_2|, & \text{if } \mathbf{q}_1, \mathbf{q}_2 \in \mathbb{S}_+, \\ |k_1(x_1 - x_2) + k_2(y_1 - y_2)|, & \text{if } \mathbf{q}_1, \mathbf{q}_2 \in \mathbb{S}_0, \\ |z_1 - z_2|, & \text{if } \mathbf{q}_1, \mathbf{q}_2 \in \mathbb{S}_-. \end{cases}$$

Taking $L = \max\{1, 2|k_1|, 2|k_2|\}$, we have $|\chi(\mathbf{q}_1) - \chi(\mathbf{q}_2)| \leq L\|\mathbf{q}_1 - \mathbf{q}_2\|_\infty$.

Suppose now that $\mathbf{q}_1 \in \mathbb{S}_+$ and $\mathbf{q}_2 \in \mathbb{S}_0$. Denote $\sigma_1 = k_1 x_1 + k_2 y_1$ and $\sigma_2 = k_1 x_2 + k_2 y_2$. Then $|\chi(\mathbf{q}_1) - \chi(\mathbf{q}_2)| = |z_1 - \sigma_2|$. We distinguish four cases, depending on the signs of z_1 and z_2.

Suppose that $z_1 \geq 0$ and $z_2 \geq 0$. From the definition of $\mathbb{S}_+$ and $\mathbb{S}_0$, we have $\sigma_1 \geq z_1$ and $|\sigma_2| \leq z_2$. Hence $z_1 - z_2 \leq z_1 - \sigma_2 \leq \sigma_1 - \sigma_2$. Suppose that $z_1 \geq 0$ and $z_2 \leq 0$. Then $|\sigma_2| \leq -z_2$ and $\sigma_1 - \sigma_2 \leq z_1 - \sigma_2 \leq z_1 + z_2 \leq z_1 - z_2$. Suppose now that $z_1 \leq 0$ and $z_2 \geq 0$. By similar arguments, $z_2 - z_1 \leq z_1 + z_2 \leq z_1 - \sigma_2 \leq \sigma_1 - \sigma_2$. Finally, when $z_1 \leq 0$ and $z_2 \leq 0$, we have $\sigma_1 - \sigma_2 \leq z_1 - \sigma_2 \leq z_1 - z_2$. In any case it follows that $|\chi(\mathbf{q}_1) - \chi(\mathbf{q}_2)| \leq L\|\mathbf{q}_1 - \mathbf{q}_2\|_\infty$.

Consider now the case $\mathbf{q}_1 \in \mathbb{S}_+$ and $\mathbf{q}_2 \in \mathbb{S}_-$. Then we have $|\chi(\mathbf{q}_1) - \chi(\mathbf{q}_2)| = |z_1 + z_2|$. Denote, as above, $\sigma_1 = k_1 x_1 + k_2 y_1$ and $\sigma_2 = k_1 x_2 + k_2 y_2$. Then by the definition of $\mathbb{S}_+$ and $\mathbb{S}_-$, it follows that $0 \leq z_1 + z_2 \leq \sigma_1 - \sigma_2$ if $z_1 z_2 \geq 0$ and $-|z_1 - z_2| \leq z_1 + z_2 \leq |z_1 - z_2|$ if $z_1 z_2 \leq 0$. Therefore, $|\chi(\mathbf{q}_1) - \chi(\mathbf{q}_2)| \leq L\|\mathbf{q}_1 - \mathbf{q}_2\|_\infty$.

Similar arguments apply to the remaining case $\mathbf{q}_1 \in \mathbb{S}_0$ and $\mathbf{q}_2 \in \mathbb{S}_-$. Thus, we conclude that $\mathbf{f}_1$ is a global Lipschitz function defined on the whole $\mathbb{S}^2$. Consequently, the degree of $\mathbf{f}$ at infinity is $n(\mathbf{f}) = 1$.

(b) follows easily from expression (2.16).

(c) follows from the expression of $\mathbf{f}_\mathbb{D}$ in the previous statement and the fact that $\chi(-z_1, -z_2, z_3) = -\chi(z_1, z_2, z_3)$.

(d) Notice that when $n(\mathbf{f}) = 1$, the vector fields $\mathbf{f}_{\mathbb{S}^2}$ and $\widetilde{\mathbf{f}}$ coincide, see expression (2.14). Consequently, $\mathbf{f}$ and $\mathbf{f}_\mathbb{D}|_{\mathrm{Int}(\mathbb{D})}$ are differentiably conjugate, see Subsection 2.9.1 for more details. □

Another expression for the function $\chi(\mathbf{z})$ defined in Proposition 3.10.1(b) is

$$\chi(\mathbf{z}) = \begin{cases} -z_3, & \text{if } z_3\,(k_1 z_1 + k_2 z_2) \le -z_3^2, \\ k_1 z_1 + k_2 z_2, & \text{if } |z_3\,(k_1 z_1 + k_2 z_2)| < z_3^2, \\ z_3, & \text{if } z_3\,(k_1 z_1 + k_2 z_2) \ge z_3^2, \end{cases} \tag{3.19}$$

as we can easily check.

In the following proposition we show that the Poincaré compactifications of two linearly conjugate fundamental systems are differentiably conjugate. We also provide an expression for the conjugacy.

Proposition 3.10.2. *Let $\dot{\mathbf{x}} = \mathbf{f}(\mathbf{x})$ and $\dot{\mathbf{x}} = \mathbf{f}^*(\mathbf{x})$ be two linearly conjugate fundamental systems, and let $\dot{\mathbf{x}} = \mathbf{f}_\mathbb{D}(\mathbf{x})$ and $\dot{\mathbf{x}} = \mathbf{f}^*_\mathbb{D}(\mathbf{x})$ be their Poincaré compactifications.*

(a) *The vector fields $\mathbf{f}_\mathbb{D}$ and $\mathbf{f}^*_\mathbb{D}$ are differentiably conjugate.*

(b) *Let M be the matrix of the conjugacy between $\mathbf{f}$ and $\mathbf{f}^*$; then*

$$\mathbf{h}(\mathbf{x}) = \frac{2}{1 - ||\mathbf{x}||^2 + \sqrt{\left(1 - ||\mathbf{x}||^2\right)^2 + 4\,||M\mathbf{x}||^2}} M\mathbf{x},$$

effects the conjugacy between $\mathbf{f}_\mathbb{D}$ and $\mathbf{f}^*_\mathbb{D}$.

Proof. Note that the function $\mathbf{h}$ given in statement (b) is a diffeomorphism defined on an open subset containing $\mathbb{D}$. So it is sufficient to prove that $D\mathbf{h}(\mathbf{x})\mathbf{f}_\mathbb{D}(\mathbf{x}) = \mathbf{f}^*_\mathbb{D}(\mathbf{h}(\mathbf{x}))$ for every $\mathbf{x} \in \mathbb{D}$.

Since $||\mathbf{x}|| = 1$ if and only if $||\mathbf{h}(\mathbf{x})|| = 1$, $\mathbf{h}$ maps $\mathrm{Int}(\mathbb{D})$ into itself and $\partial\mathbb{D}$ into itself. Thus we can divide the proof depending on whether $\mathbf{x} \in \mathrm{Int}(\mathbb{D})$, or $\mathbf{x} \in \partial\mathbb{D}$.

We begin with the case $\mathbf{x} \in \mathrm{Int}(\mathbb{D})$. Let $\mathbf{h}_\mathbb{D}$ be the conjugacy between a vector field and its Poincaré compactification, see (2.15). Since

$$\mathbf{h}_\mathbb{D}(\mathbf{x}) = \frac{1}{1 + \sqrt{1 + ||\mathbf{x}||^2}}\mathbf{x} \quad \text{and} \quad \mathbf{h}_\mathbb{D}^{-1}(\mathbf{x}) = \frac{2}{1 - ||\mathbf{x}||^2}\mathbf{x},$$

an easy computation shows that $\mathbf{h}(\mathbf{x}) = \mathbf{h}_{\mathbb{D}} \circ M \circ \mathbf{h}_{\mathbb{D}}^{-1}(\mathbf{x})$. Since both $\mathbf{h}_{\mathbb{D}}$ and M are conjugacies, $\mathbf{h}|_{\mathrm{Int}(\mathbb{D})}$ is also a conjugacy, i.e., $D\mathbf{h}(\mathbf{x})\mathbf{f}_{\mathbb{D}}(\mathbf{x}) = \mathbf{f}_{\mathbb{D}}^*(\mathbf{h}(\mathbf{x}))$ for all $\mathbf{x} \in \mathrm{Int}(\mathbb{D})$.

Now we show what happens on the boundary of $\mathbb{D}$. Take $\mathbf{x} \in \partial\mathbb{D}$, and consider $\{\mathbf{x}_n\}_{n=0}^{\infty} \subset \mathrm{Int}(\mathbb{D})$ such that $\lim\limits_{n\to\infty} \mathbf{x}_n = \mathbf{x}$. Since $\mathbf{x}_n \in \mathrm{Int}(\mathbb{D})$, we know that $D\mathbf{h}(\mathbf{x}_n)\mathbf{f}_{\mathbb{D}}(\mathbf{x}_n) = \mathbf{f}_{\mathbb{D}}^*(\mathbf{h}(\mathbf{x}_n))$ for all $n \in \mathbb{N}$. Since $\mathbf{h}$ is C^1, and $\mathbf{f}_{\mathbb{D}}$ and $\mathbf{f}_{\mathbb{D}}^*$ are continuous functions, letting $n \to \infty$ we have $D\mathbf{h}(\mathbf{x})\mathbf{f}_{\mathbb{D}}(\mathbf{x}) = \mathbf{f}_{\mathbb{D}}^*(\mathbf{h}(\mathbf{x}))$, which completes the proof. □

In the proof of Proposition 3.10.2 we have not used that $\mathbf{f}$ and $\mathbf{f}^*$ are fundamental systems. So the proposition can be stated with $\mathbf{f}$ and $\mathbf{f}^*$ vector fields satisfying the Łojasiewicz property at infinity. Furthermore, Proposition 3.10.2 may be seen as a corollary of a more general result. Namely, suppose that $\mathbf{f}$ and $\mathbf{f}^*$ are differentiably conjugate with conjugacy $\mathbf{g}$; then $\mathbf{h} = \mathbf{h}_{\mathbb{D}} \circ \mathbf{g} \circ \mathbf{h}_{\mathbb{D}}^{-1}$ is a conjugacy between $\mathbf{f}_{\mathbb{D}}|_{\mathrm{Int}(\mathbb{D})}$ and $\mathbf{f}_{\mathbb{D}}^*|_{\mathrm{Int}(\mathbb{D})}$. Moreover, if $\mathbf{g}$ satisfies $\mathbf{g}(\lambda\mathbf{x}) = \lambda^{\alpha}\mathbf{g}(\mathbf{x})$ with $\alpha > 0$, then

$$\mathbf{h}(\mathbf{x}) = \frac{2}{1 - ||\mathbf{x}||^2 + \sqrt{\left(1 - ||\mathbf{x}||^2\right)^2 + 4\,||\mathbf{g}(\mathbf{x})||^2}}\,\mathbf{g}(\mathbf{x})$$

is a differentiable function defined in an open set containing $\mathbb{D}$. Hence $\mathbf{h}$ is a differentiable conjugacy between $\mathbf{f}_{\mathbb{D}}$ and $\mathbf{f}_{\mathbb{D}}^*$.

3.11 Singular points at infinity

Let $\dot{\mathbf{x}} = \mathbf{f}_{\mathbb{D}}(\mathbf{x})$ be the Poincaré compactification of the fundamental system $\dot{\mathbf{x}} = \mathbf{f}(\mathbf{x})$, with $\mathbf{f}(\mathbf{x}) = A\mathbf{x} + \varphi(\mathbf{k}^T\mathbf{x})\mathbf{b}$. The singular points at infinity are given by the solutions of the equation $\mathbf{f}_{\mathbb{D}}|_{\partial\mathbb{D}}(\mathbf{x}) = \mathbf{0}$.

In order to study the behaviour of the flow in a neighborhood of a singular point at infinity, we define the following subsets of the Poincaré disc $\mathbb{D}$:

$$\begin{aligned}
\mathbb{D}_+ &:= \left\{\mathbf{x} \in \mathbb{D} : \mathbf{k}^T\mathbf{x} > \frac{1}{2}\left(1 - ||\mathbf{x}||^2\right)\right\},\\
\mathbb{D}_0 &:= \left\{\mathbf{x} \in \mathbb{D} : \left|\mathbf{k}^T\mathbf{x}\right| \le \frac{1}{2}\left(1 - ||\mathbf{x}||^2\right)\right\},\\
\mathbb{D}_- &:= \left\{\mathbf{x} \in \mathbb{D} : \mathbf{k}^T\mathbf{x} < -\frac{1}{2}\left(1 - ||\mathbf{x}||^2\right)\right\}.
\end{aligned}$$

Note that $\mathbb{D}_+$, $\mathbb{D}_0$ and $\mathbb{D}_-$ are contained in the images of the sets $\mathbb{S}_+ \cap \{z_3 \ge 0\}$, $\mathbb{S}_0 \cap \{z_3 \ge 0\}$ and $\mathbb{S}_- \cap \{z_3 \ge 0\}$, respectively, under the projection

$$\mathbf{p}_+(\mathbf{z}) = \frac{1}{1 + z_3}(z_1, z_2).$$

Therefore, $\mathbb{D}_+$, $\mathbb{D}_0$, and $\mathbb{D}_-$ contain the image under the conjugacy $\mathbf{h}_{\mathbb{D}}$ of the regions S_+, $L_- \cup S_0 \cup L_+$, and S_- respectively, see Figure 3.3. By definition, $\mathbb{D}_0$ is a closed set, but $\mathbb{D}_+$ and $\mathbb{D}_-$ are not closed.

For simplicity, we will denote a subset of $\mathbb{R}^2$ and its image under $\mathbf{h}_{\mathbb{D}}$ by the same letter. Thus, $L_+ = \mathbf{h}_{\mathbb{D}}(L_+)$ and $L_- = \mathbf{h}_{\mathbb{D}}(L_-)$.

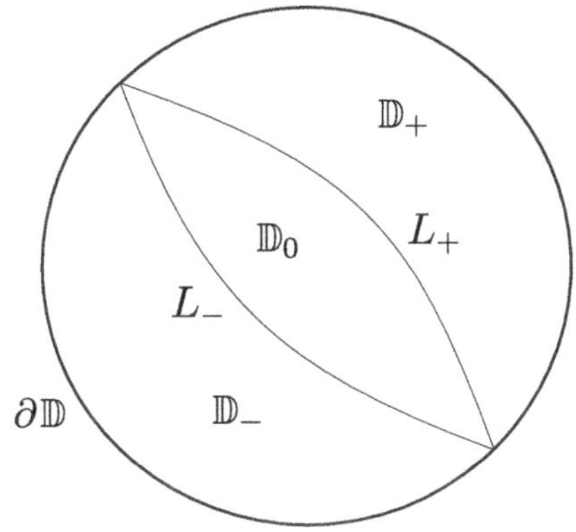

Figure 3.3: Subsets $\mathbb{D}_+$, $\mathbb{D}_0$ and $\mathbb{D}_-$ of the Poincaré disc $\mathbb{D}$.

According to Proposition 3.10.2, in order to study the singular points at infinity of a fundamental system, we can assume, with no loss of generality, that one of their fundamental matrices is given in its real Jordan normal form. Thus the rest of this section assumes that A is in real Jordan normal form.

Proposition 3.11.1. *Consider the fundamental system* $\dot{\mathbf{x}} = A\mathbf{x} + \varphi(\mathbf{k}^T\mathbf{x})\mathbf{b}$ *with parameters* (D, T, d, t) *and such that the matrix* A *is in real Jordan normal form.*

(a) *If* $t^2 - 4d < 0$, *then there are no singular points at infinity. Therefore, the infinity manifold* $\partial\mathbb{D}$ *is a periodic orbit.*

(b) *If* $t^2 - 4d = 0$ *and the matrix* A *is diagonal, then every point in the infinity manifold* $\partial\mathbb{D}$ *is a singular point.*

(c) *If* $t^2 - 4d = 0$ *and the matrix* A *is not diagonal, then there are exactly two singular points at infinity:* $\mathbf{x}_+^T = (1, 0)$ *and* $\mathbf{x}_-^T = (-1, 0)$. *Furthermore, the singular points* $\mathbf{x}_+$ *and* $\mathbf{x}_-$ *are contained in* $\partial\mathbb{D}_0$ *if and only if the first component of the vector* $\mathbf{k}$ *is zero.*

(d) *If* $t^2 - 4d > 0$, *then there are exactly four singular points at infinity:* $\mathbf{x}_+^T = (1, 0)$, $\mathbf{x}_-^T = (-1, 0)$, $\mathbf{y}_+^T = (0, 1)$ *and* $\mathbf{y}_-^T = (0, -1)$. *Furthermore, the singular points* $\mathbf{x}_+$ *and* $\mathbf{x}_-$ *are contained in* $\partial\mathbb{D}_0$ *if and only if the first component of the vector* $\mathbf{k}$ *is zero; and the singular points* $\mathbf{y}_+$ *and* $\mathbf{y}_-$ *are contained in* $\partial\mathbb{D}_0$ *if and only if the second component of the vector* $\mathbf{k}$ *is zero.*

Proof. We note that the expression of the Poincaré compactification $\mathbf{f}_{\mathbb{D}}$ restricted to the infinity manifold $\partial\mathbb{D}$ is

$$\mathbf{f}_{\mathbb{D}}(\mathbf{x}) = \begin{pmatrix} x_2^2 & -x_1x_2 \\ -x_1x_2 & x_1^2 \end{pmatrix} A\mathbf{x},$$

see Proposition 3.10.1(b).

(a) Suppose that $t^2 - 4d < 0$. Since the matrix A is in real Jordan normal form,

$$A = \begin{pmatrix} \alpha & -\beta \\ \beta & \alpha \end{pmatrix} \quad \text{with } \beta > 0.$$

Therefore,

$$\mathbf{f}_{\mathbb{D}}|_{\partial\mathbb{D}}(\mathbf{x}) = ||\mathbf{x}||^2 \begin{pmatrix} -\beta x_2 \\ \beta x_1 \end{pmatrix},$$

and $\mathbf{f}_{\mathbb{D}}|_{\partial\mathbb{D}}(\mathbf{x}) \neq \mathbf{0}$, which proves the statement.

(b) Suppose that $t^2 - 4d = 0$. Then there are two possibilities for the matrix A, but only one of them is diagonal,

$$A = \begin{pmatrix} \lambda & 0 \\ 0 & \lambda \end{pmatrix}.$$

Therefore, the expression of the Poincaré compactification in $\partial\mathbb{D}$ is

$$\mathbf{f}_{\mathbb{D}}|_{\partial\mathbb{D}}(\mathbf{x}) = \lambda \begin{pmatrix} x_1x_2^2 - x_1x_2^2 \\ -x_1^2x_2 + x_1^2x_2 \end{pmatrix},$$

which is the zero vector. Then every point in $\partial\mathbb{D}$ is a singular point.

(c) Under the assumption that A is not diagonal, we have

$$A = \begin{pmatrix} \lambda & 1 \\ 0 & \lambda \end{pmatrix} \quad \text{and} \quad \mathbf{f}_{\mathbb{D}}|_{\partial\mathbb{D}}(\mathbf{x}) = \begin{pmatrix} x_2^3 \\ -x_1x_2^2 \end{pmatrix}.$$

Hence, $\mathbf{f}_{\mathbb{D}}|_{\partial\mathbb{D}}(\mathbf{x}) = \mathbf{0}$ if and only if $x_2 = 0$ and $x_1 = \pm 1$. That is, $\mathbf{x}_+^T = (1, 0)$ and $\mathbf{x}_-^T = (-1, 0)$ are the unique singular points at infinity. Moreover, since $\mathbf{k}^T\mathbf{x}_+ = k_1$ and $\mathbf{k}^T\mathbf{x}_- = -k_1$, the singular points belong to $\partial\mathbb{D}_0$ if and only if $k_1 = 0$.

(d) Suppose that $t^2 - 4d > 0$. In this case,

$$A = \begin{pmatrix} \lambda_1 & 0 \\ 0 & \lambda_2 \end{pmatrix} \quad \text{and} \quad \mathbf{f}_{\mathbb{D}}|_{\partial\mathbb{D}}(\mathbf{x}) = (\lambda_1 - \lambda_2)\, x_1x_2 \begin{pmatrix} x_2 \\ -x_1 \end{pmatrix}.$$

Therefore, $\mathbf{f}_{\mathbb{D}}|_{\partial\mathbb{D}}(\mathbf{x}) = \mathbf{0}$ if and only if $x_1 = 0$ and $x_2 = \pm 1$, or $x_2 = 0$ and $x_1 = \pm 1$. From this we conclude that the unique singular points at infinity are $\mathbf{x}_+$, $\mathbf{x}_-$, $\mathbf{y}_+$, and $\mathbf{y}_-$. The statement follows noting that $\mathbf{k}^T\mathbf{x}_+ = k_1$, $\mathbf{k}^T\mathbf{x}_- = -k_1$, $\mathbf{k}^T\mathbf{y}_+ = k_2$, and $\mathbf{k}^T\mathbf{y}_- = -k_2$. □

In the rest of this section we will study the local phase portrait of the singular points at infinity. We start by introducing some notations.

Let $\mathbf{q}$ be a singular point at infinity of a system $\dot{\mathbf{x}} = \mathbf{f}(\mathbf{x})$; that is, $\mathbf{q} \in \partial\mathbb{D}$ and $\mathbf{f}_{\mathbb{D}}(\mathbf{q}) = \mathbf{0}$. Let $\mathbf{f}_{\mathbf{q}}$ be the vector field defined in (2.18). We recall that $\mathbf{f}_{\mathbf{q}}$ is defined on a local chart of the sphere centered at the point $\mathbf{q}$. The vector fields $\mathbf{f}_{\mathbb{D}}$ and $\mathbf{f}_{\mathbf{q}}$ are differentiably conjugate in a neighbourhood of $\mathbf{q}$ inside $\mathbb{D}$ and in a neighbourhood of $\mathbf{0}$ inside the half-plane $\{\mathbf{x} \in \mathbb{R}^2 : x_2 \geq 0\}$. Then in order to know the local phase portrait of $\mathbf{f}_{\mathbb{D}}$ at $\mathbf{q}$, it is sufficient to know the local phase portrait of $\mathbf{f}_{\mathbf{q}}$ at $\mathbf{0}$, see Figure 3.4.

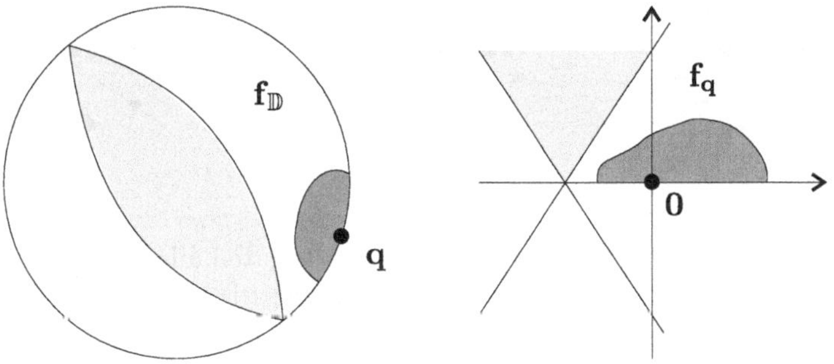

Figure 3.4: Neighbourhoods of $\mathbf{q}$ and $\mathbf{0}$ where $\mathbf{f}_{\mathbb{D}}$ and $\mathbf{f}_{\mathbf{q}}$ are differentiably conjugate.

A singular point at infinity $\mathbf{q}$ is said to be a *saddle*, a *node* or a *saddle-node*, if, for the vector field $\mathbf{f}_{\mathbf{q}}$ the singular point at the origin is a saddle, a node or a saddle-node, respectively. A singular point at infinity $\mathbf{q}$ is said to be a *non-isolated nilpotent* point or a *normally hyperbolic* point, see Figure 3.5(b) and (c), if, for the vector field $\mathbf{f}_{\mathbf{q}}$ the singular point at the origin is a non-isolated nilpotent point or a normally hyperbolic point, respectively, see Figure 3.5(a).

At this point we introduce additional concepts concerning singular points which will be needed in the following results.

We call $\mathbf{q}$ a *stable* (respectively *unstable*) *non-isolated node* if the flow of the vector field $\mathbf{f}_{\mathbf{q}}$ in a neighbourhood of the origin is topologically equivalent to the flow of $\dot{x} = -xy^2$, $\dot{y} = -y^3$ (respectively $\dot{x} = xy^2$, $\dot{y} = y^3$) in a neighbourhood of the origin, see Figure 3.5(d) and (e). The curve formed by the singular points will be called the *singular manifold* of the non-isolated node.

We call $\mathbf{q}$ a *semi-stable non-isolated node* if the flow of the vector field $\mathbf{f}_{\mathbf{q}}$ in a neighbourhood of the origin is topologically equivalent to the flow of $\dot{x} = xy$, $\dot{y} = y^2$ in a neighbourhood of the origin, see Figure 3.5(f). The curve formed by the singular points will be called the *singular manifold* of the semi-stable non-isolated node.

Given a singular point at infinity $\mathbf{q}$, in the following result we give the expression of the vector field $\mathbf{f}_{\mathbf{q}}$ in a local chart centered at $\mathbf{q}$.

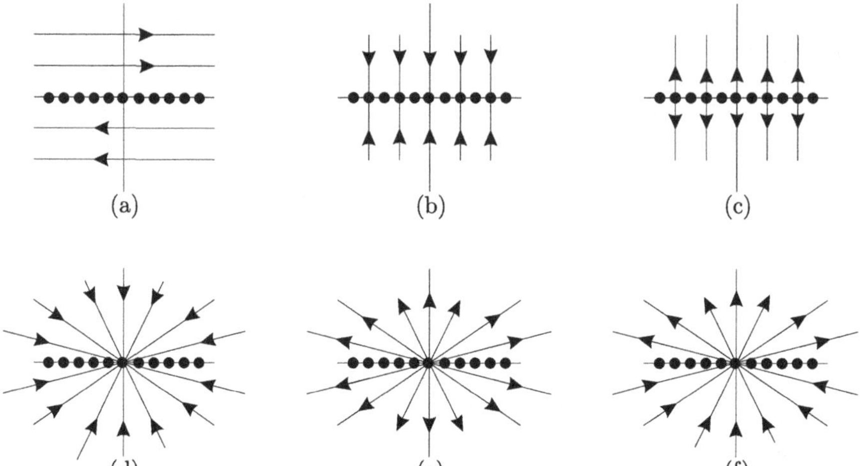

Figure 3.5: Non-isolated singular points: (a) non-isolated nilpotent; (b) stable normally hyperbolic; (c) unstable normally hyperbolic; (d) non-isolated stable node; (e) non-isolated unstable node; and (f) non-isolated semi-stable node.

Lemma 3.11.2. *Let $\mathbf{q}^T = (q_1, q_2) \in \partial\mathbb{D}$ be a singular point at infinity of the fundamental system $\dot{\mathbf{x}} = A\mathbf{x} + \varphi(\mathbf{k}^T\mathbf{x})\mathbf{b}$.*

(a) *Let χ be as in Proposition 3.10.1(b), then*

$$\mathbf{f_q}(\mathbf{x}) = M\left(A\begin{pmatrix} q_1 - x_1q_2 \\ q_2 + x_1q_1 \end{pmatrix} + \mu(\mathbf{x})\,\chi\left(\tfrac{q_1-x_1q_2}{\mu(\mathbf{x})}, \tfrac{q_2+x_1q_1}{\mu(\mathbf{x})}, \tfrac{x_2}{\mu(\mathbf{x})}\right)\mathbf{b}\right),$$

where $\mu(\mathbf{x}) = \sqrt{1+\|\mathbf{x}\|^2}$ and $M = \begin{pmatrix} -q_2 - x_1q_1 & q_1 - x_1q_2 \\ -x_2q_1 & -x_2q_2 \end{pmatrix}$.

(b) *Let $\mathbf{f}_{-\mathbf{q}}$ be the vector field defined on the local chart centered at $-\mathbf{q}$. Then $\mathbf{f_q}$ and $\mathbf{f}_{-\mathbf{q}}$ are equal.*

Proof. (a) From expression (2.18) it follows that

$$\mathbf{f_q}(\mathbf{x}) = \mu(\mathbf{x})\begin{pmatrix} -q_2 - x_1q_1 & q_1 - x_1q_2 \\ -x_2q_1 & -x_2q_2 \end{pmatrix}\mathbf{f}_{n(\mathbf{f})}\left(\tfrac{q_1-x_1q_2}{\mu(\mathbf{x})}, \tfrac{q_2+x_1q_1}{\mu(\mathbf{x})}, \tfrac{x_2}{\mu(\mathbf{x})}\right),$$

where $n(\mathbf{f})$ is the degree of $\mathbf{f}$ at infinity.

In the case of fundamental systems we have $n(\mathbf{f}) = 1$, see Proposition 3.10.1. Moreover, $\mathbf{f}_1(\mathbf{z}) = A\begin{pmatrix} z_1 \\ z_2 \end{pmatrix} + \chi(\mathbf{z})\mathbf{b}$. The statement follows immediately.

(b) The conclusion follows from statement and the fact that $\chi(-z_1, -z_2, z_3) = -\chi(z_1, z_2, z_3)$. □

Remark 3.11.3. According to Lemma 3.11.2(b), in order to study two opposite singular points at infinity, $\mathbf{q}$ and $-\mathbf{q}$, it is sufficient to study only one of them. In particular, the behaviour of the flow of the compactified vector field $\mathbf{f}_{\mathbb{D}}$ in a neighbourhood of $\mathbf{q}$ is obtained from the behaviour of the flow of $\mathbf{f}_{\mathbf{q}}$ in a non-negative y-coordinate neighbourhood of the origin. Moreover, the behaviour of the flow of the compactified vector field $\mathbf{f}_{\mathbb{D}}$ in a neighbourhood of $-\mathbf{q}$ is obtained from the behaviour of the flow of $\mathbf{f}_{\mathbf{q}}$ in a non-positive y-coordinate neighbourhood of the origin.

Remark 3.11.4. When the singular point at infinity $\mathbf{q}$ belongs to the region $\partial\mathbb{D}_+$ (respectively $\partial\mathbb{D}_-$), the vector field $\mathbf{f}_{\mathbf{q}}$ depends on the compactification of two non-homogeneous linear vector fields, see Figure 3.4. On the other hand, when $\mathbf{q}$ belongs to the central region $\partial\mathbb{D}_0$, the vector field $\mathbf{f}_{\mathbf{q}}$ depends on the compactification of the two aforementioned linear systems and on the compactification of a homogeneous one.

We will describe the behaviour of singular points at infinity separately for two groups. In the first one, using Theorem 3.11.5 to 3.11.8, we study the singular points at infinity when they belong to $\partial\mathbb{D}_+$, or $\partial\mathbb{D}_-$. In the second group, using Theorem 3.11.9 to 3.11.12, we study the singular points at infinity when they belong to $\partial\mathbb{D}_0$. Recall that in all these cases there is not loss of generality in assuming that the matrix A is in real Jordan normal form.

Theorem 3.11.5. *Consider a fundamental system* $\dot{\mathbf{x}} = A\mathbf{x} + \varphi(\mathbf{k}^T\mathbf{x})\mathbf{b}$ *with parameters* (D, T, d, t) *where* $t^2 - 4d = 0$. *Suppose that* A *is given in real Jordan normal form and is diagonal. Then every point at infinity is a singular point.*

(a) *If* $t > 0$ *(respectively,* $t < 0$*), every point in* $\partial\mathbb{D} \setminus \{\pm\mathbf{k}^\perp/\|\mathbf{k}\|\}$ *is a stable normally hyperbolic singular point (respectively, unstable normally hyperbolic singular point), and the normally hyperbolic manifold is contained in* $\partial\mathbb{D}$*, see Figure* 3.6(a).

(b) *If* $t = 0$*, every point in* $\partial\mathbb{D} \setminus \{\pm\mathbf{k}^\perp/\|\mathbf{k}\|, \pm\mathbf{b}/\|\mathbf{b}\|\}$ *is a non-isolated nilpotent singular point with the singular manifold contained in* $\partial\mathbb{D}$*. When* $T > 0$ *(respectively,* $T < 0$*), the singular points* $\pm\mathbf{b}/\|\mathbf{b}\|$ *are stable non-isolated nodes (respectively, unstable non-isolate nodes). The singular manifold of these points is contained in* $\partial\mathbb{D}$*, see Figure* 3.6(b). *When* $T = 0$*, then* $\pm\mathbf{b}/\|\mathbf{b}\| = \pm\mathbf{k}^\perp/\|\mathbf{k}\|$.

Proof. From Proposition 3.11.1(b) it follows that every point in $\partial\mathbb{D}$ is a singular point. In this theorem we are only interested in singular points at infinity which do not belong to $\mathbb{D}_0$. Therefore, since $\mathbb{D}_0 \cap \partial\mathbb{D} = \{\pm\mathbf{k}^\perp/\|\mathbf{k}\|\}$, we study only singular points in $\partial\mathbb{D} \setminus \{\pm\mathbf{k}^\perp/\|\mathbf{k}\|\}$.

Let $\mathbf{q}$ be a singular point in $\partial\mathbb{D} \setminus \{\pm\mathbf{k}^\perp/\|\mathbf{k}\|\}$ and suppose that $\mathbf{k}^T\mathbf{q} > 0$ (otherwise we consider the singular point $-\mathbf{q}$). There exists a neighbourhood U of the origin such that

$$\mathbf{k}^T\mathbf{q} + x_1\mathbf{k}^T\mathbf{q}^\perp > |x_2|,$$

for every $\mathbf{x}^T = (x_1, x_2) \in U$. Therefore, if $x_2 \geq 0$, then $(\mathbf{k}^T\mathbf{q} + x_1\mathbf{k}^T\mathbf{q}^\perp)x_2 \geq x_2^2$; and if $x_2 \leq 0$, then $(\mathbf{k}^T\mathbf{q} + x_1\mathbf{k}^T\mathbf{q}^\perp)x_2 \leq -x_2^2$. Hence,

$$\mu(\mathbf{x})\,\chi\left(\frac{q_1 - x_1q_2}{\mu(\mathbf{x})}, \frac{q_2 + x_1q_1}{\mu(\mathbf{x})}, \frac{x_2}{\mu(\mathbf{x})}\right) = |x_2|,$$

for any $\mathbf{x} \in U$, see (3.19). Thus the vector field defined in the local chart centered in $\mathbf{q}$ can be expressed as

$$\mathbf{f_q}(\mathbf{x}) = \begin{pmatrix} -q_2 - x_1q_1 & q_1 - x_1q_2 \\ -x_2q_1 & -x_2q_2 \end{pmatrix} \begin{pmatrix} \frac{t}{2}(q_1 - x_1q_2) + |x_2|\,b_1 \\ \\ \frac{t}{2}(q_2 + x_1q_1) + |x_2|\,b_2 \end{pmatrix},$$

see Lemma 3.11.2(a). Simple computations show that the system $\dot{\mathbf{x}} = \mathbf{f_q}(\mathbf{x})$ can recast as

$$\dot{x}_1 = -x_2\left(\mathbf{b}^T\mathbf{q}\,x_1 - \mathbf{b}^T\mathbf{q}^\perp\right), \quad \dot{x}_2 = -x_2\left(\mathbf{b}^T\mathbf{q}\,x_2 + \frac{t}{2}\right), \tag{3.20}$$

when $\mathbf{x} \in U$ and $x_2 \geq 0$. Notice that every point on the straight line $x_2 = 0$ is a singular point. Now we study the local phase portrait of system (3.20) at each of these singular points.

Let $\mathbf{x}(s)$, with $x_2(s) \neq 0$, be a solution of system (3.20). Performing the change of time $d\tau = x_2(s)ds$ and using the prime, instead of the dot, to denote the derivative with respect to the new time τ, it follows that

$$x_1' = -\mathbf{b}^T\mathbf{q}\,x_1 + \mathbf{b}^T\mathbf{q}^\perp, \quad x_2' = -\mathbf{b}^T\mathbf{q}\,x_2 - \frac{t}{2}. \tag{3.21}$$

(a) Suppose that $t \neq 0$. On the straight line $x_2 = 0$, the flow of system (3.21) is transversal. Returning to the original time variable, we see that the straight line $x_2 = 0$ is a stable normally hyperbolic manifold if $t > 0$, and an unstable normally hyperbolic manifold if $t < 0$.

(b) Suppose now that $t = 0$. First we consider the case $\mathbf{q} \neq \pm\mathbf{b}/\|\mathbf{b}\|$. When $\mathbf{b}^T\mathbf{q} \neq 0$, the change of time $\rho(\tau) = \mathbf{b}^T\mathbf{q}\tau$ transforms system (3.21) into the following one:

$$\frac{dx_1}{d\rho} = -x_1 + \frac{\mathbf{b}^T\mathbf{q}^\perp}{\mathbf{b}^T\mathbf{q}}, \quad \frac{dx_2}{d\rho} = -x_2.$$

Since in this case $\mathbf{b}^T\mathbf{q}^\perp \neq 0$, the origin is not a singular point; i.e., in a neighbourhood of the origin the flow is parallel. Going back through the changes of time we conclude that the straight line $x_2 = 0$ is a non-isolated nilpotent manifold of system (3.20).

When $\mathbf{b}^T\mathbf{q} = 0$, we have $\mathbf{q} = \pm\mathbf{b}^\perp/\|\mathbf{b}\|$ and $\mathbf{b}^T\mathbf{q}^\perp = \pm\|\mathbf{b}\|$. Therefore, system (3.20) can be written as $\dot{x}_1 = \pm\|\mathbf{b}\|x_2$, $\dot{x}_2 = 0$. We see that the straight line $x_2 = 0$ is a non-isolated nilpotent manifold of (3.20).

Now we consider the case $\mathbf{q} = \pm\mathbf{b}/\|\mathbf{b}\|$. Since $t = 0$, it follows that $T = \mathbf{k}^T\mathbf{b}$, see (3.12). We divide the rest of the proof in three parts, according to the sign of T. When $T > 0$, i.e., $\mathbf{k}^T\mathbf{b} > 0$, we consider the singular point at infinity $\mathbf{q} = \mathbf{b}/\|\mathbf{b}\|$. For this point we have $\mathbf{k}^T\mathbf{q} > 0$. Therefore, the behavior of the flow in a neighbourhood of $\mathbf{q}$ can be obtained from system (3.20), which can be written as $\dot{x}_1 = -\|\mathbf{b}\|x_1x_2$, $\dot{x}_2 = -\|\mathbf{b}\|x_2^2$. Then we get that the origin is a semi-stable non-isolated node and the singular manifold is $x_2 = 0$. The stable manifold is contained in the half-plane $x_2 > 0$, and the unstable one is contained in the half-plane $x_2 < 0$.

When $T < 0$, we consider the point $\mathbf{q} = -\mathbf{b}/\|\mathbf{b}\|$. Then $\mathbf{k}^T\mathbf{q} > 0$. We use again the system (3.20), which can be expressed as $\dot{x}_1 = \|\mathbf{b}\|x_1x_2$, $\dot{x}_2 = \|\mathbf{b}\|x_2^2$. Thus the origin is a semi-stable non-isolated node and the singular manifold is $x_2 = 0$. The unstable manifold is contained in the half-plane $x_2 > 0$ and the stable one in the half-plane $x_2 < 0$.

When $T = 0$, we have $\pm\mathbf{b}/\|\mathbf{b}\| = \pm\mathbf{k}^\perp/\|\mathbf{k}\|$, and the singular points $\mathbf{q} = \pm\mathbf{b}/\|\mathbf{b}\|$ belong to $\partial\mathbb{D}_0$. Therefore, we do not study them. □

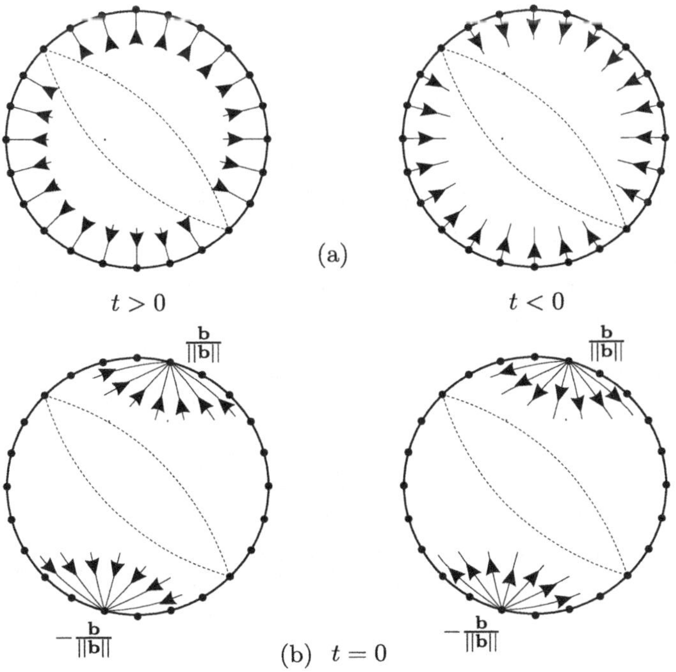

Figure 3.6: Local phase portraits of $\mathbf{f}_{\mathbb{D}}$ at the singular points at infinity when $t^2 - 4d = 0$ and A is diagonal.

In the following result we show that there are singular points at infinity such that their hyperbolic, central or singular manifolds are contained in a straight line. These straight lines play a very important role in the description of the global phase portrait. To express them we will use the coordinate system of $\mathbb{R}^2$ instead of the coordinate system of $\mathbb{D}$.

Theorem 3.11.6. *Consider a fundamental system $\dot{\mathbf{x}} = A\mathbf{x} + \varphi(\mathbf{k}^T\mathbf{x})\mathbf{b}$ with parameters (D, T, d, t), where $t^2 - 4d = 0$, and the vector $\mathbf{k} = (k_1, k_2)^T$ is such that $k_1 \neq 0$. Suppose that A is in its real Jordan normal form and is not diagonal. Then there are exactly two points at infinity, $\mathbf{x}_+$ and $\mathbf{x}_-$, and these points do not belong to $\partial\mathbb{D}_0$.*

(a) *If $t \neq 0$, then $\mathbf{x}_+$ (respectively, $\mathbf{x}_-$) is a saddle-node with the central manifold in $\partial\mathbb{D}$ and the hyperbolic manifold on the straight line $x_2 = -2\operatorname{sign}(k_1)\, b_2/t$ (respectively, $x_2 = 2\operatorname{sign}(k_1)\, b_2/t$). Moreover, when $t > 0$, the hyperbolic manifold is stable, and when $t < 0$, the hyperbolic manifold is unstable, see Figure* 3.7(a).

(b) *If $t = 0$ and $D < 0$ (respectively, $D > 0$), then a neighbourhood of $\mathbf{x}_+$ and $\mathbf{x}_-$ in $\partial\mathbb{D}$ is an elliptic sector (respectively hyperbolic sector), see Figure* 3.7(b).

(c) *If $t = 0$ and $D = 0$, then $\mathbf{x}_+$ (respectively, $\mathbf{x}_-$) is a semi-stable non-isolated node with the singular manifold on the straight line $x_2 = -\operatorname{sign}(k_1)b_1$ (respectively, $x_2 = \operatorname{sign}(k_1)b_1$), see Figure* 3.7(c).

Proof. By Theorem 3.11.1(c), there are exactly two singular points at infinity, $\mathbf{x}_+^T = (1, 0)$ and $\mathbf{x}_-^T = (-1, 0)$, which are symmetric with respect to the origin. Thus in order to study the behaviour of the flow in a neighbourhood of these singular points it is sufficient to consider only one of them. In this proof we take $\dot{\mathbf{x}} = \mathbf{f}_{\mathbf{x}_+}(\mathbf{x})$.

Since $k_1 \neq 0$, there exists a neighbourhood U of the origin such that $|k_1 + k_2x_1| > |x_2|$, for every $\mathbf{x}^T = (x_1, x_2) \in U$. Thus

$$\mu(\mathbf{x})\,\chi\left(\frac{1}{\mu(\mathbf{x})}, \frac{x_1}{\mu(\mathbf{x})}, \frac{x_2}{\mu(\mathbf{x})}\right) = \frac{k_1}{|k_1|}\,|x_2|,$$

see the expression of χ in (3.19). On the other hand, A is in real Jordan normal form, and the trace and the determinant of A satisfy $t^2 - 4d = 0$. Then, from Lemma 3.11.2(a), we get the following expression of $\mathbf{f}_{\mathbf{x}_+}$ in U:

$$\mathbf{f}_{\mathbf{x}_+}(\mathbf{x}) = \begin{pmatrix} -x_1 & 1 \\ -x_2 & 0 \end{pmatrix}\left(\begin{pmatrix} \dfrac{t}{2} + x_1 \\ \\ \dfrac{t}{2}x_1 \end{pmatrix} + \mu(\mathbf{x})\,\chi\left(\frac{1}{\mu(\mathbf{x})}, \frac{x_1}{\mu(\mathbf{x})}, \frac{x_2}{\mu(\mathbf{x})}\right)\mathbf{b}\right).$$

When $\mathbf{x} \in U$ and $x_2 \geq 0$, simple computations show that the differential system $\dot{\mathbf{x}} = \mathbf{f}_{\mathbf{x}_+}(\mathbf{x})$ can be written as

$$\dot{x}_1 = \frac{k_1b_2}{|k_1|}x_2 - \frac{k_1b_1}{|k_1|}x_1x_2 - x_1^2, \quad \dot{x}_2 = -\frac{t}{2}x_2 - x_1x_2 - \frac{k_1b_1}{|k_1|}x_2^2. \tag{3.22}$$

Now we study the origin of (3.22).

(a) If $t \neq 0$, the change of time $\tau(s) = -ts/2$ transforms system (3.22) into

$$x_1' = -2\frac{k_1 b_2}{t\,|k_1|}x_2 + 2\frac{k_1 b_1}{t\,|k_1|}x_1 x_2 + \frac{2}{t}x_1^2, \quad x_2' = x_2 + \frac{2}{t}x_1 x_2 + 2\frac{k_1 b_1}{t\,|k_1|}x_2^2,$$

where the prime denotes the derivative with respect to the variable τ. The change of variables $(z_1, z_2) \to \left(x_1 + 2\frac{k_1 b_2}{t|k_1|}x_2, x_2\right)$ transforms this system into

$$z_1' = \frac{2}{t}z_1^2 + 2\frac{k_1}{t\,|k_1|}\left(b_1 - 2\frac{b_2}{t}\right)z_1 z_2, \quad z_2' = z_2 + \frac{2}{t}z_1 z_2 + 2\frac{k_1}{t\,|k_1|}\left(b_1 - 2\frac{b_2}{t}\right)z_2^2.$$

Thus the origin is an isolated degenerate elementary singular point. Moreover, suppose that $z_2 = f(z_1)$ is the solution of the equation $z_2' = 0$ in a neighbourhood of the origin and let $X(z_1, z_2)$ be the first component of the vector field. Then the function $g(z_1) = X(z_1, f(z_1))$ has the following power series expansion:

$$g(z_1) = \frac{2}{t}z_1^2 - \frac{4}{t^2}\left(b_1 - 2\frac{b_2}{t}\right)^2 z_1^4 + O\left(z^6\right).$$

From Theorem 2.7.3(c) it follows that the origin is a degenerate elementary saddle-node with the hyperbolic manifold contained in the line $z_1 = 0$ and this manifold is unstable. Furthermore, if $t > 0$ (respectively, $t < 0$), the straight line $z_2 = 0$ contains the central manifold of the saddle-node, and this manifold is stable in the π direction (respectively, 0 direction).

Going back through the changes of variables we conclude that system (3.22) has a saddle-node at the origin with the central manifold on the x-axis and the hyperbolic manifold on the straight line $x_1 + 2k_1 b_2 x_2/(t|k_1|) = 0$. The central manifold is stable in the 0 direction and the hyperbolic manifold is stable when $t > 0$ and unstable when $t < 0$.

To write the straight line $x_1 + 2k_1 b_2 x_2/(t|k_1|) = 0$ in the coordinate system of $\mathbb{R}^2$, we have to apply the transformation $\mathbf{h}_+^{-1} \circ \mathbf{h}_{\mathbf{x}_+}^{-1}$ to this line, see (2.12) and (2.17).

(b) Suppose now that $t = 0$ and $D < 0$. From expression (3.10) we obtain $D = -k_1 b_2$. Then $k_1 b_2 > 0$. The change of time $\tau(s) = sk_1 b_2/|k_1|$ transforms (3.22) into the system

$$x_1' = x_2 - \frac{b_1}{b_2}x_1 x_2 - \frac{|k_1|}{k_1 b_2}x_1^2, \quad x_2' = -\frac{|k_1|}{k_1 b_2}x_1 x_2 - \frac{b_1}{b_2}x_2^2,$$

which has a nilpotent singular point at the origin.

Define

$$X(x_1, x_2) = -\frac{b_1}{b_2}x_1 x_2 - \frac{|k_1|}{k_1 b_2}x_1^2, \quad Y(x_1, x_2) = -\frac{|k_1|}{k_1 b_2}x_1 x_2 - \frac{b_1}{b_2}x_2^2$$

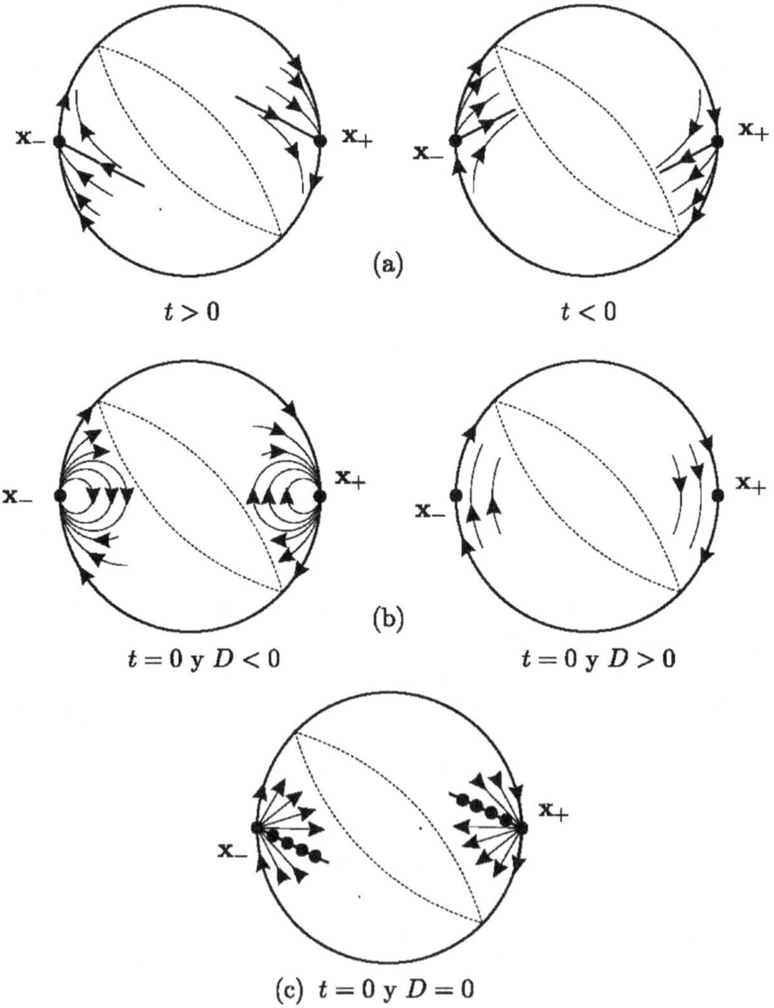

Figure 3.7: Local phase portraits of $\mathbf{f}_{\mathbb{D}}$ at the singular points at infinity when $t^2 - 4d = 0$, $k_1 \neq 0$ and A is not diagonal.

and let $x_2 = f(x_1)$ be the solution of the equation $x_2 + X(x_1, x_2) = 0$ in a neighbourhood of the origin. Then

$$f(x_1) = \frac{|k_1|}{k_1 b_2} x_1^2 + \frac{|k_1| b_1}{k_1 b_2^2} x_1^3 + O(x_1^4),$$

$$g(x_1) = Y(x_1, f(x_1)) = -\frac{1}{b_2^2} x_1^3 (1 + O(x_1)),$$

$$\Phi(x_1) = \left(\frac{\partial X}{\partial x_1} + \frac{\partial Y}{\partial x_2}\right)_{(x_1, f(x_1))} = -\frac{3|k_1|}{k_1 b_2} x_1 \left(1 + \frac{b_1}{b_2} x_1 + O(x_1^2)\right).$$

From Theorem 2.7.4(c.2) it follows that this neighbourhood of the origin is the union of a hyperbolic sector with an elliptic one. Since the axis $x_2 = 0$ is invariant under the flow, the orbits contained in $x_2 = 0$ are not in the interior of the hyperbolic sector. To determine whether the axis $x_2 = 0$ contains or not the boundary between the two sectors we need to do a blow-up in the direction $x_2 = 0$. The change of variables $x_1 = u_1$, $x_2 = u_1 u_2$ transforms the system into the system

$$u_1' = \frac{k_1 b_2}{|k_1|} u_1 u_2 - \frac{k_1 b_1}{|k_1|} u_1^2 u_2 - u_1^2, \quad u_2' = -\frac{k_1 b_2}{|k_1|} u_2^2,$$

which has a singular point at the origin and the flow of which leaves the axes invariant.

Let $\mathbf{u}(\tau) = (u_1(\tau), u_2(\tau))$ be a solution of the system such that $u_1(0) > 0$ and $u_2(0) < 0$. Since $k_1 b_2 > 0$, it follows that $u_1(\tau) > 0$ and $u_2(\tau) < 0$ in a neighbourhood of the origin. Therefore $u_1' < 0$, $u_2' < 0$, and the origin does not belong to the ω-limit set of $\mathbf{u}$.

Let $\mathbf{u}(\tau) = (u_1(\tau), u_2(\tau))$ be a solution of the system such that $u_1(0) < 0$ and $u_2(0) > 0$. Then $u_1' < 0$, $u_2' < 0$ in a neighbourhood of the origin and the origin does not belong to the α-limit set of $\mathbf{u}$.

Returning to the original variables, the quadrants $\{u_1 > 0,\ u_2 < 0\}$ and $\{u_1 < 0,\ u_2 > 0\}$ become the quadrants $\{x_1 > 0,\ x_2 < 0\}$ and $\{x_1 < 0,\ x_2 < 0\}$, respectively. Thus we conclude that the hyperbolic sector is exactly the half-plane $x_2 < 0$, and the boundary between the two sectors is the straight line $x_2 = 0$.

Similar arguments apply when $D = -k_1 b_2 > 0$. In this case we conclude that the boundary between the two sectors is the straight line $x_2 = 0$, and the hyperbolic sector is contained in the half-plane $x_2 > 0$.

(c) Suppose now that $t = 0$ and $D = 0$. Consequently, $D = -k_1 b_2$ and $b_2 = 0$. Then system (3.22) can be written as

$$\dot{x}_1 = -x_1 \left(x_1 + \frac{k_1 b_1}{|k_1|} x_2\right), \quad \dot{x}_2 = -x_2 \left(x_1 + \frac{k_1 b_1}{|k_1|} x_2\right).$$

This system has a non-isolated singular point at the origin.

The change of variables $u_1 = x_1$, $u_2 = x_1 + k_1 b_1 x_2 / |k_1|$, transforms it into the system

$$\dot{u}_1 = -u_1 u_2, \quad \dot{u}_2 = -u_2^2.$$

Thus the origin is a semi-stable non-isolated node with the singular manifold contained in $u_2 = 0$. Returning to the original variables we obtain that system (3.22) has a semi-stable non-isolated node at the origin with the singular manifold contained in the straight line $x_1 + (k_1 b_1/|k_1|)x_2 = 0$. To express this straight line in the coordinate system of $\mathbb{R}^2$ we have to apply to it the transformation $\mathbf{h}_+^{-1} \circ \mathbf{h}_{\mathbf{x}_+}^{-1}$. □

Now we study the singular points at infinity when the fundamental parameters satisfy $t^2 - 4d > 0$. In this case there exist exactly four singular points at infinity, $\mathbf{x}_+$, $\mathbf{x}_-$, $\mathbf{y}_+$, and $\mathbf{y}_-$. In the next theorem we study the local phase portrait of the compactified flow at $\mathbf{x}_+$ and $\mathbf{x}_-$. In Theorem 3.11.8 we shall study the local phase portrait of the compactified flow at $\mathbf{y}_+$ and $\mathbf{y}_-$.

Theorem 3.11.7. *Consider a fundamental system* $\dot{\mathbf{x}} = A\mathbf{x} + \varphi(\mathbf{k}^T\mathbf{x})\mathbf{b}$ *with parameters* (D, T, d, t), *where* $t^2 - 4d > 0$ *and the vector* $\mathbf{k} = (k_1, k_2)^T$ *is such that* $k_1 \neq 0$. *Suppose that A is in real Jordan normal form, and let $\lambda_1 > \lambda_2$ be its eigenvalues. Under these assumptions there exist exactly four singular points at infinity,* $\mathbf{x}_+$, $\mathbf{x}_-$, $\mathbf{y}_+$, *and* $\mathbf{y}_-$.

(a) *If $d > 0$ and $t > 0$, then $\mathbf{x}_+$ (respectively, $\mathbf{x}_-$) is a stable node. One of the characteristic directions coincides with $\partial\mathbb{D}$, and the other one coincides with the straight line $x_2 = -\operatorname{sign}(k_1)b_2/\lambda_2$ (respectively, $x_2 = \operatorname{sign}(k_1)b_2/\lambda_2$). Moreover, all the orbits except the above straight line arrive at infinity tangentially to $\partial\mathbb{D}$, see Figure* 3.8(a).

(b) *If $d > 0$ and $t < 0$, then $\mathbf{x}_+$ (respectively, $\mathbf{x}_-$) is a saddle with the stable manifold contained in $\partial\mathbb{D}$ and the unstable one contained in the straight line $x_2 = -\operatorname{sign}(k_1)b_2/\lambda_2$ (respectively, $x_2 = \operatorname{sign}(k_1)b_2/\lambda_2$), see Figure* 3.8(c).

(c) *If $d < 0$, then $\mathbf{x}_+$ (respectively, $\mathbf{x}_-$) is a stable node. One of the characteristic directions coincides with $\partial\mathbb{D}$ and the other one with the straight line $x_2 = -\operatorname{sign}(k_1)b_2/\lambda_2$ (respectively, $x_2 = \operatorname{sign}(k_1)b_2/\lambda_2$). Moreover, all the orbits arrive at infinity tangentially to the straight line, see Figure* 3.8(b).

(d) *Suppose that $d = 0$ and $t < 0$.*

(d.1) *If $D < 0$, then $\mathbf{x}_+$ (respectively, $\mathbf{x}_-$) is a stable node. One of the characteristic directions coincides with $\partial\mathbb{D}$ and the other one with the straight line $x_2 = -\operatorname{sign}(k_1)b_2/t$ (respectively, $x_2 = \operatorname{sign}(k_1)b_2/t$). Moreover, all the orbits arrive at infinity tangentially to the straight line, see Figure* 3.8(i).

(d.2) *If $D = 0$, then $\mathbf{x}_+$ (respectively, $\mathbf{x}_-$) is a stable normally hyperbolic singular point with the normally hyperbolic manifold contained in the straight line $x_2 = -\operatorname{sign}(k_1)b_2/t$ (respectively, $x_2 = \operatorname{sign}(k_1)b_2/t$), see Figure* 3.8(h).

(d.3) *If $D > 0$, then $\mathbf{x}_+$ (respectively, $\mathbf{x}_-$) is a saddle. The stable manifold is contained in $\partial\mathbb{D}$ and the unstable manifold is contained in the straight* *line* $x_2 = -\operatorname{sign}(k_1)b_2/t$ *(respectively,* $x_2 = \operatorname{sign}(k_1)b_2/t$*), see Figure* 3.8(g).

(e) *If* $d = 0$ *and* $t > 0$, *then* $\mathbf{x}_+$ *and* $\mathbf{x}_-$ *are two stable nodes such that the orbits arrive at infinity tangentially to* $\partial\mathbb{D}$ *if* $b_2 \neq 0$, *and any direction is a characteristic direction if* $b_2 = 0$. *See Figures* 3.8(d), (e) *and* (f).

Proof. To know the local phase portraits of the fundamental system at the infinite singular points $\mathbf{x}_+$ and $\mathbf{x}_-$, it is sufficient to study the system $\dot{\mathbf{x}} = \mathbf{f}_{\mathbf{x}_+}(\mathbf{x})$ in a neighbourhood of the origin, see Lemma 3.11.2(b).

Since $k_1 \neq 0$, there exists a neighbourhood U of the origin such that

$$\mu(\mathbf{x})\,\chi\left(\frac{1}{\mu(\mathbf{x})}, \frac{x_1}{\mu(\mathbf{x})}, \frac{x_2}{\mu(\mathbf{x})}\right) = \frac{k_1}{|k_1|}\,|x_2|,$$

for every $\mathbf{x} \in U$. From Lemma 3.11.2(a) and noting that A is in its real Jordan form it is easy to check that the system $\dot{\mathbf{x}} = \mathbf{f}_{\mathbf{x}_+}(\mathbf{x})$ can be written as

$$\begin{aligned} \dot{x}_1 &= (\lambda_2 - \lambda_1)\,x_1 + \frac{k_1 b_2}{|k_1|}x_2 - \frac{k_1 b_1}{|k_1|}x_1 x_2, \\ \dot{x}_2 &= -\lambda_1 x_2 - \frac{k_1 b_1}{|k_1|}x_2^2, \end{aligned} \tag{3.23}$$

when $x_2 \geq 0$.

(a) Suppose that $d > 0$ and $t > 0$. In this case the eigenvalues of A satisfy $\lambda_1 > \lambda_2 > 0$. The linear part of system (3.23) has eigenvalues $\lambda_2 - \lambda_1 < 0$ and $-\lambda_1 < 0$. Then the origin is a hyperbolic stable node. The characteristic directions coincide with the straight lines $x_2 = 0$ and $\lambda_2 x_1 + \frac{k_1 b_2}{|k_1|}x_2 = 0$. Furthermore, the orbits that are not separatrices are tangent to $x_2 = 0$ at the origin. We recall that in order to express the straight lines in the coordinates of $\mathbb{R}^2$, we have to apply the transformation $\mathbf{h}_+^{-1} \circ \mathbf{h}_{\mathbf{x}_+}^{-1}$.

(b) Suppose that $d > 0$ and $t < 0$. In this case $0 > \lambda_1 > \lambda_2$, and the origin of system (3.23) is a hyperbolic saddle. It is easy to check that the stable manifold is contained in the straight line $x_2 = 0$ and the unstable one is contained in the straight line $\lambda_2 x_1 + k_1 b_2 x_2/|k_1| = 0$.

(c) Suppose that $d < 0$. Then $\lambda_1 > 0 > \lambda_2$. The statement follows by applying the same arguments as in the proof of statement (a).

(d) Suppose that $d = 0$ and $t < 0$. In this case $0 = \lambda_1 > \lambda_2$ and system (3.23) becomes

$$\dot{x}_1 = \lambda_2 x_1 + \frac{k_1 b_2}{|k_1|}x_2 - \frac{k_1 b_1}{|k_1|}x_1 x_2, \quad \dot{x}_2 = -\frac{k_1 b_1}{|k_1|}x_2^2, \tag{3.24}$$

which has a degenerate elementary singular point at the origin.

(d.1) From (3.10) we have $D = \lambda_2 k_1 b_1$. Thus, when $D < 0$ we get $k_1 b_1 > 0$. The change of variables $u_1 = x_2$ and $u_2 = \lambda_2 x_1 + k_1 b_2 x_2/|k_1|$ transforms system (3.24) into the system

$$\dot{u}_1 = -\frac{k_1 b_1}{|k_1|}u_1^2, \quad \dot{u}_2 = \lambda_2 u_2 - \frac{k_1 b_1}{|k_1|}u_1 u_2.$$

Changing the time as $\tau(s) = \lambda_2 s$ and denoting by prime the derivative with respect to τ we obtain the system

$$u_1' = -\frac{k_1 b_1}{\lambda_2 |k_1|} u_1^2, \quad u_2' = u_2 - \frac{k_1 b_1}{\lambda_2 |k_1|} u_1 u_2,$$

which has an isolated singular point at the origin and the flow of which leaves the axes invariant. Define

$$X(u_1, u_2) = -\frac{k_1 b_1}{\lambda_2 |k_1|} u_1^2, \quad Y(u_1, u_2) = -\frac{k_1 b_1}{\lambda_2 |k_1|} u_1 u_2,$$

and let $u_2 = f(u_1)$ be the solution of the equation $u_2 + Y(u_1, u_2) = 0$ in a neighbourhood of the origin. Then $g(u_1) = X(u_1, f(u_1)) = -k_1 b_1 u_1^2/\lambda_2|k_1|$. By Theorem 2.7.3(c), the origin is a saddle-node with the unstable hyperbolic manifold on $u_1 = 0$. The central manifold of the saddle-node is contained in the line $u_2 = 0$ and it is unstable in the direction 0.

Going back through the changes of variables and time, it follows that system (3.24) has a saddle-node at the origin. The stable hyperbolic manifold of the singular point is contained in $x_2 = 0$, the central manifold is contained in $\lambda_2 x_1 + k_1 b_2 x_2/|k_1| = 0$ and it is stable in the half-plane $x_2 > 0$ and unstable in the half-plane $x_2 < 0$. Finally, the orbits arrive at the origin tangentially to the straight line.

(d.2) When $D = 0$, $k_1 b_1 = 0$, and system (3.24) becomes $\dot{x}_1 = \lambda_2 x_1 + (k_1 b_2/|k_1|)x_2$, $\dot{x}_2 = 0$. For this system the origin is a stable normally hyperbolic singular point with the normally hyperbolic manifold on the straight line $\lambda_2 x_1 + k_1 b_2 x_2/|k_1| = 0$.

(d.3) When $D > 0$, $k_1 b_1 < 0$. By applying the same arguments as those in the proof of statement (d.1) we obtain that system (3.24) has a saddle-node at the origin. The stable hyperbolic manifold of the singular point is contained in the axis $x_2 = 0$. The central manifold is contained in the straight line $\lambda_2 x_1 + k_1 b_2 x_2 |k_1| = 0$ and it is stable in the half-plane $x_2 < 0$ and unstable in $x_2 > 0$.

(e) When $d = 0$ and $t > 0$, the eigenvalues of A satisfy $\lambda_1 > \lambda_2 = 0$. Thus the linear part of system (3.23) has a hyperbolic stable node at the origin. This node is non-diagonal when $b_2 \neq 0$ and it is diagonal when $b_2 = 0$. □

Theorem 3.11.8. *Consider a fundamental system* $\dot{\mathbf{x}} = A\mathbf{x} + \varphi(\mathbf{k}^T\mathbf{x})\mathbf{b}$ *with parameters* (D, T, d, t)*, where* $t^2 - 4d > 0$ *and the vector* $\mathbf{k} = (k_1, k_2)^T$ *is such that* $k_2 \neq 0$*. Suppose that the matrix* A *is in real Jordan normal form, and let* $\lambda_1 > \lambda_2$ *be the eigenvalues of* A*. Under these assumptions there exist exactly four singular points at infinity,* $\mathbf{x}_+$*,* $\mathbf{x}_-$*,* $\mathbf{y}_+$*, and* $\mathbf{y}_-$*.*

(a) *If* $d > 0$ *and* $t > 0$*, then* $\mathbf{y}_+$ *(respectively,* $\mathbf{y}_-$*) is a saddle with the unstable manifold contained in* $\partial\mathbb{D}$ *and the stable one contained in the straight line* $x_1 = -\operatorname{sign}(k_2)b_1/\lambda_1$ *(respectively,* $x_1 = \operatorname{sign}(k_2)b_1/\lambda_1$*), see Figure* 3.8(a).

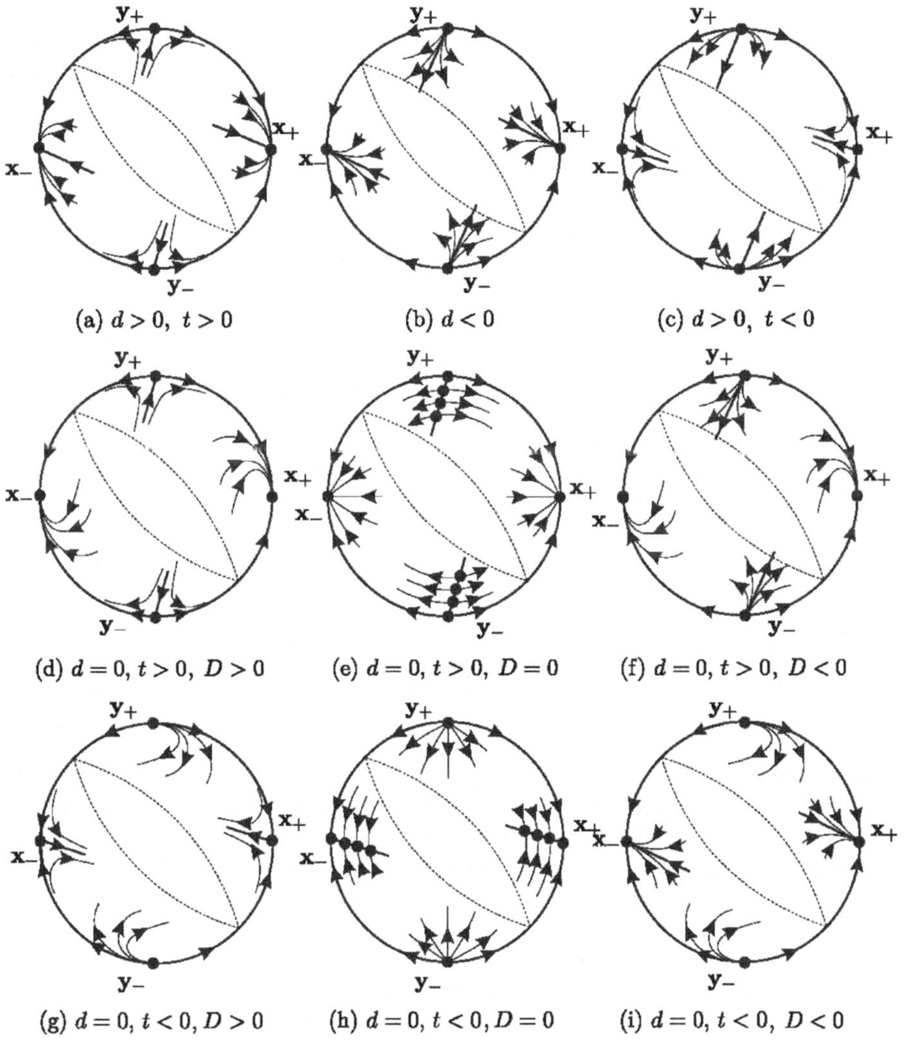

Figure 3.8: Local phase portraits of $\mathbf{f}_{\mathbb{D}}$ at the singular points at infinity when $t^2 - 4d > 0$, $k_1 \neq 0$ and $k_2 \neq 0$.

(b) *If $d > 0$ and $t < 0$, then $\mathbf{y}_+$ (respectively, $\mathbf{y}_-$) is an unstable node. One of the characteristic directions coincides with $\partial\mathbb{D}$, and the other with the straight line $x_1 = -\operatorname{sign}(k_2)b_1/\lambda_1$ (respectively, $x_1 = \operatorname{sign}(k_2)b_1/\lambda_1$). Moreover, the other orbits leave the infinity tangentially to it, see Figure* 3.8(c).

(c) *If $d < 0$, then $\mathbf{y}_+$ (respectively, $\mathbf{y}_-$) is an unstable node. One of the characteristic directions coincides with $\partial\mathbb{D}$, and the other with the straight line $x_1 = -\operatorname{sign}(k_2)b_1/\lambda_1$ (respectively, $x_1 = \operatorname{sign}(k_2)b_1/\lambda_1$). Moreover, the other orbits leave the infinity tangentially to the straight line, see Figure* 3.8(b).

(d) *If $d = 0$ and $t < 0$, then $\mathbf{y}_+$ and $\mathbf{y}_-$ are unstable nodes such that the orbits leave the infinity tangentially to it if $b_1 \neq 0$, and any direction is a characteristic direction if $b_1 = 0$, see Figures* 3.8(g), (h) *and* (i).

(e) *Suppose $d = 0$ and $t > 0$.*

(e.1) *If $D < 0$, then $\mathbf{y}_+$ (respectively, $\mathbf{y}_-$) is an unstable node. One of the characteristic directions coincides with $\partial\mathbb{D}$ and the other with the straight line $x_1 = -\operatorname{sign}(k_2)b_1/t$ (respectively, $x_1 = \operatorname{sign}(k_2)b_1/t$). Moreover, the orbits leave the infinity tangentially to the straight line, see Figure* 3.8(f).

(e.2) *If $D = 0$, then $\mathbf{y}_+$ (respectively, $\mathbf{y}_-$) is an unstable normally hyperbolic singular point. The normally hyperbolic manifold is contained in the straight line $x_1 = -\operatorname{sign}(k_2)b_1/t$ (respectively, $x_1 = \operatorname{sign}(k_2)b_1/t$), see Figure* 3.8(e).

(e.3) *If $D > 0$, then $\mathbf{y}_+$ (respectively, $\mathbf{y}_-$) is a saddle with the unstable manifold contained in $\partial\mathbb{D}$ and the stable one contained in the straight line $x_1 = -\operatorname{sign}(k_2)b_1/t$ (respectively, $x_1 = \operatorname{sign}(k_2)b_1/t$), see Figure* 3.8(d).

Proof. To determine the local phase portraits of the fundamental system at the infinite singular points $\mathbf{y}_+$ and $\mathbf{y}_-$ it is sufficient to study the system $\dot{\mathbf{x}} = \mathbf{f}_{\mathbf{y}_+}(\mathbf{x})$ in a neighbourhood of the origin, see Lemma 3.11.2(b).

Since $k_2 \neq 0$, there exists a neighbourhood U of the origin such that $|-k_1x + k_2| > |x_2|$ for every $\mathbf{x} \in U$. It is easy to check that

$$\mu(\mathbf{x})\,\chi\left(-\frac{x_1}{\mu(\mathbf{x})}, \frac{1}{\mu(\mathbf{x})}, \frac{x_2}{\mu(\mathbf{x})}\right) = \frac{k_2}{|k_2|}\,|x_2|\,.$$

Hence, if $\mathbf{x} \in U$ and $x_2 \geq 0$, the system $\dot{\mathbf{x}} = \mathbf{f}_{\mathbf{y}_+}(\mathbf{x})$ becomes

$$\begin{aligned} \dot{x}_1 &= (\lambda_1 - \lambda_2)\,x_1 - \frac{k_2 b_1}{|k_2|}x_2 - \frac{k_2 b_2}{|k_2|}x_1x_2, \\ \dot{x}_2 &= -\lambda_2 x_2 - \frac{k_2 b_2}{|k_2|}x_2^2, \end{aligned} \tag{3.25}$$

see Lemma 3.11.2(a).

(a) Suppose that $d > 0$ and $t > 0$. In this case the eigenvalues of A satisfy $\lambda_1 > \lambda_2 > 0$. The linear part of system (3.25) has eigenvalues $\lambda_1 - \lambda_2 > 0$ and $-\lambda_2 < 0$. Then, by the Hartman–Grobman theorem, the origin is a saddle. The unstable manifold of the saddle point is contained in the line $x_2 = 0$ and the stable manifold is contained in the straight line $\lambda_1 x_1 - k_2 b_1 x_2/|k_2| = 0$. To write the straight line in the coordinate system of $\mathbb{R}^2$ we apply the transformation $\mathbf{h}_+^{-1} \circ \mathbf{h}_{\mathbf{x}_+}^{-1}$.

(b) Suppose that $d > 0$ and $t < 0$. In this case the eigenvalues of A satisfy $0 > \lambda_1 > \lambda_2$. Applying again the Hartman–Grobman theorem to system (3.25), we conclude that the origin is a hyperbolic unstable node. Moreover, one of the characteristic directions of the node coincides with $x_2 = 0$, and the other with the straight line $\lambda_1 x_1 - \frac{k_2 b_1}{|k_2|} x_2 = 0$. The tangency of orbits at the origin follows from $-\lambda_2 > \lambda_1 - \lambda_2 > 0$.

(c) Suppose that $d < 0$. In this case $\lambda_1 > 0 > \lambda_2$. The statement follows similarly to statement (b). The tangency of orbits at the origin follows from $\lambda_1 - \lambda_2 > -\lambda_2$.

(d) Suppose now that $d = 0$ and $t < 0$. Thus $0 = \lambda_1 > \lambda_2$. The statement follows similarly to statement (b). The tangency of orbits at the origin depends on whether $b_1 = 0$ or $b_1 \neq 0$.

(e) Suppose that $d = 0$ and $t > 0$. Then it follows that the eigenvalues of A satisfy $\lambda_1 > \lambda_2 = 0$, and system (3.25) becomes

$$\dot{x}_1 = \lambda_1 x_1 - \frac{k_2 b_1}{|k_2|} x_2 - \frac{k_2 b_2}{|k_2|} x_1 x_2, \quad \dot{x}_2 = -\frac{k_2 b_2}{|k_2|} x_2^2. \tag{3.26}$$

(e.3) When $D > 0$, it follows from expression (3.10) that $k_2 b_2 > 0$. Thus if we replace λ_1 by λ_2, k_2 by k_1, b_2 by b_1, and b_1 by $-b_2$ in system (3.26), we obtain the system (3.24), which is studied in the proof of Theorem 3.11.7(d.1). Note that the change in time $\tau(s) = \lambda_2 s$ used in that proof preserves now the orientation. Thus, system (3.26) has a saddle-node at the origin. The unstable hyperbolic manifold of this point is contained in the line $x_2 = 0$. The central manifold is contained in the straight line $\lambda_1 x_1 - k_2 b_1 x_2/|k_2| = 0$, and it is unstable in the half-plane $x_2 > 0$ and stable in the half-plane $x_2 < 0$. Finally, the orbits which arrive at the origin are tangent to the straight line.

Statements (e.1) and (e.2) follow by the same arguments as those used in the proof of statement (e.3). □

Now let us study the local phase portraits of singular points at infinity which belong to $\mathbb{D}_0$. Vector fields in a neighbourhood of these points are formed by three different non-linear systems, see Figure 3.3. In this case, phase portraits can be drawn by studying each of these vector fields separately and by composing the respective phase portraits. Note that when the straight lines L_+ and L_- contain a characteristic direction of the singular point at infinity, more care is needed when completing the phase portrait.

Theorem 3.11.9. *Consider a fundamental system* $\dot{\mathbf{x}} = A\mathbf{x} + \varphi(\mathbf{k}^T\mathbf{x})\mathbf{b}$ *with parameters* (D, T, d, t) *where* $t^2 - 4d = 0$. *Suppose that* A *is in real Jordan normal form and is diagonal. Under these assumptions there exist exactly two singular points at infinity in* $\mathbb{D}_0$, $\pm\mathbf{k}^{\perp}/\|\mathbf{k}\|$.

(a) *If* $D > 0$, *then* $T \neq 0$. *When* $T > 0$ *(respectively* $T < 0$*) the singular points* $\pm\mathbf{k}^{\perp}/\|\mathbf{k}\|$ *are stable normally hyperbolic singular (respectively, unstable normally hyperbolic) points with the normally hyperbolic manifold contained in* $\partial\mathbb{D}$, *see Figure* 3.5(b) *and* (c).

(b) *Suppose that* $D = 0$ *and* $T > 0$.

(b.1) *If* $d = 0$, *then the phase portraits in a neighbourhood of* $\pm\mathbf{k}^{\perp}/\|\mathbf{k}\|$ *are topologically equivalent to the one shown in Figure* 3.10(b), *after reversing the orientation of the flow.*

(b.2) *If* $d \neq 0$, *then the phase portraits in a neighbourhood of* $\pm\mathbf{k}^{\perp}/\|\mathbf{k}\|$ *are topologically equivalent to the one shown in Figure* 3.10(a).

(c) *Suppose that* $D = 0$ *and* $T < 0$.

(c.1) *If* $d = 0$, *then the phase portraits in a neighbourhood of* $\pm\mathbf{k}^{\perp}/\|\mathbf{k}\|$ *are topologically equivalent to the one shown in Figure* 3.10(b).

(c.2) *If* $d \neq 0$, *then the phase portraits in a neighbourhood of* $\pm\mathbf{k}^{\perp}/\|\mathbf{k}\|$ *are topologically equivalent to the one shown in Figure* 3.10(a), *after reversing the orientation of the flow.*

(d) *If* $D = 0$ *and* $T = 0$, *then the phase portraits in a neighbourhood of* $\pm\mathbf{k}^{\perp}/\|\mathbf{k}\|$ *are topologically equivalent to the one shown in Figure 3.10(c).*

(e) *Suppose that* $D < 0$. *In this case* $t \neq 0$.

(e.1) *If* $t > 0$, *then the phase portraits in a neighbourhood of* $\pm\mathbf{k}^{\perp}/\|\mathbf{k}\|$ *are topologically equivalent to the one shown in Figure* 3.10(a).

(e.2) *If* $t < 0$, *then the phase portraits in a neighbourhood of* $\pm\mathbf{k}^{\perp}/\|\mathbf{k}\|$ *are topologically equivalent to the one shown in Figure* 3.10(a), *after reversing the orientation of the flow.*

Proof. Consider the fundamental system $\dot{\mathbf{x}} = A\mathbf{x} + \varphi(\mathbf{k}^{*T}\mathbf{x})\mathbf{b}^*$, where $\mathbf{k}^* = \mathbf{k}/\|\mathbf{k}\|$ and $\mathbf{b}^* = \|\mathbf{k}\|\,\mathbf{b}$. This system and the one in the statement of the theorem have the same fundamental matrices. By Propositions 3.6.2 and 3.10.2, their Poincaré compactifications are differentiably conjugate. Thus it is not a restriction to suppose that vector $\mathbf{k}$ satisfies $\|\mathbf{k}\| = 1$.

By Lemma 3.11.2(b), it is sufficient to study the system $\dot{\mathbf{x}} = \mathbf{f}_{\mathbf{k}^{\perp}}(\mathbf{x})$ in a neighbourhood U of the origin. Recall that we are interested only in the half-plane $x_2 \geq 0$. For simplicity of notation we use U instead of $U \cap \{(x_1, x_2) : x_2 \geq 0\}$.

From Lemma 3.11.2(a) and expression (3.19) for $\mathbf{x} \in U$ we get

$$\mathbf{f}_{\mathbf{k}^\perp}(\mathbf{x}) = \begin{cases} \mathbf{f}^-_{\mathbf{k}^\perp}(\mathbf{x}), & \text{if } x_1 \le -x_2, \\ \mathbf{f}^0_{\mathbf{k}^\perp}(\mathbf{x}), & \text{if } |x_1| \le x_2, \\ \mathbf{f}^+_{\mathbf{k}^\perp}(\mathbf{x}), & \text{if } x_1 \ge x_2, \end{cases}$$

where

$$\mathbf{f}^-_{\mathbf{k}^\perp}(\mathbf{x}) := \begin{pmatrix} -\mathbf{b}^T\mathbf{k}x_2 - \mathbf{b}^T\mathbf{k}^\perp x_1x_2 \\ -\lambda x_2 - \mathbf{b}^T\mathbf{k}^\perp x_2^2 \end{pmatrix},$$

$$\mathbf{f}^0_{\mathbf{k}^\perp}(\mathbf{x}) := \begin{pmatrix} \mathbf{b}^T\mathbf{k}x_1 + \mathbf{b}^T\mathbf{k}^\perp x_1^2 \\ -\lambda x_2 + \mathbf{b}^T\mathbf{k}^\perp x_1x_2 \end{pmatrix},$$

$$\mathbf{f}^+_{\mathbf{k}^\perp}(\mathbf{x}) := \begin{pmatrix} \mathbf{b}^T\mathbf{k}x_2 + \mathbf{b}^T\mathbf{k}^\perp x_1x_2 \\ -\lambda x_2 + \mathbf{b}^T\mathbf{k}^\perp x_2^2 \end{pmatrix}.$$

We remark that when the characteristic directions of the singular point at the origin do not coincide with the straight lines $x_2 = \pm x_1$, the local phase portrait of system $\dot{\mathbf{x}} = \mathbf{f}_{\mathbf{k}^\perp}(\mathbf{x})$ can be obtained by composition of the local phase portraits of systems $\dot{\mathbf{x}} = \mathbf{f}^-_{\mathbf{k}^\perp}(\mathbf{x})$, $\dot{\mathbf{x}} = \mathbf{f}^0_{\mathbf{k}^\perp}(\mathbf{x})$ and $\dot{\mathbf{x}} = \mathbf{f}^+_{\mathbf{k}^\perp}(\mathbf{x})$ restricted to the regions $x_1 \le -x_2$, $|x_1| \le x_2$ and $x_1 \ge x_2$, respectively (recall that $x_2 \ge 0$). Moreover, when the characteristic directions of the singular point at the origin coincide with the straight lines $x_2 = \pm x_1$, but these straight lines are invariant under the flow of the system $\dot{\mathbf{x}} = \mathbf{f}_{\mathbf{k}^\perp}(\mathbf{x})$, we can use the same argument as before to obtain the phase portrait of system $\dot{\mathbf{x}} = \mathbf{f}_{\mathbf{k}^\perp}(\mathbf{x})$ in a neighbourhood of the origin.

Note that if we change $\mathbf{b}$ to $-\mathbf{b}$ in the expression of the vector field $\mathbf{f}^+_{\mathbf{k}^\perp}$, we obtain the vector field $\mathbf{f}^-_{\mathbf{k}^\perp}$. Moreover, if we substitute $\mathbf{q} = \mathbf{k}$ in the expression for system (3.20), we obtain the expression for system $\dot{\mathbf{x}} = \mathbf{f}^-_{\mathbf{k}^\perp}(\mathbf{x})$. Thus, to understand the local phase portrait of systems $\dot{\mathbf{x}} = \mathbf{f}^+_{\mathbf{k}^\perp}(\mathbf{x})$ and $\dot{\mathbf{x}} = \mathbf{f}^-_{\mathbf{k}^\perp}(\mathbf{x})$, one can refer to the proof of Theorem 3.11.5. Next let us study the phase portrait of system $\dot{\mathbf{x}} = \mathbf{f}^0_{\mathbf{k}^\perp}(\mathbf{x})$.

Suppose that $\mathbf{b}^T\mathbf{k} \neq 0$ and $t \neq 0$. The origin is a hyperbolic singular point with linear part equal to

$$\begin{pmatrix} \mathbf{b}^T\mathbf{k} & 0 \\ 0 & -t/2 \end{pmatrix}.$$

Thus, when $\mathbf{b}^T\mathbf{k} > 0$ and $t > 0$ (respectively, $\mathbf{b}^T\mathbf{k} < 0$ and $t < 0$) the origin is a saddle point with the stable manifold (respectively, unstable manifold) contained in the line $x_1 = 0$ and with the unstable manifold (respectively, stable manifold) contained in the line $x_2 = 0$. When $\mathbf{b}^T\mathbf{k} > 0$ and $t < 0$ (respectively $\mathbf{b}^T\mathbf{k} < 0$ and $t > 0$) the origin is an unstable node (respectively, stable node) with a characteristic direction given by $x_1 = 0$ and the other given by $x_2 = 0$. Moreover, if $t/2 + \mathbf{b}^T\mathbf{k} > 0$, then the orbits are tangent to the straight line $x_1 = 0$ at the origin; if $t/2 + \mathbf{b}^T\mathbf{k} < 0$, then the orbits are tangent to $x_1 = 0$ at the origin; and if $t/2 + \mathbf{b}^T\mathbf{k} = 0$, then any direction is a characteristic direction. In the last case it is easy to check that the lines $x_2 = \pm x_1$ are invariant under the flow.

Suppose that $\mathbf{b}^T\mathbf{k} = 0$ and $t \neq 0$. The change of time $\tau(s) = -ts/2$ recasts system $\dot{\mathbf{x}} = \mathbf{f}^0_{\mathbf{k}^\perp}(\mathbf{x})$ as

$$x_1' = -2\frac{\mathbf{b}^T\mathbf{k}^\perp}{t}x_1^2, \quad x_2' = x_2 - 2\frac{\mathbf{b}^T\mathbf{k}^\perp}{t}x_1x_2,$$

which has a degenerate singular point at the origin.

If $X(x_1, x_2) = -2x_1^2\mathbf{b}^T\mathbf{k}^\perp/t$, $Y(x_1, x_2) = -2x_1x_2\mathbf{b}^T\mathbf{k}^\perp/t$, and $f(x_1)$ is the solution of the equation $x_2 + Y(x_1, x_2) = 0$ in a neighbourhood of the origin, then $X(x_1, f(x_1)) = -2x_1^2\mathbf{b}^T\mathbf{k}^\perp/t$. From Theorem 2.7.3(c) it follows that the origin is a saddle-node with the central manifold on the line $x_2 = 0$ and with the hyperbolic manifold on the line $x_1 = 0$. Moreover, when $\mathbf{b}^T\mathbf{k}^\perp > 0$ (respectively, $\mathbf{b}^T\mathbf{k}^\perp < 0$), the central manifold is unstable in the direction π (respectively 0) and the hyperbolic manifold is stable or unstable depending on whether $t > 0$ or $t < 0$, respectively.

Suppose that $\mathbf{b}^T\mathbf{k} \neq 0$ and $t = 0$. In this case the straight line $x_1 = 0$ is formed by singular points. To desingularize the system we use the change of time $d\tau = x_1(s)ds$, which transforms the system $\dot{\mathbf{x}} = \mathbf{f}^0_{\mathbf{k}^\perp}(\mathbf{x})$ into the system

$$x_1' = \mathbf{b}^T\mathbf{k} + \mathbf{b}^T\mathbf{k}^\perp x_1, \quad x_2' = \mathbf{b}^T\mathbf{k}^\perp x_2.$$

Since the origin is not a singular point, the flow is parallel in a neighbourhood of the origin. Returning to the original time variable we conclude that $x_1 = 0$ is a stable or unstable normally hyperbolic manifold depending on whether $\mathbf{b}^T\mathbf{k} < 0$ or $\mathbf{b}^T\mathbf{k} > 0$, respectively.

Finally, suppose that $\mathbf{b}^T\mathbf{k} = 0$ and $t = 0$. In this case the straight line $x_1 = 0$ is formed by singular points. The change of time $d\tau = x_1(s)ds$ transforms the system $\dot{\mathbf{x}} = \mathbf{f}^0_{\mathbf{k}^\perp}(\mathbf{x})$ into the system

$$x_1' = \mathbf{b}^T\mathbf{k}^\perp x_1, \quad x_2' = \mathbf{b}^T\mathbf{k}^\perp x_2.$$

When $\mathbf{b}^T\mathbf{k}^\perp < 0$ (respectively, $\mathbf{b}^T\mathbf{k}^\perp > 0$) the origin is a stable node (respectively, unstable node). Returning to the original time variable we conclude that the origin is a non-isolated semi-stable node with the singular manifold contained in $x_1 = 0$.

Now let us describe the local phase portrait of the system $\dot{\mathbf{x}} = \mathbf{f}_{\mathbf{k}^\perp}(\mathbf{x})$ depending on the fundamental parameters (D, T, d, t). From expressions (3.10) and (3.12) it follows that $D = t/2(t/2 + \mathbf{k}^T\mathbf{b})$ and $T = t/2 + (t/2 + \mathbf{k}^T\mathbf{b})$, respectively. Therefore $\Lambda_1 = t/2 + \mathbf{b}^T\mathbf{k}$ and $\Lambda_2 = t/2$ are the eigenvalues of the matrix $B = A + \mathbf{b}\mathbf{k}^T$.

(a) Suppose that $D > 0$. In this case the eigenvalues Λ_1 and Λ_2 have the same sign, and therefore $T = \Lambda_1 + \Lambda_2 \neq 0$. When $T > 0$, then $\Lambda_2 > 0$ and $\Lambda_1 > 0$. We have divided the proof into three parts: (a.1) $t > 0$ and $\mathbf{b}^T\mathbf{k} > 0$; (a.2) $t > 0$ and $\mathbf{b}^T\mathbf{k} = 0$; (a.3) $t > 0$, $\mathbf{b}^T\mathbf{k} < 0$. In any case $t > 0$, thus systems $\dot{\mathbf{x}} = \mathbf{f}^+_{\mathbf{k}^\perp}(\mathbf{x})$ and $\dot{\mathbf{x}} = \mathbf{f}^-_{\mathbf{k}^\perp}(\mathbf{x})$ have a stable normally hyperbolic singular point at the origin with the normal hyperbolic manifold contained in $x_2 = 0$. Moreover, the straight lines

$\mathbf{b}^T\mathbf{k}x_2 - (t/2)x_1 = C_1$ and $\mathbf{b}^T\mathbf{k}x_2 + (t/2)x_1 = C_2$ with $C_1, C_2 \in \mathbb{R}$ are invariant under the respective flows.

(a.1) When $\mathbf{b}^T\mathbf{k} > 0$, the system $\dot{\mathbf{x}} = \mathbf{f}^0_{\mathbf{k}^\perp}(\mathbf{x})$ has a hyperbolic saddle at the origin, and the stable manifold is contained in the straight line $x_1 = 0$.

(a.2) When $\mathbf{b}^T\mathbf{k} = 0$, the system $\dot{\mathbf{x}} = \mathbf{f}^0_{\mathbf{k}^\perp}(\mathbf{x})$ has a saddle-node at the origin. The hyperbolic manifold is stable and contained in the line $x_1 = 0$ and the central manifold is contained in the line $x_2 = 0$.

(a.3) When $\mathbf{b}^T\mathbf{k} < 0$ and $(t/2) + \mathbf{b}^T\mathbf{k} > 0$, the system $\dot{\mathbf{x}} = \mathbf{f}^0_{\mathbf{k}^\perp}(\mathbf{x})$ has a hyperbolic stable node at the origin. The characteristic directions coincide with the axes and any orbit is tangent to the straight line $x_2 = 0$ at the origin.

In short, the system $\dot{\mathbf{x}} = \mathbf{f}_{\mathbf{k}^\perp}(\mathbf{x})$ has a stable normal hyperbolic singular point at the origin, and the normal hyperbolic manifold is contained in $x_2 = 0$, see Figure 3.9.

The rest of the statements follow using similar arguments. □

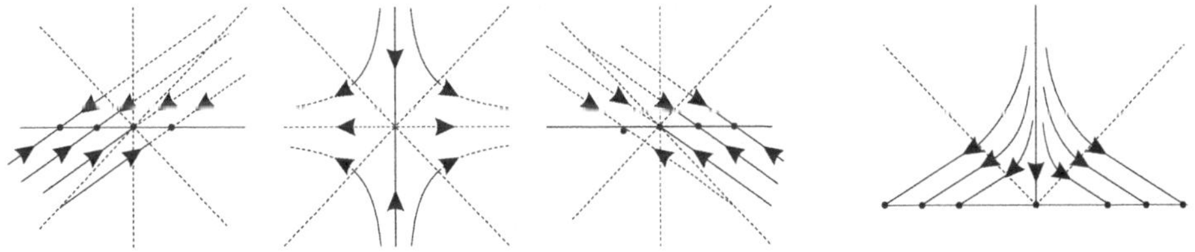

Figure 3.9: Local phase portrait of system $\dot{\mathbf{x}} = \mathbf{f}_{\mathbf{k}^\perp}(\mathbf{x})$ obtained by composing the phase portaits of systems $\dot{\mathbf{x}} = \mathbf{f}^+_{\mathbf{k}^\perp}(\mathbf{x})$, $\dot{\mathbf{x}} = \mathbf{f}^0_{\mathbf{k}^\perp}(\mathbf{x})$ and $\dot{\mathbf{x}} = \mathbf{f}^-_{\mathbf{k}^\perp}(\mathbf{x})$, when $D > 0$, $T > 0$, $t > 0$ and $\mathbf{b}^T\mathbf{k} > 0$.

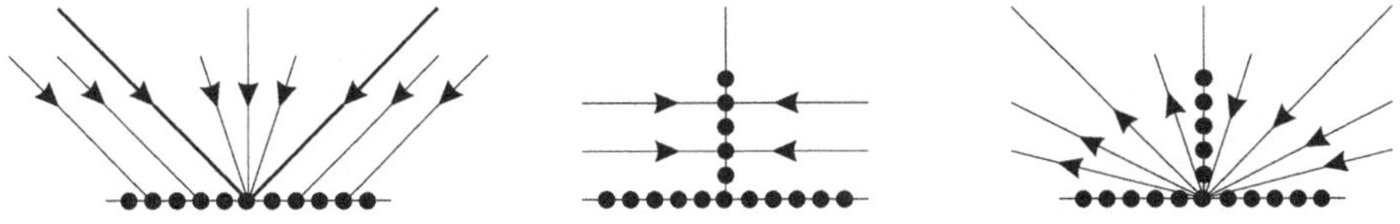

Figure 3.10: Non-isolated singular points at the infinity when $t^2 - 4d = 0$ and A is not diagonal.

Theorem 3.11.10. *Consider a fundamental system* $\dot{\mathbf{x}} = A\mathbf{x} + \varphi(\mathbf{k}^T\mathbf{x})\mathbf{b}$ *with parameters* (D, T, d, t), *where* $t^2 - 4d = 0$ *and the vector* $\mathbf{k} = (k_1, k_2)^T$ *satisfies that* $k_1 = 0$. *Suppose that matrix* A *is in real Jordan normal form and is not diagonal. Under these assumptions there are exactly two singular points at infinity* $\mathbf{x}_+$, $\mathbf{x}_-$ *and they belong to* $\mathbb{D}_0$.

(a) *If $D > 0$, then $T \neq 0$. When $T > 0$ (respectively, $T < 0$), $\mathbf{x}_+$ and $\mathbf{x}_-$ are saddle-nodes. The central manifold is contained in $\partial\mathbb{D}$ and the hyperbolic one is contained in the straight line $x_2 = 0$. Moreover, the hyperbolic manifold is stable (respectively unstable), see Figure* 3.12(a).

(b) *Suppose that $D = 0$ and $T > 0$.*

(b.1) *If $d = 0$, then the phase portraits in a neighbourhood of $\mathbf{x}_+$ and $\mathbf{x}_-$ are topologically equivalent to the one shown in Figure* 3.12(b) *after reversing the orientation of the flow.*

(b.2) *If $d \neq 0$, then $\mathbf{x}_+$ (respectively, $\mathbf{x}_-$) is a saddle-node. The central manifold is contained in $\partial\mathbb{D}$ and the hyperbolic one is contained in the straight line $x_2 = 2\,\mathrm{sign}(k_2)b_2/t$ (respectively, $x_2 = -2\,\mathrm{sign}(k_2)b_2/t$). Moreover, the hyperbolic manifold is stable, see Figure* 3.12(a).

(c) *If $D = 0$ and $T = 0$, then the phase portraits in a neighbourhood of the points $\mathbf{x}_+$ and $\mathbf{x}_-$ are topologically equivalent to the one shown in Figure* 3.12(c), (d) *or* (e) *depending on $b_1 > -1$, $b_1 = -1$, or $b_1 < -1$, respectively.*

(d) *Suppose that $D = 0$ and $T < 0$.*

(d.1) *If $d = 0$, then the phase portraits in a neighbourhood of $\mathbf{x}_+$ and $\mathbf{x}_-$ are topologically equivalent to the one shown in Figure* 3.12(b).

(d.2) *If $d \neq 0$, then $\mathbf{x}_+$ (respectively, $\mathbf{x}_-$) is a saddle-node. The central manifold is contained in $\partial\mathbb{D}$ and the hyperbolic one is contained in the straight line $x_2 = -2\,\mathrm{sign}(k_2)b_2/t$ (respectively, $x_2 = 2\,\mathrm{sign}(k_2)b_2/t$). Moreover, the hyperbolic manifold is stable, see Figure* 3.12(a).

(e) *If $D < 0$, then $t \neq 0$ and $\mathbf{x}_+$ (respectively $\mathbf{x}_-$) is a saddle-node. The central manifold is contained in $\partial\mathbb{D}$. If $t > 0$, the hyperbolic manifold is contained in the straight line $x_2 = 2\,\mathrm{sign}(k_2)b_2/t$ (respectively, $x_2 = -2\,\mathrm{sign}(k_2)b_2/t$), and it is stable, see Figure* 3.12(a); *if $t < 0$, the hyperbolic manifold is contained in the straight line $x_2 = -2k_2b_2/|k_2|t$ (respectively, $x_2 = 2k_2b_2/|k_2|t$) and it is unstable, see Figure* 3.12(a) *(note that the flow in this figure has reverse direction).*

Proof. According to the proof of Theorem 3.11.9, it suffices to describe the flow of the system $\dot{\mathbf{x}} = \mathbf{f}_{\mathbf{x}_+}(\mathbf{x})$ in a neighbourhood U of the origin, where

$$\mathbf{f}_{\mathbf{x}_+}(\mathbf{x}) := \begin{cases} \mathbf{f}^+_{\mathbf{x}_+}(\mathbf{x}), & \text{if } x_1 > x_2, \\ \mathbf{f}^0_{\mathbf{x}_+}(\mathbf{x}), & \text{if } |x_1| \leq x_2, \\ \mathbf{f}^-_{\mathbf{x}_+}(\mathbf{x}), & \text{if } x_1 < -x_2, \end{cases}$$

and

$$\mathbf{f}^+_{\mathbf{x}_+}(\mathbf{x}) := \begin{pmatrix} b_2x_2 - b_1x_1x_2 - x_1^2 \\ -(t/2)\,x_2 - x_1x_2 - b_1x_2^2 \end{pmatrix},$$

$$\mathbf{f}^-_{\mathbf{x}_+}(\mathbf{x}) := \begin{pmatrix} -b_2x_2 + b_1x_1x_2 - x_1^2 \\ -(t/2)\,x_2 - x_1x_2 + b_1x_2^2 \end{pmatrix},$$

$$\mathbf{f}^0_{\mathbf{x}_+}(\mathbf{x}) := \begin{pmatrix} b_2x_1 - (b_1+1)\,x_1^2 \\ -(t/2)\,x_2 - (b_1+1)\,x_1x_2 \end{pmatrix}.$$

Moreover, the phase portrait of the system $\dot{\mathbf{x}} = \mathbf{f}_{\mathbf{x}_+}(\mathbf{x})$ in the neighbourhood U can be obtained by the composition of the local phase portraits of systems $\dot{\mathbf{x}} = \mathbf{f}^+_{\mathbf{x}_+}(\mathbf{x})$, $\dot{\mathbf{x}} = \mathbf{f}^0_{\mathbf{x}_+}(\mathbf{x})$ and $\dot{\mathbf{x}} = \mathbf{f}^-_{\mathbf{x}_+}(\mathbf{x})$.

Changing $\mathbf{b}$ to $-\mathbf{b}$ in the expression of $\mathbf{f}^-_{\mathbf{x}_+}$, we obtain the vector field $\mathbf{f}^+_{\mathbf{x}_+}$. Moreover, the expression of the system $\dot{\mathbf{x}} = \mathbf{f}^+_{\mathbf{x}_+}(\mathbf{x})$ is equal to the one of (3.22) by taking $k_1 = 1$. Therefore, to determine the phase portraits of systems $\dot{\mathbf{x}} = \mathbf{f}^+_{\mathbf{x}_+}(\mathbf{x})$ and $\dot{\mathbf{x}} = \mathbf{f}^-_{\mathbf{x}_+}(\mathbf{x})$, we refer the reader to the proof of Theorem 3.11.6. Now let us study the phase portrait of the system $\dot{\mathbf{x}} = \mathbf{f}^0_{\mathbf{x}_+}(\mathbf{x})$ depending on the values of b_2 and t.

Suppose that $b_2 \neq 0$ and $t \neq 0$. The system $\dot{\mathbf{x}} = \mathbf{f}^0_{\mathbf{x}_+}(\mathbf{x})$ has a hyperbolic singular point at the origin, and the linear part of the system has the matrix

$$\begin{pmatrix} b_2 & 0 \\ 0 & -t/2 \end{pmatrix}.$$

Suppose that $b_2 = 0$ and $t \neq 0$. In this case the origin is a degenerated elementary singular point. When $b_1 = -1$, the straight line $x_2 = 0$ is the normally hyperbolic manifold, and it is stable for $t > 0$ and unstable for $t < 0$. When $b_1 \neq -1$, the change of time $\tau(s) = -ts/2$ transforms the system $\dot{\mathbf{x}} = \mathbf{f}^0_{\mathbf{x}_+}(\mathbf{x})$ into the system

$$x_1' = 2\frac{b_1+1}{t}x_1^2, \quad x_2' = x_2 + 2\frac{b_1+1}{t}x_1x_2.$$

Consider the functions $X(\mathbf{x}) = 2(b_1+1)x_1^2/t$ and $Y(\mathbf{x}) = 2(b_1+1)x_1x_2/t$, and let $f(x_1)$ be the solution of the equation $x_2 + Y(\mathbf{x}) = 0$ in a neighbourhood of the origin. It is easy to see that $X(x_1, f(x_1)) = 2(b_1+1)x_1^2/t$. By Theorem 2.7.3(c), the origin is a saddle-node with the central manifold contained in $x_2 = 0$ and the hyperbolic manifold contained in $x_1 = 0$. When $t > 0$ (respectively, $t < 0$) the hyperbolic manifold is stable (respectively, unstable). Moreover, when $b_1 > -1$ (respectively, $b_1 < -1$) the central manifold is stable (respectively, unstable) in the 0 direction.

Suppose that $b_2 \neq 0$ and $t = 0$. In this case the straight line $x_1 = 0$ is formed by singular points. The change of time $d\tau = x_1(s)ds$ transforms the system $\dot{\mathbf{x}} = \mathbf{f}^0_{\mathbf{x}_+}(\mathbf{x})$ into the system

$$x_1' = b_2 - (b_1+1)\,x_1, \quad x_2' = -(b_1+1)\,x_2.$$

For this system the origin is not a singular point. Thus the flow is parallel in a neighbourhood of the origin. Returning to the original time variable we conclude that the origin is a normally hyperbolic singular point, and the normally hyperbolic manifold is contained in $x_1 = 0$. Moreover, the normally hyperbolic manifold is stable when $b_2 < 0$, and unstable when $b_2 > 0$.

Suppose, finally, that $b_2 = 0$ and $t = 0$. Then we obtain the system

$$\dot{x}_1 = -(b_1+1)\,x_1^2, \quad \dot{x}_2 = -(b_1+1)\,x_1x_2.$$

When $b_1 = -1$, every point in a neighbourhood of the origin is a singular point. When $b_1 \neq -1$, if we change the time $d\tau = x_1(s)\,ds$, then we obtain that the origin is a stable node (respectively, unstable node) if $b_1 > -1$ (respectively, $b_1 < -1$). Returning to the original time variable we conclude that the origin is a semi-stable node with the singular manifold contained in $x_1 = 0$.

Now we study the local phase portrait of the system $\dot{\mathbf{x}} = \mathbf{f}_{\mathbf{x}_+}(\mathbf{x})$ depending on the fundamental parameters (D, T, d, t). From (3.10) and (3.12) it follows that $D = d + tb_2/2$ and $T = t + b_2$, and therefore the eigenvalues of the matrix $B = A + \mathbf{b}\mathbf{k}^T$ are $\Lambda_1 = t/2 + b_2$ and $\Lambda_2 = t/2$.

(a) When $D > 0$, the eigenvalues Λ_1 and Λ_2 have the same sign. Therefore, $T = \Lambda_1 + \Lambda_2 \neq 0$. Suppose that $T > 0$ (the case $T < 0$ follows by using similar arguments). Under this assumption it follows that $\Lambda_1 > 0$ and $\Lambda_2 > 0$. Thus, $t > 0$ and $t/2 + b_2 > 0$. We divide the proof into the following three cases: (a.1) $b_2 > 0$; (a.2) $b_2 = 0$; (a.3) $b_2 < 0$.

(a.1) System $\dot{\mathbf{x}} = \mathbf{f}^+_{\mathbf{x}_+}(\mathbf{x})$ has a saddle-node at the origin with the central manifold on the axis $x_2 = 0$ and the hyperbolic manifold on the straight line $x_1 + 2b_2x_2/t = 0$, see Figure 3.11. Moreover, the central manifold of the singular point is stable in the 0 direction and the hyperbolic manifold is stable.

System $\dot{\mathbf{x}} = \mathbf{f}^-_{\mathbf{x}_+}(\mathbf{x})$ has a saddle-node at the origin with the central manifold contained on $x_2 = 0$ and the hyperbolic manifold contained in $x_1 - 2b_2x_2/t = 0$. Moreover, the central manifold is stable in the 0 direction and the hyperbolic manifold is stable.

System $\dot{\mathbf{x}} = \mathbf{f}^0_{\mathbf{x}_+}(\mathbf{x})$ has a saddle at the origin with the stable and the unstable manifold contained in the lines $x_1 = 0$ and $x_2 = 0$, respectively.

Thus the neighbourhood of the origin is the union of a stable parabolic sector with a hyperbolic sector. The boundary between these sectors is contained in $x_1 = 0$, see Figure 3.11.

The remainder statements follow using similar arguments. □

Theorem 3.11.11. *Consider a fundamental system $\dot{\mathbf{x}} = A\mathbf{x} + \varphi(\mathbf{k}^T\mathbf{x})\mathbf{b}$ with parameters (D, T, d, t), where $t^2 - 4d > 0$ and the vector $\mathbf{k} = (k_1, k_2)^T$ satisfies that $k_1 = 0$. Suppose that the matrix A is in real Jordan normal form and let $\lambda_1 > \lambda_2$ be its eigenvalues. Under these assumptions there exist exactly four singular points at infinity, $\mathbf{x}_+$, $\mathbf{x}_-$, $\mathbf{y}_+$, and $\mathbf{y}_-$. Moreover, $\mathbf{x}_+$ and $\mathbf{x}_-$ belong to $\partial\mathbb{D}_0$.*

(a) *Suppose that $D > 0$. Then $T \neq 0$.*

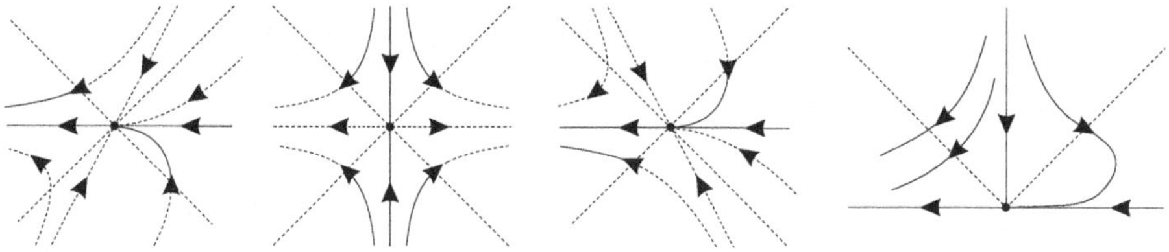

Figure 3.11: Local phase portrait of system $\dot{\mathbf{x}} = \mathbf{f}_{\mathbf{x}_+}(\mathbf{x})$ obtained by composing the phase portaits of systems $\dot{\mathbf{x}} = \mathbf{f}^+_{\mathbf{x}_+}(\mathbf{x})$, $\dot{\mathbf{x}} = \mathbf{f}^0_{\mathbf{x}_+}(\mathbf{x})$ and $\dot{\mathbf{x}} = \mathbf{f}^-_{\mathbf{x}_+}(\mathbf{x})$, when $t > 0$ and $b_2 > 0$.

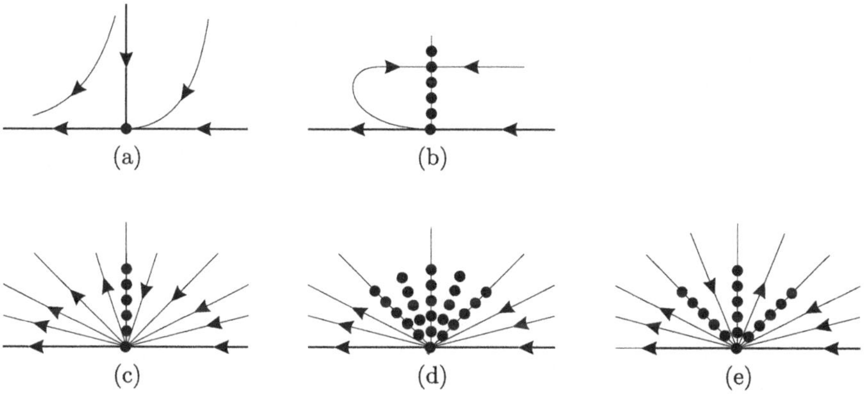

Figure 3.12: Singular points at infinity when $t^2 - 4d = 0$, matrix A is not diagonal and $k_1 = 0$.

(a.1) *If $T > 0$, then the singular points at infinity $\mathbf{x}_+$ and $\mathbf{x}_-$ are stable nodes.*

(a.2) *If $T < 0$, then the singular points at infinity $\mathbf{x}_+$ and $\mathbf{x}_-$ are saddles with the stable manifold contained in $\partial\mathbb{D}$.*

(b) *Suppose that $D = 0$ and $T > 0$.*

(b.1) *If $d = 0$ and $t < 0$, then the phase portraits in a neighbourhood of the singular points $\mathbf{x}_+$ and $\mathbf{x}_-$ are topologically equivalent to the one shown in Figure* 3.15(b) *when $b_1 \neq 0$; or in Figure* 3.15(e) *reversing the flow orientation when $b_1 = 0$.*

(b.2) *If $d = 0$ and $t > 0$ or $d \neq 0$, then $\mathbf{x}_+$ and $\mathbf{x}_-$ are stable nodes.*

(c) *Suppose that $D = 0$ and $T < 0$.*

(c.1) *If $d = 0$, then $\mathbf{x}_+$ and $\mathbf{x}_-$ are normally hyperbolic singular points with the normally hyperbolic manifold contained in the line $x_2 = 0$.*

(c.2) *If $d \neq 0$, then the phase portraits in a neighbourhood of* $\mathbf{x}_+$ *and* $\mathbf{x}_-$ *are topologically equivalent to the one shown in Figure* 3.15(a) *reversing the flow orientation.*

(d) *If $D = 0$ and $T = 0$, then the phase portraits in a neighbourhood of* $\mathbf{x}_+$ *and* $\mathbf{x}_-$ *are topologically equivalent to the one shown in Figure* 3.15(c), *when* $b_1 \neq 0$; *or in Figure* 3.15(d) *reversing the flow orientation when* $b_1 = 0$.

(e) *Suppose that $D < 0$.*

(e.1) *If $d \leq 0$ or $d > 0$ and $t > 0$, then* $\mathbf{x}_+$ *and* $\mathbf{x}_-$ *are stable nodes.*

(e.2) *If $d > 0$ and $t < 0$, then the phase portraits in a neighbourhood of* $\mathbf{x}_+$ *and* $\mathbf{x}_-$ *are topologically equivalent to the one shown in Figure* 3.15(a) *reversing the flow orientation. The boundary between the hyperbolic and the parabolic sectors is contained in the straight lines* $x_2 = \pm\, \mathrm{sign}(k_2)b_2/\lambda_2$.

(e.3) *The case $d > 0$ and $t = 0$ is not possible.*

Proof. According to the proof of Theorem 3.11.9, it is enough to study the system $\dot{\mathbf{x}} = \mathbf{f}_{\mathbf{x}_+}(\mathbf{x})$ in a neighbourhood U of the origin, where

$$\mathbf{f}_{\mathbf{x}_+}(\mathbf{x}) = \begin{cases} \mathbf{f}^+_{\mathbf{x}_+}(\mathbf{x}), & \text{if } x_1 \geq x_2, \\ \mathbf{f}^0_{\mathbf{x}_+}(\mathbf{x}), & \text{if } |x_1| < x_2, \\ \mathbf{f}^-_{\mathbf{x}_+}(\mathbf{x}), & \text{if } x_1 \leq -x_2, \end{cases}$$

and

$$\mathbf{f}^+_{\mathbf{x}_+}(\mathbf{x}) := \begin{pmatrix} (\lambda_2 - \lambda_1)x_1 + x_2 b_2 - x_1 x_2 b_1 \\ -\lambda_1 x_2 - x_2^2 b_1 \end{pmatrix},$$

$$\mathbf{f}^0_{\mathbf{x}_+}(\mathbf{x}) := \begin{pmatrix} (\lambda_2 - \lambda_1 + b_2)x_1 - b_1 x_1^2 \\ -\lambda_1 x_2 - x_1 x_2 b_1 \end{pmatrix},$$

$$\mathbf{f}^-_{\mathbf{x}_+}(\mathbf{x}) := \begin{pmatrix} (\lambda_2 - \lambda_1)x_1 - x_2 b_2 + x_1 x_2 b_1 \\ -\lambda_1 x_2 + x_2^2 b_1 \end{pmatrix}.$$

Hence, the phase portrait of the system $\dot{\mathbf{x}} = \mathbf{f}_{\mathbf{x}_+}(\mathbf{x})$ in U can be obtained by composing the local phase portraits of the systems $\dot{\mathbf{x}} = \mathbf{f}^-_{\mathbf{x}_+}(\mathbf{x})$, $\dot{\mathbf{x}} = \mathbf{f}^0_{\mathbf{x}_+}(\mathbf{x})$, and $\dot{\mathbf{x}} = \mathbf{f}^+_{\mathbf{x}_+}(\mathbf{x})$.

Changing $\mathbf{b}$ to $-\mathbf{b}$ in the expressions of $\dot{\mathbf{x}} = \mathbf{f}^+_{\mathbf{x}_+}$, we obtain the system $\dot{\mathbf{x}} = \mathbf{f}^-_{\mathbf{x}_+}$. Thus, it is enough to describe one of them. Moreover, system $\dot{\mathbf{x}} = \mathbf{f}^+_{\mathbf{x}_+}$ becomes system (3.23) by taking $k_1 = 1$. Therefore, to understand the phase portraits of both systems, $\dot{\mathbf{x}} = \mathbf{f}^-_{\mathbf{x}_+}(\mathbf{x})$ and $\dot{\mathbf{x}} = \mathbf{f}^+_{\mathbf{x}_+}(\mathbf{x})$, we refer the reader to the proof of Theorem 3.11.7.

Now let us study the phase portrait of the system $\dot{\mathbf{x}} = \mathbf{f}^0_{\mathbf{x}_+}(\mathbf{x})$. Since the matrix of its linear part is

$$\begin{pmatrix} \lambda_2 - \lambda_1 + b_2 & 0 \\ 0 & -\lambda_1 \end{pmatrix},$$

the origin is a hyperbolic singular point when $\lambda_1 \neq 0$ and $\lambda_2-\lambda_1+b_2 \neq 0$. Suppose that $\lambda_1 > 0$ (the behaviour of the singular point when $\lambda_1 < 0$ can be obtained from this case by multiplying the equation by -1 and reversing the direction of the flow). When $\lambda_2+b_2-\lambda_1 > 0$ the origin is a saddle point. Its unstable manifold is contained in the axis $x_2 = 0$ and the stable one in the axis $x_1 = 0$. When $\lambda_2+b_2-\lambda_1 < 0$ the origin is a stable node such that: for $\lambda_2+b_2 > 0$ the orbits are tangent to $x_2 = 0$ at the origin; for $\lambda_2 + b_2 < 0$ the orbits are tangent to $x_1 = 0$ at the origin; and for $\lambda_2 + b_2 = 0$ any direction is a characteristic direction.

We consider now the non-hyperbolic case. Suppose that $\lambda_1 = 0$. The change in the time variable $d\tau = x_1(s)\, ds$ transforms the system $\dot{\mathbf{x}} = \mathbf{f}^0_{\mathbf{x}_+}(\mathbf{x})$ into

$$x_1' = (\lambda_2 + b_2) - b_1 x_1, \quad x_2' = -x_2 b_1.$$

For this system the origin is not a singular point when $\lambda_2 + b_2 \neq 0$. Hence, in a neighbourhood of the origin the flow is transversal to $x_1 = 0$. Suppose that $\lambda_2 + b_2 = 0$. In this case the system has a stable node at the origin when $b_1 > 0$; an unstable node at the origin when $b_1 < 0$; or a neighbourhood of the origin is formed by singular points when $b_1 = 0$. Returning to the original time variable we obtain that: if $\lambda_2 + b_2 \neq 0$, the origin is a normally hyperbolic singular point with the normal manifold contained in $x_1 = 0$. It is stable or unstable depending on whether $\lambda_2 + b_2 < 0$ or $\lambda_2 + b_2 > 0$. Moreover, if $\lambda_2 + b_2 = 0$, then the origin is a non-isolated semi-stable node with the singular manifold contained in $x_1 = 0$.

Suppose now that $\lambda_1 \neq 0$ and $\lambda_2 + b_2 - \lambda_1 = 0$. In this case the system has a degenerate elementary singular point at the origin. With the change in the time variable $\tau(s) = -\lambda_1 s$, we obtain

$$x_1' = \frac{b_1}{\lambda_1} x_1^2, \quad x_2' = x_2 + \frac{b_1}{\lambda_1} x_1 x_2.$$

We now distinguish the cases $b_1 = 0$ and $b_1 \neq 0$. In the first one the origin is an unstable normally hyperbolic singular point with the singular manifold contained in the line $x_2 = 0$. In the second case by Theorem 2.7.3(c) the system has a saddle-node at the origin with the hyperbolic manifold contained in the line $x_1 = 0$ and with the central manifold contained in the line $x_2 = 0$. Moreover, the hyperbolic manifold is unstable and when $b_1\lambda_1 > 0$ (respectively $b_1\lambda_1 < 0$) the central manifold is stable in the π (respectively 0) direction.

Returning to the original variables we obtain the following behaviour surrounding the origin. When $b_1 = 0$, the origin is a normally hyperbolic singular point with the singular manifold on the line $x_2 = 0$. The singular manifold is stable or unstable depending on whether $\lambda_1 > 0$ or $\lambda_1 < 0$. When $b_1 \neq 0$, the origin is a saddle-node with the hyperbolic manifold on the line $x_1 = 0$ and the central manifold on the line $x_2 = 0$. The hyperbolic manifold is stable or unstable depending on $\lambda_1 > 0$ or $\lambda_1 < 0$, respectively. The central manifold is stable in the 0 or in the π direction depending on whether $b_1 > 0$ or $b_1 < 0$.

Now let us study the local phase portrait of system $\dot{\mathbf{x}} = \mathbf{f}_{\mathbf{x}_+}(\mathbf{x})$ depending on the fundamental parameters (D, T, d, t). Since $D = \lambda_1(\lambda_2 + b_2)$ and $T = \lambda_1 +$

$(\lambda_2 + b_2)$, see (3.10) and (3.12), the eigenvalues of the matrix $B = A + \mathbf{b}\mathbf{k}^T$ are $\Lambda_1 = \lambda_1$ and $\Lambda_2 = \lambda_2 + b_2$.

(a) Suppose that $D > 0$ and $T > 0$. We divide the proof of this statement into the following three parts: (a.1) $\lambda_2 + b_2 - \lambda_1 > 0$; (a.2) $\lambda_2 + b_2 - \lambda_1 = 0$; (a.3) $\lambda_2 + b_2 - \lambda_1 < 0$.

(a.1) System $\dot{\mathbf{x}} = \mathbf{f}^0_{\mathbf{x}_+}(\mathbf{x})$ has a saddle at the origin with the unstable manifold contained in the line $x_2 = 0$ and the stable manifold contained in the line $x_1 = 0$.

System $\dot{\mathbf{x}} = \mathbf{f}^+_{\mathbf{x}_+}(\mathbf{x})$ (respectively, $\dot{\mathbf{x}} = \mathbf{f}^-_{\mathbf{x}_+}(\mathbf{x})$) has a node at the origin with the characteristic directions contained in the lines $x_2 = 0$ and $\lambda_2 x_1 + b_2 x_2 = 0$ (respectively, $x_2 = 0$ and $\lambda_2 x_1 - b_2 x_2 = 0$).

In order to describe the phase portrait in a neighbourhood of the origin, we have to ensure that the characteristic directions do not coincide with the straight lines separating the domains of the different systems. Hence, since $\lambda_2 + b_2 - \lambda_1 > 0$, $\lambda_2 - \lambda_1 < 0$ and $\lambda_2 + b_2 > 0$, we have $b_2 > 0$ and $-\lambda_2/b_2 < 1$. Therefore when $\lambda_2 > 0$ the half-line $\lambda_2 x_1 + b_2 x_2 = 0$ with $x_2 \geq 0$ intersects the region $x_1 \geq x_2$ only at the origin; when $\lambda_2 < 0$ the half-line $\lambda_2 x_1 + b_2 x_2 = 0$ with $x_2 \geq 0$ is contained in the region $x_1 \geq x_2$. We conclude that the system $\dot{\mathbf{x}} = \mathbf{f}_{\mathbf{x}_+}(\mathbf{x})$ has a stable node at the origin, see Figure 3.13.

The remaining statements follow using similar arguments. □

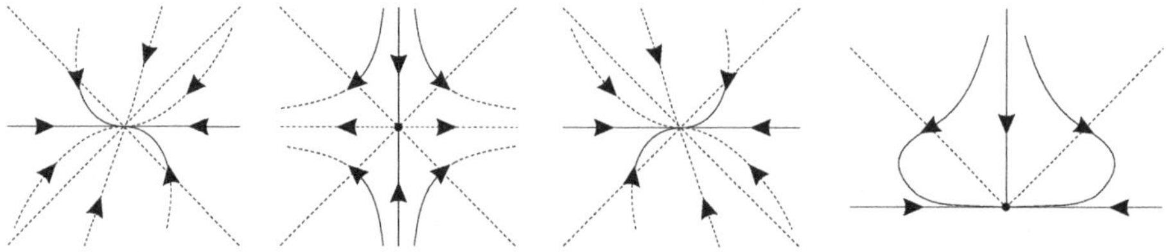

Figure 3.13: Local phase portrait of system $\dot{\mathbf{x}} = \mathbf{f}_{\mathbf{x}_+}(\mathbf{x})$ obtained by composing the phase portaits of systems $\dot{\mathbf{x}} = \mathbf{f}^+_{\mathbf{x}_+}(\mathbf{x})$, $\dot{\mathbf{x}} = \mathbf{f}^0_{\mathbf{x}_+}(\mathbf{x})$ and $\dot{\mathbf{x}} = \mathbf{f}^-_{\mathbf{x}_+}(\mathbf{x})$ when $\lambda_1 > 0$ and $\lambda_2 - \lambda_1 + b_2 > 0$.

Theorem 3.11.12. *Consider a fundamental system $\dot{\mathbf{x}} = A\mathbf{x} + \varphi(\mathbf{k}^T\mathbf{x})\mathbf{b}$ with fundamental parameters (D, T, d, t), where $t^2 - 4d > 0$ and the vector $\mathbf{k} = (k_1, k_2)^T$ satisfies $k_2 = 0$. Suppose that the matrix A is in real Jordan normal form and let $\lambda_1 > \lambda_2$ be its eigenvalues. Under these assumptions there exist exactly four singular points at infinity, $\mathbf{x}_+$, $\mathbf{x}_-$, $\mathbf{y}_+$, and $\mathbf{y}_-$. Moreover, $\mathbf{y}_+$ and $\mathbf{y}_-$ belong to $\partial\mathbb{D}_0$.*

(a) *If $D > 0$ and $T > 0$, then $\mathbf{y}_+$ and $\mathbf{y}_-$ are saddle points with the unstable manifold contained in $\partial\mathbb{D}$ and with the stable manifold contained in the line $x_1 = 0$.*

(b) *If $D > 0$ and $T < 0$, then $\mathbf{y}_+$ and $\mathbf{y}_-$ are unstable nodes.*

(c) *Suppose that* $D = 0$ *and* $T > 0$.

(c.1) *If* $d = 0$, *then* $\mathbf{y}_+$ *and* $\mathbf{y}_-$ *are unstable normal hyperbolic singular points with the normally hyperbolic manifold contained in the line* $x_1 = 0$.

(c.2) *If* $d \neq 0$, *then the phase portraits in a neighbourhood of* $\mathbf{y}_+$ *and* $\mathbf{y}_-$ *are topologically equivalent to the one shown in Figure* 3.15(a).

(d) *Suppose that* $D = 0$ *and* $T < 0$.

(d.1) *If* $d = 0$ *and* $t > 0$, *then the phase portraits in a neighbourhood of* $\mathbf{y}_+$ *and* $\mathbf{y}_-$ *are topologically equivalent to the one shown in Figure* 3.15(b) *when* $b_2 \neq 0$, *or in Figure* 3.15(e) *when* $b_2 = 0$.

(d.2) *If* $d = 0$ *and* $t < 0$ *or* $d \neq 0$, *then* $\mathbf{y}_+$ *and* $\mathbf{y}_-$ *are unstable nodes.*

(e) *If* $D = 0$ *and* $T = 0$, *then the phase portraits in a neighbourhood of* $\mathbf{y}_+$ *and* $\mathbf{y}_-$ *are topologically equivalent to the one shown in Figure* 3.15(c) *when* $b_2 \neq 0$, *or in Figure* 3.15(d) *when* $b_2 = 0$.

(f) *Suppose that* $D < 0$.

(f.1) *If* $d \leq 0$ *or* $d > 0$ *and* $t < 0$, *then* $\mathbf{y}_+$ *and* $\mathbf{y}_-$ *are unstable nodes.*

(f.2) *If* $d > 0$ *and* $t > 0$, *then the phase portraits in a neighbourhood of* $\mathbf{y}_+$ *and* $\mathbf{y}_-$ *are topologically equivalent to the one shown in Figure* 3.15(a). *The stable separatrices of* $\mathbf{y}_+$ *and* $\mathbf{y}_-$ *are contained in the straight lines* $x_2 = \pm\,\mathrm{sign}(k_1) b_1/\lambda_1$.

(f.3) *The case* $d > 0$ *and* $t = 0$ *is not possible.*

Proof. According to the proof of Theorem 3.11.9, we only have to study the system $\dot{\mathbf{x}} = \mathbf{f}_{\mathbf{y}_+}(\mathbf{x})$ in a neighbourhood U of the origin with $x_2 \geq 0$, where

$$\mathbf{f}_{\mathbf{y}_+}(\mathbf{x}) = \begin{cases} \mathbf{f}^+_{\mathbf{y}_+}(\mathbf{x}), & \text{if } x_1 \geq x_2, \\ \mathbf{f}^0_{\mathbf{y}_+}(\mathbf{x}), & \text{if } |x_1| < x_2, \\ \mathbf{f}^-_{\mathbf{y}_+}(\mathbf{x}), & \text{if } x_1 \leq -x_2, \end{cases}$$

and

$$\mathbf{f}^+_{\mathbf{y}_+}(\mathbf{x}) := \begin{pmatrix} (\lambda_1 - \lambda_2)x_1 + x_2 b_1 + x_1 x_2 b_2 \\ -\lambda_2 x_2 + x_2^2 b_2 \end{pmatrix},$$

$$\mathbf{f}^0_{\mathbf{y}_+}(\mathbf{x}) := \begin{pmatrix} (\lambda_1 + b_1 - \lambda_2)x_1 + b_2 x_1^2 \\ -\lambda_2 x_2 + x_1 x_2 b_2 \end{pmatrix},$$

$$\mathbf{f}^-_{\mathbf{y}_+}(\mathbf{x}) := \begin{pmatrix} (\lambda_1 - \lambda_2)x_1 - x_2 b_1 - x_1 x_2 b_2 \\ -\lambda_2 x_2 - x_2^2 b_2 \end{pmatrix}.$$

Changing $\mathbf{b}$ to $-\mathbf{b}$ in the expressions of $\dot{\mathbf{x}} = \mathbf{f}^+_{\mathbf{y}_+}$, we obtain the system $\dot{\mathbf{x}} = \mathbf{f}^-_{\mathbf{y}_+}$. Thus it is enough to describe one of them. Moreover, if we take $k_2 = 1$, system

$\dot{\mathbf{x}} = \mathbf{f}^+_{\mathbf{y}_+}$ coincides with system (3.25). Therefore, to draw the phase portraits of both systems $\dot{\mathbf{x}} = \mathbf{f}^-_{\mathbf{y}_+}(\mathbf{x})$ and $\dot{\mathbf{x}} = \mathbf{f}^+_{\mathbf{y}_+}(\mathbf{x})$, one can refer to the proof of Theorem 3.11.8.

Now we study the phase portrait of system $\dot{\mathbf{x}} = \mathbf{f}^0_{\mathbf{y}_+}(\mathbf{x})$. Since the matrix of its linear part is

$$\begin{pmatrix} \lambda_1 + b_1 - \lambda_2 & 0 \\ 0 & -\lambda_2 \end{pmatrix},$$

the origin is a hyperbolic singular point when $\lambda_2 \neq 0$ and $\lambda_1 + b_1 - \lambda_2 \neq 0$. Hence when $\lambda_2 > 0$ and $\lambda_1 + b_1 - \lambda_2 > 0$, the origin is a saddle point with the stable manifold tangent to $x_2 = 0$ and the unstable manifold tangent to $x_1 = 0$. When $\lambda_2 > 0$ and $\lambda_1 + b_1 - \lambda_2 < 0$, the origin is a stable node, and the orbits reach the origin tangentially to $x_2 = 0$, or to $x_1 = 0$, depending on whether $\lambda_1 + b_1 > 0$ or $\lambda_1 + b_1 < 0$. When $\lambda_1 + b_1 = 0$ any direction is a characteristic direction. The case $\lambda_2 < 0$ can be obtained from the case $\lambda_2 > 0$ by multiplying the parameters by -1 and by reversing the orientation of the flow.

Suppose that $\lambda_2 = 0$. The change of variables $d\tau = x_1(s)ds$ transforms the system $\dot{\mathbf{x}} = \mathbf{f}^0_{\mathbf{y}_+}(\mathbf{x})$ into the system

$$x_1' = \lambda_1 + b_1 + b_2 x_1, \quad x_2' = x_2 b_2.$$

For this system when $\lambda_1 + b_1 \neq 0$ the origin is not a singular point and the flow is transversal to $x_1 = 0$. On the contrary, when $\lambda_1 + b_1 = 0$ the origin can be either a stable diagonal node, an unstable diagonal node, or a non-isolated singular point, depending on whether $b_2 < 0, b_2 > 0$ or $b_2 = 0$. Returning to the original variables we obtain that the origin is a non-isolated semi-stable node with the singular manifold contained in the line $x_1 = 0$.

Suppose now that $\lambda_2 \neq 0$ and $\lambda_1 + b_1 - \lambda_2 = 0$. In this case the system $\dot{\mathbf{x}} = \mathbf{f}^0_{\mathbf{y}_+}(\mathbf{x})$ has a degenerate elementary singular point at the origin. Changing the time variable to $\tau(s) = -\lambda_2 s$, we obtain

$$x_1' = -\frac{b_2}{\lambda_2} x_1^2, \quad x_2' = x_2 - \frac{b_2}{\lambda_2} x_1 x_2.$$

When $b_2 = 0$, the origin is an unstable normally hyperbolic singular point with the singular manifold contained in the line $x_2 = 0$. According to Theorem 2.7.3(c), when $b_2 \neq 0$ the system has a saddle-node at the origin. The hyperbolic manifold is contained in the line $x_1 = 0$ and the central manifold is contained in the line $x_2 = 0$. Moreover, the hyperbolic manifold is unstable and the central manifold is stable in the 0 direction if and only if $b_2 \lambda_2 < 0$.

Returning to the original variables we conclude that when $b_2 = 0$ the origin is a normally hyperbolic singular point with the singular manifold contained in $x_2 = 0$, and this manifold is stable or unstable depending on whether $\lambda_2 > 0$ or $\lambda_2 < 0$. On the other hand, when $b_2 \neq 0$ the origin is a saddle-node with the hyperbolic manifold contained in the line $x_1 = 0$ and the central manifold contained in the line $x_2 = 0$. The hyperbolic manifold is stable or unstable depending on whether $\lambda_2 > 0$ or $\lambda_2 < 0$, and the central manifold is stable in the 0 direction if and only if $b_2\lambda_2 < 0$.

Now we study the local phase portrait of $\dot{\mathbf{x}} = \mathbf{f}_{\mathbf{y}_+}(\mathbf{x})$ depending on the fundamental parameters. Since $D = \lambda_2(\lambda_1 + b_1)$ and $T = \lambda_1 + \lambda_2 + b_1$, the eigenvalues of matrix $B = A + \mathbf{b}\mathbf{k}^T$ are $\Lambda_1 = \lambda_1 + b_1$ and $\Lambda_2 = \lambda_2$.

(a) Suppose that $D > 0$ and $T > 0$. We will consider the following three cases: (a.1) $\lambda_1 + b_1 - \lambda_2 > 0$; (a.2) $\lambda_1 + b_1 - \lambda_2 = 0$; (a.3) $\lambda_1 + b_1 - \lambda_2 < 0$.

(a.1) System $\dot{\mathbf{x}} = \mathbf{f}^0_{\mathbf{y}_+}(\mathbf{x})$ has a saddle point at the origin with the unstable manifold contained in the line $x_2 = 0$ and the stable one contained in the line $x_1 = 0$.

System $\dot{\mathbf{x}} = \mathbf{f}^+_{\mathbf{y}_+}(\mathbf{x})$ (respectively, $\dot{\mathbf{x}} = \mathbf{f}^-_{\mathbf{y}_+}(\mathbf{x})$) has a saddle point at the origin. The unstable manifold is contained in the line $x_2 = 0$ and the stable one is contained in the line $\lambda_1 x_1 - b_1 x_2 = 0$ (respectively, $\lambda_1 x_1 + b_1 x_2 = 0$).

Since $\lambda_1 > \lambda_2$ it follows that $\lambda_1 > 0$. Therefore, when $b_1 \geq 0$ the straight line $\lambda_1 x_1 - b_1 x_2 = 0$ is contained in the region $|x_2| > |x_1|$ and the system $\dot{\mathbf{x}} = \mathbf{f}_{\mathbf{y}_+}(\mathbf{x})$ has a saddle point at the origin, see Figure 3.14. Moreover, when $b_1 < 0$, then $\lambda_1/b_1 < -1$ and the straight line $\lambda_1 x_1 - b_1 x_2 = 0$ is also contained in $|x_2| > |x_1|$. Thus, the system $\dot{\mathbf{x}} = \mathbf{f}_{\mathbf{y}_+}(\mathbf{x})$ has a saddle point at the origin.

The remaining statements follow in a similar way. □

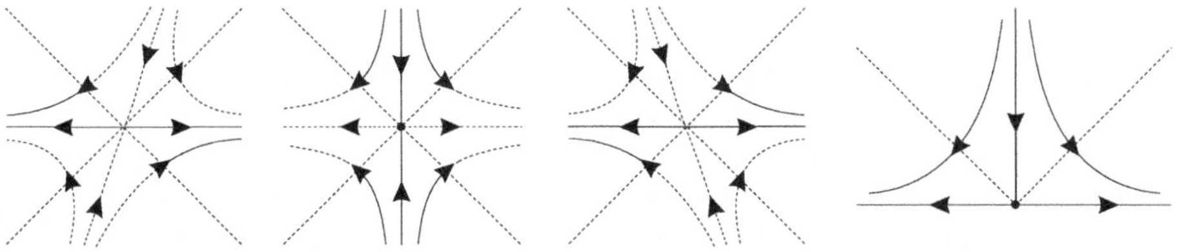

Figure 3.14: Local phase portrait of system $\dot{\mathbf{x}} = \mathbf{f}_{\mathbf{y}_+}(\mathbf{x})$ obtained by composing the phase portaits of systems $\dot{\mathbf{x}} = \mathbf{f}^+_{\mathbf{y}_+}(\mathbf{x})$, $\dot{\mathbf{x}} = \mathbf{f}^0_{\mathbf{y}_+}(\mathbf{x})$ and $\dot{\mathbf{x}} = \mathbf{f}^-_{\mathbf{y}_+}(\mathbf{x})$ when $\lambda_2 > 0$ and $\lambda_1 - \lambda_2 + b_1 > 0$.

3.12 Periodic orbits

This section is devoted to the existence and location in the phase plane of Jordan curves Γ formed by solutions. Such curves split the phase plane into two regions, one of which is denoted by Σ_Γ and is bounded. Since Σ_Γ is an invariant set, the qualitative behaviour of the flow in Σ_Γ can be obtained from the Poincaré–Bendixson Theorem.

In Lemma 3.12.1 we prove that for a fundamental system with $D \neq 0$ only three kinds of finite Jordan curves formed by solutions can exist: periodic orbits,

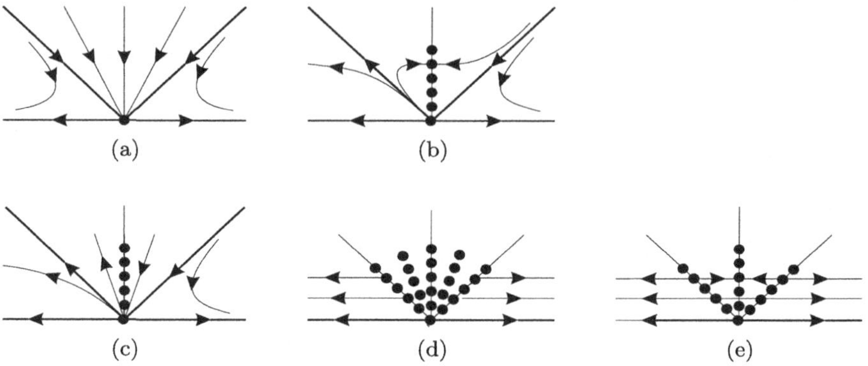

Figure 3.15: Singular points at infinity when $t^2 - 4d > 0$ and $k_2 = 0$.

homoclinic cycles and heteroclinic cycles.

Lemma 3.12.1. *Consider a fundamental system with parameters* (D, T, d, t), *where* $D \neq 0$. *If* Γ *is a Jordan curve formed by solutions, then* Γ *is a periodic orbit, a homoclinic cycle with vertex at the origin, or a heteroclinic cycle with vertices at the singular points* $\mathbf{e}_+$ *and* $\mathbf{e}_-$.

Proof. If Γ does not contain any singular point, then Γ is a periodic orbit. Suppose now that Γ contains a unique singular point $\mathbf{p}$ and a unique orbit γ. Since Γ is an invariant compact set, there exit the α- and ω-limit sets of γ and $\alpha(\gamma) = \omega(\gamma) = \mathbf{p}$. Thus Γ is a homoclinic cycle. Suppose that the vertex of the homoclinic cycle is $\mathbf{e}_+$ (the arguments are similar if we suppose that $\mathbf{e}_-$ is the vertex of Γ). Since $D \neq 0$, then $\mathbf{e}_+$ is a saddle point, see Theorem 3.9.3. It is easy to check that $\mathrm{Cl}(\Sigma_\Gamma)$ is an invariant compact and simply connected set. By applying the Poincaré–Bendixson Theorem, every orbit in Σ_Γ has an α- and an ω-limit set in $\mathrm{Cl}(\Sigma_\Gamma)$, and such limit sets are either a periodic orbit, a singular point or a separatrix cycle. In the last two cases we have singular points in Σ_Γ, which contradicts the symmetry of the vector field with respect to the origin. In the first case again we reach the same contradiction, by Cororally 2.8.4. Hence the vertex of Γ is the origin.

Suppose now that Γ contains two singular points, see Figure 3.16(a) and (b). First we will prove that the Jordan curve shown in Figure 3.16(a) is not possible; i.e. the flow on Γ must be oriented. After that we will prove that the vertices of Γ are $\mathbf{e}_+$ and $\mathbf{e}_-$.

Since $D \neq 0$ and the system has more than one singular point, Theorem 3.9.3 implies that $Dd < 0$ and the singular points are $\mathbf{0}$, $\mathbf{e}_+$ or $\mathbf{e}_-$. Moreover, any singular point is hyperbolic and the local phase portraits of $\mathbf{e}_+$ and $\mathbf{e}_-$ are identical. Finally, since $Dd < 0$, either the origin is a saddle and $\mathbf{e}_+$ and $\mathbf{e}_-$ are antisaddles, or $\mathbf{e}_+$ and $\mathbf{e}_-$ are saddle points and $\mathbf{0}$ is an antisaddle point.

Suppose that Γ is equal to the Jordan curve shown in Figure 3.16(a). If $\mathbf{e}_+$

and $\mathbf{e}_-$ are the singular points contained in Γ, by the symmetry of the vector field with respect to the origin, they are saddle points and the other separatrices of $\mathbf{e}_+$ and $\mathbf{e}_-$ split Σ_Γ into two invariant regions, A and B, see Figure 3.16(a.1). In the interior of B there are no singular points. We arrive at a contradiction by applying the Poincaré–Bendixson Theorem to this region.

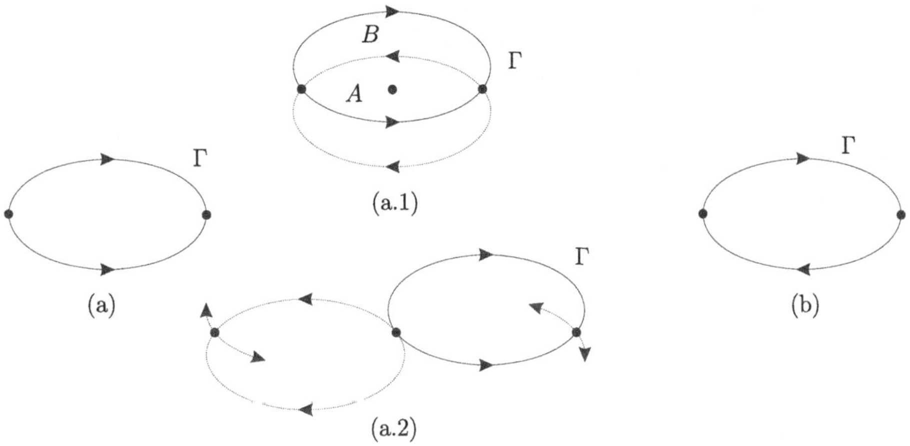

Figure 3.16: Qualitative Jordan curves formed by solutions with two singular points.

Suppose that $\mathbf{0}$ and $\mathbf{e}_+$ are the singular points belonging to Γ. There exists a Jordan curve formed by solutions Γ_-, symmetric to Γ and such that $\mathbf{0}, \mathbf{e}_- \in \Gamma_-$. It is easy to conclude that $\mathbf{e}_+$ and $\mathbf{e}_-$ are saddle points and $\mathbf{0}$ is a node, see Figure 3.16(a.2). The unstable separatrices of $\mathbf{e}_+$ and $\mathbf{e}_-$ split Σ_Γ and Σ_{Γ_-} into four invariant regions, two of them without singular points inside. We arrive to contradiction by applying the Poincaré–Bendixson Theorem to these regions. Hence, if Γ is a Jordan curve formed by two singular points and two orbits, then Γ is the curve shown in Figure 3.16(b).

In that case, since either $\mathbf{e}_+$ and $\mathbf{e}_-$ are saddles and $\mathbf{0}$ is an antisaddle, or $\mathbf{0}$ is a saddle and $\mathbf{e}_+$ and $\mathbf{e}_-$ are antisaddles, it is easy to conclude that the vertices of Γ have to be the saddle points $\mathbf{e}_+$ and $\mathbf{e}_-$. Moreover, the singular point $\mathbf{0}$ is contained in Σ_Γ. □

In the following theorem we collect some results about the existence and location of Jordan curves formed by solutions in the phase plane.

Theorem 3.12.2. *Consider a fundamental system with parameters* (D, T, d, t), *where* $D \neq 0$.

(a) *If* $T\,t > 0$, *then there are no Jordan curves formed by solutions.*

(b) *Suppose that* Γ *is a Jordan curve formed by solutions.*

(b.1) *If* $Tt < 0$, *then* $\Gamma \cap S_0 \neq \varnothing$ *and* $\Gamma \cap (S_+ \cup S_-) \neq \varnothing$.

(b.2) *If* $T \neq 0$ *and* $t = 0$, *then* $\Gamma \subset S_+ \cup L_+$ *or* $\Gamma \subset S_- \cup L_-$.

(b.3) *If* $T = 0$ *and* $t \neq 0$ *then* $\Gamma \subset L_+ \cup S_0 \cup L_-$.

Proof. Let $\mathbf{X} = (P, Q)$ be the vector field defined by the fundamental system and let Γ be a Jordan curve formed by solutions. Since the functions $\partial P/\partial x$ and $\partial Q/\partial y$ are well defined and bounded in $\Sigma_\Gamma \setminus \{L_+, L_-\}$, the divergence $\operatorname{div}(\mathbf{X}) = \partial P/\partial x + \partial Q/\partial y$ is also well defined and bounded in $\Sigma_\Gamma \setminus \{L_+, L_-\}$. Moreover,

$$\iint_{\Sigma_\Gamma} \operatorname{div}(\mathbf{X})\, dxdy = \iint_{\Sigma_\Gamma \cap S_+} \operatorname{div}(\mathbf{X})\, dxdy + \iint_{\Sigma_\Gamma \cap S_0} \operatorname{div}(\mathbf{X})\, dxdy + \iint_{\Sigma_\Gamma \cap S_-} \operatorname{div}(\mathbf{X})\, dxdy.$$

Taking into account that $\operatorname{div}(\mathbf{X}) = t$ in $S_+ \cup S_-$ and $\operatorname{div}(\mathbf{X}) = T$ in S_0, we obtain that

$$\iint_{\Sigma_\Gamma} \operatorname{div}(\mathbf{X})\, dxdy = (\mathcal{A}_+ + \mathcal{A}_-)\, t + \mathcal{A}_0 T,$$

where $\mathcal{A}_+, \mathcal{A}_-$ and $\mathcal{A}_0$ are the areas of the open regions $\Sigma_\Gamma \cap S_+$, $\Sigma_\Gamma \cap S_-$, and $\Sigma_\Gamma \cap S_0$, respectively. The rest of the proof is divided according to the type of the curve Γ: a periodic orbit, a homoclinic cycle or a heteroclinic cycle, see Lemma 3.12.1.

Suppose that Γ is a periodic orbit. By applying Green's Theorem for domains bounded by rectifiable curves [6, p. 280],

$$\iint_{\Sigma_\Gamma} \operatorname{div}(\mathbf{X})\, dxdy = \oint_\Gamma P dy - Q dx = 0.$$

Therefore, $(\mathcal{A}_+ + \mathcal{A}_-)t + \mathcal{A}_0 T = 0$ and the theorem follows easily by noting that $\mathcal{A}_+, \mathcal{A}_-$ and $\mathcal{A}_0$ are non-negative.

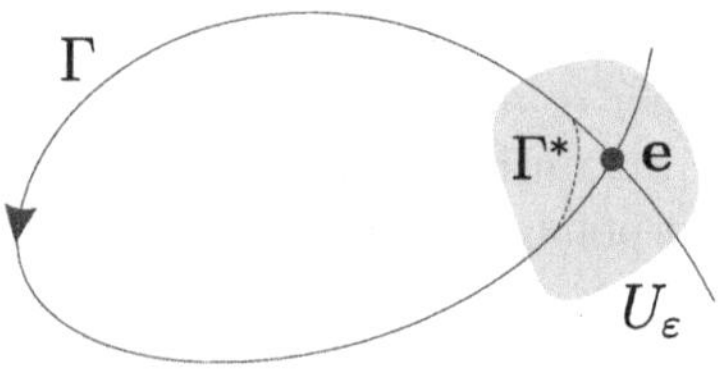

Figure 3.17: Green's Theorem for a homoclinic cycle.

Suppose now that Γ is a homoclinic cycle to a singular point $\mathbf{e}$. Since $\mathbf{X}(\mathbf{e}) = \mathbf{0}$ for every small enough $\varepsilon > 0$, there exists a neighbourhood U_ε of $\mathbf{e}$ such that $\|\mathbf{X}(\mathbf{x})\| < \varepsilon$ when $\mathbf{x} \in U_\varepsilon$. One may smooth Γ near $\mathbf{e}$ to produce a differentiable curve Γ^* such that Γ and Γ^* coincide in $\mathbb{R}^2 \setminus U_\varepsilon$ and the length of the arc where Γ and Γ^* differ is less than 1, see Figure 3.17. Applying Green's Theorem to the region bounded by Γ^* we obtain that

$$-\varepsilon < \iint_{\Sigma_{\Gamma^*}} \operatorname{div}(\mathbf{X})\, dxdy = \oint_{\Gamma^*} Pdy - Qdx < \varepsilon.$$

Therefore, $-\varepsilon < (\mathcal{A}_+^* + \mathcal{A}_-^*)t + \mathcal{A}_0^* T < \varepsilon$, where $\mathcal{A}_+^*, \mathcal{A}_-^*$ and $\mathcal{A}_0^*$ denote the areas of regions $\Sigma_{\Gamma^*} \cap S_+$, $\Sigma_{\Gamma^*} \cap S_-$, and $\Sigma_{\Gamma^*} \cap S_+$, respectively. Since $\mathcal{A}_+^*, \mathcal{A}_-^*$ and $\mathcal{A}_0^*$ tend to $\mathcal{A}_+, \mathcal{A}_-$ and $\mathcal{A}_0$, respectively, as ε tends to 0, we conclude that $(\mathcal{A}_+ + \mathcal{A}_-)t + \mathcal{A}_0 T = 0$. Similar arguments can be applied when Γ is a heteroclinic cycle. □

The ideas used in the proof of Theorem 3.12.2 are due to Lefschetz [40, pp. 238–239]. Note that these arguments can be also applied when Γ is a Jordan curve formed by solutions with any number of singular points. In this case the set of singular points contained in Γ is compact. Hence, it is always possible to find a neighbourhood U_ε such that $\|\mathbf{X}(\mathbf{x})\| < \varepsilon$ if $\mathbf{x} \in U_\varepsilon$.

3.13 Asymptotic behaviour

In this section, by using the integral expression of the solutions, we offer some preliminary results about the asymptotic behaviour of the orbits of a fundamental system.

Proposition 3.13.1. *Given a matrix $A \in L(\mathbb{R}^n)$, the following statements are equivalent.*

(a) *The eigenvalues of A have negative real part.*

(b) *There exist positive constants a, c, m, M and a non-negative constant k, such that for any $\mathbf{x}_0 \in \mathbb{R}^n$ and $s \in \mathbb{R}$ the following inequalities hold:*

$$m \left|s^k\right| e^{-as} \|\mathbf{x}_0\| \leq \|e^{As}\mathbf{x}_0\| \leq Me^{-cs}\|\mathbf{x}_0\|.$$

For a proof of this proposition see [53, p. 56].

Proposition 3.13.2. *Given a fundamental system with fundamental parameters (D, T, d, t), where $d > 0$ and $t < 0$, there exists $R > 0$ such that the ω-limit set of any orbit is contained in the ball of radius R centered at the origin.*

Proof. Take $\mathbf{x}_0 \in \mathbb{R}^2$ and let $\mathbf{x}(s)$ be the solution of the fundamental system $\dot{\mathbf{x}} = A\mathbf{x} + \varphi(\mathbf{k}^T\mathbf{x})\mathbf{b}$ such that $\mathbf{x}(0) = \mathbf{x}_0$. From expression (3.5) it follows that

$$\|\mathbf{x}(s)\| \leq \|e^{As}\mathbf{x}(0)\| + \|e^{As}\| \int_0^s \|e^{-Ar}\| \|\varphi(\mathbf{k}^T\mathbf{x})\mathbf{b}\| dr.$$

Since $d > 0$ and $t < 0$, the eigenvalues of A have negative real part. Thus, there exist positive constants c and M such that $\|e^{As}\mathbf{x}_0\| \leq Me^{-cs}\|\mathbf{x}_0\|$ for any $\mathbf{x}_0 \in \mathbb{R}^2$ and $s \in \mathbb{R}$, see Proposition 3.13.1. On the other hand, $\|\varphi(\mathbf{k}^T\mathbf{x})\mathbf{b}\| \leq \|\mathbf{b}\|$, and so

$$\begin{aligned}\|\mathbf{x}(s)\| &\leq Me^{-cs}\|\mathbf{x}(0)\| + Me^{-cs}\|\mathbf{b}\|\int_0^s \|e^{-Ar}\|dr \\ &\leq Me^{-cs}\left(\|\mathbf{x}(0)\| + \|\mathbf{b}\|\int_0^s Me^{cr}dr\right) \\ &= Me^{-cs}\left(\|\mathbf{x}(0)\| - \frac{M\|\mathbf{b}\|}{c}\right) + \frac{M^2\|\mathbf{b}\|}{c}.\end{aligned}$$

This implies that the solution $\mathbf{x}(s)$ is bounded when s tends to infinity. Therefore, the ω-limit set of this solution is contained in the ball of radius $R = M^2\|\mathbf{b}\|$. □

Proposition 3.13.3. *Given a fundamental system with fundamental parameters (D, T, d, t), where $d > 0$ and $t > 0$, there exists $R > 0$ such that the α-limit set of any orbit is contained in the ball of radius R centered at the origin.*

Proof. The change of the time variable t to $-t$ transforms the original system into a fundamental system with fundamental parameters $(D, -T, d, -t)$, see Proposition 3.7.1. The statement follows by applying Proposition 3.13.2 to this system. □

Proposition 3.13.4. *Given a fundamental system with fundamental parameters (D, T, d, t), where $d = 0$ and $t < 0$, there exist straight lines ω_1 and ω_2 which are symmetric with respect to the origin and such that $\omega_i \cap \Gamma_+ \neq \emptyset$ and $\omega_i \cap \Gamma_- \neq \emptyset$ for $i \in \{1, 2\}$. Let B_ω be the closed strip bounded by ω_1 and ω_2.*

(a) *For every solution $\mathbf{x}(s)$, there exists $s_0 > 0$ such that $\{\mathbf{x}(s) : s > s_0\} \subset B_\omega$.*

(b) *B_ω is a positively invariant set and contains every ω-limit set.*

(c) *If $D > 0$, then all orbits are positively bounded.*

Proof. Take $\mathbf{x}_0 = (x_{10}, x_{20}) \in \mathbb{R}^2$ and let $\mathbf{x}(s) = (x_1(s), x_2(s))$ be the solution of the fundamental system $\dot{\mathbf{x}} = A\mathbf{x} + \varphi(\mathbf{k}^T\mathbf{x})\mathbf{b}$ such that $\mathbf{x}(0) = \mathbf{x}_0$. From expression (3.5) it follows that

$$\mathbf{x}(s) = e^{As}\mathbf{x}(0) + \int_0^s e^{A(s-r)}\varphi\left(\mathbf{k}^T\mathbf{x}\right)\mathbf{b}\,dr.$$

Since linear maps transform straight lines into straight lines, it is not a restriction to assume that the matrix A is in real Jordan normal form. Thus

$$A = \begin{pmatrix} t & 0 \\ 0 & 0 \end{pmatrix},$$

and consequently,

$$x_1(s) = e^{ts}x_{10} + \int_0^s e^{t(s-r)}\varphi\left(\mathbf{k}^T\mathbf{x}\right)b_1 dr.$$

Since $|\varphi(\sigma)| \leq 1$ for $\sigma \in \mathbb{R}$, we conclude that

$$|x_1(s)| \leq e^{ts}\left(|x_{10}| + \frac{|b_1|}{t}\right) - \frac{|b_1|}{t}.$$

Note that the symmetric straight lines $\omega_k := \{(-1)^k(1 - |b_1|/t, x_2) : x_2 \in \mathbb{R}\}$ with $k = 1, 2$ intersect Γ_+ and Γ_-. Moreover, it is easy to check that the set B_ω has property (a).

(b) Take $\mathbf{y} = (y_1, y_2)^T \in \omega_1$. Then $y_1 = -1 + |b_1|t^{-1} < 0$ and $A\mathbf{y} = (|b_1| - t, 0)^T$, that is, the first component of the vector field is positive. We conclude that B_ω is a positively invariant set. From statement (a) it follows that B_ω contains every ω-limit set.

(d) Consider the family of segments $S_h := \{(x_1, h) : |x_1| \leq 1 - |b_1|/t\}$ for $h > 0$. Suppose that $k_2 > 0$. When h is big enough we have $k_1x_1 + k_2h > 1$, whence $S_h \subset S_+$. In the region S_+ the system is linear, with $A\mathbf{x}+\mathbf{b} = (t - |b_1| + b_1, b_2)$ and $b_2 < 0$ (note that $D = tk_2b_2 > 0$, see (3.10)). Then there exists $h_0 > 0$ such that $R_h := \{(x_1, x_2) : |x_1| \leq 1 - |b_1|/t$ and $|x_2| \leq h\}$ is a positive compact invariant set for every $h > h_0$. Moreover, $B_\omega = \bigcup_{h>h_0} R_h$, which concludes the proof.

The case $k_2 < 0$ follows by using similar arguments. □

Proposition 3.13.5. *Given a fundamental system with fundamental parameters (D, T, d, t), where $d = 0$ and $t > 0$, there exist straight lines α_1 and α_2 which are symmetric with respect to the origin and such that $\alpha_i \cap \Gamma_+ \neq \emptyset$ and $\alpha_i \cap \Gamma_- \neq \emptyset$ with $i \in \{1, 2\}$. Let B_α be the closed strip bounded by α_1 and α_2.*

(a) *For every solution $\mathbf{x}(s)$, there exists $s_0 > 0$ such that $\{\mathbf{x}(-s) : s > s_0\} \subset B_\alpha$.*

(b) *B_α is a negatively invariant set which contains every α-limit set.*

(c) *If $D > 0$, then all orbits are negatively bounded.*

Proof. The change of the time variable t to $-t$ transforms the original system into another fundamental system with fundamental parameters $(D, -T, d, -t)$, see Proposition 3.7.1. The result follows by applying Proposition 3.13.4 to this system. □

Proposition 3.13.6. *Given a fundamental system with fundamental parameters (D, T, d, t), where $d < 0$, there exist two pairs of symmetric straight lines α_1 and α_2, and ω_1 and ω_2 such that $\omega_i \cap \Gamma_+ \neq \emptyset$, $\omega_i \cap \Gamma_- \neq \emptyset$, $\alpha_i \cap \Gamma_+ \neq \emptyset$, and $\alpha_i \cap \Gamma_- \neq \emptyset$, for $i \in \{1, 2\}$. Let B_ω and B_α be the closed strips bounded by ω_1 and ω_2 and by α_1 and α_2, respectively. Then:*

(a) *For every solution $\mathbf{x}(s)$, there exists $s_0 > 0$ such that $\{\mathbf{x}(s) : s > s_0\} \subset B_\omega$ and $\{\mathbf{x}(-s) : s > s_0\} \subset B_\alpha$.*

(b) *B_α is a negatively invariant set containing every α-limit set.*

(c) *B_ω is a positively invariant set containing every ω-limit set.*

(d) $B_\alpha \cap B_\omega$ *is a compact and invariant set containing every singular point and every limit cycle.*

Proof. Take $\mathbf{x}_0 = (x_{10}, x_{20}) \in \mathbb{R}^2$ and let $\mathbf{x}(s) = (x_1(s), x_2(s))$ be the solution of the fundamental system $\dot{\mathbf{x}} = A\mathbf{x} + \varphi(\mathbf{k}^T\mathbf{x})\mathbf{b}$ such that $\mathbf{x}(0) = \mathbf{x}_0$. From expression (3.5) it follows that

$$\mathbf{x}(s) = e^{As}\mathbf{x}(0) + \int_0^s e^{A(s-r)}\varphi\left(\mathbf{k}^T\mathbf{x}\right)\mathbf{b}\,dr.$$

Since linear maps transform straight lines into straight lines, it is not a restriction to assume that the matrix A is in real Jordan normal form. Thus,

$$A = \begin{pmatrix} \lambda_1 & 0 \\ 0 & \lambda_2 \end{pmatrix}, \text{ with } \lambda_1 > 0 > \lambda_2,$$

and therefore

$$x_k(s) = e^{\lambda_k s}x_k(0) + \int_0^s e^{\lambda_k(s-r)}\varphi\left(\mathbf{k}^T\mathbf{x}\right)b_k dr,$$

for $k = 1, 2$. From this it follows that for every $s > 0$ we have

$$|x_1(-s)| \leq e^{\lambda_1(-s)}|x_1(0)| + |b_1|\int_{-s}^0 e^{\lambda_1(-s-r)}dr \leq e^{-\lambda_1 s}\left(|x_1(0)| - \frac{|b_1|}{\lambda_1}\right) + \frac{|b_1|}{\lambda_1},$$

$$|x_2(s)| \leq e^{\lambda_2 s}|x_2(0)| + |b_2|\int_0^s e^{\lambda_2(s-r)}dr \leq e^{\lambda_2 s}\left(|x_2(0)| + \frac{|b_2|}{\lambda_2}\right) - \frac{|b_2|}{\lambda_2}.$$

Hence, there exist $s_+ > 0$ and $s_- < 0$ such that $|x_2(s)| < 1 - |b_2|/\lambda_2$ when $s > s_+$, and $|x_1(s)| < 1 + |b_1|/\lambda_1$ when $s < s_-$.

Consider the symmetric straight lines $\omega_k := \{(-1)^{k+1}(x_1, 1 - |b_2|/\lambda_2) : x_1 \in \mathbb{R}\}$ and $\alpha_k = \{(-1)^{k+1}(1 + |b_1|/\lambda_1, x_2) : x_2 \in \mathbb{R}\}$, for $k = 1, 2$. It is easy to check that these straight lines intersect Γ_+ and Γ_-.

(b) Let $\mathbf{x} = (x_1, x_2)^T$ be a point on α_1 (respectively, α_2). Then $x_1 = 1 + |b_1|/\lambda_1$ which is positive (respectively, negative). Since $A\mathbf{x} + \varphi(\mathbf{k}^T\mathbf{x}) = (\lambda_1 + |b_1| + \varphi(\mathbf{k}^T\mathbf{x})b_1, \lambda_2 x_2 + \varphi(\mathbf{k}^T\mathbf{x})b_2)^T$, that is, the first coordinate of the vector field at $\mathbf{x}$ is also positive (respectively, negative), we conclude that B_α is a negatively invariant set. Moreover, by statement (a), B_α contains all the α-limit sets.

(c) The statement follows by using similar arguments to those in the proof of statement (b).

(d) Since $B_\alpha \cap B_\omega$ is a compact set, (d) follows from statements (b) and (c). □

Chapter 4

Return maps

The determination of the number and location of limit cycles of a planar differential system is one of the most difficult problems in the qualitative theory of differential equations. In the case of planar polynomial differential systems this problem is known as the second part of Hilbert's 16th problem, and it remains open even for polynomials of degree 2. In the case of fundamental systems this problem can be completely solved by using the return maps defined by the flow of the system on convenient cross sections contained in the straight lines L_+ and L_-.

Consider a fundamental system

$$\dot{\mathbf{x}} = A\mathbf{x} + \varphi\left(\mathbf{k}^T\mathbf{x}\right)\mathbf{b}, \tag{4.1}$$

with matrix $A \in L(\mathbb{R}^2)$ and vectors $\mathbf{k}, \mathbf{b} \in \mathbb{R}^2 \setminus \{\mathbf{0}\}$, and let

$$\dot{\mathbf{x}} = \begin{cases} A\mathbf{x} - \mathbf{b}, & \text{if } \mathbf{x} \in S_- \cup L_-, \\ B\mathbf{x}, & \text{if } \mathbf{x} \in L_- \cup S_0 \cup L_+, \\ A\mathbf{x} + \mathbf{b}, & \text{if } \mathbf{x} \in L_+ \cup S_+, \end{cases}$$

be the piecewise linear expression of system (4.1), where $B = A + \mathbf{b}\mathbf{k}^T$. By Theorem 3.12.2, we can distinguish three different kinds of periodic orbits Γ, depending on their location in the phase plane.

(i) Γ is contained in one of the open regions S_+, S_0, or S_- where the system is linear.

(ii) Γ intersects only one of the straight lines L_+ or L_-.

(iii) Γ intersects both the straight line L_+ and the straight line L_-.

Since the flow is linear in S_+, S_0, or S_-, periodic orbits in the class (i) appear only in the case of a linear center. Consequently, the behaviour of these orbits is well known. For this reason we restrict our attention to periodic orbits belonging either to the class (ii) or to the class (iii).

Suppose that Γ only intersects the straight line L_+. By the symmetry of the fundamental vector field with respect to the origin, the intersection point does not belong to the bounded region Σ_Γ inside Γ. Moreover, there exists another periodic orbit, denoted by Γ_-, which is symmetrical with respect to the origin to the periodic orbit Γ, and such that Γ_- only intersects the straight line L_-, see Figure 4.1(a). Thanks to the continuous dependence of the solutions of an ordinary differential equation with respect to the initial conditions, we can define a *return map* Π in a neighbourhood of one of the intersection points of Γ with L_+ by taking the cross sections contained in L_+. Similarly, we can define another return map associated to the periodic orbit Γ_- by taking the cross sections contained in L_-.

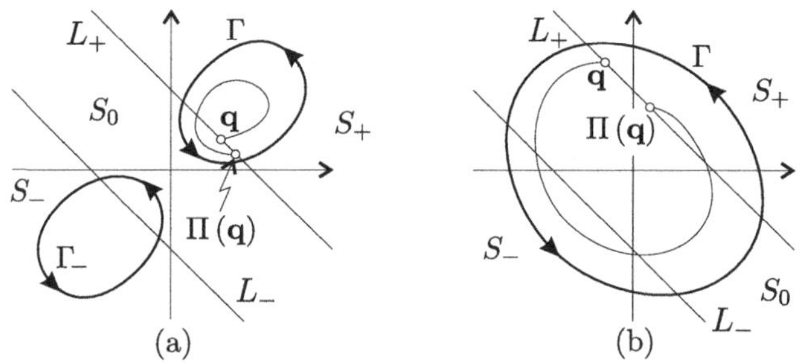

Figure 4.1: (a) Periodic orbits intersecting with only one straight line. (b) Periodic orbit intersecting with two straight lines.

Suppose now that the periodic orbit Γ intersects the two straight lines L_+ and L_-. In the same way, associated to Γ, and following the flow, we can define a *return map* Π in a neighbourhood of one of the intersection points of Γ with L_+ by taking the cross section contained in L_+, see Figure 4.1(b). Thus we reduce the study of periodic orbits in class (ii) or in class (iii) to the study of return maps defined on the straight lines L_+ and L_-.

4.1 Poincaré maps for fundamental systems

Let Γ be a periodic orbit of system (4.1) such that Γ intersects the straight lines L_+ and L_-. Let Π be the return map associated to L_+ and defined in a neighbourhood of the periodic orbit Γ. The map Π can be written as a composition of the Poincaré maps which maps points from L_+ to L_-, from L_- to L_-, from L_- to L_+ and from L_+ to L_+, see Figure 4.1(b). Similarly, if Γ is a periodic orbit intersecting only one of the straight lines, then Π can be written as a composition of the Poincaré maps which maps points either from L_+ to L_+ or from L_- to L_-, see Figure 4.1(a).

Let $\Phi(s, \mathbf{x})$ be the flow of system (4.1). We define the following subsets contained in the straight lines L_+ and L_-.

$$\begin{aligned}
\mathrm{Dom}^A_{++} &:= \{\mathbf{q} \in L_+ : \exists s_\mathbf{q} \geq 0,\ \Phi(s_\mathbf{q}, \mathbf{q}) \in L_+ \text{ and } \Phi(s, \mathbf{q}) \subset S_+\ \forall s \in (0, s_\mathbf{q})\},\\
\mathrm{Dom}^B_{+-} &:= \{\mathbf{q} \in L_+ : \exists s_\mathbf{q} > 0,\ \Phi(s_\mathbf{q}, \mathbf{q}) \in L_- \text{ and } \Phi(s, \mathbf{q}) \subset S_0\ \forall s \in (0, s_\mathbf{q})\},\\
\mathrm{Dom}^A_{--} &:= \{\mathbf{q} \in L_- : \exists s_\mathbf{q} \geq 0,\ \Phi(s_\mathbf{q}, \mathbf{q}) \in L_- \text{ and } \Phi(s, \mathbf{q}) \subset S_-\ \forall s \in (0, s_\mathbf{q})\},\\
\mathrm{Dom}^B_{-+} &:= \{\mathbf{q} \in L_- : \exists s_\mathbf{q} > 0,\ \Phi(s_\mathbf{q}, \mathbf{q}) \in L_+ \text{ and } \Phi(s, \mathbf{q}) \subset S_0\ \forall s \in (0, s_\mathbf{q})\},\\
\mathrm{Dom}^B_{++} &:= \{\mathbf{q} \in L_+ : \exists s_\mathbf{q} \geq 0,\ \Phi(s_\mathbf{q}, \mathbf{q}) \in L_+ \text{ and } \Phi(s, \mathbf{q}) \subset S_0\ \forall s \in (0, s_\mathbf{q})\},\\
\mathrm{Dom}^B_{--} &:= \{\mathbf{q} \in L_- : \exists s_\mathbf{q} \geq 0,\ \Phi(s_\mathbf{q}, \mathbf{q}) \in L_- \text{ and } \Phi(s, \mathbf{q}) \subset S_0\ \forall s \in (0, s_\mathbf{q})\}.
\end{aligned}$$

Take a point $\mathbf{q}$ in Dom^M_{kj} with $k, j \in \{+, -\}$ and $M \in \{A, B\}$. Let $\gamma(\mathbf{q})$ be the orbit through $\mathbf{q}$. The *flight time* $s_\mathbf{q}$ is defined to be the interval of time between two consecutive intersections of $\gamma(\mathbf{q})$ with the straight lines L_k and L_j, respectively. Note that during this time the orbit $\gamma(\mathbf{q})$ lies within one of the regions where the system is linear. The matrix of this linear system appears as a superscript in the name of the corresponding set.

If a set Dom^M_{jk} is non-empty, then we call $\Pi^M_{jk}(\mathbf{q}) = \Phi(s_\mathbf{q}, \mathbf{q})$ the *Poincaré map of system* (4.1) *associated to the straight lines* L_j *and* L_k.

For now it is not possible to describe precisely the decomposition of the return map Π in terms of the Poincaré maps Π^M_{jk}. This will be one of the topics in Section 4.6. Nevertheless, it is obvious that knowledge of the Poincaré maps Π^M_{jk} is essential for the knowledge of the return map Π.

After the pioneering work of Andronov [3], such Poincaré maps have been used by most authors working in this subject, see for instance [47], [24], [25] and [26]. In all these works the authors obtain, just by applying a rotation and by rescaling the original variables, a linearly conjugated fundamental system such that $L_+ = \{x = 1\}$ and $L_- = \{x = -1\}$. Hence, the expression of both the Poincaré maps and the fundamental matrices A and B are related to this new configuration.

Here we present an alternative parametrization of the straight lines L_+ and L_- in such a way that the Poincaré maps associated to them are invariant under linear transformations. That is, we can consider that the fundamental matrices A and B are in their real Jordan normal form. As it is proved in [45] these parametrization can be extended to higher dimensions, with similar consequences.

Finally, when the superscript M coincides with B, the Poincaré map Π^M_{jk} is defined by the flow of the homogeneous linear system $\dot{\mathbf{x}} = B\mathbf{x}$. Further, when M coincides with A, the Poincaré map Π^M_{jk} is defined by the flow of the non-homogeneous linear systems $\dot{\mathbf{x}} = A\mathbf{x} \pm \mathbf{b}$. Thus in Section 4.3 and Section 4.5 we study the behaviour of the Poincaré maps defined by the flow of homogeneous and non-homogeneous linear systems, respectively.

4.2 Transversality of a linear flow

Consider the linear system

$$\dot{\mathbf{x}} = A\mathbf{x}, \tag{4.2}$$

where $A \in L(\mathbb{R}^2)$. From now on, $\dot{\mathbf{q}}$ denotes the value of a vector field defined by a differential system at the point $\mathbf{q}$ of the phase plane. In particular, for the linear system (4.2) $\dot{\mathbf{q}} = A\mathbf{q}$.

Let $L = \{\mathbf{p} + \lambda\mathbf{v} : \lambda \in \mathbb{R}\}$ be a straight line which does not pass through the origin. Let $\mathbf{n}$ be the unit vector orthogonal to L such that $\mathbf{n}^T\mathbf{p} > 0$; in this case we say that $\mathbf{n}$ is *oriented in the opposite sense to the origin.* Note that choosing the vector $\mathbf{n}$ does not depend on the point $\mathbf{p}$ on L. Of course, if we take another point $\mathbf{q} = \mathbf{p} + \lambda\mathbf{v}$ in L, then $\mathbf{n}^T\mathbf{q} = \mathbf{n}^T\mathbf{p} > 0$.

The flow of system (4.2) is said to be *transversal* to the straight line L at a point $\mathbf{q}$ if $\mathbf{n}^T\dot{\mathbf{q}} \neq 0$. Otherwise, $\mathbf{q}$ is said to be a *contact point* of the flow with the straight line L. The following definitions formalize the intuitive idea of the sense of a transversal flow with respect to a straight line L. A transversal flow to L at a point $\mathbf{q} \in L$ is said to have *outside orientation* if $\mathbf{n}^T\dot{\mathbf{q}} > 0$. A transversal flow to L at a point $\mathbf{q} \in L$ is said to have *inside orientation* if $\mathbf{n}^T\dot{\mathbf{q}} < 0$. Accordingly, we define the following subsets in L

$$L^I := \left\{\mathbf{q} \in L : \mathbf{n}^T\dot{\mathbf{q}} \leq 0\right\} \text{ and } L^O := \left\{\mathbf{q} \in L : \mathbf{n}^T\dot{\mathbf{q}} \geq 0\right\}.$$

In Proposition 4.2.5 and 4.2.6 we describe the different possibilities for the sets L^I and L^O depending on the invertibility of the matrix A. Before doing this we give some technical lemmas.

Lemma 4.2.1. *Consider a linear system $\dot{\mathbf{x}} = A\mathbf{x}$ with $A \in GL(\mathbb{R}^2)$ and let L be a straight line in the phase plane which does not pass through the origin. Two points $\mathbf{p}$ and $\mathbf{q}$ in L are different if and only if $\dot{\mathbf{p}} \neq \alpha\dot{\mathbf{q}}$ for any $\alpha \in \mathbb{R}$.*

Proof. Since A is invertible, the equality $\dot{\mathbf{p}} = \alpha\dot{\mathbf{q}}$ is equivalent to the equality $\mathbf{p} = \alpha\mathbf{q}$. In this case L is a straight line through the origin, in contradiction with the hypothesis. □

Consider the unit circle $\mathbb{S}^1 = \{\mathbf{x} \in \mathbb{R}^2 : \|\mathbf{x}\| = 1\}$ and the continuous map $\Psi : \mathbb{R}^2\backslash \ker(A) \to \mathbb{S}^1$ given by

$$\Psi(\mathbf{q}) = \frac{\dot{\mathbf{q}}}{\|\dot{\mathbf{q}}\|} = \frac{A\mathbf{q}}{\|A\mathbf{q}\|}, \tag{4.3}$$

where $\ker(A)$ denotes the null space of A. The map Ψ provides information about the sense and the direction of the linear vector field defined by (4.2). For instance, if $A \in GL(\mathbb{R}^2)$, then $\Psi(\mathbb{R}^2\backslash\{\mathbf{0}\}) = \mathbb{S}^1$. This map will be particularly interesting when we restrict it to either straight lines or orbits.

Let $\alpha(\lambda)$ be a curve in the phase plane and $(1, \theta(\lambda))$ the parametrization in polar coordinates of $\Psi(\alpha(\lambda))$. We say that function $\Psi|_\alpha$ is *monotone* if $\theta(\lambda)$ is strictly monotone.

Lemma 4.2.2. *Consider a linear system* $\dot{\mathbf{x}} = A\mathbf{x}$ *with* $A \in GL(\mathbb{R}^2)$ *and let* $L = \{\mathbf{p} + \lambda\mathbf{v} : \lambda \in \mathbb{R}\}$ *be a straight line in the plane.*

(a) *If* L *does not pass through the origin, then* $\Psi|_L$ *is monotone. Furthermore* $\Psi(L)$ *is an open semi-circle of* $\mathbb{S}^1$ *with endpoints* $\Psi(\mathbf{v})$ *and* $-\Psi(\mathbf{v})$.

(b) *If* L *passes through the origin, then* $\Psi(L) = \{\Psi(\mathbf{v}), -\Psi(\mathbf{v})\}$.

Proof. (a) By Lemma 4.2.1, $\Psi|_L$ is an injective and a continuous map. Thus, $\theta(\lambda)$ is also injective and continuous. Therefore, $\theta(\lambda)$ is monotone.

Since Ψ is continuous, the set $\Psi(L)$ is a connected subset of $\mathbb{S}^1$. Hence $\Psi(L)$ is a circle arc. Moreover, by Lemma 4.2.1, if $\mathbf{w} \in \Psi(L)$, then $-\mathbf{w} \notin \Psi(L)$. Therefore, $\Psi(L)$ is contained in a semi-circle. The statement (a) follows from $\lim_{\lambda\nearrow+\infty} \Psi(\mathbf{p} + \lambda\mathbf{v}) = \Psi(\mathbf{v})$ and $\lim_{\lambda\searrow-\infty} \Psi(\mathbf{p} + \lambda\mathbf{v}) = -\Psi(\mathbf{v})$.

(b) Since the origin is contained in L, we can write $L = \{\lambda\mathbf{v} : \lambda \in \mathbb{R}\}$. The statement follows from the equality $\Psi(\lambda\mathbf{v}) = \lambda\Psi(\mathbf{v})/|\lambda|$ when $\lambda \neq 0$. □

Let $\mathbf{f} : \mathbb{R}^2 \to \mathbb{R}^2$ be a vector field and $\mathbf{v} \in \mathbb{R}^2$. The set $\{\mathbf{q} \in \mathbb{R}^2 : \mathbf{f}(\mathbf{q}) = \lambda\mathbf{v}$ with $\lambda \in \mathbb{R}\}$ is called the *isocline of* $\mathbf{f}$ *defined by the vector* $\mathbf{v}$. Therefore, the isoclines of the linear system (4.2) are the straight line through the origin, see Lemma 4.2.2(b). More precisely, we have the following result.

Lemma 4.2.3. *Set* $A \in GL(\mathbb{R}^2)$ *and* $\mathbf{v} \in \mathbb{R}^2\setminus\{\mathbf{0}\}$. *The isocline of the system* $\dot{\mathbf{x}} = A\mathbf{x}$ *defined by the vector* $\mathbf{v}$ *is the straight line through the origin* $L = \{\lambda A^{-1}\mathbf{v} : \lambda \in \mathbb{R}\}$.

Given $\mathbf{p} = (p_1, p_2)^T \in \mathbb{R}^2$ we denote by $\mathbf{p}^\perp$ the vector $(-p_2, p_1)^T$. The following properties are obvious: (a) $\mathbf{p}^\perp$ is orthogonal to $\mathbf{p}$, (b) $\|\mathbf{p}^\perp\| = \|\mathbf{p}\|$, and (c) $(\mathbf{p}^\perp)^\perp = -\mathbf{p}$.

Lemma 4.2.4. *Consider a linear system* $\dot{\mathbf{x}} = A\mathbf{x}$ *with* $A \in L(\mathbb{R}^2)$ *and let* $\mathbf{p}$ *be a point in* $\mathbb{R}^2$. *Then* $(A\dot{\mathbf{p}})^T\dot{\mathbf{p}}^\perp = -\det(A)\,\mathbf{p}^T\dot{\mathbf{p}}^\perp$.

Proof. Let t and d be the trace and the determinant of the matrix A, and let Id be the identity matrix. From the Cayley–Hamilton theorem we obtain that $A^2 - tA + d\,\mathrm{Id} = 0$. Multiplying both sides by $\mathbf{p}$ we get that $A\dot{\mathbf{p}} = t\dot{\mathbf{p}} - d\mathbf{p}$. Therefore, $(A\dot{\mathbf{p}})^T\dot{\mathbf{p}}^\perp = -d\mathbf{p}^T\dot{\mathbf{p}}^\perp$. □

Proposition 4.2.5. *Consider the linear system* $\dot{\mathbf{x}} = A\mathbf{x}$ *with* $A \in GL(\mathbb{R}^2)$ *and let* L *be a straight line in the phase plane.*

(a) *Suppose that* L *does not pass through the origin.*

(a.1) *There exists at most one contact point of the flow with* L.

(a.2) *Let* $\mathbf{p}$ *be a contact point of the flow with* L. *If* $\det(A) > 0$, *then* $L^I = \{\mathbf{p} + \lambda\dot{\mathbf{p}} : \lambda \geq 0\}$ *and* $L^O = \{\mathbf{p} + \lambda\dot{\mathbf{p}} : \lambda \leq 0\}$. *If* $\det(A) < 0$, *then* $L^I = \{\mathbf{p} + \lambda\dot{\mathbf{p}} : \lambda \leq 0\}$ *and* $L^O = \{\mathbf{p} + \lambda\dot{\mathbf{p}} : \lambda \geq 0\}$.

(a.3) *If $L^I \neq \varnothing$ and $L^O \neq \varnothing$, then there exists exactly one contact point of the flow with L.*

(a.4) *If the flow has no contact points with L, then either $L^I = L$ and $L^O = \varnothing$, or $L^I = \varnothing$ and $L^O = L$.*

(b) *If L passes through the origin, then either L is an invariant straight line (in this case any point on L is a contact point), or the origin is the unique contact point of the flow with L.*

Proof. Statement (a.1) is a consequence of Lemma 4.2.2(a).

(a.2) Since L does not pass through the origin and the matrix $A \in GL(\mathbb{R}^2)$, $\dot{\mathbf{p}} \neq \mathbf{0}$. Hence, we can write $L = \{\mathbf{p} + \lambda\dot{\mathbf{p}} : \lambda \in \mathbb{R}\}$. Then the unit orthogonal vector $\mathbf{n}$ is either $\dot{\mathbf{p}}^\perp/\|\dot{\mathbf{p}}\|$ or $-\dot{\mathbf{p}}^\perp/\|\dot{\mathbf{p}}\|$. In both cases, $(A\dot{\mathbf{p}})^T\mathbf{n} = -\det(A)\,\mathbf{p}^T\mathbf{n}$, see Lemma 4.2.4.

Take $\mathbf{q} \in L$ such that $\mathbf{q} \neq \mathbf{p}$. Then we can write $\mathbf{q} = \mathbf{p} + \lambda\dot{\mathbf{p}}$ with $\lambda \neq 0$, and therefore

$$\mathbf{n}^T\dot{\mathbf{q}} = \mathbf{n}^T A\mathbf{q} = \mathbf{n}^T\dot{\mathbf{p}} + \lambda\mathbf{n}^T A\dot{\mathbf{p}} = \lambda\mathbf{n}^T A\dot{\mathbf{p}} = -\lambda\det(A)\mathbf{n}^T\mathbf{p}.$$

Since $\mathbf{n}^T\mathbf{p} > 0$, the orientation of the flow on the straight line L depends on the sign of λ and the statement follows.

(a.3) Consider the continuous function $f(\lambda) = \mathbf{n}^T A(\mathbf{q} + \lambda\mathbf{v})$ defined on the straight line $L = \{\mathbf{q} + \lambda\mathbf{v} : \lambda \in \mathbb{R}\}$. Clearly f satisfies that $f(\lambda_1) \leq 0$ when $\mathbf{q} + \lambda_1\mathbf{v} \in L^I$ and $f(\lambda_2) \geq 0$ when $\mathbf{q} + \lambda_2\mathbf{v} \in L^O$. Therefore, there exists a λ^* such that $f(\lambda^*) = 0$. That is, $\mathbf{q} + \lambda^*\mathbf{v}$ is a contact point of the flow with L. The uniqueness of the contact point follows from statement (a.1).

(a.4) Consider the function f defined in the proof of (a.3). If there are no singular points on L, then either $f(\lambda) > 0$ or $f(\lambda) < 0$, which proves the statement.

(b) Since L is a straight line through the origin, we can write $L = \{\lambda\mathbf{v} : \lambda \in \mathbb{R}\}$. Thus the origin is a contact point of the flow with L. Suppose that there exists another such contact point $\mathbf{q}$. Then $\mathbf{q} = \lambda\mathbf{v}$ with $\lambda \neq 0$, and $A\mathbf{q} = \alpha\mathbf{v}$ with $\alpha \in \mathbb{R}$. Therefore, $A\mathbf{v} = (\alpha/\lambda)\mathbf{v}$, i.e., $\mathbf{v}$ is an eigenvector of A and L is invariant under the flow. □

Given a linear system $\dot{\mathbf{x}} = A\mathbf{x}$ with $\det(A) \neq 0$ and a straight line L which does not pass through the origin, in Proposition 4.2.5 we have proved that if there exists a contact point of the flow with L, we can split the straight line into the two half-lines L^I and L^O in such a way that the flow over L^I and L^O has opposite sense. For a treatment of a more general case we refer the reader to [45]. Under a natural restriction, in the next proposition we prove that even when $\det(A) = 0$, the straight line L can also be divided into the two half-lines L^I and L^O. From now on we denote by $\det(\mathbf{v}, \mathbf{w})$ the determinant of the matrix whose columns are the vectors $\mathbf{v}$ and $\mathbf{w}$ in $\mathbb{R}^2$.

Proposition 4.2.6. *Consider a planar linear system $\dot{\mathbf{x}} = A\mathbf{x}$ with $\det(A) = 0$ and such that A is not the zero matrix. Let L be a non-invariant straight line which does* *not pass through the origin and let* $\mathbf{n}$ *be the unit orthogonal vector to* L *oriented in the opposite sense to the origin.*

(a) *If* $\mathbf{p}$ *is a contact point of the flow with* L*, then* $\det(A\mathbf{n}^\perp, \mathbf{n}^\perp) \neq 0$.

(b) *If* $\mathbf{p}$ *is a contact point of the flow with* L*, then* $\mathbf{p}$ *is a singular point.*

(c) *There exists at most one contact point of the flow with* L.

(d) *Let* $\mathbf{p}$ *be a contact point of the flow with* L*. If* $\det(A\mathbf{n}^\perp, \mathbf{n}^\perp) > 0$*, then* $L^I = \{\mathbf{p} + \lambda\mathbf{n}^\perp : \lambda \leq 0\}$ *and* $L^O = \{\mathbf{p} + \lambda\mathbf{n}^\perp : \lambda \geq 0\}$*. If* $\det(A\mathbf{n}^\perp, \mathbf{n}^\perp) < 0$*, then* $L^I = \{\mathbf{p} + \lambda\mathbf{n}^\perp : \lambda \geq 0\}$ *and* $L^O = \{\mathbf{p} + \lambda\mathbf{n}^\perp : \lambda \leq 0\}$.

(e) *If* $L^I \neq \varnothing$ *and* $L^O \neq \varnothing$*, then there exists exactly one contact point of the flow with* L.

Proof. (a) It is clear that $L = \{\mathbf{p}+\lambda\mathbf{n}^\perp : \lambda \in \mathbb{R}\}$. Suppose that $\det(A\mathbf{n}^\perp, \mathbf{n}^\perp) = 0$; that is, $\mathbf{n}^T A\mathbf{n}^\perp = 0$. For any $\lambda \in \mathbb{R}$ it follows that $\mathbf{n}^T A(\mathbf{p} + \lambda\mathbf{n}^\perp) = 0$. Hence L is an invariant straight line, which contradicts our assumptions.

(b) Let $\mathbf{p}$ be a contact point of the flow with L. Then $\dot{\mathbf{p}} = \lambda_0\mathbf{n}^\perp$ with $\lambda_0 \in \mathbb{R}$ and $(A\dot{\mathbf{p}})^T\dot{\mathbf{p}}^\perp = -\lambda_0^2(A\mathbf{n}^\perp)^T\mathbf{n} = -\lambda_0^2\det(A\mathbf{n}^\perp, \mathbf{n}^\perp)$. From Lemma 4.2.4 we have that $(A\dot{\mathbf{p}})^T\dot{\mathbf{p}}^\perp = -\det(A)\,\mathbf{p}^T\dot{\mathbf{p}}^\perp = 0$, and so $\lambda_0 = 0$ and $\dot{\mathbf{p}} = \mathbf{0}$.

(c) Suppose that $\mathbf{p}$ and $\mathbf{q}$ are two different contact points of the flow with L. Thus we can write $L = \{\mathbf{p} + \lambda(\mathbf{p} - \mathbf{q}) : \lambda \in \mathbb{R}\}$. Since $A\mathbf{p} = A\mathbf{q} = \mathbf{0}$, see statement (b), it follows that $L = \ker(A)$, which contradicts that L does not pass through the origin. Therefore the contact point is unique.

(d) Since $\mathbf{p}$ is a contact point of the flow with L, we can write $L = \{\mathbf{p}+\lambda\mathbf{n}^\perp : \lambda \in \mathbb{R}\}$. The statement follows as a consequence of the equality $\mathbf{n}^T A(\mathbf{p} + \lambda\mathbf{n}^\perp) = \lambda\det(A\mathbf{n}^\perp, \mathbf{n}^\perp)$ for $\lambda \in \mathbb{R}$.

(e) Take $\mathbf{q}_1 \in L^I$ and $\mathbf{q}_2 \in L^O$. Then $L = \{\mathbf{q}_1 + \lambda(\mathbf{q}_2 - \mathbf{q}_1) : \lambda \in \mathbb{R}\}$. The continuous function $f(\lambda) = \mathbf{n}^T A\mathbf{q}_1 + \lambda\mathbf{n}^T A(\mathbf{q}_2 - \mathbf{q}_1)$ defined on $\mathbb{R}$ satisfies that $f(0) < 0$ and $f(1) > 0$. Thus, there exists $\lambda_0 \in (0,1)$ such that $\mathbf{p} = \mathbf{q}_1+\lambda_0(\mathbf{q}_2-\mathbf{q}_1)$ is a contact point of the flow with L. The uniqueness follows from the statement (c). □

In Propositions 4.2.5 and 4.2.6 we prove that the flow defined by a planar linear system is transversal to any non-invariant straight line in the phase plane, except in a contact point $\mathbf{p}$, in case that it exists. In this case the contact point splits the straight line into two half-lines such that the flow has opposite sense in each of them.

Let L be a straight line not passing through the origin. Then L splits the phase plane into two half-planes S_0 and S, where S_0 is the one containing the origin. Let $\gamma(\mathbf{p})$ be the orbit through the contact point $\mathbf{p}$. In the next result we prove that in a neighbourhood of $\mathbf{p}$ the orbit $\gamma(\mathbf{p})$ is contained in one of the half-planes $S_0 \cup L$ or $S \cup L$. In Proposition 4.2.10 we prove that, under some restrictions, this behaviour is not only local, but also global.

Proposition 4.2.7. *Consider a linear system* $\dot{\mathbf{x}} = A\mathbf{x}$ *with* $A \in GL(\mathbb{R}^2)$. *Let* L *be a straight line in the phase plane not passing through the origin,* $\mathbf{p}$ *a contact point of the flow with* L, *and* $\mathbf{x}(s)$ *the solution of the system such that* $\mathbf{x}(0) = \mathbf{p}$. *Define* $f(s) = \mathbf{n}^T \Psi(\mathbf{x}(s))$, $s \in \mathbb{R}$.

(a) *If* $\det(A) > 0$, *then there exists* $\varepsilon > 0$ *such that* $\{\mathbf{x}(s) : s \in (-\varepsilon, \varepsilon)\} \subset S_0 \cup L$ *and* $f(s)$ *is strictly decreasing in* $(-\varepsilon, \varepsilon)$.

(b) *If* $\det(A) < 0$, *then there exists* $\varepsilon > 0$ *such that* $\{\mathbf{x}(s) : s \in (-\varepsilon, \varepsilon)\} \subset S \cup L$ *and* $f(s)$ *is strictly increasing in* $(-\varepsilon, \varepsilon)$.

Proof. Since any rotation transforms L into a straight line L^* not passing through the origin and S_0 into the connected component of $\mathbb{R}^2 \setminus L^*$ containing the origin, we can assume without loss of generality that $L = \{x_2 = b\}$ with $b > 0$. Hence $\mathbf{p} = (p_1, b)^T$ and $\dot{\mathbf{p}} = (\dot{p}_1, 0)^T$ with $\dot{p}_1 \neq 0$, see Figure 4.2.

By the Inverse Function Theorem, there exist a neighbourhood I of p_1 and a differentiable function $\tau : I \to (-\delta, \delta)$ such that $\tau(x_1(s)) = s$, where $\mathbf{x}(s) = (x_1(s), x_2(s))^T$ and $\mathbf{x}(0) = \mathbf{p}$. To simplify the notation, we define $x_2 : I \to \mathbb{R}$ by $x_2(x_1) = x_2(\tau(x_1))$. It is clear that x_2 is a differentiable function and $\{\mathbf{x}(s) : s \in (-\delta, \delta)\} = \{(x_1, x_2(x_1)) : x_1 \in I\}$. Hence, $\mathbf{x}(s)$ is locally contained either in S_0 or in S if and only if $x_2(x_1)$ has a local maximum or minimum at $x_1 = p_1$, see Figure 4.2. We compute the sign of $d^2x_2/dx_1^2\big|_{x_1=p_1}$ in order to distinguish these two situations.

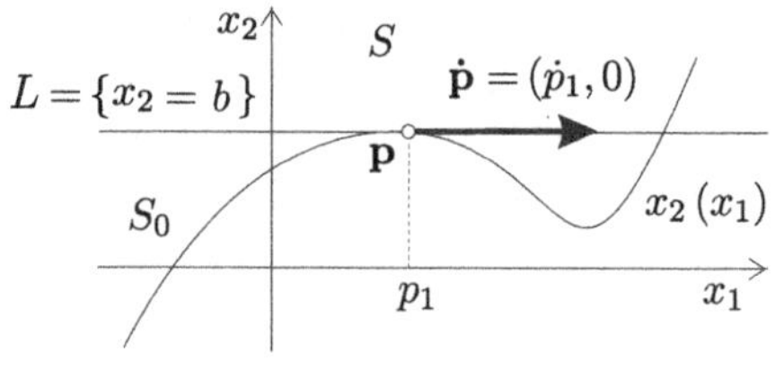

Figure 4.2: Graph of the function $x_2(x_1)$.

Let $(a_{ij})_{1\le i,j\le 2}$ be the coefficients of the matrix A. Then $\dot{x}_1 = a_{11}x_1 + a_{12}x_2$, $\dot{x}_2 = a_{21}x_1 + a_{22}x_2$, and

$$\frac{d^2x_2}{dx_1^2} = \frac{d}{dx_1}\left(\frac{dx_2}{dx_1}\right) = \frac{x_1\dfrac{dx_2}{dx_1} - x_2}{\left(a_{11}x_1 + a_{12}x_2\right)^2}\det(A).$$

Therefore

$$\left.\frac{d^2x_2}{dx_1^2}\right|_{x_1=p_1} = \frac{-b}{\left(\dot{p}_1\right)^2}\det(A). \tag{4.4}$$

(a) Suppose that $\det(A) > 0$. In this case $x_2(x_1)$ has a local maximum at $x_1 = p_1$. Hence, there exists $\varepsilon > 0$ such that $\{\mathbf{x}(s) : s \in (-\varepsilon, \varepsilon)\} \subset S_0$.

Since the function $f(s)$ defined in the proposition can be expressed as

$$f(s) = \mathbf{n}^T \frac{\dot{\mathbf{x}}(s)}{||\dot{\mathbf{x}}(s)||} = \frac{\dot{x}_1}{|\dot{x}_1|} \frac{dx_2/dx_1}{\sqrt{1+(dx_2/dx_1)^2}},$$

where $\mathbf{n} = (0,1)^T$ and $\dot{\mathbf{x}} = \dot{x}_1(1, dx_2/dx_1)$ in $(-\varepsilon, \varepsilon)$, we have that

$$\frac{df}{ds} = \frac{\dot{x}_1}{|\dot{x}_1|}\dot{x}_1 \frac{d}{dx_1}\left(\frac{dx_2/dx_1}{\sqrt{1+(dx_2/dx_1)^2}}\right) = \frac{d^2x_2}{dx_1^2}\frac{|\dot{x}_1|}{\left(1+(dx_2/dx_1)^2\right)^{\frac{3}{2}}}.$$

From expression (4.4) it follows that $df/ds < 0$ in $(-\varepsilon, \varepsilon)$.

Statement (b) follows by using similar arguments. □

Lemma 4.2.8. *Consider a planar linear system $\dot{\mathbf{x}} = A\mathbf{x}$ whose matrix A has only real eigenvalues. Let L and γ be respectively a straight line in the phase plane passing through the origin and an orbit of the system. If γ intersects L at more than one point, then $\gamma \subset L$ and L is invariant under the flow.*

Proof. Suppose that $\det(A) = 0$. In this case the orbits of the linear system are either singular points or they are contained in straight lines, see Subsection 2.5.3, and the lemma is obvious.

Suppose now that $\det(A) \neq 0$. Since linear transformations map straight lines into straight lines, we can assume without loss of generality that A is given in its real Jordan normal form. Therefore, A is either

$$\text{(i)} \begin{pmatrix} \lambda_1 & 0 \\ 0 & \lambda_2 \end{pmatrix} \text{ with } \lambda_1 \geq \lambda_2, \text{ or (ii)} \begin{pmatrix} \lambda & 1 \\ 0 & \lambda \end{pmatrix} \text{ with } \lambda \in \mathbb{R}.$$

Suppose that $\mathbf{p}$ and $\mathbf{q}$ are two points in $L \cap \gamma$. Since $\mathbf{p}, \mathbf{q} \in L$, then $\mathbf{q} = r\mathbf{p}$. Moreover, since $\mathbf{p}, \mathbf{q} \in \gamma$, then $\mathbf{q} = e^{As_0}\mathbf{p}$ with $s_0 > 0$. Therefore,

$$\left(r\,\mathrm{Id} - e^{As_0}\right)\mathbf{p} = \mathbf{0}. \tag{4.5}$$

Suppose that A is given by expression (i). Then by (4.5), either $p_1 = 0$, $p_2 = 0$, or $\lambda_1 = \lambda_2$, where $\mathbf{p} = (p_1, p_2)^T$. In any case we obtain that $\gamma \subset L$.

Suppose that A is given by expression (ii). Then by (4.5), $p_2 = 0$ and therefore γ is also contained in L. □

Using the map Ψ defined in expression (4.3) in the next result we study the set of directions of a linear vector field restricted to an orbit.

Lemma 4.2.9. *Consider a linear system $\dot{\mathbf{x}} = A\mathbf{x}$ with $A \in GL(\mathbb{R}^2)$ and let γ be an orbit.*

(a) *If all the eigenvalues of the matrix A are real and γ is not contained in a straight line, then $\Psi|_\gamma$ is monotone and $\Psi(\gamma)$ is contained in an open semicircle.*

(b) *If all the eigenvalues of the matrix A have real part equal to zero and the orbit γ is not a singular point, then $\Psi(\gamma) = \mathbb{S}^1$.*

Proof. (a) Let $\mathbf{p}$ and $\mathbf{q}$ be two points in γ. Since A is an invertible matrix, if $\Psi(\mathbf{p}) = \Psi(\mathbf{q})$, then $\mathbf{p} = \lambda\mathbf{q}$, where $\lambda = \|A\mathbf{p}\|/\|A\mathbf{q}\|$. Therefore, γ intersects a straight line L which passes through the origin in two different points. From Lemma 4.2.8 it follows that $\gamma \subset L$, which contradicts our assumptions. Therefore, the map $\Psi|_\gamma$ is injective.

Since $\Psi(-\mathbf{p}) = -\Psi(\mathbf{p})$, the same argument that we used before can be applied to show that, if $\Psi(\mathbf{p}) \in \Psi(\gamma)$, then $-\Psi(\mathbf{p}) \notin \Psi(\gamma)$. Therefore, $\Psi(\gamma)$ is contained in an open semi-circle.

(b) The proof follows straightforward by the representation $\gamma = \{e^{sA}\mathbf{p} : s \in \mathbb{R}\}$, where e^{sA} is the composition of a rotation of angle βs and a homothetic transformation. □

Proposition 4.2.10. *Consider a linear system $\dot{\mathbf{x}} = \mathbf{A}\mathbf{x}$ with $A \in GL(\mathbb{R}^2)$ and such that all the eigenvalues of A are real. Let L be a straight line in the plane which does not pass through the origin, $\mathbf{p}$ a contact point of the flow with L, and $\gamma(\mathbf{p})$ the orbit through $\mathbf{p}$.*

(a) *If* $\det(A) > 0$, *then* $\gamma \subset S_0 \cup L$.

(b) *If* $\det(A) < 0$, *then* $\gamma \subset S \cup L$.

Proof. (a) Let $\mathbf{x}(s)$ be the solution of the system such that $\mathbf{x}(0) = \mathbf{p}$. Since $\det(A) > 0$ by Proposition 4.2.7(a), there exists $\varepsilon > 0$ such that $\{\mathbf{x}(s) : s \in (-\varepsilon, \varepsilon)\} \subset S_0 \cup L$ and the function $f(s) = \mathbf{n}^T\Psi(\mathbf{x}(s))$ is continuous and strictly decreasing in $(-\varepsilon, \varepsilon)$.

Suppose that there exists $s_1 > 0$ such that $\mathbf{x}(s_1) \in L^O$, i.e., $f(s_1) > 0$. Since $f(0) = 0$ and f is decreasing in a neighbourhood of $s = 0$, there exists $s^* \in (0, s_1)$ such that $f(s^*) = 0$. This implies that $\Psi(\mathbf{x}(0)) = \pm\Psi(\mathbf{x}(s^*))$, in contradiction with Lemma 4.2.9(a). Similar arguments can be applied if we suppose that there exists $s_1 < 0$ such that $\mathbf{x}(s_1) \in L^I$. Therefore, $\gamma \subset S_0 \cup L$.

Statement (b) follows by using similar arguments. □

4.3 Poincaré maps of homogeneous linear systems

Consider the planar homogeneous linear system

$$\dot{\mathbf{x}} = A\mathbf{x} \tag{4.6}$$

and let L_+ and L_- be two different straight lines in the phase plane which are symmetric with respect to the origin. Note that then L_+ and L_- do not pass through the origin. Moreover, L_+ and L_- split the phase plane into three regions. We denote by S_0 the open strip containing the origin, and by S_+ and S_- the half-planes bounded by L_+ and L_-, respectively.

Using the expression of the flow of the linear system (4.6) and arguments similar to those employed in Subsection 4.1, we can define on L_+ and on L_- the following subsets:

$$\begin{aligned}
\mathrm{Dom}_{++} &:= \{\mathbf{q} \in L_+ : \exists s_\mathbf{q} > 0 \text{ such that } e^{s_\mathbf{q} A}\mathbf{q} \in L_+, \text{ and either} \\
&\qquad e^{sA}\mathbf{q} \subset S_+ \text{ or } e^{sA}\mathbf{q} \subset S_0 \ \forall s \in (0, s_\mathbf{q})\} \cup CP_+, \\
\mathrm{Dom}_{+-} &:= \{\mathbf{q} \in L_+ : \exists s_\mathbf{q} > 0 \text{ such that } e^{s_\mathbf{q} A}\mathbf{q} \in L_-, \text{ and} \\
&\qquad e^{sA}\mathbf{q} \subset S_0 \ \forall s \in (0, s_\mathbf{q})\}, \\
\mathrm{Dom}_{--} &:= \{\mathbf{q} \in L_- : \exists s_\mathbf{q} > 0 \text{ such that } e^{s_\mathbf{q} A}\mathbf{q} \in L_-, \text{ and either} \\
&\qquad e^{sA}\mathbf{q} \subset S_- \text{ or } e^{sA}\mathbf{q} \subset S_0 \ \forall s \in (0, s_\mathbf{q})\} \cup CP_-, \\
\mathrm{Dom}_{-+} &:= \{\mathbf{q} \in L_- : \exists s_\mathbf{q} > 0 \text{ such that } e^{s_\mathbf{q} A}\mathbf{q} \in L_+, \text{ and} \\
&\qquad e^{sA}\mathbf{q} \subset S_0 \ \forall s \in (0, s_\mathbf{q})\},
\end{aligned} \tag{4.7}$$

where CP_+ and CP_- are empty sets or consist of the contact points of the flow with the straight lines L_+ or L_-, respectively.

If for some $j, k \in \{+, -\}$ we have that $\mathrm{Dom}_{jk} \neq \varnothing$, then we define the *Poincaré map* Π_{jk} *of the homogeneous linear system* (4.6) *associated to the straight lines* L_j *and* L_k as $\Pi_{jk} : \mathrm{Dom}_{jk} \to L_k$, $\Pi_{jk}(\mathbf{q}) = e^{s_\mathbf{q} A}\mathbf{q}$.

Remark 4.3.1. We remark that the Poincaré maps Π_{jk}^M defined in Subsection 4.1 are the same as the corresponding one defined above. Thus, in order to study the maps Π_{jk}^M it is sufficient to study the maps Π_{jk}.

In Proposition 4.3.3 we present some results on the domain of the Poincaré maps. Necessary and sufficient conditions for the existence of these maps are given in Proposition 4.3.4. But first we prove a technical lemma.

Lemma 4.3.2. *Consider a planar linear system* $\dot{\mathbf{x}} = A\mathbf{x}$ *with* A *not the zero matrix. Let* L_+ *and* L_- *be two symmetric straight lines in the plane and let* Dom_{jk} *be the sets defined in* (4.7). *If for some* $j, k \in \{+, -\}$ *the set* $\mathrm{Dom}_{jk} \neq \varnothing$, *then there exists exactly one contact point* $\mathbf{p}_+$ *of the flow with* L_+. *In this case* $\mathbf{p}_- = -\mathbf{p}_+$ *is the unique contact point of the flow with* L_-.

Proof. Suppose that there exists a point $\mathbf{q}_1$ in Dom_{++}. (The other cases follow in a similar way.) By definition, there exist $s_{\mathbf{q}_1} > 0$ and $\mathbf{q}_2 \in L_+$ such that $\mathbf{q}_2 = e^{s_{\mathbf{q}_1} A}\mathbf{q}_1$; i.e., $\mathbf{q}_2 = \Pi_{++}(\mathbf{q}_1)$. Moreover, either $e^{sA}\mathbf{q}_1 \subset S_+$ or $e^{sA}\mathbf{q}_1 \subset S_0$ for every $s \in (0, s_{\mathbf{q}_1})$. Therefore, the flow at the points $\mathbf{q}_1$ and $\mathbf{q}_2$ has opposite sense. Thus $L_+^I \neq \varnothing$ and $L_+^O \neq \varnothing$. The lemma follows from Propositions 4.2.5(a.3) or 4.2.6(e), depending on whether $\det(A) \neq 0$ or $\det(A) = 0$. □

Proposition 4.3.3. *Consider a planar linear system* $\dot{\mathbf{x}} = A\mathbf{x}$ *with* A *not the zero matrix. Let* L_+ *and* L_- *be two symmetric straight lines in the plane and let* Dom_{jk} *be the sets defined in* (4.7)*) Suppose that* $\mathrm{Dom}_{jk} \neq \varnothing$ *for every* $j, k \in \{+, -\}$.

(a) *If* $\det(A) > 0$, *then*

$$\begin{aligned}
\Pi_{++} &: \mathrm{Dom}_{++} \subset L_+^O \longrightarrow L_+^I,\\
\Pi_{+-} &: \mathrm{Dom}_{+-} \subset L_+^I \longrightarrow L_-^O,\\
\Pi_{--} &: \mathrm{Dom}_{--} \subset L_-^O \longrightarrow L_-^I,\\
\Pi_{-+} &: \mathrm{Dom}_{-+} \subset L_-^I \longrightarrow L_+^O.
\end{aligned}$$

(b) *If* $\det(A) = 0$, *then* $\mathrm{Dom}_{++} = \{\mathbf{p}_+\}$, $\mathrm{Dom}_{--} = \{\mathbf{p}_-\}$, $\Pi_{++}(\mathbf{p}_+) = \mathbf{p}_+$, $\Pi_{--}(\mathbf{p}_-) = \mathbf{p}_-$, *and*

$$\begin{aligned}
\Pi_{+-} &: \mathrm{Dom}_{+-} \subset L_+^I \longrightarrow L_-^O,\\
\Pi_{-+} &: \mathrm{Dom}_{-+} \subset L_-^I \longrightarrow L_+^O.
\end{aligned}$$

(c) *If* $\det(A) < 0$, *then*

$$\begin{aligned}
\Pi_{++} &: \mathrm{Dom}_{++} \subset L_+^I \longrightarrow L_+^O,\\
\Pi_{+-} &: \mathrm{Dom}_{+-} \subset L_+^I \longrightarrow L_-^O,\\
\Pi_{--} &: \mathrm{Dom}_{--} \subset L_-^I \longrightarrow L_-^O,\\
\Pi_{-+} &: \mathrm{Dom}_{-+} \subset L_-^I \longrightarrow L_+^O.
\end{aligned}$$

Proof. (a) Since $\mathrm{Dom}_{++} \neq \varnothing$ and $\mathrm{Dom}_{--} \neq \varnothing$, there exist a contact point $\mathbf{p}_+$ of the flow with the straight line L_+, and a contact point $\mathbf{p}_-$ of the flow with the straight line L_-, see Lemma 4.3.2. These points split L_+ and L_- into the respective half-lines L_+^I, L_+^O, L_-^I and L_-^O, see Proposition 4.2.5.

Let $\mathbf{x}(s)$ be the solution of the system such that $\mathbf{x}(0) = \mathbf{p}_+$. In Proposition 4.2.7(a) we proved that there exists $\varepsilon > 0$ such that $\mathbf{x}(s) \subset S_0 \cup L_+$ if $s \in (-\varepsilon, \varepsilon)$. Thanks to the continuous dependence of the solutions of a differential equation on the initial conditions, we conclude that if $\mathbf{q}_1 \in \mathrm{Dom}_{++}$ and $\mathbf{q}_2 = \Pi_{++}(\mathbf{q}_1)$, then the flow at $\mathbf{q}_1$ has outward sense and the flow at $\mathbf{q}_2$ has inward sense, see Figure 4.3(a). Therefore $\mathbf{q}_1 \in L_+^O$ and $\mathbf{q}_2 \in L_+^I$.

Take now a point $\mathbf{q}_3$ in Dom_{+-} and let $\mathbf{q}_4 = \Pi_{+-}(\mathbf{q}_3)$. By the definition of the set Dom_{+-}, the orbit through $\mathbf{q}_3$ is contained during the flight time in S_0. Therefore, $\mathbf{q}_1 \in L_+^I$ and $\mathbf{q}_2 \in L_-^O$, see Figure 4.3(a).

Using the symmetry of the linear vector field with respect to the origin, we obtain the result about the Poincaré maps Π_{--} and Π_{-+}.

(b) Let $\mathbf{p}_+$ be the unique contact point of the flow with L_+, see Lemma 4.3.2. By Proposition 4.2.6(b), $\mathbf{p}_+$ is a singular point. Thus, $\mathbf{p}_+ \in \mathrm{Dom}_{++}$ and $\Pi_{++}(\mathbf{p}_+) = \mathbf{p}_+$. Similarly, there exists a unique contact point $\mathbf{p}_-$ with the straight line L_- which is a singular point and $\Pi_{--}(\mathbf{p}_-) = \mathbf{p}_-$

Since $\det(A) = 0$, the flow evolves in parallel straight lines. Suppose that L_+ is not invariant under the flow. Then every orbit intersects L_+ in at most one point. Therefore, $\text{Dom}_{++} = \{\mathbf{p}_+\}$ and $\text{Dom}_{--} = \{\mathbf{p}_-\}$. Similar arguments to those used in the proof of statement (a) show that $\Pi_{+-} : \text{Dom}_{+-} \subset L_+^I \longrightarrow L_-^O$ and $\Pi_{-+} : \text{Dom}_{-+} \subset L_-^I \longrightarrow L_+^O$.

Suppose now that L_+ is an invariant straight line. Then clearly Dom_{+-} and Dom_{-+} are empty sets, which contradicts our assumptions.

Statement (c) follows by using similar arguments to those used in the proof of statement (a). □

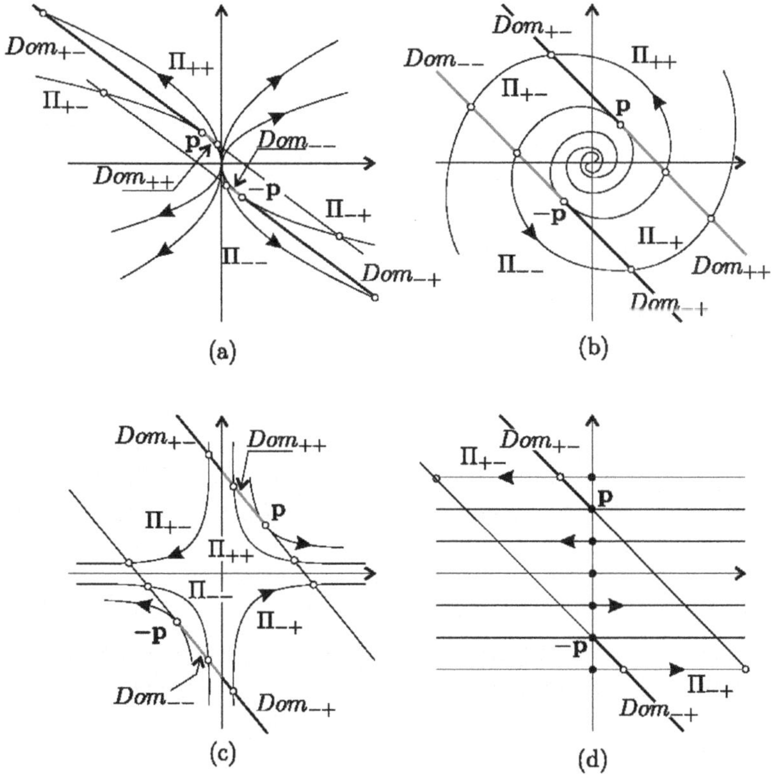

Figure 4.3: Domain of the Poincaré maps when $\det(A) > 0$ (a) and (b); when $\det(A) < 0$ (c); and when $\det(A) = 0$ (d).

In the following result we present necessary and sufficient conditions for the existence of the Poincaré maps Π_{jk} with $j, k \in \{+, -\}$.

Proposition 4.3.4. *Consider a planar linear system* $\dot{\mathbf{x}} = A\mathbf{x}$ *with non-zero matrix* A. *Let* L_+ *and* L_- *be two symmetric straight lines in the plane. The Poincaré maps associated to* L_+ *and* L_- *are defined if and only if the flow of the system has a unique contact point with* L_+.

Proof. Suppose that the Poincaré maps are defined. By definition, this implies that Dom_{jk} are not empty. Therefore, the flow has a unique contact point $\mathbf{p}_+$ with L_+ and a unique contact point $\mathbf{p}_-$ with L_-, see Lemma 4.3.2.

Conversely, suppose now that the flow has a unique contact point $\mathbf{p}_+$ with L_+. Then $\mathbf{p}_+$ lies in Dom_{++} and the Poincaré map Π_{++} is defined at it and satisfies that $\Pi_{++}(\mathbf{p}_+) = \mathbf{p}_+$. Moreover, by the symmetry of the vector field with respect to the origin, $\mathbf{p}_- = -\mathbf{p}_+$ is a contact point of the flow with L_-. Therefore, $\mathbf{p}_- \in \mathrm{Dom}_{--}$ and $\Pi_{--}(\mathbf{p}_-) = \mathbf{p}_-$.

It remains to prove that Π_{+-} and Π_{-+} are also defined. We have divided the proof according to the sign of $d = \det(A)$.

Suppose that $d > 0$. The contact points $\mathbf{p}_+$ and $\mathbf{p}_-$ split the straight lines L_+ and L_- into the half-lines L_+^I, L_+^O, L_-^I and L_-^O respectively. Let γ be the orbit through $\mathbf{p}_+$ and suppose that the eigenvalues of A have non-zero imaginary part. Since γ cannot be contained in S_0, where S_0 is the connected component of $\mathbb{R}^2 \setminus L_+$ containing the origin, we conclude that either $\gamma^+ \cap L_-^O \neq \varnothing$ or $\gamma^- \cap L_-^I \neq \varnothing$. Thanks to the continuous dependence of the solutions of a differential equation on the initial conditions, Π_{+-} and Π_{-+} are defined. Suppose now that the eigenvalues of A are real. By Proposition 4.2.10, it follows that $\gamma \subset S_0 \cup L_+$. Consider now the following two cases: (i) $\gamma \cap L_- \neq \varnothing$ and (ii) $\gamma \cap L_- = \varnothing$.

(i) In this case either γ^+ intersects L_-^O or γ^- intersects L_-^I. Suppose that $\gamma^+ \cap L_-^O \neq \varnothing$. The continuous dependence of the solutions of a differential equation on the initial conditions implies that the Poincaré map Π_{+-} is defined. Now suppose that $\gamma^+ \cap L_-^I \neq \varnothing$. In a similar way we obtain that Π_{-+} is defined.

(ii) Since $\gamma \subset S_0 \cup L_+$ and $\gamma \cap L_- = \varnothing$, the orbit γ lies in the strip bounded by L_+ and L_-. Moreover, since the system is linear and the eigenvalues of A are real numbers, either γ^+ or γ^- is not bounded. Suppose that γ^+ is not bounded. Let $\mathbf{x}(s)$ be the solution of the system such that $\mathbf{x}(0) = \mathbf{p}_+$. Then $\lim\limits_{s \nearrow +\infty} \Psi(\mathbf{x}(s)) = \Psi(\mathbf{p}_+) = \Psi(\mathbf{x}(0))$, which contradicts Lemma 4.2.9. Therefore γ intersects L_-. As we have seen in (i), this implies that Π_{+-} and Π_{-+} are defined.

Suppose that $d < 0$, i.e., the origin is a saddle point. Since $\mathbf{p}_+$ is a contact point of the flow with L_+, we can write $L_+ = \{\mathbf{p}_+ + \lambda \dot{\mathbf{p}}_+ : \lambda \in \mathbb{R}\}$. Suppose that L_+ is parallel to any separatrix of the saddle. Thus, $\dot{\mathbf{p}}_+$ is an eigenvector of A. Let $\lambda_1 \neq 0$ be the corresponding eigenvalue. Set $\mathbf{q} = \mathbf{p}_+ - \lambda_1^{-1} \dot{\mathbf{p}}_+ \in L_+$. Then $A\mathbf{q} = \mathbf{0}$ which implies that $\mathbf{q} = 0$. Consequently, the straight line L_+ passes through the origin, which contradicts our assumptions. Therefore, L_+ and L_- intersect both separatrices of the origin. By the continuous dependence of the solutions of a differential equation on the initial conditions, it follows that Π_{+-} and Π_{-+} are defined, see Figure 4.3(c).

Suppose now that $d = 0$. Since $\mathbf{p}_+$ is the unique contact point of the flow with the straight line L_+, we have that $L_+ \cap \ker(A) = \{\mathbf{p}_+\}$ and the orbits are contained in straight lines transversal to L_+ and L_-. It is easy to conclude that Π_{+-} and Π_{-+} are well defined, see Figure 4.3(d). □

4.3.1 Poincaré maps π_{jk}

Suppose that the Poincaré maps Π_{jk} of the homogeneous linear system (4.6) associated to the symmetric straight lines L_+ and L_- are defined. Hence, there exists a unique contact point $\mathbf{p}$ of the flow with L_+, see Proposition 4.3.4. Let $\mathbf{n}$ be the unit vector orthogonal to L_+ oriented in the opposite sense with respect to the origin, i.e., $\mathbf{n}^T\mathbf{p} > 0$. Then $L_+ = \{\mathbf{p} + \lambda\mathbf{v} : \lambda \in \mathbb{R}\}$ and $L_- = \{-\mathbf{p} + \lambda\mathbf{v} : \lambda \in \mathbb{R}\}$, where $\mathbf{v} = \dot{\mathbf{p}}$ or $\mathbf{v} = \mathbf{n}^\perp$ depending on whether $\det(A) \neq 0$ or $\det(A) = 0$.

If $\det(A) > 0$, then by Proposition 4.2.5(a.2),

$$\begin{aligned} L_+^I &= \{\mathbf{p} + a\dot{\mathbf{p}} : a \geq 0\}, & L_+^O &= \{\mathbf{p} - a\dot{\mathbf{p}} : a \geq 0\}, \\ L_-^I &= \{-\mathbf{p} - a\dot{\mathbf{p}} : a \geq 0\}, & L_-^O &= \{-\mathbf{p}+a\dot{\mathbf{p}} : a \geq 0\}, \end{aligned} \tag{4.8}$$

and if $\det(A) < 0$, then

$$\begin{aligned} L_+^I &= \{\mathbf{p} - a\dot{\mathbf{p}} : a \geq 0\}, & L_+^O &= \{\mathbf{p} + a\dot{\mathbf{p}} : a \geq 0\}, \\ L_-^I &= \{-\mathbf{p} + a\dot{\mathbf{p}} : a \geq 0\}, & L_-^O &= \{-\mathbf{p} - a\dot{\mathbf{p}} : a \geq 0\}. \end{aligned} \tag{4.9}$$

Suppose now that $\det(A) = 0$. If $\det(A\mathbf{n}^\perp, \mathbf{n}^\perp) < 0$, then by Proposition 4.2.6(d),

$$\begin{aligned} L_+^I &= \left\{\mathbf{p} + a\mathbf{n}^\perp : a \geq 0\right\}, & L_+^O &= \left\{\mathbf{p} - a\mathbf{n}^\perp : a \geq 0\right\}, \\ L_-^I &= \left\{-\mathbf{p} - a\mathbf{n}^\perp : a \geq 0\right\}, & L_-^O &= \left\{-\mathbf{p}+a\mathbf{n}^\perp : a \geq 0\right\}, \end{aligned} \tag{4.10}$$

and if $\det(A\mathbf{n}^\perp, \mathbf{n}^\perp) > 0$, then

$$\begin{aligned} L_+^I &= \left\{\mathbf{p} - a\mathbf{n}^\perp : a \geq 0\right\}, & L_+^O &= \left\{\mathbf{p} + a\mathbf{n}^\perp : a \geq 0\right\}, \\ L_-^I &= \left\{-\mathbf{p} + a\mathbf{n}^\perp : a \geq 0\right\}, & L_-^O &= \left\{-\mathbf{p}-a\mathbf{n}^\perp : a \geq 0\right\}. \end{aligned} \tag{4.11}$$

Using this parametrization of L_+ and L_- we can associate to any point $\mathbf{q}$ on L_+ and on L_- a unique value $a \geq 0$, called the *coordinate of* $\mathbf{q}$. The following statements are obvious:

(i) The unique points on L_+ and on L_- with coordinate equal to 0 are the contact points $\mathbf{p}$ and $-\mathbf{p}$;

(ii) two symmetric points in L_+ and in L_- have equal coordinates.

Take $\mathbf{q}_1 \in L_j$ and $\mathbf{q}_2 \in L_k$ such that $\mathbf{q}_2 = \Pi_{jk}(\mathbf{q}_1)$, where $j, k \in \{+, -\}$. Let a_1 and a_2 be the coordinates of $\mathbf{q}_1$ and $\mathbf{q}_2$, respectively. The *Poincaré map* π_{jk} is defined by $a_2 = \pi_{jk}(a_1)$. To know the qualitative behaviour of the Poincaré map π_{jk} with $j.k \in \{+, -\}$, is equivalent to know the qualitative behaviour of the Poincaré map Π_{jk}. The following results present some important properties of the Poincaré maps π_{jk}.

Lemma 4.3.5. *Consider a planar linear system $\dot{\mathbf{x}} = A\mathbf{x}$ with non-zero matrix A. Let L_+ and L_- be two symmetric straight lines in the plane. Suppose that the Poincaré maps π_{jk} with $j, k \in \{+, -\}$ are defined. Then:*

(a) *π_{++} and π_{--} coincide.*

(b) *π_{+-} and π_{-+} coincide.*

(c) *The Poincaré maps π^*_{jk} associated to the flow of the linear system $\dot{\mathbf{x}} = -A\mathbf{x}$ and the straight lines L_+ and L_- are defined, and they satisfy $\pi^*_{jk} = \pi^{-1}_{jk}$.*

(d) *π_{jk} are analytic functions. Moreover, the inverse of these functions are also analytic functions.*

Proof. Statements (a) and (b) follow immediately by using the symmetry of the vector field with respect to the origin and by noting that the coordinates of two symmetric points on L_+ and L_- are equal.

(c) The change of time variable $\tau = -s$ transforms the linear system $\dot{\mathbf{x}} = A\mathbf{x}$ into the linear system $\dot{\mathbf{x}} = -A\mathbf{x}$. Therefore, the orbits of the two systems coincide but have opposite orientation. We conclude that the Poincaré maps π^*_{jk} associated to the system $\dot{\mathbf{x}} = -A\mathbf{x}$ are defined and they coincide with the Poincaré maps π^{-1}_{jk}.

(d) The statement follows by noting that the flight time s is an analytic function of the initial condition $\mathbf{x}$ and that the flow $\Phi(s, \mathbf{x})$ of the linear system is also an analytic function of its arguments s and $\mathbf{x}$, see Subsection 2.7.2 for more details. □

In order to study the qualitative behaviour of all the Poincaré maps π_{jk} associated to a planar linear system it is enough to consider the maps π_{++} and π_{+-}, see Lemma 4.3.5(a) and (b). Moreover, we can restrict ourselves to the case when $\operatorname{trace}(A) \geq 0$. The case when $\operatorname{trace}(A) < 0$ then follows by taking π^{-1}_{++} and π^{-1}_{+-}, see Lemma 4.3.5(c).

Another important property of the Poincaré maps π_{jk} is that they are invariant by linear changes of coordinates. Hence, we can consider that the matrix of the linear system is expressed in the most convenient way, for instance in its real Jordan form. A generalization of this fact can be found in [45]. Before proving this, we show that the half-lines L^I_+, L^O_+, L^I_- and L^O_- are invariant under linear changes of coordinates.

Given a matrix $M \in GL(\mathbb{R}^2)$ and a subset L in the plane, we denote by ML the set $\{M\mathbf{q} : \mathbf{q} \in L\}$.

Lemma 4.3.6. *Let A be a non-zero matrix, $M \in GL(\mathbb{R}^2)$ and L a straight line in the plane. If $\mathbf{p}$ is a contact point of the flow of the planar system $\dot{\mathbf{x}} = A\mathbf{x}$ with L, then $\mathbf{p}^* = M\mathbf{p}$ is a contact point of the flow of the system $\dot{\mathbf{x}} = A^*\mathbf{x}$ with the straight line $L^* = ML$, where $A^* = MAM^{-1}$. Moreover, $\dot{\mathbf{p}}^* = M\dot{\mathbf{p}}$, $L^{*I} = ML^I$ and $L^{*O} = ML^O$.*

Proof. Since the linear change of coordinates $\mathbf{y} = M\mathbf{x}$ transforms the system $\dot{\mathbf{x}} = A\mathbf{x}$ into the system $\dot{\mathbf{y}} = A^*\mathbf{y}$, it is clear that $\mathbf{p}^* = M\mathbf{p}$ is a contact point of the flow of the new system with the straight line L^*. Moreover, $\dot{\mathbf{p}}^* = A^*\mathbf{p}^* = MA\mathbf{p} = M\dot{\mathbf{p}}$. The contact points $\mathbf{p}$ and $\mathbf{p}^*$ divide the straight lines L and L^* in two half-lines L^I, L^O and L^{*I}, L^{*O}, respectively. In order to show that $L^{*I} = ML^I$

and $L^{*O} = ML^{*O}$ we carry out the proof depending on the invertibility of the matrix A.

Suppose that $d = \det(A) > 0$ (the case $d < 0$ follows in a similar way). Hence $\det(A^*) > 0$. From Proposition 4.2.5(a.2) we have $L^I = \{\mathbf{p} + \lambda\dot{\mathbf{p}} : \lambda \geq 0\}$, $L^O = \{\mathbf{p} + \lambda\dot{\mathbf{p}} : \lambda \leq 0\}$, $L^{*I} = \{\mathbf{p}^* + \lambda\dot{\mathbf{p}}^* : \lambda \geq 0\}$ and $L^{*O} = \{\mathbf{p}^* + \lambda\dot{\mathbf{p}}^* : \lambda \leq 0\}$. Therefore, $L^{*I} = ML^I$ and $L^{*O} = ML^O$.

Suppose now that $d = 0$. Let $\mathbf{n}$ and $\mathbf{n}^*$ be the unit orthogonal vectors to L and L^*, respectively, which are oriented in the direction opposite to the origin. Since

$$\mathbf{p}^{*T}\left(M\mathbf{n}^\perp\right)^\perp = -\det\left(M\mathbf{p}, M\mathbf{n}^\perp\right) = -\det\left(M\right)\mathbf{p}^T\mathbf{n},$$

we have that

$$\mathbf{n}^* = -\operatorname{sign}\left(\det\left(M\right)\right)\frac{\left(M\mathbf{n}^\perp\right)^\perp}{||M\mathbf{n}^\perp||}, \quad \mathbf{n}^{*\perp} = \operatorname{sign}\left(\det\left(M\right)\right)\frac{M\mathbf{n}^\perp}{||M\mathbf{n}^\perp||}$$

and

$$\det\left(A^*\mathbf{n}^{*\perp}, \mathbf{n}^{*\perp}\right) = \frac{\det\left(M\right)}{||M\mathbf{n}^\perp||}\det\left(A\mathbf{n}^\perp, \mathbf{n}^\perp\right).$$

Suppose that $\det(A\mathbf{n}^\perp, \mathbf{n}^\perp) > 0$ (the case $\det(A\mathbf{n}^\perp, \mathbf{n}^\perp) < 0$ follows similarly). Then $L^I = \{\mathbf{p} + \lambda\mathbf{n}^\perp : \lambda \leq 0\}$ and $L^O = \{\mathbf{p} + \lambda\mathbf{n}^\perp : \lambda \geq 0\}$, see Proposition 4.2.6(d). If $\det(M) > 0$, then $\det(A^*\mathbf{n}^{*\perp}, \mathbf{n}^{*\perp}) > 0$. From Proposition 4.2.6(d) we obtain that $L^{*I} = \{\mathbf{p}^* + \lambda\mathbf{n}^{*\perp} : \lambda \leq 0\}$ and $L^{*O} = \{\mathbf{p}^* + \lambda\mathbf{n}^{*\perp} : \lambda \geq 0\}$. Therefore, $L^{*I} = ML^I$ and $L^{*O} = ML^O$. If $\det(M) < 0$, then $\det(A^*\mathbf{n}^{*\perp}, \mathbf{n}^{*\perp}) < 0$ and $L^{*I} = \{\mathbf{p}^* + \lambda\mathbf{n}^{*\perp} : \lambda \geq 0\}$ and $L^{*O} = \{\mathbf{p}^* + \lambda\mathbf{n}^{*\perp} : \lambda \leq 0\}$. In this case we also conclude that $L^{*I} = ML^I$ and $L^{*O} = ML^O$. □

Proposition 4.3.7. *Consider a linear system $\dot{\mathbf{x}} = A\mathbf{x}$ with $A \in GL(\mathbb{R}^2)$ and let L_+ and L_- be two symmetric straight lines in the plane. Suppose that the Poincaré maps π_{jk} associated to L_+ and L_- are defined. If $M \in GL(\mathbb{R}^2)$, then the Poincaré maps are invariant under the change of coordinates $\mathbf{y} = M\mathbf{x}$.*

Proof. Since the Poincaré maps π_{jk} are defined, there exists a contact point $\mathbf{p}$ of the flow with L_+, see Proposition 4.3.4. Thus $\mathbf{p}^* = M\mathbf{p}$ is the contact point of the flow of the system $\dot{\mathbf{x}} = A^*\mathbf{x}$ with the straight line $L_+^* = ML_+$, where $A^* = MAM^{-1}$, see Lemma 4.3.6. By Proposition 4.3.4, the Poincaré maps π_{jk}^* associated to the flow of the system $\dot{\mathbf{x}} = A^*\mathbf{x}$ and the straight lines L_+^* and $L_-^* = ML_-$ are defined. We will prove that $\pi_{+k}^*(a) = \pi_{+k}(a)$ for $k \in \{+, -\}$.

Consider a value a in the domain of the map π_{++}. Hence, there exist a point $\mathbf{q}_1$ in L_+^O and a point $\mathbf{q}_2$ in L_+^I such that $\mathbf{q}_1 = \mathbf{p} - a\dot{\mathbf{p}}$, $\mathbf{q}_2 = \mathbf{p} + \pi_{++}(a)\,\dot{\mathbf{p}}$ and $\mathbf{q}_2 = \Pi_{++}(\mathbf{q}_1)$. Suppose that $\det(A) > 0$ (the case when $\det(A) < 0$ follows in a similar way). Then $L_+^{*O} = ML_+^O$ and $L_+^{*I} = ML_+^I$, see Lemma 4.3.6. Therefore $\mathbf{q}_1^* = M\mathbf{q}_1 = \mathbf{p}^* - aM\dot{\mathbf{p}} \in L_+^{*O}$, $\mathbf{q}_2^* = M\mathbf{q}_2 = \mathbf{p}^* + \pi_{++}(a)M\dot{\mathbf{p}} \in L_+^{*I}$ and $\mathbf{q}_2^* = \Pi_{++}^*(\mathbf{q}_1^*)$. Noting that $\dot{\mathbf{p}}^* = A^*\mathbf{p}^* = M\dot{\mathbf{p}}$, it follows that $\pi_{++}^*(a) = \pi_{++}(a)$.

Consider now that a belongs to the domain of the map π_{+-}. Using similar arguments it can be proved that $\pi_{+-}^*(a) = \pi_{+-}(a)$. □

Next we prove that when $\det(A) = 0$ the Poincaré map π_{+-} is not invariant under linear changes of coordinates. In this case, however, the new Poincaré map π^*_{+-} is just a scaled version of π_{+-}.

Proposition 4.3.8. *Consider a planar linear system $\dot{\mathbf{x}} = A\mathbf{x}$ with non-zero matrix A such that $\det(A) = 0$. Let L_+ and L_- be two symmetric straight lines in the plane. Suppose that the Poincaré map π_{+-} associated to L_+ and L_- is defined. If $M \in GL(\mathbb{R}^2)$, then the Poincaré map π^*_{+-} associated to the flow of the system $\dot{\mathbf{x}} = A^*\mathbf{x}$, where $A^* = MAM^{-1}$, and the straight lines ML_+ and ML_- are defined. Moreover there exists $K > 0$ such that $\pi^*_{+-}(Ka) = K\pi_{+-}(a)$.*

Proof. Since π_{+-} is defined, there exists a contact point $\mathbf{p}$ of the flow with L_+. Then $\mathbf{p}^* = M\mathbf{p}$ is a contact point of the flow of the system $\dot{\mathbf{x}} = A^*\mathbf{x}$ with $L^*_+ = ML_+$. We conclude that the Poincaré map π^*_{+-} is well defined, see Proposition 4.3.4.

Suppose that $\det(A\mathbf{n}^\perp, \mathbf{n}^\perp) > 0$ (similar arguments can be applied in the case $\det(A\mathbf{n}^\perp, \mathbf{n}^\perp) < 0$). Take $a \geq 0$ belonging to the domain of π_{+-}. There exist $\mathbf{q}_1 \in L^I_+$ and $\mathbf{q}_2 \in L^O_-$ such that $\mathbf{q}_1 = \mathbf{p} - a\mathbf{n}^\perp$ and $\mathbf{q}_2 = -\mathbf{p} - \pi_{+-}(a)\mathbf{n}^\perp$. Since $L^{*I}_+ = ML^I_+$ and $L^{*O}_- = ML^O_-$, the points $\mathbf{q}^*_1 = M\mathbf{q}_1$ and $\mathbf{q}^*_2 = M\mathbf{q}_2$ satisfy

$$\mathbf{q}^*_1 = \mathbf{p}^* - \operatorname{sign}(\det(M))\, a^*\mathbf{n}^{*\perp},$$
$$\mathbf{q}^*_2 = -\mathbf{p}^* - \operatorname{sign}(\det(M))\, \pi^*_{+-}(a^*)\, \mathbf{n}^{*\perp}.$$

We conclude that

$$\operatorname{sign}(\det(M))\, a^*\mathbf{n}^{*\perp} = aM\mathbf{n}^\perp,$$
$$\operatorname{sign}(\det(M))\, \pi^*_{+-}(a^*)\, \mathbf{n}^{*\perp} = \pi_{+-}(a)\, M\mathbf{n}^\perp.$$

Since

$$\mathbf{n}^{*\perp} = \operatorname{sign}(\det(M))\, \frac{M\mathbf{n}^\perp}{||M\mathbf{n}^\perp||},$$

setting $K = \|M\mathbf{n}^\perp\| > 0$ it follows that $a^* = Ka$ and $\pi^*_{+-}(a^*) = K\,\pi_{+-}(a)$. Consequently, $\pi^*_{+-}(Ka) = K\,\pi_{+-}(a)$, as claimed. □

As we have proved in Propositions 4.3.7 and 4.3.8, in order to study the qualitative behaviour of the Poincaré maps π_{jk} we can assume that the matrix of the linear system is given in real Jordan normal form. Looking at the implicit expressions for the Poincaré maps in Proposition 4.3.11, it is easy to understand the simplification in computations that this argument introduces.

Corollary 4.3.9. *Consider a planar linear system $\dot{\mathbf{x}} = A\mathbf{x}$ such that the trace t and the determinant d of the matrix A satisfy $t^2 - 4d \neq 0$. Let L_+ and L_- be two symmetric straight lines in the plane and suppose that the Poincaré maps π_{jk} associated to L_+ and L_- are defined. Then π_{jk} are analytic functions with respect to the parameters t and d.*

Proof. Without loss of generality we can assume that the matrix A is in its real Jordan normal form, see Propositions 4.3.7 and 4.3.8. Since $t^2 - 4d \neq 0$, the coefficients of the Jordan matrix A are analytic functions with respect to the parameters t and d. The corollary follows from the differentiable dependence of the solutions of a linear differential system on parameters. □

4.3.2 Existence of the Poincaré maps

In the next result we characterize the existence of the Poincaré maps π_{jk}.

Theorem 4.3.10. *Consider a planar linear system* $\dot{\mathbf{x}} = A\mathbf{x}$ *with non-zero matrix* A*. Let* L_+ *and* L_- *be two symmetric straight lines in the plane. The following statements are equivalent.*

(a) *The Poincaré maps* π_{jk} *associated to* L_+ *and* L_- *are defined.*

(b) *There exists a unique contact point of the flow of the linear system with* L_+*.*

(c) *If* $\mathbf{v}$ *is a vector parallel to the straight line* L_+*, then* $\mathbf{v}$ *and* $A\mathbf{v}$ *are linearly independent, i.e.,* $\det(A\mathbf{v}, \mathbf{v}) \neq 0$.

Proof. The equivalence between statements (a) and (b) is exactly Proposition 4.3.4. Then it remains to prove that statements (b) and (c) are equivalent.

Let $\mathbf{v}$ be a vector parallel to L_+ and $\mathbf{q}$ a point in L_+. The straight lines L_+ and L_- can be expressed as $L_+ = \{\mathbf{q} + \lambda\mathbf{v} : \lambda \in \mathbb{R}\}$ and $L_- = \{-\mathbf{q} + \lambda\mathbf{v} : \lambda \in \mathbb{R}\}$. We divide the proof into two cases, depending on whether $\det(A) = 0$ or $\det(A) \neq 0$.

Suppose that $\det(A) \neq 0$. By Lemma 4.2.3, the straight line $L = \{\lambda A^{-1}\mathbf{v} : \lambda \in \mathbb{R}\}$ is the isocline defined by $\mathbf{v}$. There exists a unique contact point of the flow with L_+ if and only if L_+ and L intersect at a unique point, i.e., $\det(A^{-1}\mathbf{v}, \mathbf{v}) \neq 0$ or, equivalently, $\det(A\mathbf{v}, \mathbf{v}) \neq 0$.

Suppose now that $\det(A) = 0$. Since A is not the zero matrix, if there exists a unique contact point of the flow with L_+, then L_+ is a non-invariant straight line. Suppose that $\det(A\mathbf{v}, \mathbf{v}) = 0$. This implies that $A\mathbf{v} = \alpha\mathbf{v}$ for some $\alpha \in \mathbb{R}$, and so L_+ is an invariant straight line, which contradicts our assumption. Therefore, $\det(A\mathbf{v}, \mathbf{v}) \neq 0$. Conversely, if $\det(A\mathbf{v}, \mathbf{v}) \neq 0$, then the straight lines L_+ and $\ker(A)$ are not parallel. Therefore L_+ and $\ker(A)$ intersect at a unique point $\mathbf{p}$, which is a singular point. Moreover, $L_+ = \{\mathbf{p} + \lambda\mathbf{v} : \lambda \in \mathbb{R}\}$ is a non-invariant straight line. By Proposition 4.2.6(c), $\mathbf{p}$ is the unique contact point of the flow with L_+. □

4.3.3 Implicit equations of the Poincaré maps π_{jk}

We now present some implicit equations for π_{jk} depending on the sign of the determinant of the matrix A.

Proposition 4.3.11. *Consider a planar linear system* $\dot{\mathbf{x}} = A\mathbf{x}$ *with non-zero matrix* A. *Let* L_+ *and* L_- *be two symmetric straight lines in the plane and let* $\mathbf{p}$ *be a contact point of the flow with* L_+.

(a) *If* $\det(A) > 0$, *then there exist two analytic functions,* $s_{++}(a) \geq 0$ *and* $s_{+-}(a) > 0$, *such that*

$$
\begin{aligned}
(\mathrm{Id} + \pi_{++}(a)\, A)\, \mathbf{p} &= e^{s_{++}(a)A}\, (\mathrm{Id} - aA)\, \mathbf{p}, \\
(-\mathrm{Id} + \pi_{+-}(a)\, A)\, \mathbf{p} &= e^{s_{+-}(a)A}\, (\mathrm{Id} + aA)\, \mathbf{p}.
\end{aligned}
$$

(b) *If* $\det(A) < 0$, *then there exist two analytic functions,* $s_{++}(a) \geq 0$ *and* $s_{+-}(a) > 0$, *such that*

$$
\begin{aligned}
(\mathrm{Id} + \pi_{++}(a)\, A)\, \mathbf{p} &= e^{s_{++}(a)A}\, (\mathrm{Id} - aA)\, \mathbf{p}, \\
(-\mathrm{Id} - \pi_{+-}(a)\, A)\, \mathbf{p} &= e^{s_{+-}(a)A}\, (\mathrm{Id} - aA)\, \mathbf{p}.
\end{aligned}
$$

(c) *Suppose that* $\det(A) = 0$. *There exists an analytic function* $s_{+-}(a) > 0$ *such that*

(c.1) *if* $\det(A\mathbf{n}^\perp, \mathbf{n}^\perp) > 0$, *then* $-\mathbf{p} - \pi_{+-}(a)\mathbf{n}^\perp = e^{s_{+-}(a)A}(\mathbf{p} - a\mathbf{n}^\perp)$,

(c.2) *if* $\det(A\mathbf{n}^\perp, \mathbf{n}^\perp) < 0$, *then* $-\mathbf{p} + \pi_{+-}(a)\mathbf{n}^\perp = e^{s_{+-}(a)A}(\mathbf{p} + a\mathbf{n}^\perp)$.

Proof. (a) Since $\det(A) > 0$ it follows that $\mathrm{Dom}_{++} \subset L_+^O$ and $\Pi_{++}(\mathrm{Dom}_{++}) \subset L_+^I$, see Proposition 4.3.3(a). Take $\mathbf{q}_1 \in L_+^O$ and $\mathbf{q}_2 \in L_+^I$ such that $\mathbf{q}_2 = \Pi_{++}(\mathbf{q}_1)$, and let a be the coordinate of $\mathbf{q}_1$. Then $\pi_{++}(a)$ is the coordinate of $\mathbf{q}_2$, i.e., $\mathbf{q}_1 = \mathbf{p} - a\dot{\mathbf{p}}$ and $\mathbf{q}_2 = \mathbf{p} + \pi_{++}(a)\dot{\mathbf{p}}$, see (4.8).

On the other hand, since $\mathbf{q}_2 = \Pi_{++}(\mathbf{q}_1)$, there exists an analytic function $s_{\mathbf{q}_1}$ (the flight time) such that $\mathbf{q}_2 = e^{s_{\mathbf{q}_1}A}\mathbf{q}_1$. Therefore, setting $s_{++}(a) = s_{\mathbf{q}_1}$ we obtain that $\mathbf{p} + \pi_{++}(a)\dot{\mathbf{p}} = e^{s_{++}(a)A}(\mathbf{p} - a\dot{\mathbf{p}})$. The statement follows by using the fact that $\dot{\mathbf{p}} = A\mathbf{p}$.

The remaining statements can be proved in a similar way. □

4.4 Qualitative behaviour of the maps π_{jk}

Suppose that the Poincaré maps π_{++} and π_{+-} associated to the flow of a planar homogeneous linear system $\dot{\mathbf{x}} = A\mathbf{x}$ and two symmetric straight lines in the plane L_+ and L_- are defined. In this section we study the qualitative behaviour of the maps π_{++} and π_{+-} depending on the values of $t = \mathrm{trace}(A)$ and $d = \det(A)$.

Since π_{++} and π_{+-} are invariant under linear changes of coordinates when $d \neq 0$, see Proposition 4.3.7, and their qualitative behaviour does not change when $d = 0$, see Proposition 4.3.8, we can consider without loss of generality that A is in real Jordan normal form. Accordingly, this section is divided in subsections, each devoted to one of the real Jordan normal forms.

Moreover, we need to study only the behaviour of π_{++} and π_{+-} when $t > 0$: the case $t < 0$ can be obtained from the case $t > 0$ by Lemma 4.3.5(c). Accordingly, in each of the following subsections we summarize the qualitative behaviour of π_{++} and π_{+-} in two propositions and two corollaries. We deal first with the case $t > 0$.

Throughout this section we will write $f'(a)$ and $f''(a)$ to denote the first and the second derivative of a function $f(a)$ with respect to a.

4.4.1 Diagonal node: $d > 0$ and $t^2 - 4d > 0$

It is known that in this case the matrix A has two real and distinct eigenvalues $\lambda_1 = (t + \sqrt{t^2 - 4d})/2$ and $\lambda_2 = (t - \sqrt{t^2 - 4d})/2$, and the real Jordan normal form of A is

$$\begin{pmatrix} \lambda_1 & 0 \\ 0 & \lambda_2 \end{pmatrix}.$$

Proposition 4.4.1. *Consider $A \in GL(\mathbb{R}^2)$ such that $d > 0$, $t > 0$ and $t^2 - 4d > 0$. Then the eigenvalues of A satisfy $\lambda_1 > \lambda_2 > 0$. Let π_{++} be the Poincaré map defined by the flow of the linear system $\dot{\mathbf{x}} = A\mathbf{x}$ and associated to two symmetric straight lines in the plane, L_+ and L_-. Then the following holds:*

(a) $\pi_{++} : [0, \lambda_1^{-1}) \to [0, +\infty)$, $\pi_{++}(0) = 0$, $\lim_{a \nearrow \lambda_1^{-1}} \pi_{++}(a) = +\infty$ *and* $\pi_{++}(a) > a$ *in* $(0, \lambda_1^{-1})$.

(b) *If* $a \in (0, \lambda_1^{-1})$, *then* $\pi'_{++}(a) > 1$ *and* $\lim_{a \searrow 0} \pi'_{++}(a) = 1$.

(c) *If* $a \in (0, \lambda_1^{-1})$, *then* $\pi''_{++}(a) > 0$.

(d) *The graph of* π_{++} *has a vertical asymptote at* $a = \lambda_1^{-1}$.

(e) π_{++} *is implicitly defined by the expression*

$$\left(\frac{2 + \pi_{++}(a)\left(t - \sqrt{t^2 - 4d}\right)}{2 - a\left(t - \sqrt{t^2 - 4d}\right)} \right)^{\frac{t + \sqrt{t^2 - 4d}}{t - \sqrt{t^2 - 4d}}} = \frac{2 + \pi_{++}(a)\left(t + \sqrt{t^2 - 4d}\right)}{2 - a\left(t + \sqrt{t^2 - 4d}\right)}.$$

(f) *The qualitative behaviour of the graph of* π_{++} *is shown in Figure* 4.5(a).

Proof. By Proposition 4.3.11(a), the map π_{++} satisfies that

$$(\mathrm{Id} + \pi_{++}(a) A)\,\mathbf{p} = e^{s_{++}(a)A}\,(\mathrm{Id} - aA)\,\mathbf{p},$$

where $\mathbf{p} = (p_1, p_2)^T$ is the contact point of the flow of the linear system with L_+, $a \geq 0$, $b = \pi_{++}(a) \geq 0$ and $s = s_{++}(a) \geq 0$.

Without loss of generality we can assume that the matrix A is in real Jordan normal form. Thus the coordinate axes are invariant under the flow. Since $\mathbf{p}$ is the contact point and L_+ does not pass through the origin, $p_1 \neq 0$ and $p_2 \neq 0$. Therefore the map $b = \pi_{++}(a)$ is determined by the system

$$1 + b\lambda_1 = (1 - a\lambda_1)\, e^{\lambda_1 s}, \quad 1 + b\lambda_2 = (1 - a\lambda_2)\, e^{\lambda_2 s}, \tag{4.12}$$

and the inequalities $a \geq 0$, $b \geq 0$ and $s \geq 0$.

(a) Note that $s = 0$, $a = 0$ and $b = 0$ is a solution of system (4.12). Hence, $\pi_{++}(0) = 0$. Furthermore, since we are interested in solutions with $b \geq 0$, it follows that $1 - a\lambda_1 > 0$ and $1 - a\lambda_2 > 0$, i.e., $a < \lambda_1^{-1} < \lambda_2^{-1}$. Therefore, the domain of π_{++} is contained in $[0, \lambda_1^{-1})$.

Now we find the solutions of system (4.12) such that $s > 0$. Multiplying the first equation by λ_2, the second one by λ_1 and substracting the first from the second, we obtain the following parametric equations of π_{++}:

$$\begin{aligned} a\,(s) &= \frac{\lambda_1 \left(1 - e^{\lambda_2 s}\right) - \lambda_2 \left(1 - e^{\lambda_1 s}\right)}{d\,(e^{\lambda_1 s} - e^{\lambda_2 s})}, \\ b\,(s) &= \frac{(\lambda_2 - \lambda_1)\, e^{ts} - \lambda_2 e^{\lambda_2 s} + \lambda_1 e^{\lambda_1 s}}{d\,(e^{\lambda_2 s} - e^{\lambda_1 s})}, \end{aligned} \tag{4.13}$$

where $t = \lambda_1 + \lambda_2$ and $d = \lambda_1 \lambda_2$. Since $\lambda_1 > \lambda_2$, functions $a(s)$ and $b(s)$ are defined and differentiable in $(0, +\infty)$. Moreover, $a(s) > 0$ in $(0, +\infty)$, $\lim\limits_{s \searrow 0} b(s) = 0$ and $\lim\limits_{s \searrow 0} a(s) = 0$.

Differentiating with respect to s in (4.12) and isolating da/ds and db/ds we obtain

$$\frac{da}{ds} = b \frac{\lambda_1 - \lambda_2}{e^{\lambda_1 s} - e^{\lambda_2 s}}, \quad \frac{db}{ds} = a \frac{\lambda_1 - \lambda_2}{e^{\lambda_1 s} - e^{\lambda_2 s}} e^{ts}.$$

Since $a(s) > 0$ in $(0, +\infty)$, we have $b'(s) > 0$. Hence, $b(s) > 0$ and consequently $a'(s) > 0$. Finally, from (4.13) it follows that $\lim\limits_{s \nearrow +\infty} a(s) = \lambda_1^{-1}$ and $\lim\limits_{s \nearrow +\infty} b(s) = +\infty$. We conclude that $\pi_{++} : [0, \lambda_1^{-1}) \to [0, +\infty)$ and $\lim\limits_{a \nearrow \lambda_1^{-1}} \pi_{++}(a) = +\infty$.

Note that in order to finish the proof of the statement (a) it remains to verify that $\pi_{++}(a) > a$ in $[0, \lambda_1^{-1})$. This inequality will be shown in the proof of the statement (b) below.

(d) The statement follows by noting that

$$\lim_{a \nearrow \lambda_1^{-1}} \frac{db}{da} = \lim_{a \nearrow \lambda_1^{-1}} \frac{a e^{ts}}{b} = +\infty.$$

(e) Isolating s in (4.12) we obtain

$$\left(\frac{1 + b\lambda_1}{1 - a\lambda_1} \right)^{\frac{\lambda_2}{\lambda_1}} = \frac{1 + b\lambda_2}{1 - a\lambda_2}, \tag{4.14}$$

which is an implicit expression of $\pi_{++}(a)$. The expression of the statement (e) is obtained by substituting $\lambda_1 = (t + \sqrt{t^2 - 4d})/2$ and $\lambda_2 = (t - \sqrt{t^2 - 4d})/2$ in (4.14)

(b) and (c) Differentiating expression (4.14) with respect to a and isolating db/da we have that

$$\pi'_{++}(a) = \frac{db}{da} = \frac{a}{b} \frac{1 + bt + b^2 d}{1 - at + a^2 d}. \tag{4.15}$$

The qualitative behaviour of the parabolas $1+bt+b^2d$ and $1-at+a^2d$ is represented in Figure 4.4. It is easy to conclude that $\pi'_{++}(a) > 0$ in $(0, \lambda_1^{-1})$. Moreover, from (4.15) and (4.13) it follows that

$$\lim_{a \searrow 0} \pi'_{++}(a) = \lim_{s \searrow 0} \frac{a(s)}{b(s)} = \lim_{s \searrow 0} \frac{\lambda_2 \left(1 - e^{\lambda_1 s}\right) - \lambda_1 \left(1 - e^{\lambda_2 s}\right)}{(\lambda_2 - \lambda_1) e^{ts} - \lambda_2 e^{\lambda_2 s} + \lambda_1 e^{\lambda_1 s}}.$$

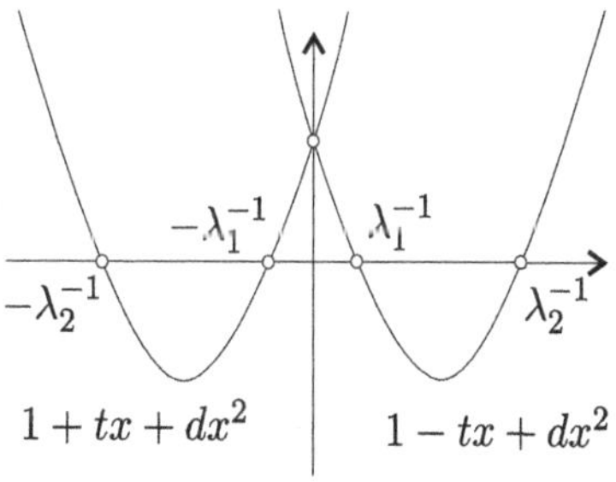

Figure 4.4: Qualitative behaviour of $1 + t\,x + d\,x^2$ and $1 - t\,x + d\,x^2$ when $d > 0$, $t > 0$ and $t^2 - 4d > 0$.

Hence, by applying l'Hôpital's rule twice we have

$$\lim_{a \searrow 0} \pi'_{++}(a) = \frac{d(\lambda_2 - \lambda_1)}{t^2(\lambda_2 - \lambda_1) - \lambda_2^3 + \lambda_1^3} = 1.$$

On the other hand, differentiating with respect to a in (4.15) we obtain

$$\pi''_{++}(a) = \frac{d}{da}\left(\frac{db}{da}\right) = \frac{(b-a)(b+a)}{ab^2(1 - at + a^2 d)} \frac{db}{da}. \tag{4.16}$$

Note that it remains to be proved that $\pi_{++}(a) > a$ in $(0, \lambda_1^{-1})$ because in this case $\pi''_{++}(a) > 0$, which will finish the proof.

Suppose that there exists a value a_0 in $(0, \lambda_1^{-1})$ such that $\pi_{++}(a_0) = a_0$. The function $g(a) = \pi_{++}(a) - a$ is continuously defined in $[0, a_0]$ and analytic in $(0, a_0)$. Moreover, $g(0) = g(a_0) = 0$. Thus, there exists $\xi \in (0, a_0)$ such that $g'(\xi) = 0$ or, equivalently, $\pi'_{++}(\xi) = 1$. The function

$$h(a) = \begin{cases} 1, & \text{if } a = 0, \\ \pi'_{++}(a), & \text{if } a \in (0, \xi], \end{cases}$$

is continuous in $[0,\xi]$, differentiable in $(0,\xi)$ and satisfies that $h(0) = h(\xi) = 1$. Hence, there exists $a_1 \in (0,\xi)$ such that $h'(a_1) = 0$, equivalently, $\pi''_{++}(a_1) = 0$. By (4.16), this implies that $\pi_{++}(a_1) = a_1$, equivalently, $g(a_1) = 0$. Using this argument repeatedly we obtain a strictly decreasing sequence $\{a_n\}_{n=0}^{+\infty}$ such that $g(a_k) = 0$ for any $k \geq 0$, which contradicts the fact that g is a non-identically zero analytic function. Then $\pi_{++}(a) \neq a$ in $(0, \lambda_1^{-1})$. Noting that $\lim_{a \nearrow \lambda_1^{-1}} \pi_{++}(a) = +\infty$, see statement (a), it follows that $\pi_{++}(a) > a$ in $(0, \lambda_1^{-1})$. □

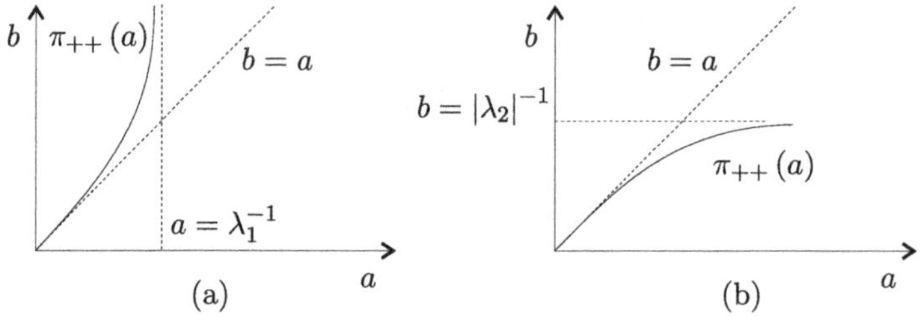

Figure 4.5: Qualitative behaviour of the Poincaré map π_{++} for the parameters (a) $d > 0$, $t > 0$ and $t^2 - 4d > 0$, (b) $d > 0$, $t < 0$ and $t^2 - 4d > 0$.

Corollary 4.4.2. *Consider $A \in GL(\mathbb{R}^2)$ such that $d > 0$, $t < 0$ and $t^2 - 4d > 0$. Then the eigenvalues of A satisfy $0 > \lambda_1 > \lambda_2$. Let π_{++} be the Poincaré map defined by the flow of the linear system $\dot{\mathbf{x}} = A\mathbf{x}$ associated to two symmetric straight lines in the plane, L_+ and L_-. Then:*

(a) $\pi_{++} : [0, +\infty) \to [0, |\lambda_2|^{-1})$, $\pi_{++}(0) = 0$, $\lim_{a \nearrow +\infty} \pi_{++}(a) = |\lambda_2|^{-1}$ *and* $\pi_{++}(a) < a$ *in* $(0, |\lambda_2|^{-1})$.

(b) *If* $a \in (0, +\infty)$, *then* $0 < \pi'_{++}(a) < 1$ *and* $\lim_{a \searrow 0} \pi'_{++}(a) = 1$.

(c) *If* $a \in (0, +\infty)$, *then* $\pi''_{++}(a) < 0$.

(d) *The graph of* π_{++} *has a horizontal asymptote at* $b = |\lambda_2|^{-1}$.

(e) π_{++} *is implicitly defined by the expression*

$$\left(\frac{2 + \pi_{++}(a)\left(t - \sqrt{t^2 - 4d}\right)}{2 - a\left(t - \sqrt{t^2 - 4d}\right)} \right)^{\frac{t + \sqrt{t^2-4d}}{t - \sqrt{t^2-4d}}} = \frac{2 + \pi_{++}(a)\left(t + \sqrt{t^2 - 4d}\right)}{2 - a\left(t + \sqrt{t^2 - 4d}\right)}.$$

(f) *The qualitative behaviour of the graph of* π_{++} *is shown in Figure* 4.5(b).

Proof. The proof follows straightforward by noting that the map π_{++} is the inverse of the map described at Proposition 4.4.1, see Lemma 4.3.5(c). □

Proposition 4.4.3. *Consider $A \in GL(\mathbb{R}^2)$ such that $d > 0$, $t > 0$ and $t^2 - 4d > 0$. Then the eigenvalues of A satisfy $\lambda_1 > \lambda_2 > 0$. Let π_{+-} be the Poincaré map defined by the flow of the linear system $\dot{\mathbf{x}} = A\mathbf{x}$ and associated to two symmetric straight lines in the plane L_+ and L_-. Then:*

(a) *There exists a value $b^* > \lambda_2^{-1}$ such that $\pi_{+-} : [0, +\infty) \to [b^*, +\infty)$, $\pi_{+-}(0) = b^*$, $\lim\limits_{a \nearrow +\infty} \pi_{+-}(a) = +\infty$, and $\pi_{+-}(a) > a$ in $[0, +\infty)$.*

(b) *If $a \in (0, +\infty)$, then $0 < \pi'_{+-}(a) < 1$ and $\lim\limits_{a \searrow 0} \pi'_{+-}(a) = 0$.*

(c) *If $a \in [0, +\infty)$, then $\pi''_{+-}(a) > 0$.*

(d) *The straight line $b = a + 2t/d$ is an asymptote of the graph of π_{+-} when a tends to $+\infty$.*

(e) *π_{+-} is implicitly defined by the expression*

$$\left(\frac{\pi_{+-}(a) \left(t - \sqrt{t^2 - 4d}\right) - 2}{a \left(t - \sqrt{t^2 - 4d}\right) + 2} \right)^{\frac{t + \sqrt{t^2 - 4d}}{t - \sqrt{t^2 - 4d}}} = \frac{\pi_{+-}(a) \left(t + \sqrt{t^2 - 4d}\right) - 2}{a \left(t + \sqrt{t^2 - 4d}\right) + 2}.$$

(f) *The qualitative behaviour of the graph of π_{+-} is shown in Figure 4.6(a).*

Proof. Applying arguments similar to those in the proof of Proposition 4.4.1, we obtain that $b = \pi_{+-}(a)$ is determined by the system

$$-1 + b\lambda_1 = (1 + a\lambda_1)\, e^{\lambda_1 s}, \quad -1 + b\lambda_2 = (1 + a\lambda_2)\, e^{\lambda_2 s}, \tag{4.17}$$

and the inequalities $a \geq 0$, $b \geq 0$ and $s > 0$.

(a) and (d) Since $a \geq 0$, from (4.17) it follows that $-1 + b\lambda_1 > 0$ and $-1 + b\lambda_2 > 0$, i.e., $b > \lambda_2^{-1} > \lambda_1^{-1}$. Thus the image of π_{+-} is contained in $(\lambda_1^{-1}, +\infty)$.

Multiplying the first equation in (4.17) by λ_2, the second one by λ_1 and substracting the second equation from the first, we obtain the parametric equations of π_{+-}:

$$\begin{aligned} a(s) &= \frac{\lambda_2 \left(1 + e^{\lambda_1 s}\right) - \lambda_1 \left(1 + e^{\lambda_2 s}\right)}{d \left(e^{\lambda_2 s} - e^{\lambda_1 s}\right)}, \\ b(s) &= \frac{(\lambda_2 - \lambda_1)\, e^{ts} + \lambda_2 e^{\lambda_2 s} - \lambda_1 e^{\lambda_1 s}}{d \left(e^{\lambda_2 s} - e^{\lambda_1 s}\right)}. \end{aligned} \tag{4.18}$$

The auxiliary function $f(s) = \lambda_2(1 + e^{\lambda_1 s}) - \lambda_1(1 + e^{\lambda_2 s})$ satisfies that $f'(s) = d(e^{\lambda_1 s} - e^{\lambda_2 s}) > 0$, $f(0) = 2(\lambda_2 - \lambda_1) < 0$ and $\lim\limits_{s \nearrow +\infty} f(s) = +\infty$. Consequently, there exists a unique value $s^* > 0$ such that $f(s^*) = 0$, equivalently, $a(s^*) = 0$. Hence, if $s \in (0, s^*)$, then $a(s) > 0$ and $b(s) > 0$.

Differentiating with respect to s in (4.17) and isolating da/ds and db/ds, we obtain that

$$\frac{da}{ds}=\frac{e^{\lambda_2 s}(1+a\lambda_2)-e^{\lambda_1 s}(1+a\lambda_1)}{e^{\lambda_1 s}-e^{\lambda_2 s}},\quad \frac{db}{ds}=a\frac{(\lambda_2-\lambda_1)e^{ts}}{e^{\lambda_1 s}-e^{\lambda_2 s}}.$$

Since $da/ds<0$ in $(0,s^*]$, $\lim\limits_{s\searrow 0}a(s)=+\infty$ and $a(s^*)=0$, the domain of definition of $\pi_{+-}(a)$ is $[0,+\infty)$. Moreover, since $db/ds<0$ in $(0,s^*]$ and $\lim\limits_{s\searrow 0}b(s)=+\infty$, the image of $\pi_{+-}(a)$ is contained in $[b^*,+\infty)$ where $b^*=b(s^*)>\lambda_2^{-1}$.

Finally from expression (4.18) we have $\lim\limits_{a\nearrow+\infty}\pi_{+-}(a)/a=\lim\limits_{s\searrow 0}b/a=1$ and

$$b(s)-a(s)=\frac{(\lambda_2-\lambda_1)(e^{ts}-1)+\lambda_2\left(e^{\lambda_2 s}-e^{\lambda_1 s}\right)+\lambda_1\left(e^{\lambda_2 s}-e^{\lambda_1 s}\right)}{d\left(e^{\lambda_2 s}-e^{\lambda_1 s}\right)}>0.$$

From this expression it follows that $b(s)>a(s)$, i.e., $\pi_{+-}(a)>a$. Moreover, by applying l'Hôpital's rule twice we have $\lim\limits_{a\nearrow+\infty}(b-a)=2t/d$. We conclude that the straight line $b=a+2t/d$ is an asymptote of the graph of $\pi_{+-}(a)$ when a tends to $+\infty$.

(e) Isolating s in (4.17) we obtain an implicit expression of π_{+-}

$$\left(\frac{b\lambda_1-1}{a\lambda_1+1}\right)^{\frac{\lambda_2}{\lambda_1}}=\frac{b\lambda_2-1}{a\lambda_2+1}. \tag{4.19}$$

The statement follows by substituting the values of λ_1 and λ_2 in (4.19).

(b) and (c) Differentiating with respect to a in expression (4.19) and isolating db/da yields

$$\pi'_{+-}(a)=\frac{db}{da}=\frac{a}{b}\frac{1-bt+b^2d}{1+at+a^2d}, \tag{4.20}$$

where the parabolas $1-bt+b^2d$ and $1+at+a^2d$ are shown in Figure 4.4. Since $a\geq 0$ and $b>\lambda_2^{-1}$, then $\pi'_{+-}(a)>0$ in $[0,+\infty)$ and $\lim\limits_{a\searrow 0}\pi'_{+-}(a)=0$.

From expression (4.20) we compute

$$\pi''_{+-}(a)=\frac{(b-a)(b+a)}{ab^2(1+at+a^2d)}\frac{db}{da}>0,$$

which proves the proposition. □

Corollary 4.4.4. *Consider $A\in GL(\mathbb{R}^2)$ such that $d>0$, $t<0$, and $t^2-4d>0$. Then the eigenvalues of A satisfy $0>\lambda_1>\lambda_2$. Let π_{+-} be the Poincaré map defined by the flow of the linear system $\dot{\mathbf{x}}=A\mathbf{x}$ and associated to two symmetric straight lines in the plane L_+ and L_-. Then:*

(a) *There exists a value $a^*>|\lambda_1|^{-1}$ such that $\pi_{+-}:[a^*,+\infty)\longrightarrow[0,+\infty)$, $\pi_{+-}(a^*)=0$, $\lim\limits_{a\nearrow+\infty}\pi_{+-}(a)=+\infty$, and $\pi_{+-}(a)<a$ in $[a^*,+\infty)$.*

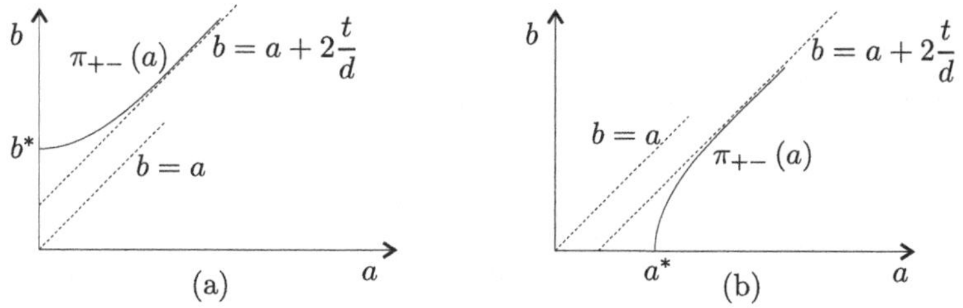

Figure 4.6: Qualitative behaviour of the Poincaré map π_{+-} for the parameters (a) $d > 0$, $t > 0$, and $t^2 - 4d > 0$, (b) $d > 0$, $t < 0$, and $t^2 - 4d > 0$.

(b) *If* $a \in (a^*, +\infty)$, *then* $\pi'_{+-}(a) > 1$ *and* $\lim_{a \searrow a^*} \pi'_{+-}(a) = +\infty$.

(c) *If* $a \in (a^*, +\infty)$, *then* $\pi''_{+-}(a) < 0$.

(d) *The straight line* $b = a + 2t/d$ *is an asymptote for the graph of* π_{+-} *when* a *tends to* $+\infty$.

(e) π_{+-} *is implicitly defined by the expression of Proposition* 4.4.3(e).

(f) *The qualitative behaviour of the graph of* π_{+-} *is shown in Figure* 4.6(b).

Proof. The proof follows easily by using that π_{+-} is the inverse map of the one described in Proposition 4.4.3, see Lemma 4.3.5(c). □

4.4.2 Non-diagonal node: $d > 0$ and $t^2 - 4d = 0$

When $t^2 - 4d = 0$, the matrix A has two real eigenvalues which are equal, $\lambda_1 = \lambda_2 = \lambda$. In this case there exist two different real Jordan normal forms for the matrix A, one diagonal and the other non-diagonal. In the diagonal case it is easy to check the non-existence of contact points of the flow with any straight line not passing through the origin. This implies the non-existence of Poincaré maps, see Theorem 4.3.10. Thus we only need to consider the non-diagonal case, that is,

$$A = \begin{pmatrix} \lambda & 1 \\ 0 & \lambda \end{pmatrix} \quad \text{with } \lambda = \frac{t}{2}.$$

Lemma 4.4.5. *Consider the function* $\psi_1 : \mathbb{R}^2 \to \mathbb{R}$ *given by* $\psi_1(x, y) = 1 + e^{xy}(1 - xy)$. *The qualitative behaviour of the graph of* $\psi_1(x, y_0)$ *is shown in Figure* 4.7, *depending on whether* $y_0 > 0$ (a) *or* $y_0 < 0$ (b).

Proof. Since $\partial\psi_1/\partial x|_{(x,y_0)} = -xy_0^2 e^{xy_0}$ and $\partial^2\psi_1/\partial x^2|_{(x,y_0)} = -y_0(1 + y_0)e^{xy_0}$, when $y_0 > 0$ the unique critical value of $\psi_1(x, y_0)$ is $x = 0$ which is a maximum. Moreover, $\lim_{x \nearrow -\infty} \psi_1(x, y_0) = 1$, $\lim_{x \nearrow +\infty} \psi_1(x, y_0) = -\infty$, and $\psi_1(0, y_0) = 2$. This proves the lemma. The case $y_0 < 0$ follows by noting that $\psi_1(-x, y) = \psi_1(x, -y)$. □

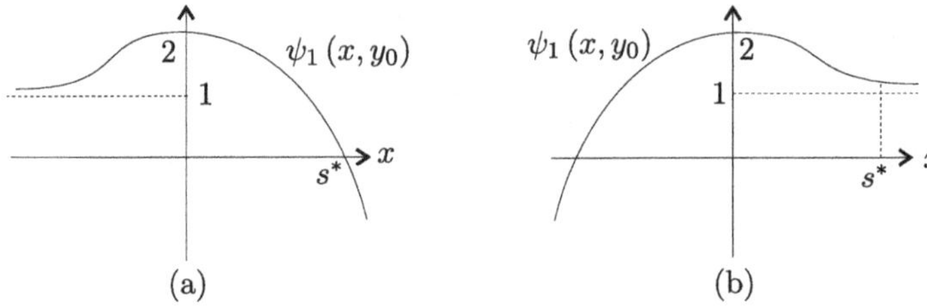

Figure 4.7: Qualitative behaviour of $\psi_1(x, y_0) = 1 + e^{xy_0}(1 - xy_0)$ when $y_0 > 0$ (a) and when $y_0 < 0$ (b).

Proposition 4.4.6. *Consider $A \in GL(\mathbb{R}^2)$ such that $d > 0$, $t > 0$, and $t^2 - 4d = 0$. Then the eigenvalues of A satisfy that $\lambda_1 = \lambda_2 = \lambda$. Assume that A is not diagonalizable and let π_{++} be the Poincaré map defined by the flow of the linear system $\dot{\mathbf{x}} = A\mathbf{x}$ and associated to two symmetric straight lines in the plane L_+ and L_-. Then:*

(a) $\pi_{++} : [0, \lambda^{-1}) \to [0, +\infty)$, $\pi_{++}(0) = 0$, $\lim_{a \nearrow \lambda^{-1}} \pi_{++}(a) = +\infty$, *and* $\pi_{++}(a) > a$ *in* $(0, \lambda^{-1})$.

(b) *If* $a \in (0, \lambda^{-1})$, *then* $\pi'_{++}(a) > 1$ *and* $\lim_{a \searrow 0} \pi'_{++}(a) = 1$.

(c) *If* $a \in (0, \lambda^{-1})$, *then* $\pi''_{++}(a) > 0$.

(d) *The graph of* π_{++} *has a vertical asymptote at* $a = \lambda^{-1}$.

(e) π_{++} *is implicitly defined by the expression*

$$\frac{t\,\pi_{++}(a) + 2}{2 - at} = e^{\frac{2t(\pi_{++}(a) + a)}{(t\,\pi_{++}(a) + 2)(2 - at)}}.$$

(f) *The qualitative behaviour of the graph of* π_{++} *is shown in Figure* 4.5(a).

Proof. By Proposition 4.3.11(a),

$$(\mathrm{Id} + \pi_{++}(a)\,A)\,\mathbf{p} = e^{sA}\,(\mathrm{Id} - aA)\,\mathbf{p},$$

where $\mathbf{p} = (p_1, p_2)^T$ is the contact point of the flow with L_+, $a \geq 0$, $\pi_{++}(a) \geq 0$, and $s \geq 0$.

Without loss of generality we assume that the matrix A is in real Jordan normal form. Thus $x_2 = 0$ is a straight line invariant under the flow. Since $\mathbf{p}$ is the contact point and L_+ does not pass through the origin, $p_2 \neq 0$. Therefore, the map $b = \pi_{++}(a)$ is implicitly defined by the system

$$1 + b\lambda = (1 - a\lambda)\,e^{\lambda s}, \quad b = (-a + s\,(1 - a\lambda))\,e^{\lambda s}, \tag{4.21}$$

and the inequalities $a \geq 0$, $b \geq 0$ and $s \geq 0$.

(a) and (d) Arguments similar to those used in the proof of Proposition 4.4.1 show that the domain of definition of π_{++} is contained in $[0, \lambda_1^{-1})$ and $\pi_{++}(0) = 0$.

We are now interested in the solutions of (4.21) such that $s > 0$. Multiplying the second equation by λ and substituting $b\lambda$ in the first one, we obtain the following parametric expression of π_{++}:

$$a(s) = \frac{-1 + e^{-\lambda s} + \lambda s}{\lambda^2 s}, \quad b(s) = \frac{-1 - \lambda s + e^{\lambda s}}{\lambda^2 s}. \tag{4.22}$$

Note that both functions are positive and differentiable in $(0, +\infty)$.

Differentiating in (4.21) with respect to s and isolating da/ds, we obtain that $da/ds = s^{-1} b(s) e^{-\lambda s} > 0$. Moreover, since $\lim\limits_{s \nearrow +\infty} a(s) = \lambda^{-1}$, the domain of definition of $\pi_{++}(a)$ is $[0, \lambda^{-1})$.

On the other hand, since $\lim\limits_{s \nearrow +\infty} b(s) = +\infty$, we obtain $\lim\limits_{a \nearrow \lambda^{-1}} \pi_{++}(a) = +\infty$. Therefore, the straight line $a = \lambda^{-1}$ is an asymptote of the graph of $\pi_{++}(a)$.

Finally, since $b(s) - a(s) > 0$ (note that the function $f(s) = e^{\lambda s} - e^{-\lambda s} - 2\lambda s$ satisfies that $f(0) = 0$ and $f'(s) > 0$ when $s > 0$), it follows that $\pi_{++}(a) > a$ in $[0, \lambda^{-1})$.

(e) The statement follows by isolating s in the first equation of system (4.21) and substituting it in the second one.

(b) and (c) Differentiating in (4.22) with respect to s we obtain that

$$\frac{da}{ds} = \frac{\psi_1(s, -\lambda) - 2}{(\lambda s)^2}, \quad \frac{db}{ds} = \frac{\psi_1(s, \lambda) - 2}{(\lambda s)^2}, \tag{4.23}$$

where $\psi_1(s, \lambda) = 1 + e^{s\lambda}(1 - s\lambda)$. Hence

$$\frac{d\pi_{++}}{da} = \frac{db/ds}{da/ds} = \frac{\psi_1(s, \lambda) - 2}{\psi_1(s, -\lambda) - 2} > 0,$$

see Lemma 4.4.5. By applying l'Hôpital's rule,

$$\lim_{a \searrow 0} \pi'_{++}(a) = \lim_{s \searrow 0} \frac{db/ds}{da/ds} = \lim_{s \searrow 0} \frac{e^{\lambda s}(1 - \lambda s) - 1}{e^{-\lambda s}(1 + \lambda s) - 1} = 1.$$

Differentiating $d\pi_{++}/da$ with respect to a, it follows that

$$\frac{d^2\pi_{++}}{da^2} = \frac{d}{ds}\left(\frac{\psi_1(s, \lambda) - 2}{\psi_1(s, -\lambda) - 2}\right)\frac{1}{\frac{da}{ds}} = -\frac{(b - a)}{(\psi_1(s, -\lambda) - 2)^2}\frac{1}{\frac{da}{ds}},$$

which proves that $\pi''_{++}(a) > 0$ in $(0, \lambda^{-1})$. □

Corollary 4.4.7. *Consider $A \in GL(\mathbb{R}^2)$ such that $d > 0$, $t < 0$, and $t^2 - 4d = 0$. Then the eigenvalues of A satisfy $\lambda_1 = \lambda_2 = \lambda < 0$. Assume that A is not diagonalizable and let π_{++} be the Poincaré map defined by the flow of the linear system $\dot{\mathbf{x}} = A\mathbf{x}$ and associated to two symmetric straight lines in the plane L_+ and L_-. Then:*

(a) $\pi_{++} : [0, +\infty) \to [0, |\lambda|^{-1})$, $\pi_{++}(0)=0$, $\lim_{a\nearrow+\infty} \pi_{++}(a)=|\lambda|^{-1}$, *and* $\pi_{++}(a)<a$ *in* $(0, +\infty)$.

(b) *If* $a \in (0, +\infty)$, *then* $0 < \pi'_{++}(a) < 1$ *and* $\lim_{a\searrow 0} \pi'_{++}(a) = 1$.

(c) *If* $a \in (0, +\infty)$, *then* $\pi''_{++}(a) < 0$ *in* $(0, +\infty)$.

(d) *The straight line* $b = |\lambda|^{-1}$ *is a horizontal asymptote of the graph of* π_{++} *when* a *tends to* $+\infty$.

(e) π_{++} *is implicitly defined by the expression of Proposition* 4.4.6(e).

(f) *The qualitative behaviour of the graph of* π_{++} *is shown in Figure* 4.5(b).

Proof. The proof follows directly by using that the Poincaré map π_{++} is the inverse of the map described in Proposition 4.4.6. □

Proposition 4.4.8. *Consider* $A \in GL(\mathbb{R}^2)$ *such that* $d > 0$, $t > 0$, *and* $t^2 - 4d = 0$. *Then the eigenvalues of* A *satisfy* $\lambda_1 = \lambda_2 = \lambda > 0$. *Assume that* A *is not diagonalizable and let* π_{+-} *be the Poincaré map defined by the flow of the linear system* $\dot{\mathbf{x}} = A\mathbf{x}$ *and associated to two symmetric straight lines in the plane* L_+ *and* L_-. *Then:*

(a) *There exists a value* $b^* > \lambda^{-1}$ *such that* $\pi_{+-} : [0, +\infty) \to [b^*, +\infty)$, $\pi_{+-}(0) = b^*$, $\lim_{a\nearrow+\infty} \pi_{+-}(a) = +\infty$, *and* $\pi_{+-}(a) > a$ *in* $(0, +\infty)$.

(b) *If* $a \in (0, +\infty)$, *then* $0 < \pi'_{+-}(a) < 1$ *and* $\lim_{a\searrow 0} \pi'_{+-}(a) = 0$.

(c) *If* $a \in (0, +\infty)$, *then* $\pi''_{+-}(a) > 0$.

(d) *The straight line* $b = a + 2t/d$ *is an asymptote of the graph of* π_{+-} *when* a *tends to* $+\infty$.

(e) π_{+-} *is implicitly defined by the expression*

$$\frac{t\,\pi_{+-}(a) - 2}{2 + a\,t} = e^{\frac{2t\left(\pi_{+-}(a)+a\right)}{(t\,\pi_{+-}(a)-2)(2+t\,a)}}.$$

(f) *The qualitative behaviour of the graph of* π_{+-} *is shown in Figure* 4.6(a).

Proof. Arguments similar to those used in the proof of Proposition 4.4.6 show that $b = \pi_{+-}(a)$ is implicitly determined by the system

$$1 - b\lambda = -(1 + a\lambda)\,e^{\lambda s}, \quad -b = -(a + s\,(1 + a\lambda))\,e^{\lambda s}, \tag{4.24}$$

and the inequalities $a \geq 0$, $b \geq 0$ and $s > 0$.

(a) Let $s^* > 0$ be the zero of the function $\psi_1(s, \lambda)$ defined in Lemma 4.4.5(a), and set $b^* = s^* e^{\lambda s^*}$. It is easy to check that $a = 0$, $b = b^*$ and $s = s^*$ is a solution of system (4.24), i.e., $\pi_{+-}(0) = b^* > \lambda^{-1}$.

Solving (4.24) we obtain the parametric equations of π_{+-}:

$$a(s) = \frac{1 + e^{-\lambda s} - \lambda s}{\lambda^2 s} \quad \text{and} \quad b(s) = \frac{1 + \lambda s + e^{\lambda s}}{\lambda^2 s}, \tag{4.25}$$

where $a(s)$ and $b(s)$ are differentiable functions in $(0, +\infty)$. Differentiating these functions with respect to s it follows that

$$\frac{da}{ds} = -\frac{\psi_1(s, -\lambda)}{\lambda^2 s^2} < 0 \quad \text{and} \quad \frac{db}{ds} = -\frac{\psi_1(s, \lambda)}{\lambda^2 s^2} < 0,$$

see Lemma 4.4.5. From this we have that $a(s)$ and $b(s)$ are strictly decreasing functions in $(0, s^*)$. Since $\lim_{s \searrow 0} a(s) = +\infty$, the domain of definition of π_{+-} is $[0, +\infty)$.

From (4.25) it follows that $\lim_{s \searrow 0} b(s) = +\infty$, hence $\lim_{a \nearrow +\infty} \pi_{+-}(a) = +\infty$ and $b(s) > a(s)$ in $(0, s^*)$. Therefore, $\pi_{+-}(a) > a$ in $[0, +\infty)$.

(d) From expression (4.25) we obtain

$$\lim_{s \searrow 0} \frac{b}{a} = 1 \quad \text{and} \quad \lim_{s \searrow 0} (b - a) = \frac{4}{\lambda} = 2\frac{t}{d},$$

which proves the statement.

Statement (e) follows easily by isolating s in (4.24).

(b) and (c) Since $a'(s) < 0$ and $b'(s) < 0$, we get $\pi'_{+-}(a) > 0$, $\lim_{a \searrow 0} \pi'_{+-}(a) = 0$, and

$$\pi''_{+-}(a) = \frac{d}{ds}\left(\frac{\psi_1(s, \lambda)}{\psi_1(s, -\lambda)}\right)\frac{1}{\frac{da}{ds}} = -\frac{\lambda^2 s(b - a)}{\frac{da}{ds}\psi_1(s, -\lambda)^2} > 0,$$

which proves the statement. □

Corollary 4.4.9. *Consider $A \in GL(\mathbb{R}^2)$ such that $d > 0$, $t < 0$, and $t^2 - 4d = 0$. Then the eigenvalues of A satisfy $\lambda_1 = \lambda_2 = \lambda < 0$. Assume that A is not diagonalizable and let π_{+-} be the Poincaré map defined by the flow of the linear system $\dot{\mathbf{x}} = A\mathbf{x}$ and associated to two symmetric straight lines in the plane L_+ and L_-. Then:*

(a) *There exists a value $a^* > |\lambda|^{-1}$ such that $\pi_{+-} : [a^*, +\infty) \to [0, +\infty)$, $\pi_{+-}(0) = 0$, $\lim_{a \nearrow +\infty} \pi_{+-}(a) = +\infty$ and $\pi_{+-}(a) < a$ in $(a^*, +\infty)$.*

(b) *If $a \in (a^*, +\infty)$, then $\pi'_{+-}(a) > 1$ and $\lim_{a \searrow a^*} \pi'_{+-}(a) = +\infty$.*

(c) *If $a \in (a^*, +\infty)$, then $\pi''_{+-}(a) < 0$.*

(d) *The straight line $b = a + 2t/d$ is an asymptote of the graph of π_{+-} when a tends to $+\infty$.*

(e) *π_{+-} is implicitly defined by the expression of Proposition 4.4.8(e).*

(f) *The qualitative behaviour of the graph of* π_{+-} *is shown in Figure* 4.6(b).

Proof. The proof follows easily by using that the Poincaré map π_{+-} is the inverse of the map described in the Proposition 4.4.8. $\square$

4.4.3 Center and focus: $t^2 - 4d < 0$

In this subsection we assume that the matrix A has two complex eigenvalues $\lambda = \alpha + i\beta$ and $\bar{\lambda}$, where $\alpha = t/2$ and $\beta = \sqrt{4d - t^2}/2$. In this case the real Jordan normal form of A is

$$\begin{pmatrix} \alpha & -\beta \\ \beta & \alpha \end{pmatrix} \quad \text{with } \beta \neq 0.$$

Hence, when $t = 0$, i.e., $\alpha = 0$, the singular point at the origin is a center, otherwise it is a focus.

Lemma 4.4.10. *Consider the function* $\psi_2 : \mathbb{R}^2 \to \mathbb{R}$ *given by* $\psi_2(x, y) = 1 - e^{xy}(\cos(x) - y\sin(x))$. *The qualitative behaviour of* $\psi_2(x, y_0)$ *in* $(-\infty, 2\pi]$ *is shown in Figure* 4.8(a) *when* $y_0 > 0$ *and* (b) *when* $y_0 < 0$.

Proof. Since $\partial\psi_2/\partial x|_{(x,y)} = (1 + y^2)e^{xy}\sin(x)$, the critical values of ψ_2 are $x_k = k\pi$, where $k \in \mathbb{Z}$. From $\partial^2\psi_2/\partial x^2\big|_{(x,y)} = (1 + y^2)e^{xy}(\cos(x) + y\sin(x))$ and assuming that $y_0 > 0$ it follows that ψ_2 has a local minimum at x_k for k even or has a local maximum at x_k for k odd.

On the other hand, when $p \in \mathbb{Z}$ and $p < 0$ we have $\psi_2(x_{2p}, y_0) = 1 - e^{2p\pi y_0} > 0$, and consequently $\psi_2(x, y_0) > 0$ when $x < 0$. Moreover, $\psi_2(0, y_0) = 0$, $\psi_2(\pi, y_0) = 1 + e^{\pi y_0} > 0$, and $\psi_2(2\pi, y_0) = 1 - e^{2\pi y_0} < 0$. Therefore, there exists a unique zero τ^* in $(0, 2\pi)$ which proves the lemma for $y_0 > 0$. The case $y_0 < 0$ follows by using the fact that $\psi_2(-x, y) = \psi_2(x, -y)$. $\square$

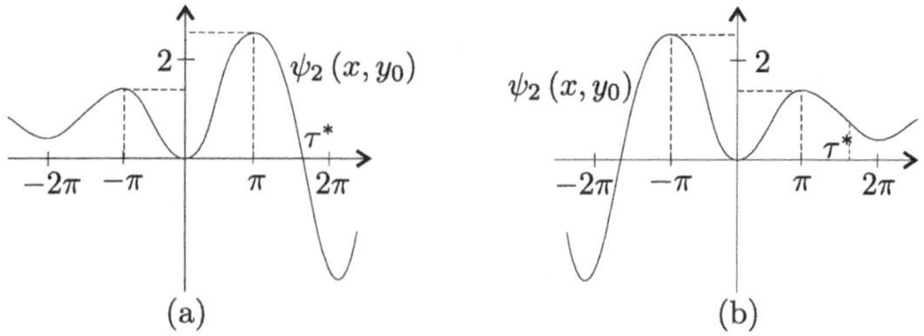

Figure 4.8: Qualitative behaviour of $\psi_2\,(x, y_0) = 1 - e^{xy_0}\,(\cos\,(x) - y_0 \sin\,(x))$; (a) when $y_0 > 0$, (b) when $y_0 < 0$.

Proposition 4.4.11. *Consider* $A \in GL(\mathbb{R}^2)$ *such that* $d > 0$, $t \geq 0$, *and* $t^2 - 4d < 0$. *Let* π_{++} *be the Poincaré map defined by the flow of the linear system* $\dot{\mathbf{x}} = A\mathbf{x}$ *and associated to two symmetric straight lines in the plane* L_+ *and* L_-. *Then:*

(a) *If* $t > 0$, *then* $\pi_{++} : [0, +\infty) \to [0, +\infty)$, $\pi_{++}(0) = 0$, $\lim_{a \nearrow +\infty} \pi_{++}(a) = +\infty$, *and* $\pi_{++}(a) > a$ *in* $(0, +\infty)$.

(a.1) *If* $a \in (0, +\infty)$, *then* $\pi'_{++}(a) > 1$ *and* $\lim_{a \searrow 0} \pi'_{++}(a) = 1$.

(a.2) *If* $a \in (0, +\infty)$, *then* $\pi''_{++}(a) > 0$.

(a.3) *Set* $\gamma = t/\sqrt{4d - t^2}$. *The straight line* $b = e^{\gamma\pi}a - t(1 + e^{\gamma\pi})/d$ *is an asymptote of the graph of* π_{++} *when* a *tends to* $+\infty$.

(a.4) π_{++} *is implicitly defined by the expression*

$$\frac{1 + t\pi_{++}(a) + d\pi_{++}(a)^2}{1 - ta + da^2} = e^{2\gamma \arctan\left(\frac{(a+\pi_{++}(a))\beta}{(\pi_{++}(a)-a)\alpha+1-a\pi_{++}(a)d}\right)}.$$

(a.5) *The qualitative behaviour of the graph of the map* π_{++} *is shown in Figure 4.9(a).*

(b) *If* $t = 0$, *then* π_{++} *is the identity in* $[0, +\infty)$.

Proof. Let $\mathbf{p} = (p_1, p_2)^T$ be the contact point of the flow with L_+. By Proposition 4.3.11(a),

$$(\mathrm{Id} + \pi_{++}(a) A)\, \mathbf{p} = e^{sA} (\mathrm{Id} - aA)\, \mathbf{p},$$

where $a \geq 0$, $\pi_{++}(a) \geq 0$ and $s \geq 0$. Since L_+ does not pass through the origin, we conclude that $p_1 \neq 0$ or $p_2 \neq 0$, which implies that $b = \pi_{++}(a)$ is defined by the system

$$\begin{array}{rcl} 1 + b\alpha & = & e^{\alpha s} \left\{ \cos(\beta s) + a \left[\beta \sin(\beta s) - \alpha \cos(\beta s) \right] \right\}, \\ b\beta & = & e^{\alpha s} \left\{ \sin(\beta s) - a \left[\alpha \sin(\beta s) + \beta \cos(\beta s) \right] \right\}, \end{array} \tag{4.26}$$

and the inequalities $a \geq 0$, $b \geq 0$ and $s \geq 0$.

(a) Since $s = 0$, $a = 0$ and $b = 0$ is a solution of system (4.26), we have $\pi_{++}(0) = 0$. Furthermore, if $a = a_0$, $b = b_0$ and $s = s_0$ is a solution of (4.26), then s_0 is the flight time between the points $\mathbf{q}_1 = \mathbf{p} - a\dot{\mathbf{p}}$ and $\mathbf{q}_2 = \mathbf{p} + b\dot{\mathbf{p}}$, see Section 4.3. Thus βs_0 is the angle between $\mathbf{q}_1$ and $\mathbf{q}_2$, and consequently $\beta s \in [0, \pi)$.

Define $\tau = \beta s$ and $\gamma = \alpha/\beta$. Solving system (4.26) with $\tau \in (0, \pi)$ we obtain the following parametric equations of π_{++}:

$$a(\tau) = \frac{\beta e^{-\gamma\tau}}{d \sin(\tau)} \psi_2(\tau, \gamma) \quad \text{and} \quad b(\tau) = \frac{\beta e^{\gamma\tau}}{d \sin(\tau)} \psi_2(\tau, -\gamma), \tag{4.27}$$

where ψ_2 is the function described in Lemma 4.4.10.

Since $\lim_{\tau \nearrow \pi} a(\tau) = +\infty$ and $\lim_{\tau \nearrow \pi} b(\tau) = +\infty$, the domain of definition of π_{++} is $[0, +\infty)$ and $\lim_{a \nearrow +\infty} \pi_{++}(a) = +\infty$. Moreover, when $\tau \in (0, \pi)$ we have

$$b(\tau) - a(\tau) = \frac{\beta}{d \sin(\tau)} \left(e^{\gamma\tau} - e^{-\gamma\tau} - 2\gamma \sin(\tau) \right) > 0,$$

and therefore $\pi_{++}(a) > a$ in $(0, +\infty)$.

(a.3) From expression (4.27) it follows that

$$\lim_{a \nearrow +\infty} \frac{\pi_{++}(a)}{a} = \lim_{\tau \nearrow \pi} \frac{b(\tau)}{a(\tau)} = \lim_{\tau \nearrow \pi} e^{2\gamma\tau} \frac{\psi_2(\tau, -\gamma)}{\psi_2(\tau, \gamma)} = e^{\gamma\pi}.$$

Hence, applying l'Hôpital's rule we obtain

$$\lim_{a \nearrow +\infty} \pi_{++}(a) - e^{\gamma\pi} a = \lim_{\tau \nearrow \pi} b(\tau) - e^{\gamma\pi} a(\tau) = \frac{-t(1 + e^{\gamma\pi})}{d},$$

and therefore the straight line $b = e^{\gamma\pi} a - t(1 + e^{\gamma\pi})/d$ is an asymptote of the graph of $\pi_{++}(a)$.

(a.4) Adding the two equations squared in system (4.26) and dividing them we obtain

$$1 + tb + db^2 = e^{2\gamma\tau}\left(1 - ta + da^2\right),$$
$$\tan(\tau) = \frac{(a + b)\beta}{(b - a)\alpha + 1 - abd},$$

which proves the statement.

(a.1) and (a.2) Differentiating in (4.27) with respect to τ it follows that

$$\frac{da}{d\tau} = \frac{\beta}{d\sin^2(\tau)}\psi_2(\tau, -\gamma) \quad \text{and} \quad \frac{db}{d\tau} = \frac{\beta}{d\sin^2(\tau)}\psi_2(\tau, \gamma).$$

Thus, $\pi'_{++}(a) = \psi_2(\tau, \gamma)/\psi_2(\tau, -\gamma) > 0$ and $\lim\limits_{a \searrow 0} \pi'_{++}(a) = 1$, see Lemma 4.4.10. Moreover,

$$\pi''_{++}(a) = \frac{d}{d\tau}\left(\frac{db}{da}\right)\frac{1}{\frac{da}{d\tau}} = \frac{2d\left(1 + \gamma^2\right)\sin^3(\tau)}{\beta\psi_2(\tau, -\gamma)^3}\left(\sinh(\gamma\tau) - \gamma\sin(\tau)\right).$$

Since $\sinh(\gamma\tau) > \gamma\sin(\tau)$ when $\tau \in (0, \pi)$, we conclude that $\pi''_{++}(a) > 0$ in the interval $(0, +\infty)$.

(b) Since $t = 0$, we obtain that $\gamma = \alpha/\beta = 0$. Applying expression (4.27), it follows that $\pi_{++}(a) = a$ in $(0, +\infty)$. The statement follows by noting that $a = 0$, $b = 0$ and $s = 0$ is also a solution of (4.26). □

Corollary 4.4.12. *Consider $A \in GL(\mathbb{R}^2)$ such that $d > 0$, $t < 0$, and $t^2 - 4d < 0$. Let π_{++} be the Poincaré map defined by the flow of the linear system $\dot{\mathbf{x}} = A\mathbf{x}$ and associated to two symmetric straight lines in the plane L_+ and L_-. Then:*

(a) *$\pi_{++} : [0, +\infty) \to [0, +\infty)$, $\pi_{++}(0) = 0$, $\lim\limits_{a \nearrow +\infty} \pi_{++}(a) = +\infty$, and $\pi_{++}(a) < a$ in $(0, +\infty)$.*

(b) *If $a \in (0, +\infty)$, then $0 < \pi'_{++}(a) < 1$ and $\lim\limits_{a \nearrow 0} \pi'_{++}(a) = 1$.*

(c) *If $a \in (0, +\infty)$, then $\pi''_{++}(a) < 0$.*

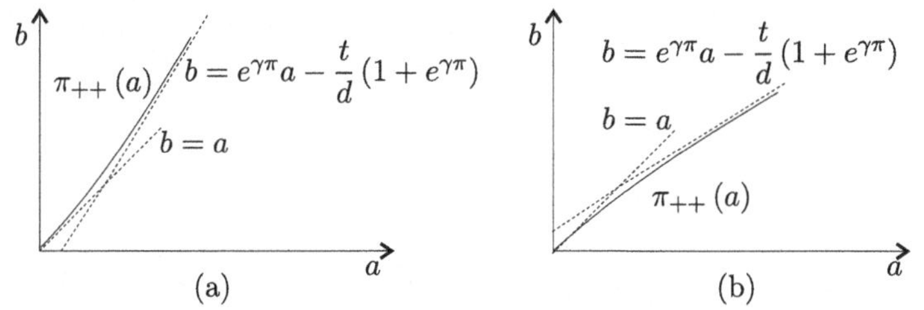

Figure 4.9: Qualitative behaviour of the Poincaré map π_{++}; (a) when $t^2 - 4d < 0$ and $t > 0$, (b) when $t^2 - 4d < 0$ and $t < 0$.

(d) *Set* $\gamma = t/\sqrt{4d - t^2}$. *The straight line* $b = e^{\gamma\pi}a - t(1+e^{\gamma\pi})/d$ *is an asymptote of the graph of* π_{++} *when* a *tends to* $+\infty$.

(e) π_{++} *is implicitly defined by the expression of Proposition* 4.4.11(e).

(f) *The qualitative behaviour of the graph of* π_{++} *is shown in Figure* 4.9(b).

Proof. The proof follows directly by using that π_{++} is the inverse of the map described in Proposition 4.4.11. □

Proposition 4.4.13. *Consider* $A \in GL(\mathbb{R}^2)$ *such that* $d > 0$, $t \geq 0$, *and* $t^2 - 4d < 0$. *Let* π_{+-} *be the Poincaré map defined by the flow of the linear system* $\dot{\mathbf{x}} = A\mathbf{x}$ *and associated to two symmetric straight lines in the plane* L_+ *and* L_-. *Then:*

(a) *If* $t > 0$, *then there exists a value* $b^* > 0$ *such that* $\pi_{+-} : [0, +\infty) \to [b^*, +\infty)$, $\pi_{+-}(0) = b^*$, $\lim_{a \nearrow +\infty} \pi_{+-}(a) = +\infty$, *and* $\pi_{+-}(a) > a$ *in* $(0, +\infty)$.

(a.1) *If* $a \in (0, +\infty)$, *then* $0 < \pi'_{+-}(a) < 1$ *and* $\lim_{a \searrow 0} \pi'_{+-}(a) = 0$.

(a.2) *If* $a \in (0, +\infty)$, *then* $\pi''_{+-}(a) > 0$.

(a.3) *The straight line* $b = a + 2t/d$ *is an asymptote of the graph of* π_{+-} *when* a *tends to* $+\infty$.

(a.4) π_{+-} *is implicitly defined by the expression*

$$\frac{1 - t\pi_{+-}(a) + d\pi_{+-}(a)^2}{1 + ta + da^2} = e^{2\gamma \arctan\left(\frac{(a+\pi_{+-}(a))\beta}{(\pi_{+-}(a)-a)\alpha - 1 + a\pi_{+-}(a)d}\right)}.$$

(a.5) *The qualitative behaviour of the graph of the map* π_{+-} *is shown in Figure* 4.6(a).

(b) *If* $t = 0$, *then the map* π_{+-} *is the identity function in* $[0, +\infty)$.

Proof. Arguments similar to those used in the proof of Proposition 4.4.11 imply that π_{+-} is determined by the system

$$\begin{aligned} -1+b\alpha &= e^{\alpha s}\left\{\cos(\beta s)+a\left[\alpha\cos(\beta s)-\beta\sin(\beta s)\right]\right\}, \\ b\beta &= e^{\alpha s}\left\{\sin(\beta s)+a\left[\alpha\sin(\beta s)+\beta\cos(\beta s)\right]\right\}, \end{aligned} \tag{4.28}$$

and the inequalities $a \geq 0$, $b \geq 0$ and $s > 0$.

(a) and (a.3) Following the proof of Proposition 4.4.11, we conclude that $\tau = \beta s$ belongs to $(0, \tau^*]$, where τ^* is the unique zero of the function $2 - \psi_2(\tau, \gamma)$ in $(0, \pi)$, see Lemma 4.4.10 and Figure 4.10.

Solving system (4.28) for $\tau \in (0, \tau^*]$ and $\gamma = \alpha/\beta$ we obtain the parametric equations of π_{+-}:

$$a(\tau) = \frac{\beta e^{-\tau\gamma}}{d\sin(\tau)}\left(2-\psi_2(\tau,\gamma)\right) \quad \text{and} \quad b(\tau) = \frac{\beta e^{\tau\gamma}}{d\sin(\tau)}\left(2-\psi_2(\tau,-\gamma)\right). \tag{4.29}$$

From these equations we conclude that $\lim_{\tau\searrow 0} a(\tau) = +\infty$ and $\lim_{\tau\searrow 0} b(\tau) = +\infty$ which implies that the domain of definition of $\pi_{+-}(a)$ is $[0, +\infty)$; $\lim_{a\nearrow+\infty} \pi_{+-}(a) = +\infty$;

$$b(\tau) - a(\tau) = \frac{1}{d\sin(\tau)}\left(2\alpha\sin(\tau) + \beta\left(e^{\gamma\tau} - e^{-\gamma\tau}\right)\right) > 0 \text{ if } \tau \in (0, \tau^*],$$

that is, $\pi_{+-}(a) > a$ in $(0, +\infty)$; $\lim_{a\nearrow+\infty} \pi_{+-}(a)/a = \lim_{\tau\searrow 0} b(\tau)/a(\tau) = 1$; and by applying l'Hôpital's rule,

$$\lim_{a\nearrow+\infty} \pi_{+-}(a) - a = \lim_{\tau\searrow 0} b(\tau) - a(\tau) = 2t/d,$$

which implies that $b = a + 2t/d$ is an asymptote of the graph of $\pi_{+-}(a)$.

Statement (a.4) follows by arguments similar to those used in the proof of Proposition 4.4.11(e).

(a.1) and (a.2) Differentiating in (4.29) with respect to τ it follows that

$$\left.\frac{da}{d\tau}\right|_{\tau} = \beta\frac{\psi_2(\tau,-\gamma)-2}{d\sin^2(\tau)} < 0 \quad \text{and} \quad \left.\frac{db}{d\tau}\right|_{\tau} = \beta\frac{\psi_2(\tau,\gamma)-2}{d\sin^2(\tau)} < 0.$$

Hence, $\pi'_{+-}(a) > 0$, $\lim_{a\searrow 0} \pi'_{+-}(a) = \lim_{\tau\nearrow\tau^*} (\psi_2(\tau,\gamma)-2)/(\psi_2(\tau,-\gamma)-2) = 0$ and $\lim_{a\nearrow+\infty} \pi'_{+-}(a) = \lim_{\tau\searrow 0} (\psi_2(\tau,\gamma)-2)/(\psi_2(\tau,-\gamma)-2) = 1$. Moreover, taking into account that $db/d\tau < 0$ the image of π_{+-} is contained in $[b^*, +\infty)$, where $b^* = b(\tau^*)$.

Computing the second derivative of π_{+-} with respect to a we have that

$$\pi''_{+-}(a) = \frac{d}{d\tau}\left(\frac{db}{da}\right)\frac{1}{\frac{da}{d\tau}} = \frac{2d\sin^3(\tau)\left(1+\gamma^2\right)}{\beta\left(2-\varphi(-\tau,\gamma)\right)^3}\left(\sinh(\gamma\tau)+\gamma\sin(\tau)\right).$$

Since $0 < \tau \leq \tau^* < \pi$, we conclude that $\pi''_{+-}(a) > 0$.

Statement (b) follows in much the same way as Proposition 4.4.13(b). □

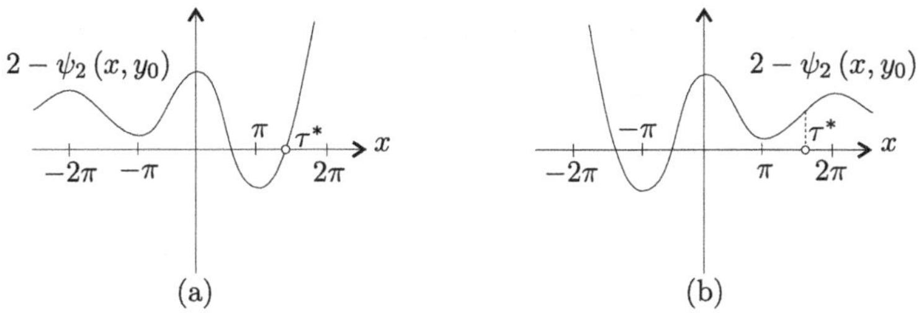

Figure 4.10: Qualitative behaviour of the function $2-\psi_2(x,y_0)$; (a) when $y_0>0$, (b) when $y_0<0$.

Corollary 4.4.14. *Consider $A \in GL(\mathbb{R}^2)$ such that $d>0$, $t<0$, and $t^2-4d<0$. Let π_{+-} be the Poincaré map defined by the flow of the linear system $\dot{\mathbf{x}} = A\mathbf{x}$ and associated to two symmetric straight lines in the plane L_+ and L_-. Then:*

(a) *There exists a value $a^*>0$ such that $\pi_{+-}:[a^*,+\infty) \to [0,+\infty)$, $\pi_{+-}(a^*)=0$, and $\lim_{a\nearrow+\infty} \pi_{+-}(a) = +\infty$. Moreover, $\pi_{+-}(a) < a$ in $(a^*,+\infty)$.*

(b) *If $a \in (0,+\infty)$, then $\pi'_{+-}(a) > 1$ and $\lim_{a\searrow a^*} \pi'_{+-}(a) = +\infty$.*

(c) *If $a \in (0,+\infty)$, then $\pi''_{+-}(a) < 0$.*

(d) *The straight line $b = a + 2t/d$ is an asymptote of the graph of π_{+-} when a tends to $+\infty$.*

(e) *π_{+-} is implicitly defined by the expression in Proposition 4.4.13(a.4).*

(f) *The qualitative behaviour of the graph of π_{+-} is shown in Figure 4.6(b).*

Proof. The proof follows directly by using that the Poincaré map π_{+-} is the inverse of the map described in Proposition 4.4.13. □

4.4.4 Saddle: $d<0$

In this subsection we consider the case where the matrix A has two real eigenvalues with different sign; that is, $\lambda_1 > 0 > \lambda_2$, where $\lambda_1 = (t+\sqrt{t^2-4d})/2$ and $\lambda_2 = (t-\sqrt{t^2-4d})/2$. The real Jordan normal form of A is

$$A = \begin{pmatrix} \lambda_1 & 0 \\ 0 & \lambda_2 \end{pmatrix}.$$

Proposition 4.4.15. *Consider $A \in GL(\mathbb{R}^2)$ such that $d<0$ and $t \geq 0$. Then the eigenvalues of A satisfy $\lambda_1 > 0 > \lambda_2$. Let π_{++} be the Poincaré map defined by the flow of the linear system $\dot{\mathbf{x}} = A\mathbf{x}$ and associated to two symmetric straight lines in the plane L_+ and L_-. Then:*

(a) *If* $t > 0$, *then* $\pi_{++} : [0, \lambda_1^{-1}) \to [0, |\lambda_2|^{-1})$, $\pi_{++}(0) = 0$, $\lim_{a \nearrow \lambda_1^{-1}} \pi_{++}(a) = |\lambda_2|^{-1}$, *and* $\pi_{++}(a) > a$ *in* $(0, \lambda_1^{-1})$.

(a.1) *If* $a \in (0, \lambda_1^{-1})$, *then* $\pi'_{++}(a) > 1$. *Furthermore,* $\lim_{a \searrow 0} \pi'_{++}(a) = 1$ *and* $\lim_{a \nearrow \lambda_1^{-1}} \pi'_{++}(a) = +\infty$.

(a.2) *If* $a \in (0, \lambda_1^{-1})$, *then* $\pi''_{++}(a) > 0$.

(a.3) *The graph of* π_{++} *has a vertical asymptote at* $a = \lambda_1^{-1}$.

(a.4) π_{++} *is implicitly defined by the expression*

$$\left(\frac{2 + \pi_{++}(a)\left(t - \sqrt{t^2 - 4d}\right)}{2 - a\left(t - \sqrt{t^2 - 4d}\right)} \right)^{\frac{t+\sqrt{t^2-4d}}{t-\sqrt{t^2-4d}}} = \frac{2 + \pi_{++}(a)\left(t + \sqrt{t^2 - 4d}\right)}{2 - a\left(t + \sqrt{t^2 - 4d}\right)}.$$

(a.5) *The qualitative behaviour of the graph of* π_{++} *is shown in Figure* 4.12(a).

(b) *If* $t = 0$, *then* π_{++} *is the identity in* $[0, \lambda_1^{-1})$.

Proof. Arguments similar to those in the proof of Proposition 4.4.1 show that the map $b = \pi_{++}(a)$ is defined by the system

$$1 + b\lambda_1 = e^{\lambda_1 s}(1 - a\lambda_1), \quad 1 + b\lambda_2 = e^{\lambda_2 s}(1 - a\lambda_2), \tag{4.30}$$

and the inequalities $a \geq 0$, $b \geq 0$ and $s \geq 0$.

(a) As in the proof of Proposition 4.4.1(a), from system (4.30) we obtain the following information: $\pi_{++}(0) = 0$; the domain of definition of π_{++} is contained in $(0, \lambda_1^{-1})$; and the parametric equations of π_{++} are

$$\begin{aligned} a(s) &= \frac{\lambda_2\left(1 - e^{\lambda_1 s}\right) - \lambda_1\left(1 - e^{\lambda_2 s}\right)}{d\left(e^{\lambda_2 s} - e^{\lambda_1 s}\right)}, \\ b(s) &= \frac{(\lambda_2 - \lambda_1)e^{ts} + \lambda_1 e^{\lambda_1 s} - \lambda_2 e^{\lambda_2 s}}{d\left(e^{\lambda_2 s} - e^{\lambda_1 s}\right)}. \end{aligned} \tag{4.31}$$

Note that the functions $a(s)$ and $b(s)$ are differentiable in $s \in (0, +\infty)$.

Differentiating in (4.30) with respect to s and isolating da/ds and db/ds we obtain that

$$\frac{da}{ds} = b(s)\frac{\lambda_1 - \lambda_2}{e^{\lambda_1 s} - e^{\lambda_2 s}} \quad \text{and} \quad \frac{db}{ds} = 1 - b(s)\frac{\lambda_1 e^{\lambda_2 s} - \lambda_2 e^{\lambda_1 s}}{e^{\lambda_1 s} - e^{\lambda_2 s}}.$$

Hence, since $\lim_{s \searrow 0} b(s) = 0$ and $\lim_{s \nearrow +\infty} b(s) = |\lambda_2|^{-1}$, see expression (4.30), we conclude that $b(s) \geq 0$ and $a'(s) > 0$ in $(0, +\infty)$. Now using that $\lim_{s \searrow 0} a(s) = 0$ and

$\lim_{s \nearrow +\infty} a(s) = \lambda_1^{-1}$ it follows that the domain of definition of π_{++} is $[0, \lambda_1^{-1})$ and $\lim_{a \nearrow \lambda_1^{-1}} \pi_{++}(a) = |\lambda_2|^{-1}$.

We will prove the inequality $\pi_{++}(a) > a$ at the end of the present proof.

(a.4) Isolating s in each of the equations of system (4.30) we obtain the following implicit expression for the map $b = \pi_{++}(a)$:

$$\frac{1 + b\lambda_2}{1 - a\lambda_2} = \left(\frac{1 - a\lambda_1}{1 + b\lambda_1}\right)^{-\frac{\lambda_2}{\lambda_1}}. \tag{4.32}$$

(a.1), (a.2) and (a.3) Differentiating in (4.32) with respect to a and isolating db/da we obtain that

$$\pi'_{++}(a) = \frac{db}{da} = \frac{a}{b}\frac{1 + bt + b^2 d}{1 - at + a^2 d}. \tag{4.33}$$

The behaviour of the parabolas $1 + bt + b^2 d$ and $1 - at + a^2 d$ is shown in Figure 4.11. From this it is easy to conclude that $\pi'_{++}(a) > 0$ when $a \in (0, \lambda_1^{-1})$ and $\lim_{a \nearrow \lambda_1^{-1}} \pi'_{++}(a) = +\infty$. Furthermore, since $\lim_{a \nearrow \lambda_1^{-1}} \pi_{++}(a) = |\lambda_2|^{-1}$, the graph of π_{++} has a vertical asymptote at $a = \lambda_1^{-1}$.

By applying l'Hôpital's rule we obtain

$$\lim_{s \searrow 0} \frac{b(s)}{a(s)} = \lim_{s \searrow 0} \frac{(\lambda_2 - \lambda_1) e^{ts} + \lambda_1 e^{\lambda_1 s} - \lambda_2 e^{\lambda_2 s}}{\lambda_2 (1 - e^{\lambda_1 s}) - \lambda_1 (1 - e^{\lambda_2 s})} = 1,$$

which implies that $\lim_{a \searrow 0} \pi'_{++}(a) = 1$, see expression (4.33).

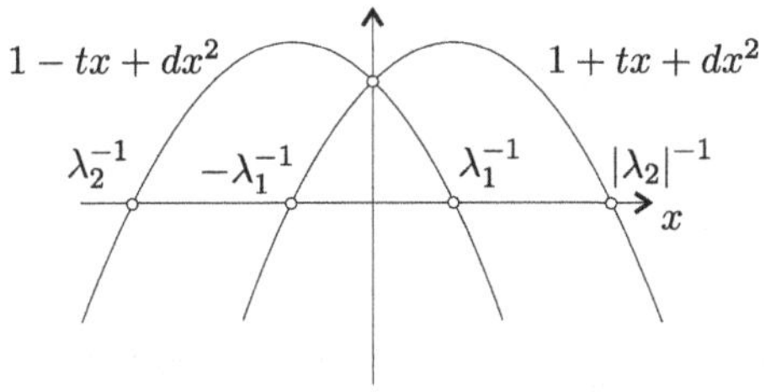

Figure 4.11: Qualitative behaviour of the parabolas $1 + tx + dx^2$ and $1 - tx + dx^2$ when the parameter $d < 0$.

Differentiating expression (4.33) with respect to a we get

$$\pi''_{++}(a) = \frac{db}{da}\frac{(b - a)(b + a)}{ab^2 (1 - at + a^2 d)}.$$

Hence, $\pi''_{++}(a) > 0$ if $\pi_{++}(a) > a$; $\pi''_{++}(a) = 0$ if $\pi_{++}(a) = a$; and $\pi''_{++}(a) < 0$ if $\pi_{++}(a) < a$. From this we conclude that $\pi_{++}(a) \neq a$, see the end of the proof of Proposition 4.4.1(e) for more details. Moreover, since $\lim_{a \nearrow \lambda_1^{-1}} \pi_{++}(a) = |\lambda_2|^{-1} > \lambda_1^{-1}$, we have $\pi_{++}(a) > a$ and $\pi''_{++}(a) > 0$.

(b) If $t = 0$, then $\lambda_1 = -\lambda_2$ and $\pi_{++}(a) = a$, see (4.31). □

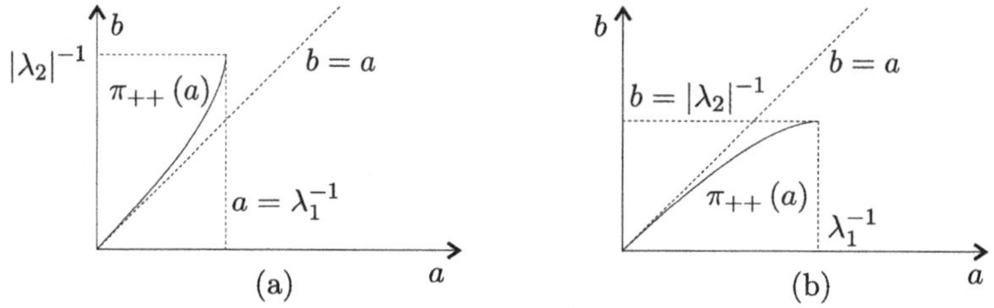

Figure 4.12: Qualitative behaviour of the Poincaré map π_{++} (a) when $d < 0$ and $t > 0$, (b) when $d < 0$ and $t < 0$.

Corollary 4.4.16. *Consider $A \in GL(\mathbb{R}^2)$ such that $d < 0$ and $t < 0$. Then the eigenvalues of A satisfy $\lambda_1 > 0 > \lambda_2$. Let π_{++} be the Poincaré map defined by the flow of the linear system $\dot{\mathbf{x}} = A\mathbf{x}$ and associated to two symmetric straight lines in the plane L_+ and L_-. Then:*

(a) $\pi_{++} : [0, \lambda_1^{-1}) \to [0, |\lambda_2|^{-1})$, $\pi_{++}(0) = 0$, $\lim_{a \nearrow \lambda_1^{-1}} \pi_{++}(a) = |\lambda_2|^{-1}$, *and* $\pi_{++}(a) < a$ *in* $(0, \lambda_1^{-1})$.

(b) *If* $a \in (0, \lambda_1^{-1})$, *then* $0 < \pi'_{++}(a) < 1$ *and* $\lim_{a \searrow 0} \pi'_{++}(a) = 1$.

(c) *If* $a \in (0, \lambda_1^{-1})$, *then* $\pi''_{++}(a) < 0$.

(d) *The straight line* $b = |\lambda_2|^{-1}$ *is a horizontal asymptote of the graph of* π_{++} *when* a *tends to* $+\infty$.

(e) π_{++} *is implicitly defined by the expression in Proposition* 4.4.15(a.4).

(f) *The qualitative behaviour of the graph of* π_{++} *is represented in Figure* 4.12(b).

Proof. The proof follows directly by using that the Poincaré map π_{++} is the inverse of the map described in Proposition 4.4.15. □

Corollary 4.4.17. *Consider $A \in GL(\mathbb{R}^2)$ such that $d < 0$ and let $\lambda_1 > 0 > \lambda_2$ be the eigenvalues of A. Suppose that the flow of the linear system $\dot{\mathbf{x}} = A\mathbf{x}$ has a contact point $\mathbf{p}$ with a straight line L not passing through the origin. Then L intersects with the stable and the unstable subspaces of the origin at $\mathbf{p} - \lambda_1^{-1}\dot{\mathbf{p}}$ and $\mathbf{p} + |\lambda_2|^{-1}\dot{\mathbf{p}}$, respectively.*

Proof. Since L does not pass through the origin, $\mathbf{p}$ is the unique contact point of the flow with L, i.e., L is not invariant under the flow, see Proposition 4.2.5(a). Thus L intersects the stable and the unstable subspaces of the origin at points $\mathbf{u}$ and $\mathbf{v}$, respectively. Moreover, the point $\mathbf{p}$ splits L into the two half-lines $L^I = \{\mathbf{p} + \lambda\dot{\mathbf{p}} : \lambda \leq 0\}$ and $L^O = \{\mathbf{p} + \lambda\dot{\mathbf{p}} : \lambda \geq 0\}$. It is clear that $\mathbf{u} \in L^I$ and $\mathbf{v} \in L^O$. Let $a_0 > 0$ and $b_0 > 0$ be the coordinates of the points $\mathbf{u}$ and $\mathbf{v}$, respectively. From the continuous dependence of the solutions of a linear differential system on the initial conditions, it follows that $\lim_{a \nearrow a_0} \pi_{++}(a) = b_0$. Therefore, the statement is a consequence of Proposition 4.4.15(a) and Corollary 4.4.16(a). □

Proposition 4.4.18. *Consider $A \in GL(\mathbb{R}^2)$ such that $d < 0$ and $t \geq 0$. Then the eigenvalues of A satisfy $\lambda_1 > 0 > \lambda_2$. Let π_{+-} be the Poincaré map defined by the flow of the linear system $\dot{\mathbf{x}} = A\mathbf{x}$, associated to two symmetric straight lines in the plane L_+ and L_-. Then:*

(a) *If $t > 0$, then $\pi_{+-} : (\lambda_1^{-1}, +\infty) \to (|\lambda_2|^{-1}, +\infty)$, $\lim_{a \searrow \lambda_1^{-1}} \pi_{+-}(a) = |\lambda_2|^{-1}$, and $\lim_{a \nearrow +\infty} \pi_{+-}(a) = +\infty$.*

  (a.1) *If $a \in (\lambda_1^{-1}, +\infty)$, then $\pi'_{+-}(a) > 1$, $\lim_{a \searrow \lambda_1^{-1}} \pi'_{+-}(a) = +\infty$, and $\lim_{a \nearrow +\infty} \pi'_{+-}(a) = 1$.*

  (a.2) *If $a \in (\lambda_1^{-1}, +\infty)$, then $\pi''_{+-}(a) < 0$.*

  (a.3) *The straight line $b = a - 2t/d$ is an asymptote of the graph of π_{+-} when a tends to $+\infty$.*

  (a.4) *π_{+-} is implicitly defined by the expression*

$$\left(\frac{\pi_{+-}(a)\left(t - \sqrt{t^2 - 4d}\right) - 2}{a\left(t - \sqrt{t^2 - 4d}\right) + 2}\right)^{\frac{t+\sqrt{t^2-4d}}{t-\sqrt{t^2-4d}}} = \frac{\pi_{+-}(a)\left(t + \sqrt{t^2 - 4d}\right) - 2}{a\left(t + \sqrt{t^2 - 4d}\right) + 2}.$$

  (a.5) *The qualitative behaviour of the graph of π_{+-} is shown in Figure* 4.13(a).

(b) *If $t = 0$, then π_{+-} is the identity map in $(\lambda_1^{-1}, +\infty)$.*

Proof. Arguments similar to those used in the proof of Proposition 4.4.3 show that the map $b = \pi_{+-}(a)$ is determined by the system

$$1 + b\lambda_1 = (-1 + a\lambda_1)\, e^{\lambda_1 s}, \quad 1 + b\lambda_2 = (-1 + a\lambda_2)\, e^{\lambda_2 s}, \tag{4.34}$$

and the inequalities $a \geq 0$, $b \geq 0$ and $s > 0$. From this we conclude that $a > \lambda_1^{-1}$ and $b > |\lambda_2|^{-1}$.

By solving system (4.34) for $s > 0$ we get the parametric equations of π_{+-}:

$$\begin{aligned} a(s) &= \frac{\lambda_2\left(1 + e^{\lambda_1 s}\right) - \lambda_1\left(1 + e^{\lambda_2 s}\right)}{d\left(e^{\lambda_1 s} - e^{\lambda_2 s}\right)}, \\ b(s) &= \frac{(\lambda_2 - \lambda_1)\, e^{ts} + \lambda_2 e^{\lambda_2 s} - \lambda_1 e^{\lambda_1 s}}{d\left(e^{\lambda_1 s} - e^{\lambda_2 s}\right)}. \end{aligned} \tag{4.35}$$

Hence, since $\lim_{s\searrow 0} a(s) = +\infty$, $\lim_{s\searrow 0} b(s) = +\infty$, $\lim_{s\nearrow +\infty} a(s) = \lambda_1^{-1}$, and $\lim_{s\nearrow +\infty} b(s) = |\lambda_2|^{-1}$, we conclude that $\pi_{+-}(a)$ is defined in $(\lambda_1^{-1}, +\infty)$, $\lim_{a\nearrow\lambda_1^{-1}} \pi_{+-}(a) = |\lambda_2|^{-1}$ and $\lim_{a\nearrow +\infty} \pi_{+-}(a) = +\infty$.

From (4.35) we obtain that $\lim_{a\nearrow +\infty} \pi_{+-}(a)/a = \lim_{s\searrow 0} b(s)/a(s) = 1$. Moreover by applying l'Hôpital's rule we get $\lim_{a\nearrow +\infty} b(s) - a(s) = -2t/d$. Therefore, $b = a - 2t/d$ is an asymptote of the graph of π_{+-}.

The implicit expression of π_{+-}

$$\left(\frac{-1 + a\lambda_1}{1 + b\lambda_1}\right)^{-\frac{\lambda_2}{\lambda_1}} = \frac{1 + b\lambda_2}{-1 + a\lambda_2}, \tag{4.36}$$

follows from system (4.34). Differentiating in (4.36) with respect to a and isolating db/da yields

$$\frac{db}{da} = \frac{a}{b}\frac{1 + bt + b^2 d}{1 - at + a^2 d}, \tag{4.37}$$

where the graphs of the parabolas are qualitatively depicted in Figure 4.11. Thus, it is easy to conclude that $\pi'_{+-}(a) > 1$, $\lim_{a\searrow\lambda_1^{-1}} \pi'_{+-}(a) = +\infty$, and

$$\pi''_{+-}(a) = \frac{db}{da}\frac{(b-a)(b+a)}{ab^2(1 - at + a^2 d)}.$$

Therefore, $\pi''_{+-}(a) < 0$. For more details, see the proof of Proposition 4.4.3 □

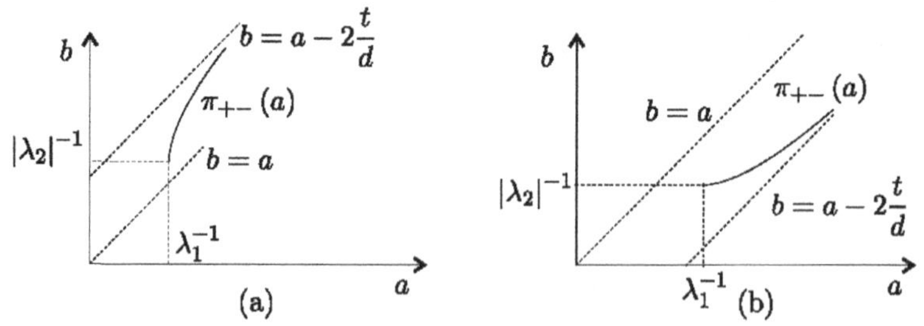

Figure 4.13: Qualitative behaviour of the Poincaré map π_{+-} (a) when $d < 0$ and $t > 0$, (b) when $d < 0$ and $t < 0$.

Corollary 4.4.19. *Consider $A \in GL(\mathbb{R}^2)$ such that $d < 0$ and $t < 0$. Then the eigenvalues of A satisfy $\lambda_1 > 0 > \lambda_2$. Let π_{+-} be the Poincaré map defined by the flow of the linear system $\dot{\mathbf{x}} = A\mathbf{x}$ and associated to two symmetric straight lines in the plane L_+ and L_-. Then:*

(a) $\pi_{+-} : (\lambda_1^{-1}, +\infty) \to (|\lambda_2|^{-1}, +\infty)$, $\lim_{a \searrow \lambda_1^{-1}} \pi_{+-}(a) = |\lambda_2|^{-1}$, $\lim_{a \nearrow +\infty} \pi_{+-}(a) = +\infty$, *and* $\pi_{+-}(a) < a$ *in the domain* $(\lambda_1^{-1}, +\infty)$.

(b) *If* $a \in (\lambda_1^{-1}, +\infty)$, *then* $0 < \pi'_{+-}(a) < 1$. *Furthermore,* $\lim_{a \searrow \lambda_1^{-1}} \pi'_{+-}(a) = 0$ *and* $\lim_{a \nearrow +\infty} \pi'_{+-}(a) = 1$.

(c) *If* $a \in (\lambda_1^{-1}, +\infty)$, *then* $\pi''_{+-}(a) > 0$.

(d) *The straight line* $b = a - 2t/d$ *is an asymptote of the graph of* π_{+-} *when* a *tends to* $+\infty$.

(e) π_{+-} *is implicitly defined by the expression in Proposition* 4.4.18(a.4).

(f) *The qualitative behaviour of the graph of* π_{+-} *is shown in Figure* 4.13(b).

Proof. The proof follows directly by using that the Poincaré map π_{+-} is the inverse of the map described in the Proposition 4.4.18, see Lemma 4.3.5(c). □

4.4.5 Degenerate node: $d = 0$

We suppose now that the matrix A has two real eigenvalues, one being equal to 0 and the other one equal to t. Hence, A has two different real Jordan normal forms. When $t \neq 0$, then the real Jordan normal form of A is

$$A = \begin{pmatrix} t & 0 \\ 0 & 0 \end{pmatrix},$$

while when $t = 0$, the real Jordan normal form of A is

$$A = \begin{pmatrix} 0 & 1 \\ 0 & 0 \end{pmatrix}.$$

Note that we do not consider the case where A is the zero matrix.

In any case, the behaviour of the Poincaré map π_{++} is trivial, see Proposition 4.3.3(b). Therefore we restrict our attention to the Poincaré map π_{+-}.

Proposition 4.4.20. *Consider* $A \in L(\mathbb{R}^2)$ *not the zero matrix and such that* $d = 0$. *Let* π_{+-} *be the Poincaré map defined by the flow of the linear system* $\dot{\mathbf{x}} = A\mathbf{x}$ *and associated to two symmetric straight lines in the plane* L_+ *and* L_-. *Then:*

(a) *If* $t > 0$, *then there exists a value* $b^* > 0$ *such that* $\pi_{+-} : [0, +\infty) \to [b^*, +\infty)$ *and* $\pi_{+-}(a) = a + b^*$.

(b) *If* $t = 0$, *then* π_{+-} *is the identity map in* $[0, +\infty)$.

(c) *If* $t < 0$, *then there exist a value* $b^* > 0$ *such that* $\pi_{+-} : [b^*, +\infty) \to [0, +\infty)$ *and* $\pi_{+-}(a) = a - b^*$.

Proof. Let $\mathbf{n} = (n_1, n_2)^T$ be the unit orthogonal vector to L_+ oriented in the direction opposite to the origin and suppose that $\det(A\mathbf{n}^\perp, \mathbf{n}^\perp) > 0$. The case $\det(A\mathbf{n}^\perp, \mathbf{n}^\perp) < 0$ follows by using similar arguments. Let $\mathbf{p} = (p_1, p_2)^T$ be the contact point of the flow with L_+. From Proposition 4.3.11(c) it follows that the map π_{+-} satisfies

$$-\mathbf{p} - \pi_{+-}(a)\,\mathbf{n}^\perp = e^{sA}\left(\mathbf{p} - a\mathbf{n}^\perp\right).$$

Therefore $\mathbf{p} \in \ker(A) \setminus \{\mathbf{0}\}$, see Proposition 4.2.6(b).

(a) Without loss of generality we can consider that the matrix A is in real Jordan normal form. Thus, $p_1 = 0$, $p_2 \neq 0$ and the map $b = \pi_{+-}(a)$ is implicitly defined by the system

$$bn_2 = e^{ts}an_2, \quad bn_1 = an_1 - 2p_2,$$

and the inequalities $a \geq 0$, $b \geq 0$ and $s > 0$. From $\det(A\mathbf{n}^\perp, \mathbf{n}^\perp) = -tn_1n_2 > 0$ we obtain that $n_1n_2 < 0$. Moreover, $\mathbf{p}^T\mathbf{n} = p_2n_2 > 0$ implies that $p_2/n_1 < 0$. Therefore, $\pi_{+-}(a) = a - 2p_2/n_1$ for $a \geq 0$.

(b) Without loss of generality we can consider that the matrix A is in real Jordan normal form. Thus, $p_1 \neq 0$, $p_2 = 0$ and the map $b = \pi_{+-}(a)$ is implicitly defined by the system

$$-p_1 + bn_2 = p_1 + an_2 - san_1, \quad -bn_1 = -an_1,$$

and the inequalities $a \geq 0$, $b \geq 0$ y $s > 0$. Since $\mathbf{p}^T\mathbf{n} = p_1n_1 > 0$, we obtain that $n_1 \neq 0$ and $\pi_{+-}(a) = a$.

(c) The proof follows by using the fact that the Poincaré map π_{+-} is the inverse of the map described in the statement (a), see Proposition 4.3.5(c). □

4.5 Poincaré maps of non-homogeneous linear systems

To finish our study about the Poincaré maps defined by the flow of a fundamental system and associated to two symmetric straight lines L_+ and L_-, we need to analyze the Poincaré maps defined by the flow of a non-homogeneous linear system

$$\dot{\mathbf{x}} = A\mathbf{x} + \mathbf{b}, \tag{4.38}$$

where $A \in L(\mathbb{R}^2)$ and $\mathbf{b} \in \mathbb{R}^2 \setminus \{\mathbf{0}\}$, and associated to a straight line not passing through the origin. We denote by L_+ this straight line and by S_0 and S_+ the half-planes bounded by L_+, where S_0 is the half-plane containing the origin. As before, $\mathbf{n}$ denotes the unit orthogonal vector to the straight line which is oriented in the direction opposite to the origin.

Since L_+ does not pass through the origin it can be split into the two subsets, L_+^I and L_+^O. The set $CP_+ = L_+^I \cap L_+^O$ is formed by the contact points of the flow of the non-homogeneous linear system (4.38) with L_+. Define in L_+ the subset

$$\begin{aligned}\mathrm{Dom}_{++} := &\left\{\mathbf{q} \in L_+^O : \exists s_{\mathbf{q}} \geq 0,\ \Phi\left(s_{\mathbf{q}}, \mathbf{q}\right) \in L_+^I \text{ and } \Phi\left(s, \mathbf{q}\right) \subset S_+ \ \forall s \in \left(0, s_{\mathbf{q}}\right)\right\}\\ &\cup CP_+.\end{aligned}$$

When $\mathrm{Dom}_{++} \neq \varnothing$, we can define *the Poincaré map of the non-homogeneous linear system* (4.38) *associated to the straight line* L_+ by $\Pi_{++} : \mathrm{Dom}_{++} \subset L_+^O \to L_+^I$ with $\Pi_{++}(\mathbf{q}) = \Phi(s_\mathbf{q}, \mathbf{q})$.

Suppose that the flow of system (4.38) has a unique contact point $\mathbf{p}$ with L_+. Since $\mathbf{p} \in L_+^I$ and $\mathbf{p} \in L_+^O$, $\Pi_{++}(\mathbf{p}) = \mathbf{p}$. In the particular case $\mathrm{Dom}_{++} = CP_+$ we say that *the behaviour of* Π_{++} *is trivial.*

Note that the Poincaré map Π_{++} here defined corresponds with the Poincaré map Π_{++}^A defined by the flow of a fundamental system and associated with the straight line L_+, see Section 4.1. The study of the qualitative behaviour of the map Π_{++} is divided into two subsections depending on the invertibility of the matrix A. Thus, in Subsection 4.5.1 we deal with the case $\det(A) \neq 0$ and in Subsection 4.5.2 with the case $\det(A) = 0$.

4.5.1 Non-homogeneous linear systems with $A \in GL(\mathbb{R}^2)$

Suppose that A is invertible. Then we can consider the point $\mathbf{e}_+ = -A^{-1}\mathbf{b}$. The translation $\mathbf{y} = \mathbf{x} - \mathbf{e}_+$ transforms the system (4.38) into the homogeneous linear system

$$\dot{\mathbf{y}} = A\mathbf{y}, \tag{4.39}$$

and the straight line L_+ into the straight line L_+^*. Thus, if $\mathbf{e}_+ \notin L_+$, then L_+^* does not pass through the origin. Define $L_-^* = \{-\mathbf{q} : \mathbf{q} \in L_+^*\}$. Note that the homogeneous linear system (4.39) and the straight lines L_+^* and L_-^* fulfill the conditions of Section 4.3. Therefore, if the Poincaré map Π_{++} associated to the flow of (4.38) and the straight line L_+ is defined, then it induces a Poincaré map associated to the flow of the homogeneous linear system (4.39) and the straight line L_+^*. Moreover, the converse statement is also true. Let Π_{jk}^* denote the Poincaré maps induced by the translation above. Hence, we have the following result.

Proposition 4.5.1. *Consider a matrix* $A \in GL(\mathbb{R}^2)$, *a vector* $\mathbf{b} \in \mathbb{R}^2 \setminus \{\mathbf{0}\}$ *and the point* $\mathbf{e}_+ = -A^{-1}\mathbf{b}$. *Let* L_+ *be a straight line in the plane not passing through the origin and such that* $\mathbf{e}_+ \notin L_+$. *The Poincaré map* Π_{++} *associated to the flow of the non-homogeneous linear system* $A\mathbf{x} + \mathbf{b}$ *and the straight line* L_+ *is defined if and only if there exists a unique contact point of the flow with* L_+. *In this case* L_+ *is a non-invariant straight line and* L_+^I *and* L_+^O *are non-empty half-lines.*

Proof. The statement is a consequence of Proposition 4.3.4. □

It is clear that the behaviour of the map Π_{++} can be obtained from the behaviour of the map Π_{++}^*. Furthermore, this last map can be expressed as a composition of the Poincaré maps considered in Section 4.3.

Depending on whether $\mathbf{e}_+ \in S_+$ or $\mathbf{e}_+ \in S_0$, the translation $\mathbf{y} = \mathbf{x} - \mathbf{e}_+$ preserves or reverses the orientation of the flow on L_+ and L_+^*, see Figure 4.14. Thus when $\mathbf{e}_+ \in S_0$ the translation $\mathbf{y} = \mathbf{x} - \mathbf{e}_+$ transforms the half-lines L_+^I and L_+^O into the half-lines L_+^{*I} and L_+^{*O}, respectively. When $\mathbf{e}_+ \in S_+$, the translation transforms L_+^I into L_+^{*O} and L_+^O into L_+^{*I}. This implies the following result.

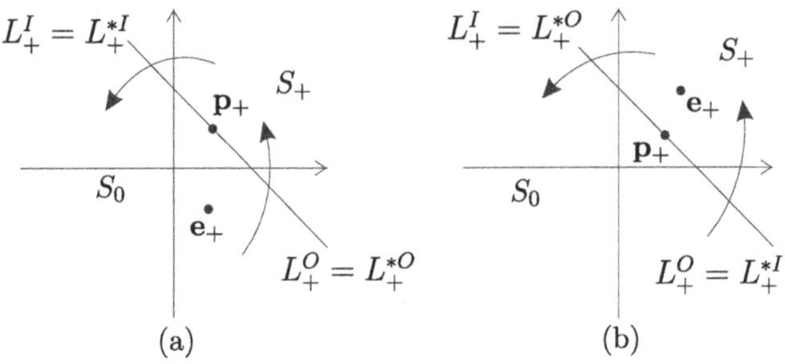

Figure 4.14: Relation between the half-lines L_+^I, L_+^O, L_+^{*I} and L_+^{*O} depending on (a) $\mathbf{e}_+ \in S_0$, (b) $\mathbf{e}_+ \in S_+$.

Proposition 4.5.2. *Consider a matrix $A \in GL(\mathbb{R}^2)$, a vector $\mathbf{b} \in \mathbb{R}^2 \setminus \{\mathbf{0}\}$ and the point $\mathbf{e}_+ = -A^{-1}\mathbf{b}$. Let L_+ be a straight line in the plane not passing through the origin and such that $\mathbf{e}_+ \notin L_+$. Suppose that the Poincaré map Π_{++} associated to the flow of the system $\dot{\mathbf{x}} = A\mathbf{x}+\mathbf{b}$ and the straight line L_+ is defined, and let Π_{++}^* be the Poincaré map induced by the translation $\mathbf{y} = \mathbf{x} - \mathbf{e}_+$.*

(a) *Suppose that $\mathbf{e}_+ \in S_0$.*

  (a.1) *If* $\det(A) > 0$, *then Π_{++}^* is the Poincaré map defined in Section 4.3.*

  (a.2) *If* $\det(A) < 0$, *then Π_{++}^* is trivial.*

(b) *Suppose that $\mathbf{e}_+ \in S_+$.*

  (b.1) *Assume that* $\det(A) > 0$. *If $t^2 - 4d \geq 0$, then Π_{++}^* is trivial. If $t^2 - 4d < 0$, then Π_{++}^* coincides with the composition of the Poincaré maps $\Pi_{-+} \circ \Pi_{--} \circ \Pi_{+-}$ where Π_{-+}, Π_{--} and Π_{+-} are the maps defined in Section 4.3.*

  (b.2) *If* $\det(A) < 0$, *then Π_{++}^* is the Poincaré map defined in Section 4.3.*

Proof. (a) Since $\mathbf{e}_+ \in S_0$, the translation $\mathbf{y} = \mathbf{x} - \mathbf{e}_+$ transforms the half-lines L_+^I and L_+^O into L_+^{*I} and L_+^{*O}, respectively. Moreover, it is easy to check that the domain Dom_{++} of Π_{++}^* is contained in L_+^{*O}.

When $\det(A) > 0$, the Poincaré map defined by a linear flow with the domain contained in L_+^{*O} is the map defined in Section 4.3; this proves statement (a.1).

When $\det(A) < 0$, the domain of Π_{++}^* is contained in the half-line L_+^{*I}, see Proposition 4.3.3(c). Thus Π_{++}^* is defined in the intersection of the half-lines L_+^{*I} and L_+^{*O}, i.e., in the contact point. Applying again this argument it follows that the image of Π_{++}^* is also the contact point. Therefore the behaviour of Π_{++}^* is trivial, as claimed in (a.2).

(b) Suppose that $\mathbf{e}_+ \in S_+$. The translation $\mathbf{y} = \mathbf{x} - \mathbf{e}_+$ transforms the half-lines L_+^I and L_+^O into L_+^{*O} and L_+^{*I}, respectively. By Proposition 4.3.3(a), there are no Poincaré maps defined on L_+^I which have the image contained on L_+^O. Thus either the behaviour of Π_{++}^* is trivial, or $\Pi_{++}^* = \Pi_{-+} \circ \Pi_{--} \circ \Pi_{+-}$. In this last case the orbits have to surround the origin. Therefore, if $t^2 - 4d \geq 0$, then Π_{++}^* is trivial, see Lemma 4.2.9(a). If $t^2 - 4d < 0$, then Π_{++}^* is non-trivial. Thus $\Pi_{++}^* = \Pi_{-+} \circ \Pi_{--} \circ \Pi_{+-}$, which proves the statement (b.1).

If $\det(A) < 0$, then by Proposition 4.3.3(c) we have two possibilities: (i) Π_{++}^* is the map Π_{++}; (ii) $\Pi_{++}^* = \Pi_{-+} \circ \Pi_{--} \circ \Pi_{+-}$. In this last case the orbits have to surround the origin, which proves the statement (b.2). □

In order to study the Poincaré maps defined by the flow of a non-homogeneous linear system with regular matrix and associated to a straight line not passing through the origin, we restrict our attention to the Poincaré maps defined by the flow of a homogeneous linear system, see Proposition 4.5.2. When $\det(A) > 0$, the behaviour of the map Π_{++}^* depends on whether $\mathbf{e}_+ \in S_0$, or $\mathbf{e}_+ \in S_+$ and $t^2 - 4d < 0$. To distinguish between these two situations we denote the map $\Pi_{-+} \circ \Pi_{--} \circ \Pi_{+-}$ by $\widetilde{\Pi}_{++}$.

Thus, the behaviour of the Poincaré maps defined by a non-homogeneous linear flow and associated to a straight line which does not pass through the origin is determined by π_{++} and $\widetilde{\pi}_{++} = \pi_{-+} \circ \pi_{--} \circ \pi_{+-}$. There are two different ways of studying the map $\widetilde{\pi}_{++}$: one by using the well know information about maps π_{+-} and π_{++}, see Section 4.3, and the other by obtaining a new expression for $\widetilde{\pi}_{++}$. The latter is adopted here.

Note that $\widetilde{\Pi}_{++} = \Pi_{-+} \circ \Pi_{--} \circ \Pi_{+-}$ satisfies that $\widetilde{\Pi}_{++} : L_+^{*I} \to L_+^{*O}$ where $L_+^{*I} = \{\mathbf{p} + a\dot{\mathbf{p}} : a \geq 0\}$ and $L_+^{*O} = \{\mathbf{p} - a\dot{\mathbf{p}} : a \geq 0\}$, see (4.8). Thus the map $\widetilde{\pi}_{++}$ is implicitly defined by the equation

$$\mathbf{p} - \widetilde{\pi}_{++}(a)\,\dot{\mathbf{p}} = e^{As}(\mathbf{p} + a\dot{\mathbf{p}}) \tag{4.40}$$

and the inequalities $a \geq 0$, $b = \widetilde{\pi}_{++}(a) \geq 0$ and $s \geq 0$.

4.5.2 Non-homogeneous linear systems with $A \notin GL(\mathbb{R}^2)$

Suppose now that the matrix A of the non-homogeneous linear system (4.38) is singular, that is $\det(A) = 0$. To describe the Poincaré maps defined by the flow of this system and associated to a straight line in the plane, we distinguish different situations. First we consider the case where A is the zero matrix. Then $\dot{\mathbf{x}} = \mathbf{b}$ where $\mathbf{b} \in \mathbb{R}^2 \setminus \{\mathbf{0}\}$. Therefore the orbits of this system are contained in straight lines and consequently the behaviour of Π_{++} is trivial.

Suppose now that system (4.38) has a singular point $\mathbf{e}$, i.e., $A\mathbf{e} + \mathbf{b} = \mathbf{0}$. The translation $\mathbf{y} = \mathbf{x} - \mathbf{e}$ transforms (4.38) into the homogeneous system $\dot{\mathbf{y}} = A\mathbf{y}$ and the Poincaré map Π_{++} into the Poincaré map Π_{++}^*. By Proposition 4.3.3(b), the behaviour of the map Π_{++}^* is trivial. Hence the map Π_{++} associated to the system (4.38) is also trivial. For this reason, we restrict our attention to the Poincaré maps defined by the flow of a non-homogeneous linear system (4.38) without singular points and such that the matrix A is singular, but not the zero matrix.

Let L_+ be a straight line in the plane which is not invariant under the flow. In this subsection we define a coordinate system onr L_+ and a map $\widetilde{\pi}_{++}$ which is invariant under linear transformations so that we can reduce the study of Π_{++} to the study of $\widetilde{\pi}_{++}$.

Let $\mathbf{n}$ be the unit orthogonal vector to L_+ which is oriented in the direction opposite to the origin. We define the following subsets in L_+:

$$L_+^I := \{\mathbf{q} \in L_+ : \mathbf{n}^T\dot{\mathbf{q}} \leq 0\} \quad \text{and} \quad L_+^O := \{\mathbf{q} \in L_+ : \mathbf{n}^T\dot{\mathbf{q}} \geq 0\},$$

where $\dot{\mathbf{q}} = A\mathbf{q} + \mathbf{b}$.

Proposition 4.5.3. *Consider a singular non-zero matrix $A \in L(\mathbb{R}^2)$ and a vector $\mathbf{b} \in \mathbb{R}^2 \setminus \{\mathbf{0}\}$. Let L_+ be a straight line in the plane which is not invariant under the flow of the system $\dot{\mathbf{x}} = A\mathbf{x} + \mathbf{b}$. Suppose that this non-homogeneous linear system has no singular points. Then:*

(a) *The flow of the system has at most a contact point with L_+.*

(b) *If $L_+^I \neq \varnothing$ and $L_+^O \neq \varnothing$, then the system has exactly one contact point with L_+.*

(c) *Suppose that there exists a contact point $\mathbf{p}$ of the flow with the straight line L_+ and let $\mathbf{n}$ be as above. Then $\det(A\mathbf{n}^\perp, \mathbf{b}) \neq 0$. Moreover, if $\det(A\mathbf{n}^\perp, \mathbf{b}) > 0$, then $L_+^I = \{\mathbf{p}+\lambda\dot{\mathbf{p}} : \lambda \leq 0\}$ and $L_+^O = \{\mathbf{p}+\lambda\dot{\mathbf{p}} : \lambda \geq 0\}$, and if $\det(A\mathbf{n}^\perp, \mathbf{b}) < 0$, then $L_+^I = \{\mathbf{p}+\lambda\dot{\mathbf{p}} : \lambda \geq 0\}$ and $L_+^O = \{\mathbf{p}+\lambda\dot{\mathbf{p}} : \lambda \leq 0\}$.*

(d) *The Poincaré map Π_{++} is defined if and only if the flow of the system has a unique contact point with L_+.*

Proof. (a) Suppose that the flow of the system has two contact points with L_+, $\mathbf{p}_1$ and $\mathbf{p}_2$. Thus $L_+ = \{\mathbf{p}_1 + \lambda(\mathbf{p}_1 - \mathbf{p}_2) : \lambda \in \mathbb{R}\}$, $A\mathbf{p}_1 + \mathbf{b} = \alpha_1(\mathbf{p}_1 - \mathbf{p}_2)$ and $A\mathbf{p}_2 + \mathbf{b} = \alpha_2(\mathbf{p}_1 - \mathbf{p}_2)$. From this it follows that $A(\mathbf{p}_1 - \mathbf{p}_2) = \alpha(\mathbf{p}_1 - \mathbf{p}_2)$, which implies that L_+ is an invariant straight line, in contradiction with our hypothesis. Therefore there exists at most one contact point with L_+.

(b) Take a point $\mathbf{p}$ in L_+. We can write $L_+ = \{\mathbf{p} + \lambda\mathbf{n}^\perp : \lambda \in \mathbb{R}\}$. Consider the auxiliar function $f(\lambda) = \mathbf{n}^T[A(\mathbf{p}+\lambda\mathbf{n}^\perp)+\mathbf{b}]$. Since $L_+^I \neq \varnothing$ and $L_+^O \neq \varnothing$ there exist λ_1 and λ_2 such that $f(\lambda_1) \leq 0$ and $f(\lambda_2) \geq 0$. The existence of a contact point follows from the continuity of the function f. The uniqueness of the contact point is a consequence of statement (a).

(c) Let $\mathbf{p}$ be a contact point of the flow with L_+. Since the system has no singular points, $\dot{\mathbf{p}} \neq \mathbf{0}$. From this we conclude that $L_+ = \{\mathbf{p}+\lambda\dot{\mathbf{p}} : \lambda \in \mathbb{R}\}$, that the vector $\mathbf{n}$ is equal to either $\dot{\mathbf{p}}^\perp/\|\dot{\mathbf{p}}\|$ or $-\dot{\mathbf{p}}^\perp/\|\dot{\mathbf{p}}\|$, and that $\mathbf{n}^T(A\dot{\mathbf{p}}) = \det(A\mathbf{n}^\perp, \mathbf{b})$.

Consider now a different point $\mathbf{q}$ in L_+. Then $\mathbf{q} = \mathbf{p} + \lambda\dot{\mathbf{p}}$ with $\lambda \neq 0$. Therefore

$$\mathbf{n}^T\dot{\mathbf{q}} = \mathbf{n}^T(A\mathbf{p} + \mathbf{b}) + \lambda\mathbf{n}^T A\dot{\mathbf{p}} = \lambda \det\left(A\mathbf{n}^\perp, \mathbf{b}\right).$$

If we suppose that $\det(A\mathbf{n}^\perp, \mathbf{b}) = 0$, then since $\mathbf{n}^T\dot{\mathbf{q}} = 0$ for every point $\mathbf{q} \in L_+$, we conclude that L_+ is an invariant straight line, which contradicts the hypothesis.

Therefore, $\det(A\mathbf{n}^{\perp}, \mathbf{b}) \neq 0$. The expression of L_+^I and L_+^O follows from the above equation.

(d) Suppose that the map Π_{++} is defined. Then the half-straight lines $L_+^I \neq \varnothing$ and $L_+^O \neq \varnothing$. By statement (b), the flow of the system has exactly one contact point with the straight line L_+.

Conversely, if the flow has exactly one contact point $\mathbf{p}$ with L_+, then $\mathbf{p}$ splits L_+ into the half-lines L_+^I and L_+^O. Let $\gamma(\mathbf{p})$ be the orbit through the point $\mathbf{p}$. By the continuous dependence of the solutions of a linear differential system with respect to the initial conditions we conclude the existence of a Poincaré map in a neighbourhood of $\gamma(\mathbf{p})$. □

If the flow of the non-homogeneous linear system (4.38) has a contact point $\mathbf{p}$ with the straight line L_+, then we can associate a value $a \geq 0$, called *coordinate*, to any point on L_+, see Proposition 4.5.3(c). Thus if $\det(A\mathbf{n}^{\perp}, \mathbf{b}) > 0$, then

$$L_+^I = \{\mathbf{p} - a\dot{\mathbf{p}} : a \geq 0\} \quad \text{and} \quad L_+^O = \{\mathbf{p} + a\dot{\mathbf{p}} : a \geq 0\},$$

while if $\det(A\mathbf{n}^{\perp}, \mathbf{b}) < 0$, then

$$L_+^I = \{\mathbf{p} + a\dot{\mathbf{p}} : a \geq 0\} \quad \text{and} \quad L_+^O = \{\mathbf{p} - a\dot{\mathbf{p}} : a \geq 0\}.$$

Let Π_{++} be the Poincaré map defined by the flow of system (4.38) and associated to the straight line L_+, and let $\mathbf{q}_1$ and $\mathbf{q}_2$ be two points on L_+ such that $\mathbf{q}_2 = \Pi_{++}(\mathbf{q}_1)$. We denote by $\widetilde{\pi}_{++}$ the map which transforms the coordinate of $\mathbf{q}_1$ into the coordinate of $\mathbf{q}_2$. In this way we can reduce the study of the behaviour of $\widetilde{\pi}_{++}$ to the study of the behaviour of Π_{++}.

By using similar arguments to those used in the homogeneous case we obtain the next result which we present without proof.

Lemma 4.5.4. *Consider a singular matrix $A \in L(\mathbb{R}^2)$ and a vector $\mathbf{b} \in \mathbb{R}^2 \setminus \{\mathbf{0}\}$. Let L_+ be a straight line in the plane which does not pass through the origin. Suppose that the Poincaré map $\widetilde{\pi}_{++}$ associated to the flow of the non-homogeneous linear system $\dot{\mathbf{x}} = A\mathbf{x} + \mathbf{b}$ and the straight line L_+ is defined. Then:*

(a) *$\widetilde{\pi}_{++}^*$ associated to the flow of the system $\dot{\mathbf{x}} = -A\mathbf{x} - \mathbf{b}$ and the straight line L_+ is defined and satisfies that $\widetilde{\pi}_{++}^* = \widetilde{\pi}_{++}^{-1}$.*

(b) *$\widetilde{\pi}_{++}$ depends analytically on the parameter $t = \operatorname{trace}(A)$.*

(c) *$\widetilde{\pi}_{++}$ is an analytic function of its argument and its inverse is also analytic.*

By Lemma 4.5.4, in order to determine the qualitative behaviour of the Poincaré map $\widetilde{\pi}_{++}$ defined by the system (4.38), it is enough to consider the case $\operatorname{trace}(A) \geq 0$. The case $\operatorname{trace}(A) < 0$ follows by considering $\widetilde{\pi}_{++}^{-1}$.

Lemma 4.5.5. *Consider a non-zero matrix $A \in L(\mathbb{R}^2)$, a vector $\mathbf{b} \in \mathbb{R}^2 \setminus \{\mathbf{0}\}$, and a regular matrix $M \in GL(\mathbb{R}^2)$ such that $\det(M) > 0$. Let L be a straight line in the plane and $\mathbf{p}$ be a contact point of the flow of the system $\dot{\mathbf{x}} = A\mathbf{x} + \mathbf{b}$ with L.*

Then $\mathbf{p}^* = M\mathbf{p}$ *is a contact point of the flow of the system* $\dot{\mathbf{x}} = A^*\mathbf{x} + \mathbf{b}^*$ *with* $L^* = ML$, *where* $A^* = MAM^{-1}$ *and* $\mathbf{b}^* = M\mathbf{b}$. *Moreover,* $\dot{\mathbf{p}}^* = M\dot{\mathbf{p}}$, $L^{*I} = ML^I$ *and* $L^{*O} = ML^O$.

Proof. The linear change of coordinates $\mathbf{y} = M\mathbf{x}$ transforms the system $\dot{\mathbf{x}} = A\mathbf{x}+\mathbf{b}$, the straight line L, and the contact point $\mathbf{p}$ into the system $\dot{\mathbf{y}} = A^*\mathbf{y}+\mathbf{b}^*$, the straight line $L^* = ML$, and the contact point $\mathbf{p}^* = M\mathbf{p}$, respectively.

Let $\mathbf{n}$ and $\mathbf{n}^*$ be the unit orthogonal vectors to L and L^*, respectively. Then

$$\det\left(A^*\mathbf{n}^{*\perp}, \mathbf{b}^*\right) = \frac{\det(M)}{||M\mathbf{n}^\perp||} \det\left(A\mathbf{n}^\perp, \mathbf{b}\right).$$

See the proof of Lemma 4.3.6 for more details.

Assume that $\det(A\mathbf{n}^\perp, \mathbf{b}) > 0$ (the case $\det(A\mathbf{n}^\perp, \mathbf{b}) < 0$ is treated in a similar way). In this case $L^I_+ = \{\mathbf{p}-a\dot{\mathbf{p}} : a \geq 0\}$ and $L^O_+ = \{\mathbf{p}+a\dot{\mathbf{p}} : a \geq 0\}$, see Proposition 4.5.3(c). On the other hand, since $\det(M) > 0$, we obtain that $\det(A^*\mathbf{n}^{*\perp}, \mathbf{b}^*) > 0$. Hence, $L^{*I} = \{\mathbf{p}^*-a\dot{\mathbf{p}}^* : a \geq 0\}$ and $L^{*O} = \{\mathbf{p}^*+a\dot{\mathbf{p}}^* : a \geq 0\}$, i.e., $L^{*I} = ML^I$ and $L^{*O} = ML^O$. □

Proposition 4.5.6. *Consider a singular matrix* $A \in L(\mathbb{R}^2)$ *and a vector* $\mathbf{b} \in \mathbb{R}^2\backslash\{\mathbf{0}\}$. *Let* L_+ *be a straight line in the plane which does not pass through the origin. Suppose that the Poincaré map* $\widetilde{\pi}_{++}$ *associated to the flow of* $\dot{\mathbf{x}} = A\mathbf{x} + \mathbf{b}$ *and to the straight line* L_+ *is defined. If* $M \in GL(\mathbb{R}^2)$ *such that* $\det(M) > 0$, *then* $\widetilde{\pi}_{++}$ *is invariant under the change of coordinates* $\mathbf{y} = M\mathbf{x}$.

Proof. Suppose that $\det(A\mathbf{n}^\perp, \mathbf{b}) > 0$ (the case $\det(A\mathbf{n}^\perp, \mathbf{b}) < 0$ is treated similarly). By Lemma 4.5.5, $\mathbf{p}^* = M\mathbf{p}$ is the contact point of the flow of the system $\dot{\mathbf{y}} = A^*\mathbf{y} + \mathbf{b}^*$ with the straight line $L^*_+ = ML_+$, where $A^* = MAM^{-1}$ and $\mathbf{b}^* = M\mathbf{b}$. Moreover, $\mathbf{p}^*$ splits L^*_+ into the two half-lines $L^{*I}_+ = ML^I_+$ and $L^{*O}_+ = ML^O_+$.

Since $\widetilde{\pi}_{++}$ is well defined, there exist a coordinate $a \geq 0$, a point $\mathbf{q}_1$ in L^O_+, and a point $\mathbf{q}_2$ in L^I_+ such that $\mathbf{q}_2 = \Pi_{++}(\mathbf{q}_1)$, $\mathbf{q}_1 = \mathbf{p} + a\dot{\mathbf{p}}$ and $\mathbf{q}_2 = \mathbf{p} - \widetilde{\pi}_{++}(a)\,\dot{\mathbf{p}}$. Therefore, $M\mathbf{q}_1 \in L^{*O}_+$, $M\mathbf{q}_2 \in L^{*I}_+$ and $M\mathbf{q}_1 = \Pi^*_{++}(M\mathbf{q}_2)$, where $M\mathbf{q}_1 = M\mathbf{p} + aM\dot{\mathbf{p}}$ and $M\mathbf{q}_2 = M\mathbf{p} - \widetilde{\pi}_{++}(a)M\dot{\mathbf{p}}$. This proves the proposition. □

By the last proposition, we can assume that the matrix A is given in real Jordan normal form. In other case, since $\det(A) = 0$, we can always transform the matrix A into its real Jordan normal form by an orientation-preserving change of coordinates.

4.5.3 Qualitative behaviour of the Poincaré map $\widetilde{\pi}_{++}$

In this subsection we study the qualitative behaviour of the Poincaré map defined by the flow of the non-homogeneous linear system $\dot{\mathbf{x}} = A\mathbf{x} + \mathbf{b}$. Recall that if $d = \det(A) \neq 0$, then the Poincaré map above is equal to one of the Poincaré maps defined by a linear flow, see Proposition 4.5.2 and Section 4.4. Only when

$t^2-4d<0$ and the singular point $\mathbf{e}_+$ belongs to S_+ the Poincaré map denoted by $\widetilde{\pi}_{++}$ does not coincide to any of the Poincaré maps defined by a linear flow. In this case such Poincaré maps can be expressed as a composition of the Poincaré maps studied in Section 4.4; that is, $\widetilde{\pi}_{++}=\pi_{-+}\circ\pi_{--}\circ\pi_{+-}$. Proposition 4.5.7 and Corollary 4.5.8 describe the behaviour of the map $\widetilde{\pi}_{++}$ when $d\neq 0$, $t^2-4d<0$ and $t\geq 0$ or $t<0$, respectively.

In the degenerate case $d=0$ there exists two possibilities for the Poincaré map $\widetilde{\pi}_{++}$. If the non-homogeneous linear system $\dot{\mathbf{x}}=A\mathbf{x}+\mathbf{b}$ has a singular point, then the behaviour of $\widetilde{\pi}_{++}$ is trivial, see Subsection 4.5.2. If the non-homogeneous system has no singular points, then $\widetilde{\pi}_{++}$ cannot be reduced to any of the Poincaré maps associated to the homogeneous case. In Proposition 4.5.9 and Corollary 4.5.10 we will describe the behaviour of the map $\widetilde{\pi}_{++}$ when $d=0$, and $t\geq 0$ and $t<0$, respectively.

Proposition 4.5.7. *Consider a matrix $A\in GL(\mathbb{R}^2)$ with parameters $d>0$, $t\geq 0$ and $t^2-4d<0$, and a vector $\mathbf{b}\in\mathbb{R}^2\setminus\{\mathbf{0}\}$. Let $\widetilde{\pi}_{++}$ be the Poincaré map defined by the flow of the system $\dot{\mathbf{x}}=A\mathbf{x}+\mathbf{b}$ and associated to a straight line L_+ which does not pass through the origin. Then:*

(a) *If $t>0$, then there exist a value $b^*>0$ such that $\widetilde{\pi}_{++}:[0,+\infty)\to[b^*,+\infty)$, $\widetilde{\pi}_{++}(0)=b^*$, $\lim\limits_{a\nearrow+\infty}\widetilde{\pi}_{++}(a)=+\infty$, and $\widetilde{\pi}_{++}(a)>a$ in $(0,+\infty)$.*

  (a.1) *If $a\in(0,+\infty)$, then $\widetilde{\pi}'_{++}(a)>0$ and $\lim\limits_{a\searrow 0}\widetilde{\pi}'_{++}(a)=0$.*

  (a.2) *If $a\in(0,+\infty)$, then $\widetilde{\pi}''_{++}(a)>0$.*

  (a.3) *The straight line $b=e^{\gamma\pi}a+t(1+e^{\gamma\pi})/d$ is an asymptote of the graph of $\widetilde{\pi}_{++}$ when a tends to $+\infty$, where $\gamma=t/\sqrt{4d-t^2}$.*

  (a.4) *$\widetilde{\pi}_{++}$ is implicitly defined by the expression*

$$\frac{1-t\widetilde{\pi}_{++}(a)+d\widetilde{\pi}_{++}(a)^2}{1+ta+da^2}=e^{2\gamma\arctan\left(\frac{(a+\widetilde{\pi}_{++}(a))\beta}{(\widetilde{\pi}_{++}(a)-a)\alpha-1+ad\widetilde{\pi}_{++}(a)}\right)}.$$

  (a.5) *The qualitative behaviour of the graph of $\widetilde{\pi}_{++}$ is depicted in Figure 4.15(a).*

(b) *If $t=0$, then $\widetilde{\pi}_{++}$ is the identity in $[0,+\infty)$.*

Proof. By Proposition 4.5.2(b.1), the map $\widetilde{\pi}_{++}$ can be expressed as a composition of the maps π_{-+}, π_{--} and π_{+-}, i.e., $\widetilde{\pi}_{++}=\pi_{-+}\circ\pi_{--}\circ\pi_{+-}$. Therefore, the statements (a), (a.1) and (a.2) are consequences of Propositions 4.4.11 and 4.4.13.

(a.3) We assume that A is in real Jordan normal form, see Proposition 4.3.7. From (4.40) it follows that the map $b=\widetilde{\pi}_{++}(a)$ is defined by the system

$$\begin{aligned}1-b\alpha &= e^{\alpha s}\left\{\cos(\beta s)+a\left[\alpha\cos(\beta s)-\beta\sin(\beta s)\right]\right\},\\ -b\beta &= e^{\alpha s}\left\{\sin(\beta s)+a\left[\alpha\sin(\beta s)+\beta\cos(\beta s)\right]\right\},\end{aligned}\tag{4.41}$$

and the inequalities $a \geq 0$, $b \geq 0$ and $s \geq 0$, where $\alpha = t/2 \geq 0$ and $\beta = \sqrt{4d - t^2}/2 > 0$.

Consider the change of the time variable $\tau = \beta s$ and take $\gamma = \alpha/\beta$. Isolating a and b in (4.41), we obtain the parametric equations of $\widetilde{\pi}_{++}$,

$$\begin{aligned} a(\tau) &= \frac{\beta\cos(\tau) - \alpha\sin(\tau) - \beta e^{-\gamma\tau}}{d\sin(\tau)}, \\ b(\tau) &= \frac{\alpha\sin(\tau) + \beta\cos(\tau) - \beta e^{\gamma\tau}}{d\sin(\tau)}. \end{aligned} \tag{4.42}$$

Since A is in real Jordan normal form, τ is the angle covered by the solution during the flight time s. Hence, we conclude that $\tau \in (\pi, \tau^*)$, where $\tau^* < 2\pi$.

From (4.42) it can be proved that $\lim_{\tau \searrow \pi} a(\tau) = +\infty$ and

$$\lim_{a \nearrow +\infty} \frac{\widetilde{\pi}_{++}(a)}{a} = \lim_{\tau \searrow \pi} \frac{b}{a} = \frac{1 + e^{\gamma\pi}}{1 + e^{-\gamma\pi}} = e^{\gamma\pi}.$$

By applying l'Hôpital's rule it is easy to check that $\lim_{a \nearrow +\infty} (\widetilde{\pi}_{++}(a) - e^{\gamma\pi} a) = t(1 + e^{\gamma\pi})/d$, which implies that the straight line $b = e^{\gamma\pi} a + t(1 + e^{\gamma\pi})/d$ is an asymptote of the graph of $\widetilde{\pi}_{++}(a)$.

(a.4) For more details, see the proof of Proposition 4.4.11(a.4).

(b) If $t = 0$, then $\alpha = 0$ and $\gamma = 0$. Therefore the statement follows from equation (4.42). □

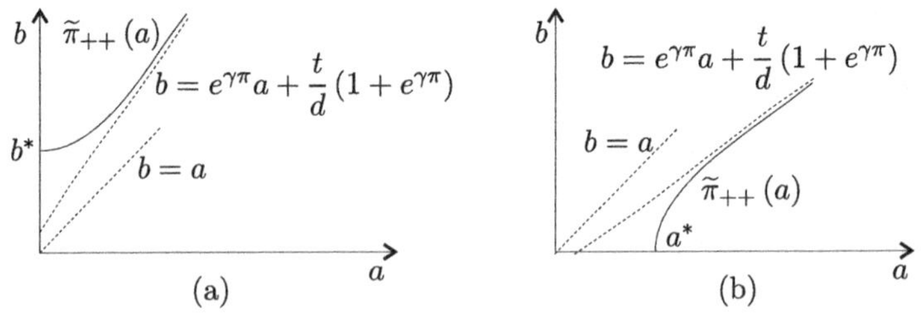

Figure 4.15: Qualitative behaviour of the Poincaré map $\widetilde{\pi}_{++}$ (a) when $t^2 - 4d < 0$ and $t > 0$, (b) when $t^2 - 4d < 0$ and $t < 0$.

Corollary 4.5.8. *Consider a matrix $A \in GL(\mathbb{R}^2)$ with parameters $d > 0$, $t < 0$ and $t^2 - 4d < 0$, and a vector $\mathbf{b} \in \mathbb{R}^2 \setminus \{\mathbf{0}\}$. Let $\widetilde{\pi}_{++}$ be the Poincaré map defined by the flow of the system $\dot{\mathbf{x}} = A\mathbf{x} + \mathbf{b}$ and associated to a straight line which does not pass through the origin. Then:*

(a) *There exists a value $a^* > 0$ such that $\widetilde{\pi}_{++} : [a^*, +\infty) \to [0, +\infty)$, $\widetilde{\pi}_{++}(a^*) = 0$, and $\lim_{a \nearrow +\infty} \widetilde{\pi}_{++}(a) = +\infty$. Moreover, $\widetilde{\pi}_{++}(a) < a$ in $(a^*, +\infty)$.*

(b) *If* $a \in (a^*, +\infty)$, *then* $\widetilde{\pi}'_{++}(a) > 0$ *and* $\lim_{a \searrow a^*} \widetilde{\pi}'_{++}(a) = +\infty$.

(c) *If* $a \in (a^*, +\infty)$, *then* $\widetilde{\pi}''_{++}(a) < 0$.

(d) *The straight line* $b = e^{\gamma\pi}a + t(1+e^{\gamma\pi})/d$ *is an asymptote of the graph of* $\widetilde{\pi}_{++}$ *when* a *tends to* $+\infty$, *where* $\gamma = t/\sqrt{4d - t^2}$.

(e) $\widetilde{\pi}_{++}$ *is implicitly defined by the expression of the Proposition* 4.5.7(e).

(f) *The qualitative behaviour of the graph of* $\widetilde{\pi}_{++}$ *is depicted in Figure* 4.15(b).

Proof. The proof follows straightforward by using the fact that the map $\widetilde{\pi}_{++}$ is the inverse of the map described in Proposition 4.5.7. □

Proposition 4.5.9. *Consider a non-zero singular matrix* $A \in L(\mathbb{R}^2)$ *and a vector* $\mathbf{b} \in \mathbb{R}^2 \setminus \{\mathbf{0}\}$. *Let* L_+ *be a non-invariant straight line which does not pass through the origin and* $\mathbf{n}$ *be the unit orthogonal vector to* L_+ *oriented in the direction opposite to the origin. Suppose that the non-homogeneous linear system* $\dot{\mathbf{x}} = A\mathbf{x}+\mathbf{b}$ *has no singular points and that the Poincaré map* $\widetilde{\pi}_{++}$ *associated to the flow of the system and to the straight line* L_+ *is defined. Then:*

(a) *If* $\det(A\mathbf{n}^\perp, \mathbf{b}) > 0$, *then the domain of* $\widetilde{\pi}_{++}$ *is* $a = 0$, *and* $\widetilde{\pi}_{++}(0) = 0$.

(b) *If* $\det(A\mathbf{n}^\perp, \mathbf{b}) < 0$ *and* $t > 0$, *then* $\widetilde{\pi}_{++} : [0, t^{-1}) \to [0, +\infty)$, $\widetilde{\pi}_{++}(0) = 0$, $\lim_{a \nearrow t^{-1}} \widetilde{\pi}_{++}(a) = +\infty$, *and* $\widetilde{\pi}_{++}(a) > a$ *in* $(0, t^{-1})$.

(b.1) *If* $a \in (0, t^{-1})$, *then* $\widetilde{\pi}'_{++}(a) > 1$ *and* $\lim_{a \searrow 0} \widetilde{\pi}'_{++}(a) = 1$.

(b.2) *If* $a \in (0, t^{-1})$, *then* $\widetilde{\pi}''_{++}(a) > 0$.

(b.3) $\widetilde{\pi}_{++}$ *is implicitly defined by the expression*

$$1 + t\widetilde{\pi}_{++}(a) = (1 - at)\, e^{t(a+\widetilde{\pi}_{++}(a))}.$$

(b.4) *The graph of* $\widetilde{\pi}_{++}$ *has a vertical asymptote at* $a = t^{-1}$.

(b.5) *The qualitative behaviour of the graph of* $\widetilde{\pi}_{++}$ *is depicted in Figure* 4.16(a).

(c) *If* $\det(A\mathbf{n}^\perp, \mathbf{b}) < 0$ *and* $t = 0$, *then the map* $\widetilde{\pi}_{++}$ *is the identity in* $[0, +\infty)$.

Proof. By Proposition 4.5.6, we can assume that the matrix A is in real Jordan normal form, i.e.,

$$\text{(i) } A = \begin{pmatrix} t & 0 \\ 0 & 0 \end{pmatrix} \quad \text{or} \quad \text{(ii) } A = \begin{pmatrix} 0 & 1 \\ 0 & 0 \end{pmatrix},$$

depending on whether $t \neq 0$ or $t = 0$. Therefore, the flow $\Phi(s, \mathbf{x})$ of the non-homogeneous linear system $\dot{\mathbf{x}} = A\mathbf{x} + \mathbf{b}$ is given by

$$\text{(i) } \begin{pmatrix} e^{ts}\left(x_1 + \frac{b_1}{t}\right) - \frac{b_1}{t} \\ x_2 + b_2 s \end{pmatrix} \quad \text{or} \quad \text{(ii) } \begin{pmatrix} \frac{b_2}{2}s^2 + (x_2 + b_1)\, s + x_2 \\ x_2 + b_2 s \end{pmatrix},$$

where $\mathbf{b} = (b_1, b_2)^T$ with $b_2 \neq 0$ and $\mathbf{x} = (x_1, x_2)^T$. Note that $b_2 \neq 0$ because the system has no singular points.

Since $\widetilde{\pi}_{++}$ is well defined, the flow of the system has a unique contact point $\mathbf{p} = (p_1, p_2)^T$ with the straight line L_+, see Proposition 4.5.3(d).

(a) Suppose that $\det(A\mathbf{n}^\perp, \mathbf{b}) > 0$. By Proposition 4.5.3(d), the contact point $\mathbf{p}$ splits L_+ into the two half-lines, L_+^I and L_+^O. Moreover, we obtain that $L_+^I = \{\mathbf{p} - a\dot{\mathbf{p}} : a \geq 0\}$, $L_+^O = \{\mathbf{p}+a\dot{\mathbf{p}} : a \geq 0\}$, and the map $\widetilde{\pi}_{++}$ takes coordinates of points of L_+^O into coordinates of points of L_+^I. Thus $\widetilde{\pi}_{++}$ is implicitly defined by the equation

$$\mathbf{p} - \widetilde{\pi}_{++}(a)\,\dot{\mathbf{p}} = \Phi(s, \mathbf{p} + a\dot{\mathbf{p}})$$

and the inequalities $a \geq 0$, $\widetilde{\pi}_{++}(a) \geq 0$ and $s \geq 0$. Substituting in this equation the expression of the flow in (i) and (ii) we obtain two systems. It is easy to check that in both cases the second equation of these systems is given by $\widetilde{\pi}_{++}(a) = -(s+a)$, which implies that $\widetilde{\pi}_{++}$ is only defined in $a = 0$ and $\widetilde{\pi}_{++}(0) = 0$.

(b) Assume that $\det(A\mathbf{n}^\perp, \mathbf{b}) < 0$ and $t < 0$. Arguments similar to those in the proof of statement (a) show that the map $\widetilde{\pi}_{++}$ is implicitly defined by the equation

$$\mathbf{p} + \widetilde{\pi}_{++}(a)\,\dot{\mathbf{p}} = \Phi(s, \mathbf{p} - a\dot{\mathbf{p}})$$

and the inequalities $a \geq 0$, $\widetilde{\pi}_{++}(a) \geq 0$ and $s \geq 0$. Since $t > 0$, substituting in this equation the expression of the flow corresponding to the case (i) we obtain that

$$\begin{aligned} p_1 + \widetilde{\pi}_{++}(a)(tp_1 + b_1) &= e^{ts}\left(p_1 - a(tp_1 + b_1) + \frac{b_1}{t}\right) - \frac{b_1}{t}, \\ \widetilde{\pi}_{++}(a)\, b_2 &= -ab_2 + sb_2. \end{aligned} \tag{4.43}$$

Note that the second equation is now $\widetilde{\pi}_{++}(a) = s - a$, which does not imply that $\widetilde{\pi}_{++}$ is only defined in $a = 0$. It is easy to check that $x_1 = -b_1/t$ is an invariant straight line of the flow of the system. Therefore, $tp_1 + b_1 \neq 0$. Applying this to the first equation in (4.43) we get

$$1 + t\widetilde{\pi}_{++}(a) = e^{t(\widetilde{\pi}_{++}(a)+a)}(1 - ta). \tag{4.44}$$

Since $t > 0$, from (4.44) it follows that $\widetilde{\pi}_{++}(0) = 0$ and the domain of definition of $\widetilde{\pi}_{++}$ is contained in $[0, t^{-1})$.

Introduce now the auxiliary function $\psi_3(x) = (1 + tx)e^{-tx}$. Then expression (4.44) can be written as $\psi_3(\widetilde{\pi}_{++}(a)) = \psi_3(-a)$. The following properties of the function ψ_3 can be easily verified: ψ_3 is defined on $\mathbb{R}$; ψ_3 is strictly increasing on $(-\infty, 0)$ and strictly decreasing on $(0, +\infty)$; $\psi_3(0) = 1$; $\lim\limits_{x \searrow -\infty} \psi_3(x) = -\infty$; $\lim\limits_{x \nearrow +\infty} \psi_3(x) = 0$, and $\psi_3(-t^{-1}) = 0$. We deduce that the domain of $\widetilde{\pi}_{++}$ is $[0, t^{-1})$, that $\lim\limits_{a \nearrow t^{-1}} \widetilde{\pi}_{++}(a) = +\infty$, and that $\widetilde{\pi}_{++}(a) \neq a$ in $[0, t^{-1})$, which implies that $\widetilde{\pi}_{++}(a) > a$ in $[0, t^{-1})$.

Differentiating expression (4.44) with respect to a we obtain that $\widetilde{\pi}'_{++}(a) = ae^{at}/(\widetilde{\pi}_{++}(a)e^{-t\widetilde{\pi}_{++}(a)})$. Define now a new auxiliary function by $\psi_4(x) = xe^{tx} -$

$\widetilde{\pi}_{++}(x)e^{-t\widetilde{\pi}_{++}(x)}$. Since $\psi_4(0)=0$ and $\psi_4'(x)>0$ in $(0,t^{-1})$, it follows that $\psi_4(x)>0$ in $(0,t^{-1})$. From this we conclude that $\widetilde{\pi}'_{++}(a)>1$ in $(0,t^{-1})$.

Differentiating twice expression (4.44) with respect to a we obtain

$$\widetilde{\pi}''_{++}(a)=\frac{e^{t(a+2\widetilde{\pi}_{++}(a))}}{(\widetilde{\pi}_{++}(a))^3}\left[(1+at)(\widetilde{\pi}_{++}(a))^2e^{-t\widetilde{\pi}_{++}(a)}-a^2e^{ta}(1-t\widetilde{\pi}_{++}(a))\right].$$

Suppose now that there exists $a^*\in(0,t^{-1})$ such that $\widetilde{\pi}''_{++}(a^*)=0$; that is,

$$(1+a^*t)(\widetilde{\pi}_{++}(a^*))^2e^{-t\widetilde{\pi}_{++}(a^*)}=(a^*)^2e^{ta^*}(1-t\widetilde{\pi}_{++}(a^*)).$$

This yields that $\widetilde{\pi}_{++}(a^*)<t^{-1}$ and $(1+a^*t)e^{-t\widetilde{\pi}_{++}(a^*)}\le e^{ta^*}(1-t\widetilde{\pi}_{++}(a^*))$. Expressing this last inequality in terms of the auxiliary function ψ_3 gives $\psi_3(a^*)\le\psi_3(-\widetilde{\pi}_{++}(a^*))$, which contradicts the fact that ψ_3 is strictly decreasing in $(0,+\infty)$ and $\widetilde{\pi}_{++}(a)>a$. Therefore, we have $\widetilde{\pi}''_{++}(a)\neq 0$ if $a\in(0,t^{-1})$. Finally, since $\lim\limits_{a\nearrow t^{-1}}\widetilde{\pi}''_{++}(a)>0$, we get that $\widetilde{\pi}''_{++}(a)>0$ if $a\in(0,t^{-1})$.

From $\widetilde{\pi}'_{++}(a)>1$ in $(0,t^{-1})$ it follows that $\lim\limits_{a\searrow 0}\widetilde{\pi}'_{++}(a)=L\ge 1$. Applying l'Hôpital's rule we obtain that

$$L=\lim_{a\searrow 0}\widetilde{\pi}'_{++}(a)=\lim_{a\searrow 0}\left(\frac{1}{\widetilde{\pi}'_{++}(a)}\frac{e^{at}(1+at)}{e^{-t\widetilde{\pi}_{++}(a)}(1-t\widetilde{\pi}_{++}(a))}\right)=\frac{1}{L},$$

which implies that $L=1$.

(c) Arguments similar to those in the proof of statement (a) show that when $\det(A\mathbf{n}^\perp,\mathbf{b})<0$ and $t=0$, the map $\widetilde{\pi}_{++}$ is implicitly defined by the system

$$\begin{aligned}\widetilde{\pi}_{++}(a)(p_2+b_1)&=\tfrac{b_2}{2}s^2+[b_1+p_2-ab_2]\,s-a(p_2+b_1),\\ \widetilde{\pi}_{++}(a)\,b_2&=-ab_2+sb_2.\end{aligned}$$

Isolating s in the second equation and substituting it in the first one we obtain that $\widetilde{\pi}_{++}(a)=a$, which finishes the proof. □

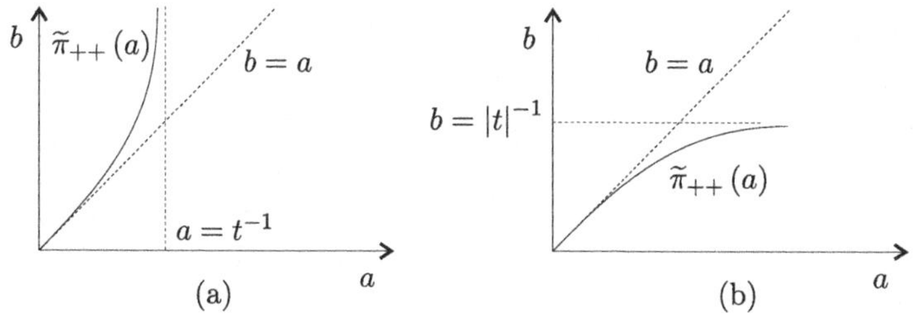

Figure 4.16: Qualitative behaviour of the Poincaré map $\widetilde{\pi}_{++}$ (a) when $d=0$ and $t>0$, (b) $d=0$ and $t<0$.

Corollary 4.5.10. *Consider a non-zero singular matrix $A \in L(\mathbb{R}^2)$ and a vector $\mathbf{b} \in \mathbb{R}^2 \setminus \{\mathbf{0}\}$. Let L_+ be a non-invariant straight line which does not pass through the origin and let $\mathbf{n}$ be the unit orthogonal vector to L_+ oriented in the direction opposite to the origin. Suppose that the non-homogeneous linear system $\dot{\mathbf{x}} = A\mathbf{x}+\mathbf{b}$ has no singular points and that the Poincaré map $\widetilde{\pi}_{++}$ associated to the flow and to L_+ is defined. Then:*

(a) *If* $\det(A\mathbf{n}^{\perp}, \mathbf{b}) < 0$ *and* $t < 0$, *then* $\widetilde{\pi}_{++} : [0, +\infty) \to [0, |t|^{-1})$, $\widetilde{\pi}_{++}(0) = 0$, $\lim_{a \nearrow +\infty} \widetilde{\pi}_{++}(a) = |t|^{-1}$, *and* $\widetilde{\pi}_{++}(a) < a$ *in* $(0, +\infty)$.

(b) *If* $a \in (0, +\infty)$, *then* $0 < \widetilde{\pi}'_{++}(a) < 1$ *and* $\lim_{a \searrow 0} \widetilde{\pi}'_{++}(a) = 1$.

(c) *If* $a \in (0, +\infty)$, *then* $\widetilde{\pi}''_{++}(a) > 0$.

(d) $\widetilde{\pi}_{++}$ *is implicitly defined by the expression of Proposition* 4.5.9(b.3).

(e) *The graph of* $\widetilde{\pi}_{++}$ *has a horizontal asymptote at* $b = |t|^{-1}$.

(f) *The qualitative behaviour of the graph of* $\widetilde{\pi}_{++}$ *is shown in Figure* 4.16(b).

Proof. The corollary follows by using that $\widetilde{\pi}_{++}$ is the inverse of the map described in the Proposition 4.5.9(b), see Lemma 4.5.4(a). □

4.6 Return maps of fundamental systems

In this section we describe the return map Π defined by the flow of a fundamental system and associated to the straight line L_+ as the composition of the Poincaré maps Π_{jk}^M, where $j, k \in \{+, -\}$ and $M \in \{A, B\}$, see Section 4.1. Those Poincaré maps have been reduced to the Poincaré maps defined by the flow of a homogeneous and a non-homogeneous linear system (Π_{jk} and $\widetilde{\Pi}_{++}$), and they have been studied in detail in Sections 4.3 and 4.5 via the so-defined Poincaré maps π_{jk} with $j, k \in \{+, -\}$ and $\widetilde{\pi}_{++}$. From now on we use the superscript A or B to identify which of the two linear systems define the Poincaré maps π_{jk}. Thus π_{jk}^B are the Poincaré maps defined by the homogeneous linear system $\dot{\mathbf{x}} = B\mathbf{x}$, and π_{jk}^A are the Poincaré maps defined by the non-homogeneous linear system $\dot{\mathbf{x}} = A\mathbf{x} \pm \mathbf{b}$.

Consider the fundamental system

$$\dot{\mathbf{x}} = A\mathbf{x} + \varphi\left(\mathbf{k}^T\mathbf{x}\right)\mathbf{b}, \tag{4.45}$$

with $A \in L(\mathbb{R}^2)$ and $\mathbf{k}, \mathbf{b} \in \mathbb{R}^2 \setminus \{\mathbf{0}\}$. Take $L_+ = \{\mathbf{x} \in \mathbb{R}^2 : \mathbf{k}^T\mathbf{x} = 1\}$ and $L_- = \{\mathbf{x} \in \mathbb{R}^2 : \mathbf{k}^T\mathbf{x} = -1\}$. Since L_+ and L_- does not pass through the origin, we can split them into the subsets L_+^I, L_+^O, L_-^I and L_-^O. Since we know when the flow $\Phi(s, \mathbf{x})$ of the fundamental system (4.45) has a contact point with the straight line L_+, these subsets are half-lines.

Consider the subset

$$\mathrm{Dom} := \left\{\mathbf{q} \in L_+^I : \exists s_{\mathbf{q}} > 0, \Phi\left(s_{\mathbf{q}}, \mathbf{q}\right) \in L_+^I \text{ and } \forall s \in \left(0, s_{\mathbf{q}}\right), \Phi\left(s, \mathbf{q}\right) \notin L_+^I\right\}.$$

When Dom $\neq \varnothing$, we define *the return map associated to the flow of the fundamental system* (4.45) *and to the half-line* L_+^I by $\Pi : \text{Dom} \subset L_+^I \to L_+^I$ and $\Pi(\mathbf{q}) = \Phi(s_{\mathbf{q}}, \mathbf{q})$. It is easy to check that if Dom $\neq \varnothing$, then two or more of the sets Dom_{++}, Dom_{+-}, Dom_{--} and Dom_{-+}, see (4.7), are not empty. This implies that there exists exactly one contact point $\mathbf{p}$ of the flow of the fundamental system with L_+, see Lemma 4.3.2. By the symmetry of the vector field with respect to the origin, $-\mathbf{p}$ is the contact point of the flow with L_-. The existence of contact points on L_+ and L_- allows us to associate a non-negative number to any points on L_+ and L_- called the coordinate of the point. For more details, see Subsections 4.3.1 and 4.5.2. We define the *return map* π as the map that transforms the coordinate of $\mathbf{q}$ into the coordinate of $\Pi(\mathbf{q})$. As usual, we restrict our attention to the return map π instead of Π.

In the following result we express π as the composition of the Poincaré maps π_{jk}^A, $\widetilde{\pi}_{++}^A$ and π_{jk}^B. Given a map f we denote by f^2 the map $f \circ f$.

Theorem 4.6.1. *Consider a fundamental system* $\dot{\mathbf{x}} = A\mathbf{x} + \varphi(\mathbf{k}^T\mathbf{x})\mathbf{b}$ *with fundamental matrices* (A, B) *and parameters* (D, T, d, t). *Then:*

(a) *The return map* π *is defined if and only if the flow of the system has exactly one contact point with the straight line* L_+.

(b) π *is defined if an only if* $\det(A\mathbf{k}^\perp, \mathbf{k}^\perp) \neq 0$.

(c) *If* $D > 0$, $d \neq 0$ *and* π *is defined, then* $\pi = (\pi_{++}^A \circ \pi_{+-}^B)^2$.

(d) *If* $D > 0$, $d = 0$ *and* π *is defined, then* $\pi = (\widetilde{\pi}_{++}^A \circ \pi_{+-}^B)^2$.

(e) *Suppose that* $D < 0$, $d > 0$ *and the return map* π *is defined. Let* $\Lambda_1 > 0 > \Lambda_2$ *be the eigenvalues of the matrix* B. *If* $t^2 - 4d < 0$, *then*

$$\pi(a) = \begin{cases} \left(\widetilde{\pi}_{++}^A \circ \pi_{++}^B\right)(a), & \text{if } a \in \left[0, \Lambda_1^{-1}\right), \\ \left(\widetilde{\pi}_{++}^A \circ \pi_{+-}^B\right)^2(a), & \text{if } a \in \left(\Lambda_1^{-1}, +\infty\right). \end{cases}$$

If $t^2 - 4d \geq 0$, *then the domain of* π *is* $a = 0$ *and* $\pi(0) = 0$.

(f) *If* $D < 0$ *and* $d \leq 0$, *then the domain of* π *is* $a = 0$ *and* $\pi(0) = 0$.

(g) π *is analytic with respect to* a.

Proof. (a) As we have seen, the existence of the return map π implies the existence of exactly one contact point. Thus it remains to prove only the converse. Suppose that the flow of the system has exactly one contact point $\mathbf{p}$ with L_+. Then $\Pi(\mathbf{p}) = \mathbf{p}$ and the return map is defined.

(b) Since the fundamental system in S_0 is $\dot{\mathbf{x}} = B\mathbf{x}$, the existence of a contact point of the flow with L_+ is equivalent to $\det(B\mathbf{k}^\perp, \mathbf{k}^\perp) \neq 0$, see Theorem 4.3.10. The statement follows by noting that $B = A + \mathbf{b}\mathbf{k}^T$ and $\det(A\mathbf{k}^\perp, \mathbf{k}^\perp) = \det(B\mathbf{k}^\perp, \mathbf{k}^\perp)$.

(c) Suppose that $D > 0$ and $d > 0$ (when $d < 0$, the same arguments can be applied to prove the statement). In this case the flow of the system has a contact point with L_+, see statement (a). Hence the Poincaré maps π^B_{+-} and π^B_{-+} are defined, see Proposition 4.3.4. Moreover, π^B_{+-} maps coordinates of points in L^I_+ into coordinates of points in L^O_-; and π^B_{-+} maps coordinates of points in L^I_- into coordinates of points in L^O_+, see Proposition 4.3.3(a).

Suppose that $\mathbf{e}_+ = -A^{-1}\mathbf{b} \in S_+$. Then the fundamental system has a singular point in S_+, which contradicts Theorem 3.9.3(a). Therefore, $\mathbf{e}_+ \notin S_+$. By Proposition 4.5.1, this implies that π^A_{++} and π^A_{--} are well defined. Moreover, π^A_{++} maps coordinates from L^O_+ to L^I_+, and π^A_{--} maps coordinates from L^O_- to L^I_-, see Proposition 4.5.2(a).

We conclude that there exists exactly one return map π defined in L^I_+. Moreover, this map can be expressed as a composition of the Poincaré maps π^M_{jk}, i.e., $\pi = \pi^A_{++} \circ \pi^B_{-+} \circ \pi^A_{--} \circ \pi^B_{+-}$. The statement follows by noting that $\pi^A_{++} = \pi^A_{--}$ and $\pi^B_{+-} = \pi^B_{-+}$, see Lemma 4.3.5(a) and (b).

(d) The statement follows by arguments similar to those used in the proof of (c).

(e) Since $D < 0$, it is easy to conclude that the Poincaré maps π^B_{jk} with $j,k \in \{+,-\}$ are well defined, see the proof of statement (a) for more details. Moreover, π^B_{++} maps coordinates from L^I_+ to L^O_+ and π^B_{+-} maps coordinates from L^I_+ to L^O_-.

Since $d > 0$, $\mathbf{e}_+ = -A^{-1}\mathbf{b}$ is a singular point contained in S_+, see Theorem 3.9.3(b). Therefore, the Poincaré maps $\widetilde{\pi}^A_{++}$ and $\widetilde{\pi}^A_{--}$ are well defined, see Proposition 4.5.1.

Suppose that $t^2 - 4d \geq 0$. By Proposition 4.5.2(a), the behaviour of the map $\widetilde{\pi}^A_{++}$ is trivial and so is the behaviour of the return map π. Suppose that $t^2 - 4d < 0$. In this case $\widetilde{\pi}^A_{++}$ maps coordinates from L^O_+ to L^I_+. We have two possibilities for the return map π: either $\pi = \widetilde{\pi}^A_{++} \circ \pi^B_{++}$, or $\pi = \widetilde{\pi}^A_{++} \circ \pi^B_{-+} \circ \widetilde{\pi}^A_{--} \circ \pi^B_{+-}$. The domain of both of these maps follows from Propositions 4.4.15 and 4.4.18 and from Corollaries 4.4.16 and 4.4.19. The statement follows by noting that $\widetilde{\pi}^A_{++} = \widetilde{\pi}^A_{--}$ and $\pi^B_{+-} = \pi^B_{-+}$.

(f) Arguments similar to those used in the proof of the statement (d) show that the Poincaré maps π^B_{jk} are well defined when $D < 0$.

Suppose that $d < 0$. From Theorem 3.9.3(a) it follows that $\mathbf{e}_+ \notin S_+$. Moreover, the Poincaré map π^A_{++} is defined, see Proposition 4.5.1. The behaviour of π^A_{++} follows from Proposition 4.5.2(a.2).

Suppose $d = 0$. Without loss of generality, we can assume that

$$\text{(i) } A = \begin{pmatrix} t & 0 \\ 0 & 0 \end{pmatrix} \quad \text{or} \quad \text{(ii) } A = \begin{pmatrix} 0 & 1 \\ 0 & 0 \end{pmatrix}.$$

The case where A is the zero matrix is not considered because this would imply that $D = 0$, see (3.10).

Since $\mathbf{n} = \mathbf{k}/\|\mathbf{k}\|$ is the unit orthogonal vector to L_+ oriented in the direction opposite to the origin, in case (i) we obtain that $\det(A\mathbf{n}^\perp, \mathbf{b}) = -tk_2b_2 = -D > 0$ (see expression (3.10)), and in case (ii) we obtain that $\det(A\mathbf{n}^\perp, \mathbf{b}) = -D > 0$. By Proposition 4.5.9(a), it follows that the behavior of $\widetilde{\pi}^A_{++}$ is trivial.

(g) If π is defined, then we can write π as a composition of analytic maps, which proves the statement. □

From the definition of the return map π it follows that a periodic orbit intersecting L_+ and/or L_- is associated to a fixed point of π. Thus, any fixed point of π is associated to a periodic orbit intersecting L_+ and/or L_-. Further, the existence of periodic orbits which are not contained into one of the regions S_+, S_0 or S_- is equivalent to the existence of a fixed point of the return map π. Moreover, any isolated fixed point of π is associated to a periodic orbit, which is a limit cycle, Γ. When Γ is hyperbolic, we can obtain its stability from Theorem 2.7.5. In the case that Γ is not a hyperbolic limit cycle we have the following proposition.

Proposition 4.6.2. *Consider a fundamental system* $\dot{\mathbf{x}} = A\mathbf{x} + \varphi(\mathbf{k}^T\mathbf{x})\mathbf{b}$ *and let* a^* *be a fixed point of the return map* π *associated to a periodic orbit* Γ.

(a) *If there exists* $\varepsilon > 0$ *such that* $|\pi'(a)| < 1$ *in* $(a^*, a^* + \varepsilon)$, *then* Γ *is an outside asymptotically stable periodic orbit.*

(b) *If there exists* $\varepsilon > 0$ *such that* $|\pi'(a)| < 1$ *in* $(a^* - \varepsilon, a^*)$, *then* Γ *is an inside asymptotically stable periodic orbit.*

(c) *If there exists* $\varepsilon > 0$ *such that* $|\pi'(a)| > 1$ *in* $(a^*, a^* + \varepsilon)$, *then* Γ *is an outside asymptotically unstable periodic orbit.*

(d) *If there exists* $\varepsilon > 0$ *such that* $|\pi'(a)| > 1$ *in* $(a^* - \varepsilon, a^*)$, *then* Γ *is an inside asymptotically unstable periodic orbit.*

Proof. (a) Suppose that $|\pi'(a)| < 1$ in $(a^*, a^* + \varepsilon)$. The statement follows by noting that the sequence $a_{n+1} = \pi(a_n)$ with $a_0 \in (a^*, a^* + \varepsilon)$ is contained in $(a^*, a^* + \varepsilon)$ and tends to a^*.

The remaining statements follows in a similar way. □

Lamerey map

For a fundamental system with fundamental matrices (A, B) and parameters $D > 0$ and $d \neq 0$, the *Lamerey maps* are defined by

$$g_A(a) = \pi_{++}^{-A}(a) - \pi_{+-}^{B}(a) \quad \text{and} \quad g_B(a) = \pi_{++}^{A}(a) - \pi_{+-}^{-B}(a). \tag{4.46}$$

When $D > 0$ and $d = 0$, the *Lamerey maps* are defined by

$$g_A(a) = \widetilde{\pi}_{++}^{-A}(a) - \pi_{+-}^{B}(a) \quad \text{and} \quad g_B(a) = \widetilde{\pi}_{++}^{A}(a) - \pi_{+-}^{-B}(a). \tag{4.47}$$

When $D < 0$, $d > 0$ and $t^2 - 4d < 0$, the *Lamerey maps* are defined by

$$g_A(a) = \begin{cases} \widetilde{\pi}_{++}^{-A}(a) - \pi_{++}^{B}(a), & \text{if } a \in [0, \Lambda_1^{-1}), \\ \widetilde{\pi}_{++}^{-A}(a) - \pi_{+-}^{B}(a), & \text{if } a \in (\Lambda_1^{-1}, +\infty), \end{cases}$$
$$\text{and} \tag{4.48}$$
$$g_B(a) = \begin{cases} \widetilde{\pi}_{++}^{A}(a) - \pi_{++}^{-B}(a), & \text{if } a \in [0, \Lambda_1^{-1}), \\ \widetilde{\pi}_{++}^{A}(a) - \pi_{+-}^{-B}(a), & \text{if } a \in (\Lambda_1^{-1}, +\infty), \end{cases}$$

where $\Lambda_1 > 0 > \Lambda_2$ are the eigenvalues of the matrix B. In the next result we prove that the zeros of the Lamerey maps are the fixed points of the return map π, and therefore they are associated to periodic orbits that are not contained in one of the regions S_+, S_0, or S_-.

Lemma 4.6.3. *Consider a fundamental system with matrices (A, B) and fundamental parameters (D, T, d, t).*

(a) *The Lamerey maps are defined if and only if the flow of the system has a contact point with the straight line L_+.*

(b) *The Lamerey map g_A (respectively g_B) has a zero at a_0 if and only if a_0 is a fixed point of the return map π.*

Proof. Statement (a) follows from Theorem 4.6.1(a).

(b) Suppose that $D > 0$ and $d \neq 0$. The remaining cases can be proved by using similar arguments. Consider the Lamerey map g_A (similar considerations apply to g_B as well).

Suppose that a_0 is a zero of g_A. Then $\pi_{++}^{-A}(a_0) = \pi_{+-}^{B}(a_0)$. Since $\pi_{++}^{-A} = (\pi_{++}^{A})^{-1}$ by Lemma 4.3.5(c), we have that $(\pi_{++}^{A} \circ \pi_{+-}^{B})(a_0) = a_0$, and consequently $(\pi_{++}^{A} \circ \pi_{+-}^{B})^2(a_0) = a_0$. By Theorem 4.6.1(b), this means that a_0 is a fixed point of the return map π.

Suppose now that a_0 is a fixed point of π, i.e., $(\pi_{++}^{A} \circ \pi_{+-}^{B})^2(a_0) = a_0$. By the symmetry of the flow with respect to the origin it can be concluded that $(\pi_{++}^{A} \circ \pi_{+-}^{B})(a_0) = a_0$. The statement follows by using that $\pi_{++}^{-A} = (\pi_{++}^{A})^{-1}$. □

A graphical tool to study the existence and the location of the zeros of the Lamerey maps g_A or g_B is provided by the so-called *Lamerey diagrams*, see [3], p. 161. These diagrams depict the graph of the Poincaré maps which define the Lamerey map. For instance, when $D > 0$ and $d \neq 0$, we draw the graph of π_{++}^{A} and π_{+-}^{B}. Thus, a_0 is a zero of g_A or g_B if and only if a_0 is an intersecting point of the graph of π_{++}^{A} and π_{+-}^{B}.

Proper fundamental systems

In Theorem 4.6.1(b) we characterize the existence of return maps in fundamental systems by $\det(A\mathbf{k}^{\perp}, \mathbf{k}^{\perp}) \neq 0$. From this equation we obtain an interesting relationship between the existence of the return map and ideas originating in control theory, and in particular in the concept of observability of a control system. For an introduction to control theory we refer the reader to [36]. More specific results can be found in [32] and [35].

A control system is composed of a *steady state equation* (usually a differential equation) $\dot{\mathbf{x}} = \mathbf{f}(\mathbf{x}, \mathbf{u})$, which depends on the states $\mathbf{x}(s)$ of the system and on the inputs $\mathbf{u}(s)$, and of an *output function* $\mathbf{y} = \mathbf{g}(\mathbf{x}, \mathbf{u})$, which also depends on the states and on the inputs.

A control system is said to be *observable* if there exists a time $s_1 > 0$ such that any initial state $\mathbf{x}(s_0)$ can be distinguished from another state $\mathbf{x}_0$ by knowing the input $\mathbf{u}(s)$ and the output $\mathbf{y}(s)$ in the interval $[s_0, s_1]$. That is, the state of an observable control system is determined by the input and the output. This property is essentially for applications: if a system is not observable (we cannot obtain the state of the system from the outputs), then it cannot be controled (we cannot steer the system from one state to another).

In the case when the steady state equation is a fundamental system we obtain that observability is equivalent to the existence of the return map π. This result is presented in Theorem 4.6.5. As usual, the gradient of a function $\mathbf{h}$ is denoted by $\nabla \mathbf{h}$.

Proposition 4.6.4. *The control system* $\dot{\mathbf{x}} = \mathbf{f}(\mathbf{x}) + \mathbf{g}(\mathbf{x})\mathbf{u}$, $y = \mathbf{h}(x)\mathbf{u}$ *is observable if and only if the matrix*

$$K = \begin{pmatrix} \nabla \mathbf{h} \\ \mathbf{f}^T \nabla \mathbf{h} \end{pmatrix}$$

has maximal rank.

A proof of this result can be found in [32], p. 135. From this proposition we obtain that if the control system $\dot{\mathbf{x}} = A\mathbf{x} + \varphi(\mathbf{k}^T\mathbf{x})\mathbf{b}$, $y = \mathbf{x}^T\mathbf{k}$ is observable, then the determinant of

$$K = K(A, \mathbf{k}) = \begin{pmatrix} \mathbf{k}^T \\ \mathbf{k}^T A \end{pmatrix}$$

does not vanish. This implies that the flow of the fundamental system $\dot{\mathbf{x}} = A\mathbf{x} + \varphi(\mathbf{k}^T\mathbf{x})\mathbf{b}$ has a contact point with the straight line $L_+ = \{\mathbf{x} \in \mathbb{R}^2 : \mathbf{k}^T\mathbf{x} = 1\}$, see Theorem 4.3.10. Note that the definition of observable control system depends on the output $\mathbf{u}^T\mathbf{h}(\mathbf{x})$ and we work only with the steady state equation. Following Komuro [37] and Chua [59], we introduce the notion of *proper fundamental system.* A fundamental system $\dot{\mathbf{x}} = A\mathbf{x} + \varphi(\mathbf{k}^T\mathbf{x})\mathbf{b}$ is said to be *proper* if $\det(K(A, \mathbf{k})) \neq 0$. Thus, any control system with a proper fundamental system as steady state equation and output $y = \mathbf{k}^T\mathbf{x}$ is observable. Therefore, we obtain the following result.

Theorem 4.6.5. *Given a fundamental system* $\dot{\mathbf{x}} = A\mathbf{x} + \varphi(\mathbf{k}^T\mathbf{x})\mathbf{b}$, *the following statements are equivalent.*

(a) *The flow of the fundamental system has exactly one contact point with* L_+.

(b) *The vectors* $A\mathbf{k}^\perp$ *and* $\mathbf{k}^\perp$ *are independent, i.e.,* $\det(A\mathbf{k}^\perp, \mathbf{k}^\perp) \neq 0$.

(c) *The vectors* $B\mathbf{k}^{\perp}$ *and* $\mathbf{k}^{\perp}$ *are independent, i.e.,* $\det(B\mathbf{k}^{\perp}, \mathbf{k}^{\perp}) \neq 0$.

(d) *The fundamental system is proper.*

(e) *The control system* $\dot{\mathbf{x}} = A\mathbf{x} + \varphi(\mathbf{k}^T\mathbf{x})\mathbf{b}$*;* $y = \mathbf{k}^T\mathbf{x}$ *is observable.*

An important property of proper fundamental systems is that they admit a canonical form, called *proper form*, which involves only the fundamental parameters. The extension of this relationship to higher dimension is developed in [45]. Another interesting approach to the proper form of a fundamental system in higher dimensions can be found in [13]. In that work the authors are using an equivalent expression, called the Van der Pol–Duffing form of a fundamental system.

Proposition 4.6.6. *For any proper fundamental system* $\dot{\mathbf{x}} = A\mathbf{x} + \varphi(\mathbf{k}^T\mathbf{x})\mathbf{b}$ *with fundamental parameters* (D, T, d, t) *there exists a change of coordinates* $\mathbf{y} = M\mathbf{x}$ *with* $\mathbf{y} = (y_1, y_2)^T$*, which transforms the system into the system*

$$\dot{\mathbf{y}} = \begin{pmatrix} 0 & -d \\ 1 & t \end{pmatrix} \mathbf{y} + \varphi(y_2) \begin{pmatrix} d - D \\ T - t \end{pmatrix}.$$

Proof. In [13] and [59] the authors prove that for a proper fundamental systems there exists a change of coordinates $\mathbf{y} = M\mathbf{x}$ such that

$$MAM^{-1} = \begin{pmatrix} 0 & -d \\ 1 & t \end{pmatrix}.$$

The result follows by noting that $\mathbf{k}^T M^{-1} = (0,1)^T$ and $M\mathbf{b} = (d - D, T - t)^T$. □

Since limit cycles are important in applications, some authors pay attention to fundamental systems exhibiting such orbits, see [55]. We have seen that the existence of limit cycles in fundamental systems implies the existence of the return map π. By Theorem 4.6.5, fundamental systems with limit cycles are contained in the class of proper fundamental systems.

4.7 Fundamental parameter space

In this section we show that *proper systems are dense in the family of fundamental systems.* We find an algebraic manifold W in the parameter space $\mathbb{R}^4$ such that any fundamental system with fundamental parameters $(D, T, d, t) \notin W$ is proper. The manifold W is called the *Whitney umbrella* and has important properties from a geometric viewpoint.

As a consequence of Theorem 4.6.5 we obtain a relationship between proper systems and the real Jordan normal form of the fundamental matrices. We recall that the vector $\mathbf{k}^{\perp} = (-k_2, k_1)^T$ is parallel to the straight lines L_+ and L_-, see (3.6) and (3.7).

Lemma 4.7.1. *Consider a fundamental system* $\dot{\mathbf{x}} = A\mathbf{x} + \varphi(\mathbf{k}^T\mathbf{x})\mathbf{b}$ *with vector* $\mathbf{k} = (k_1, k_2)^T$ *and fundamental matrices* (A, B). *Suppose that either* A *or* B *is given in its real Jordan normal form. Then:*

(a) *If* A *or* B *is equal to* $\begin{pmatrix} \alpha & -\beta \\ \beta & \alpha \end{pmatrix}$ *with* $\beta \neq 0$, *then the system is proper.*

(b) *If* A *or* B *is equal to* $\begin{pmatrix} \lambda & 1 \\ 0 & \lambda \end{pmatrix}$, *then the system is proper if and only if* $k_1 \neq 0$.

(c) *If* A *or* B *is equal to* $\begin{pmatrix} \lambda & 0 \\ 0 & \lambda \end{pmatrix}$, *then the system is not proper.*

(d) *If* A *or* B *is equal to* $\begin{pmatrix} \lambda_1 & 0 \\ 0 & \lambda_2 \end{pmatrix}$ *with* $\lambda_1 > \lambda_2$, *then the system is proper if and only if* $k_1 \neq 0$ *and* $k_2 \neq 0$.

Proof. Suppose that the matrix A is in real Jordan normal form. The same arguments can be applied if B is in real Jordan normal form.

Easy computations show that: $\det(A\mathbf{k}^\perp, \mathbf{k}^\perp) = -\beta\|\mathbf{k}^\perp\|^2$ in the case (a), $\det(A\mathbf{k}^\perp, \mathbf{k}^\perp) = k_1^2$ in the case (b); $\det(A\mathbf{k}^\perp, \mathbf{k}^\perp) = 0$ in the case (c); and $\det(A\mathbf{k}^\perp, \mathbf{k}^\perp) = k_1k_2(\lambda_1 - \lambda_2)$ in the case (d). The lemma follows from Theorem 4.6.5. □

We start by locating in the fundamental parameter space $\mathbb{R}^4$ the region where the real Jordan normal forms of the fundamental matrices A and B are not determined by the fundamental parameters.

Define the three-dimensional vector subspace

$$W^B := \{(D, T, d, t) \in \mathbb{R}^4 : T^2 - 4D = 0\}.$$

Consider a fundamental system with fundamental parameters $(D, T, d, t) \in W^B$ and fundamental matrices (A, B). Since the fundamental parameters are invariant under linear changes of coordinates, we can assume without loss of generality that B is in real Jordan normal form. Therefore, if B is diagonal, then

$$\begin{aligned} B &= \begin{pmatrix} T/2 & 0 \\ 0 & T/2 \end{pmatrix}, \\ A &= \begin{pmatrix} -k_1b_1 + T/2 & -k_2b_1 \\ -k_1b_2 & -k_2b_2 + T/2 \end{pmatrix}, \end{aligned}$$

and the fundamental parameter $d = \frac{T}{2}(t - \frac{T}{2})$. If B is not diagonal, then

$$\begin{aligned} B &= \begin{pmatrix} T/2 & 1 \\ 0 & T/2 \end{pmatrix}, \\ A &= \begin{pmatrix} -k_1b_1 + T/2 & 1 - k_2b_1 \\ -k_1b_2 & -k_2b_2 + T/2 \end{pmatrix} \end{aligned}$$

and the fundamental parameter $d = \frac{T}{2}(t - \frac{T}{2}) + k_1 b_2$.

Consider now the family of algebraic surfaces $F=\{F_c^B\}_{c\in\mathbb{R}}$ in $\mathbb{R}^4$ given by

$$F_c^B = \left\{ \left(\frac{T^2}{4}, T, \frac{T}{2}\left(t - \frac{T}{2}\right) + c, t \right) : (T,t) \in \mathbb{R}^2 \right\}. \tag{4.49}$$

It is clear that W^B is foliated by F_c^B. Moreover, given a fundamental system with fundamental matrices (A, B) and fundamental parameters $(D, T, d, t) \in F_c^B$ with $c \neq 0$, the real Jordan form of the matrix B is not diagonal. If $c \neq 0$, i.e., $(D, T, d, t) \in F_0^B$, we cannot decide if the real Jordan normal form of B is diagonal or not.

In a similar way, the family of algebraic surfaces $F^A = \{F_c^A\}_{c\in\mathbb{R}}$ given by

$$F_c^A = \left\{ \left(\frac{t}{2}\left(T - \frac{t}{2}\right) + c, T, \frac{t^2}{4}, t \right) : (T,t) \in \mathbb{R}^2 \right\},$$

foliates the vector subspace $W^A := \{(D, T, d, t) \in \mathbb{R}^4 : t^2 - 4d = 0\}$. Moreover, only when the fundamental parameters lie in F_0^A, there exists an uncertainty in the real Jordan form of the matrix A. The following lemma summarizes these facts.

Lemma 4.7.2. (a) *Consider a fundamental system with matrices (A, B) and parameters (D, T, d, t) in F_c^B with $c \neq 0$ (respectively, F_c^A). Then the real Jordan form of the matrix B (respectively, A) is not diagonal.*

(b) *Take (D, T, d, t) in F_0^B (respectively, in F_0^A). Then there exist fundamental systems with matrices (A, B) and parameters (D, T, d, t) such that the real Jordan normal form of the matrix B (respectively, A) is not diagonal. There exist fundamental systems with matrices (A, B) and parameters (D, T, d, t) such that the real Jordan normal form of the matrix B (respectively, A) is diagonal.*

Recall that for a fundamental system with parameters (D, T, d, t) and matrices (A, B) the eigenvalues of A are $\lambda_1 = (t+\sqrt{t^2 - 4d})/2$ and $\lambda_2 = (t-\sqrt{t^2 - 4d})/2$, and the eigenvalues of B are $\Lambda_1 = (T+\sqrt{T^2 - 4D})/2$ and $\Lambda_2 = (T-\sqrt{T^2 - 4D})/2$. If λ_1 and λ_2 are real, then $\lambda_1 \geq \lambda_2$, and if Λ_1 and Λ_2 are real, then $\Lambda_1 \geq \Lambda_2$.

Consider the three dimensional algebraic manifolds

$$\begin{aligned} W_1 &:= \left\{(D, T, d, t) \in \mathbb{R}^4 : T^2 - 4D \geq 0,\ d = \Lambda_1 (t - \Lambda_1)\right\}, \\ W_2 &:= \left\{(D, T, d, t) \in \mathbb{R}^4 : T^2 - 4D \geq 0,\ d = \Lambda_2 (t - \Lambda_2)\right\}, \\ W &:= \left\{(D, T, d, t) \in \mathbb{R}^4 : (2(d - D) - T(t - T))^2 = (t - T)\left(T^2 - 4D\right)\right\} \end{aligned} \tag{4.50}$$

and the surface

$$F^* := \left\{(D, T, d, t) \in \mathbb{R}^4 : T^2 - 4D \geq 0,\ D = d \text{ and } T = t\right\}.$$

Note that $W = W_1 \cup W_2$. These manifolds are represented in Figure 4.17.

Given a fundamental system with matrices (A, B) and fundamental parameters $(D, T, d, t) \in W_1$, one has that Λ_1 is a common eigenvalue of the matrices A and B. Hence, either $\Lambda_1 = \lambda_1$ or $\Lambda_1 = \lambda_2$. Similarly, if $(D, T, d, t) \in W_2$, then either $\Lambda_2 = \lambda_1$ or $\Lambda_2 = \lambda_2$. Finally, if $(D, T, d, t) \in F^*$, then $\Lambda_1 = \lambda_1$ and $\Lambda_2 = \lambda_2$.

Lemma 4.7.3. *Let W, W_1, W_2 and F^* be the manifolds defined in (4.50). Then:*

(a) $W \subset \{(D, T, d, t) \in \mathbb{R}^4 : T^2 - 4D \geq 0$ *and* $t^2 - 4d \geq 0\}$.

(b) $W_1 \cap W_2 = F_0^B \cup F^*$.

(c) $F_0^A \cap F_0^B \subset F^*$.

(d) $F_0^A \subset W$.

Proof. (a) Suppose that $(D, T, d, t) \in W$. Since $W = W_1 \cup W_2$, from (4.50) it follows that $t^2 - 4d = (t - 2\Lambda_1)^2$ if $(D, T, d, t) \in W_1$, or $t^2 - 4d = (t - 2\Lambda_2)^2$ if $(D, T, d, t) \in W_2$. In both cases we have $t^2 - 4d \geq 0$.

(b) First we prove that $W_1 \cap W_2 \subseteq F_0^B \cup F^*$. Take $(D, T, d, t) \in W_1 \cap W_2$ and suppose that $T^2 - 4D = 0$. Thus $d = T(t - T/2)/2$, see (4.50). From expression (4.49) we conclude that $(D, T, d, t) \in F_0^B$ and consequently $(D, T, d, t) \in F_0^B \cup F^*$. Suppose now that $T^2 - 4D > 0$. Then (4.50) yields $d = \Lambda_1(t - \Lambda_1) = \Lambda_2(t - \Lambda_2)$, which implies $t = \Lambda_1 + \Lambda_2 = T$ (note that $\Lambda_1 > \Lambda_2$). Substituting $t = T$ in $d = \Lambda_1(t - \Lambda_1)$ it follows that $d = D$. Therefore, $(D, T, d, t) \in F_0^B \cup F^*$.

We now prove that $W_1 \cap W_2 \supseteq F_0^B \cup F^*$. Suppose that $(D, T, d, t) \in F_0^B$. Hence $T^2 - 4D = 0$, $\Lambda_1 = \Lambda_2 = T/2$ and $d = T(t - T/2)/2$, which implies $(D, T, d, t) \in W_1 \cap W_2$. Otherwise, if $(D, T, d, t) \in F^*$, then $T^2 - 4D \geq 0$, $t = T$, $d = D$ and consequently $\Lambda_1 = \lambda_1$ and $\Lambda_2 = \lambda_2$. Thus, $(D, T, d, t) \in W_1 \cap W_2$.

(c) Take $(D, T, d, t) \in F_0^A \cap F_0^B$. Since $(D, T, d, t) \in F_0^A$, it holds that $D = t(T - t/2)/2$, and since $(D, T, d, t) \in F_0^B$, it follows that $D = T^2/4$. Thus we obtain that $(T - t)^2 = 0$, or equivalently $T = t$. Therefore, $D = d$ and $(D, T, d, t) \subset F^*$.

(d) If $(D, T, d, t) \in F_0^A$, then $D = t(T - t/2)/2$ and $d = t^2/4 = t(t - t/2)/2$. Thus $t/2$ is a common eigenvalue of A and B, and consequently $(D, T, d, t) \in W$. In particular, if $t/2 > (T - t/2)$, i.e., $t > T$, then $(D, T, d, t) \in W_1$; and if $t/2 < (T - t/2)$, then $(D, T, d, t) \in W_2$. Finally, if $t = T$, then $d = D$ and $(D, T, d, t) \in W_1 \cap W_2$. □

Given a fundamental system such that one of the fundamental matrices A or B is in real Jordan normal form, in order to know whether the system is proper or not we need to study when the components of the vector $\mathbf{k} = (k_1, k_2)^T$ vanish, see Lemma 4.7.1. In the following result we obtain necessary conditions.

Lemma 4.7.4. *Consider a fundamental system $\dot{\mathbf{x}} = A\mathbf{x} + \varphi(\mathbf{k}^T\mathbf{x})\mathbf{b}$ with parameters (D, T, d, t) and such that the matrix A is in real Jordan form.*

(a) *If $t^2 - 4d \geq 0$ and $k_1 = 0$, then $T^2 - 4D \geq 0$.*

(b) *If $t^2 - 4d > 0$ and $k_2 = 0$, then $T^2 - 4D \geq 0$.*

Proof. (a) Suppose that $t^2 - 4d \geq 0$. Since the matrix A is in real Jordan form, $\mathbf{k}^\perp = (-k_2, 0)^T$ is an eigenvector of A with eigenvalue $\lambda_1 \in \mathbb{R}$. Moreover, $B = A + \mathbf{b}\mathbf{k}^T$. Thus $\mathbf{k}^\perp$ is also an eigenvector of the matrix B with eigenvalue λ_1. This implies that $T^2 - 4D \geq 0$.

Statement (b) can be proved by using similar arguments. In this case note that $t^2 - 4d = 0$ does not imply that $(0,1)^T$ is an eigenvector of the matrix A. □

Lemma 4.7.5. *Consider a fundamental system* $\dot{\mathbf{x}} = A\mathbf{x} + \varphi(\mathbf{k}^T\mathbf{x})\mathbf{b}$ *with parameters* (D, T, d, t) *satisfying that* $T^2 - 4D \geq 0$ *and such that the matrix* A *is in real Jordan normal form. Then;*

(a) *If* A *is diagonal and* $(D, T, d, t) \notin W$, *then* $k_1 \neq 0$ *and* $k_2 \neq 0$.

(b) *If* $t^2 - 4d = 0$, *the matrix* A *is non-diagonal, and* $(D, T, d, t) \notin W$, *then* $k_1 \neq 0$.

(c) *Suppose that* $(D, T, d, t) \in W_1 \setminus W_2$. *If* $k_1 = 0$, *then* $2\Lambda_1 \geq t$. *If* $k_2 = 0$, *then* $2\Lambda_1 \leq t$.

(d) *Suppose that* $(D, T, d, t) \in W_2 \setminus W_1$. *If* $k_1 = 0$, *then* $2\Lambda_2 \geq t$. *If* $k_2 = 0$, *then* $2\Lambda_2 \leq t$.

Proof. Let (A, B) be the fundamental matrices of the system.

(a) Since the matrix A is diagonal, $(1,0)^T$ and $(0,1)^T$ are the eigenvectors of A. Thus, if either $k_1 = 0$ or $k_2 = 0$, then $\mathbf{k}^\perp$ is an eigenvector of A. Moreover, since $B = A + \mathbf{b}\mathbf{k}^T$ the vector $\mathbf{k}^\perp$ is also an eigenvector of B with the same eigenvalue. Hence, $(D, T, d, t) \in W$, which contradicts the hypothesis. Therefore $k_1 \neq 0$ and $k_2 \neq 0$.

Statement (b) follows by using similar arguments.

(c) Suppose that $(D, T, d, t) \in W_1 \setminus W_2$. Hence, $d = \Lambda_1(t - \Lambda_1)$. Therefore, Λ_1 and $t - \Lambda_1$ are the eigenvalues of the matrix A. If we suppose that $t^2 - 4d = 0$, then $\Lambda_1 = t - \Lambda_1$ and $t = 2\Lambda_1$, which proves the statement.

Suppose now that $t^2 - 4d > 0$. In this case

$$A = \begin{pmatrix} \lambda_1 & 0 \\ 0 & \lambda_2 \end{pmatrix} \text{ with } \lambda_1 > \lambda_2, \text{ and } B = \begin{pmatrix} \lambda_1 + k_1 b_1 & k_2 b_1 \\ k_1 b_2 & \lambda_2 + k_2 b_2 \end{pmatrix}.$$

Assume that $k_1 = 0$. From the above expression it follows that λ_1 and $\lambda_2 + k_2 b_2$ are the eigenvalues of the matrix B. Since $t - \Lambda_1$ and Λ_1 are the eigenvalues of A suppose that $\lambda_1 = t - \Lambda_1$ and $\lambda_2 = \Lambda_1$. Then $\Lambda_1 = \Lambda_1 + k_2 b_2$. From this we obtain that $k_2 b_2 = 0$ and therefore $D = d$ and $T = t$. Consequently, $d = \Lambda_2(t - \Lambda_2)$ and $(D, T, d, t) \in W_2$, which contradicts the hypothesis. Therefore, $\lambda_1 = \Lambda_1$ and $\lambda_2 = t - \Lambda_1$ with $\lambda_1 > \lambda_2$ and the statement follows.

Statement (d) follows by using arguments similar to those in statement (a). □

Theorem 4.7.6. (a) *Every fundamental system with parameters* $(D, T, d, t) \notin W$ *is proper.*

(b) *Every fundamental system with parameters* $(D, T, d, t) \in F^*$ *is not proper.*

(c) *Consider* $(D, T, d, t) \in W \setminus F^*$. *There exist fundamental systems with parameters* (D, T, d, t) *which are proper. There exist fundamental systems with parameters* (D, T, d, t) *which are not proper.*

Proof. Since fundamental parameters and contact points are invariant under linear changes of coordinates, we can assume that either the matrix A or the matrix B is given in its real Jordan normal form.

(a) Given a fundamental system with parameters $(D, T, d, t) \notin W$ we distinguish three cases. First we suppose that $T^2 - 4D < 0$ or $t^2 - 4d < 0$. Then from Lemma 4.7.1(a) we conclude that the system is proper. Suppose now that $T^2 - 4D \geq 0$ and that the matrix A is diagonal. In this case the components of the vector $\mathbf{k}$ satisfy that $k_1 \neq 0$ and $k_2 \neq 0$, see Lemma 4.7.5(a). Since $(D, T, d, t) \notin F_0^A$, see Lemma 4.7.3(d), we have $t^2 - 4d > 0$ and therefore the system is proper, see Lemma 4.7.1(d). Suppose now that $T^2 - 4D \geq 0$ and that the matrix A is non-diagonal. Hence, $t^2 - 4d = 0$ and $k_1 \neq 0$, see Lemma 4.7.5(b). By Lemma 4.7.1(d), we conclude that the system is proper.

(b) Given a fundamental system with $(D, T, d, t) \in F^*$ we have that $T^2 - 4D \geq 0$, $T = t$ and $D = d$. Consider the following cases. Suppose that $T^2 - 4D = 0$ and the matrix B is diagonal. Therefore, the system is not proper, see Lemma 4.7.1(c). Suppose now that B is not diagonal. Hence, the matrices B and A are the ones described in (4.49), and therefore, $t = T - \mathbf{k}^T\mathbf{b}$ and $d = D - T\mathbf{k}^T\mathbf{b}/2 + k_1 b_2$. Since $T = t$ and $D = d$, it is clear that $\mathbf{k}^T\mathbf{b} = 0$ and $k_1 b_2 = 0$. Thus, if $k_1 \neq 0$, then $b_2 = 0$ and $b_1 = 0$, in contradiction with $\mathbf{b} \neq \mathbf{0}$. Therefore, $k_1 = 0$ and the system is not proper, see Lemma 4.7.1(b).

Suppose that $T^2 - 4D > 0$. From expression (3.10) it follows that $D - d = \Lambda_1 k_2 b_2 + \Lambda_2 k_1 b_1$ and $T - t = k_1 b_1 + k_2 b_2$. Since $D = d$, $T = t$ and $\Lambda_1 > \Lambda_2$, we obtain that $k_1 b_1 = 0$ and $k_2 b_2 = 0$. From this we deduce that if $k_1 \neq 0$, then $b_1 = 0$ and $b_2 \neq 0$, which implies that $k_2 = 0$. In a similar way, if $k_2 \neq 0$, then $k_1 = 0$. Therefore, $k_1 k_2 = 0$ and the system is not proper, see Lemma 4.7.1(d).

(c) Consider $(D, T, d, t) \in W \setminus F^*$. Suppose that the parameters (D, T, d, t) lie in $F_0^B \setminus F^*$. Then $T \neq t$, $D = T^2/4$ and $d = T(t - T/2)/2$, see (4.49). We define the following matrices:

$$B = \begin{pmatrix} T/2 & 0 \\ 0 & T/2 \end{pmatrix} \text{ and } A = \begin{pmatrix} -(T-t) + T/2 & 0 \\ 0 & T/2 \end{pmatrix}.$$

Thus $D = \det(B)$, $T = \operatorname{trace}(B)$, $d = \det(A)$ and $t = \operatorname{trace}(A)$. Since A and B satisfy the condition of Proposition 3.6.1, there exists a fundamental system with parameters (D, T, d, t) and fundamental matrices (A, B). By Lemma 4.7.1(c), we conclude that this system is non-proper.

We define now the following matrices:

$$B = \begin{pmatrix} T/2 & 1 \\ 0 & T/2 \end{pmatrix} \text{ and } A = \begin{pmatrix} -(T-t) + T/2 & 1 \\ 0 & T/2 \end{pmatrix}.$$

Similar arguments prove that there exists a fundamental system with parameters (D,T,d,t) and fundamental matrices (A,B). In this case the vectors $\mathbf{k}=(k_1,k_2)^T$ and $\mathbf{b}=(b_1,b_2)^T$ satisfy that $k_1b_1 = T-t \neq 0$. Consequently, $k_1 \neq 0$, and by Lemma 4.7.1(b), the system is proper.

Finally, we suppose that the parameters (D,T,d,t) are contained in $W\backslash(W_1\cap W_2)$. If $(D,T,d,t)\in W_1\backslash W_2$ (similar arguments apply when $(D,T,d,t)\in W_1\backslash W_2$), then we obtain that $T^2-4D>0$ and $d=\Lambda_1(t-\Lambda_1)$, see expression (4.50). In this case we define the following matrices:

$$B=\begin{pmatrix} \Lambda_1 & 0 \\ 0 & \Lambda_2 \end{pmatrix} \text{ and } A(\beta)=\begin{pmatrix} \Lambda_1-(T-t) & \beta \\ 0 & \Lambda_2 \end{pmatrix} \text{ with } \beta\in\mathbb{R}.$$

It is clear that for any $\beta\in\mathbb{R}$ there exists a fundamental system with parameters (D,T,d,t) and fundamental matrices $A(\beta)$ and B, see Proposition 3.6.1. Moreover the vectors $\mathbf{k}$ and $\mathbf{b}$ satisfy that $k_1b_1=T-t\neq 0$, $-k_2b_1=\beta$ and $b_2=0$. From the first equality we obtain that $k_1\neq 0$ and from the others we obtain that $k_2\neq 0$ if and only if $\beta\neq 0$. Therefore the system is proper if and only if $\beta\neq 0$, see Lemma 4.7.1(d). □

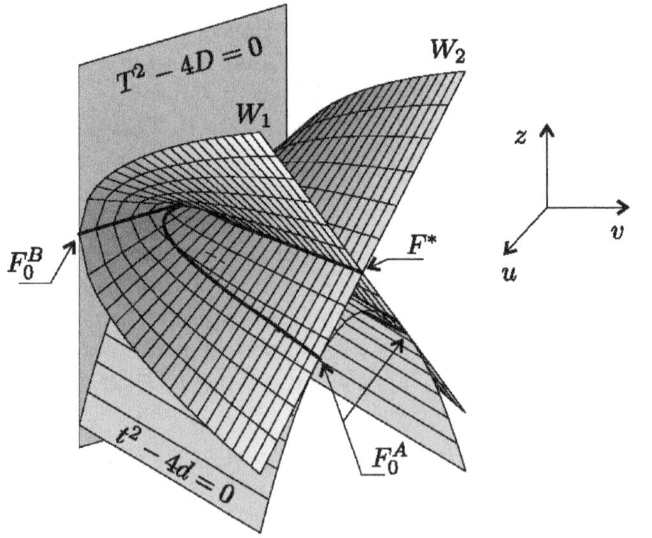

Figure 4.17: Intersection with $t=t_0$ of the manifold W in the coordinates $u=t-T$, $v=T^2-4D$ and $z=2\,(d-D)+T\,(T-t)$.

Set $t=t_0$. By using the change of coordinates $u=t-T$, $v=T^2-4D$ and $z=2(d-D)+T(T-t)$, the manifold W can be expressed as $z^2=u^2v$, see expression (4.50). This algebraic manifold is called the *Whitney umbrella* which has important properties from the geometric point of view [28]. For instance the origin is a *pinch point* also called a *branch point* or a *Whitney singularity.*

The intersection of the manifold W with the space $t = t_0$ is represented in Figure 4.17. The intersection of the manifolds $T^2 - 4D = 0$ and $t^2 - 4d = 0$ with $t = t_0$ is given by the surfaces $v = 0$ and $z = \left(v + u^2\right)/2$. In Figure 4.17 we represent also the intersection of the surfaces F_0^A, F_0^B and F^* with $t = t_0$. From this we can see graphically all statements of the Lemma 4.7.3. Finally, it is easy to check that W can be expressed as

$$W = \left\{(D, T, d, t) \in \mathbb{R}^4 : (d - D)^2 - (t - T)(Td - Dt) = 0\right\}.$$

Chapter 5

Phase portraits

Let $\dot{\mathbf{x}} = \mathbf{f}(\mathbf{x})$ be a fundamental system and let $\dot{\mathbf{x}} = \mathbf{f}_{\mathbb{D}}(\mathbf{x})$ be its Poincaré compactification. In this chapter we describe all the global phase portraits of the compactified system $\dot{\mathbf{x}} = \mathbf{f}_{\mathbb{D}}(\mathbf{x})$ and we study how they vary with the parameters (D, T, d, t). Using the topological equivalence relation, we present the bifurcation set in the fundamental parameter space $\mathbb{R}^4$.

In the study of the phase portrait of the compactified system we note two facts. First, the behaviour of the compactified flow of $\dot{\mathbf{x}} = \mathbf{f}_{\mathbb{D}}(\mathbf{x})$ on the boundary $\partial\mathbb{D}$ of the Poincaré disc corresponds to the behaviour of the flow of the fundamental system $\dot{\mathbf{x}} = \mathbf{f}(\mathbf{x})$ at infinity. This study has been the goal of Section 3.11. Second, the flow of the compactified system in the interior Σ_∞ of the Poincaré disc is differentiably equivalent to the flow of the fundamental system in $\mathbb{R}^2$. Hence, the local phase portrait at the singular points in Σ_∞ (the finite singular points) can be obtained from Theorems 3.9.3 and 3.9.5. Moreover, to study the limit cycles and their local phase portraits, we use the return map defined by the flow of the fundamental system with respect to one of the symmetric straight lines L_+ or L_-. This return map has been the main objective of Chapter 4.

Each of the sections of this chapter presents different local phase portraits obtained by fixing the sign of the parameters D and T. We start by defining some subsets in the parameter space (D, T, d, t) and introduce the notation used in the rest of the chapter.

5.1 Introduction

For any fixed $D, T \in \mathbb{R}^2$ we denote by $\Pi_{D,T}$ the plane

$$\Pi_{D,T} := \left\{(D, T, d, t) : (t, d) \in \mathbb{R}^2\right\}$$

in the parameter space. When no confusion can arise we denote by (t, d) the points of $\Pi_{D,T}$.

Let E be a subset of $\mathbb{R}^4$. We denote by $E|_{D,T}$ the intersection of E with the plane $\Pi_{D,T}$. Thus $W|_{D,T} = W_1|_{D,T} \cup W_2|_{D,T}$ denotes the intersection of the plane $\Pi_{D,T}$ with the manifold W, see (4.50). Note that the sets $W_1|_{D,T}$ and $W_2|_{D,T}$ are two different straight lines when $T^2-4D>0$, and they represent the same straight line when $T^2-4D=0$.

When no confusion can arise, the elements of the set $E|_{D,T}$ will be denoted by a two-dimensional vector (t,d) instead of the four-dimensional one (D,T,d,t). In fact we will use both notations indistinguishably. Note that the change in the order of the coordinates when we pass from four to two coordinates is only a matter of convention. If $E|_{D,T}$ does not depend essentially on the parameters D and T, then we preserve the name E for the set $E \cap \Pi_{D,T}$. Thus the intersection with the plane $\Pi_{D,T}$ of the following regions in the parameter space

$$\begin{aligned}
\mathcal{C}_1^1 &:= \left\{(D,T,d,t) \in \mathbb{R}^4 : t^2-4d>0,\, t>0,\, d>0\right\},\\
\mathcal{C}_1^2 &:= \left\{(D,T,d,t) \in \mathbb{R}^4 : t^2-4d<0,\, t>0\right\},\\
\mathcal{C}_2^1 &:= \left\{(D,T,d,t) \in \mathbb{R}^4 : t^2-4d<0,\, t<0\right\},\\
\mathcal{C}_2^2 &:= \left\{(D,T,d,t) \in \mathbb{R}^4 : t^2-4d>0,\, t<0,\, d>0\right\},\\
\mathcal{C}_3 &:= \left\{(D,T,d,t) \in \mathbb{R}^4 : t<0,\, d<0\right\},\\
\mathcal{C}_4 &:= \left\{(D,T,d,t) \in \mathbb{R}^4 : t>0,\, d<0\right\},
\end{aligned} \tag{5.1}$$

the hyperplanes

$$\begin{aligned}
\mathcal{SN}_\infty &:= \left\{(D,T,d,t) \in \mathbb{R}^4 : t^2-4d=0\right\},\\
\mathcal{N} &:= \left\{(D,T,d,t) \in \mathbb{R}^4 : d=0\right\},
\end{aligned}$$

and the plane

$$\mathcal{O} := \left\{(D,T,d,t) \in \mathbb{R}^4 : t=0,\, d=0\right\},$$

will be denoted by the same symbol, that is, $\mathcal{C}_1^1$, $\mathcal{C}_1^2$, $\mathcal{C}_2^1$, $\mathcal{C}_2^2$, $\mathcal{C}_3$, $\mathcal{C}_4$, $\mathcal{SN}_\infty$, $\mathcal{N}$, and $\mathcal{O}$, respectively. The manifold $\mathcal{SN}_\infty$ is easily located in Figure 4.17. Moreover, the straight lines $W_1|_{D,T}$ and $W_2|_{D,T}$ are tangent to $\mathcal{SN}_\infty$ at the points

$$\mathcal{VB}_1|_{D,T} := W_1|_{D,T} \cap \mathcal{SN}_\infty, \quad \mathcal{VB}_2|_{D,T} := W_2|_{D,T} \cap \mathcal{SN}_\infty, \tag{5.2}$$

where $F_0^A = \mathcal{VB}_1 \cup \mathcal{VB}_2$. Finally, $W_1|_{D,T}$ and $W_2|_{D,T}$ intersect at the point $(t,d) = (T,D)$ belonging to F^*, see (4.50).

Given a fundamental system $\dot{\mathbf{x}} = A\mathbf{x} + \varphi(\mathbf{k}^T\mathbf{x})\mathbf{b}$ with fundamental parameters $(D,T,d,t) \in \mathcal{SN}_\infty$, it follows from Lemma 4.7.2(c) that if $(t,d) \notin \mathcal{VB}_1|_{D,T} \cup \mathcal{VB}_2|_{D,T}$, then the real Jordan normal form of the matrix A is not diagonal. Otherwise, i.e., if $(t,d) \in \mathcal{VB}_1|_{D,T} \cup \mathcal{VB}_2|_{D,T}$, then the fundamental parameters do not determine if the real Jordan normal form of A is diagonal or not, see Lemma 4.7.2(d).

5.2 The case $D > 0$ and $T < 0$

Consider a fundamental system $\dot{\mathbf{x}} = A\mathbf{x} + \varphi(\mathbf{k}^T\mathbf{x})\mathbf{b}$ with parameters (D, T, d, t) and fundamental matrices (A, B). Thus we have the matrix $B = A + \mathbf{k}^T\mathbf{b}$ and the vectors $\mathbf{k}, \mathbf{b} \in \mathbb{R}^2 \setminus \{\mathbf{0}\}$. In this section we assume that $D > 0$ and $T < 0$.

Preliminary results about the number and the local phase portraits at the singular points of the fundamental systems with $D > 0$ and $T < 0$ can be found in [1]. The work of Llibre and Sotomayor [44] presents the different global phase portraits in the plane and the bifurcation set of such systems. In this section we complete this work by considering the behaviour at infinity and by describing the phase portraits in the Poincaré disc and the bifurcations of the singular points at infinity.

5.2.1 Proper fundamental systems

By Theorem 4.7.6, a fundamental system with parameters $D > 0$, $T < 0$ and $(t, d) \notin W|_{D,T} = W_1|_{D,T} \cup W_2|_{D,T}$ is a proper fundamental system. Moreover, if $t = T$ and $d = D$ or $(t, d) \in \mathcal{VB}_1|_{D,T} \cup \mathcal{VB}_2|_{D,T}$, then the system is not proper.

In Figure 5.1 we represent the sets $W|_{D,T}$, $\mathcal{VB}_1|_{D,T}$ and $\mathcal{VB}_2|_{D,T}$ when $T^2 - 4D > 0$. If $T^2 - 4D = 0$, then $W_1|_{D,T} = W_2|_{D,T}$ and $\mathcal{VB}_1|_{D,T} = \mathcal{VB}_2|_{D,T}$. If $T^2 - 4D < 0$, then $W|_{D,T} = \varnothing$.

Suppose that $T^2 - 4D \geq 0$ and let $\Lambda_1 \geq \Lambda_2$ be the eigenvalues of the matrix B. Define the half-lines $W_1^*|_{D,T} = W_1|_{D,T} \cap \{t \leq 2\Lambda_1\}$ and $W_2^*|_{D,T} = W_2|_{D,T} \cap \{t \leq 2\Lambda_2\}$. If we suppose that the matrix A is in real Jordan normal form, then from Lemma 4.7.5 we obtain that (t, d) belongs to either $W_1^*|_{D,T}$ or $W_2^*|_{D,T}$ and $k_1 = 0$.

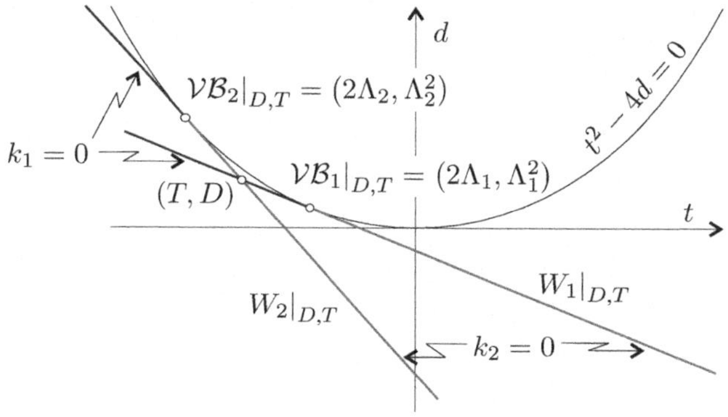

Figure 5.1: Straight lines $W_1|_{D,T}$ and $W_2|_{D,T}$ when $D > 0, T < 0$ and $T^2 - 4D > 0$.

5.2.2 Singular points

We start with the study of the singular points of the compactified vector field $\dot{\mathbf{x}} = \mathbf{f}_{\mathbb{D}}(\mathbf{x})$ in the interior of the Poincaré disc Σ_∞. The following result is a corollary of Theorem 3.9.3.

Proposition 5.2.1. *Consider a fundamental system with parameters $D > 0$ and $T < 0$; thus the origin is an asymptotically stable hyperbolic singular point.*

(a) *If $d \geq 0$, then the unique singular point is the origin and it is a focus if $T^2 - 4D < 0$ and a node if $T^2 - 4D \geq 0$.*

(b) *If $d < 0$, then there exist exactly three singular points: the origin (with the same possible cases as in statement* (a)*) and two saddle points $\mathbf{e}_+$ and $\mathbf{e}_-$.*

5.2.3 Behaviour at infinity

Given a fundamental system with fundamental matrices (A, B) and parameters $t^2 - 4d = 0$, we recall that if $(t, d) \notin \mathcal{VB}_1|_{D,T} \cup \mathcal{VB}_2|_{D,T}$, then the real Jordan normal form of the matrix A is not diagonal, see Lemmas 4.7.2 and 4.7.3. Moreover, if $(t, d) \in \mathcal{VB}_1|_{D,T} \cup \mathcal{VB}_2|_{D,T}$, then the real Jordan normal form of A can be diagonal or not.

Proposition 5.2.2. *Consider a fundamental system with parameters $D > 0$ and $T \leq 0$.*

(a) *If $t^2 - 4d < 0$, then the system has a periodic orbit at infinity.*

(b) *If $t^2 - 4d = 0$ and the real Jordan form of A is diagonal, then infinity is an unstable normally hyperbolic manifold. We remark that this occurs only if $(t, d) \in \mathcal{VB}_1|_{D,T} \cup \mathcal{VB}_2|_{D,T}$.*

(c) *Suppose that $t^2 - 4d = 0$ and the real Jordan normal form of A is not diagonal. Then there exist exactly two singular points at infinity, $\mathbf{x}_+ \in \partial\mathbb{D}_+$ and $\mathbf{x}_- \in \partial\mathbb{D}_-$.*

  (c.1) *If $t \neq 0$, then $\mathbf{x}_+$ (respectively, $\mathbf{x}_-$) is a saddle-node with center manifold contained in $\partial\mathbb{D}$. The hyperbolic manifold is stable when $t > 0$ and unstable when $t < 0$.*

  (c.2) *If $t = 0$, then a neighbourhood of $\mathbf{x}_+$ and $\mathbf{x}_-$ in $\mathbb{D}$ is formed by a hyperbolic sector.*

(d) *Suppose that $t^2 - 4d > 0$. Then there exist exactly four singular points at infinity, $\mathbf{x}_+, \mathbf{y}_+ \in \partial\mathbb{D}_+$ and $\mathbf{x}_-$, $\mathbf{y}_- \in \partial\mathbb{D}_-$.*

  (d.1) *If $d \geq 0$ and $t > 0$, then $\mathbf{x}_+$ and $\mathbf{x}_-$ are stable nodes, and $\mathbf{y}_+$ and $\mathbf{y}_-$ are saddle points with the unstable manifold contained in $\partial\mathbb{D}$.*

  (d.2) *If $d \geq 0$ and $t < 0$, then $\mathbf{x}_+$ and $\mathbf{x}_-$ are saddle points with the stable manifold contained in $\partial\mathbb{D}$, and $\mathbf{y}_+$ and $\mathbf{y}_-$ are unstable nodes.*

(d.3) *If $d < 0$, then $\mathbf{x}_+$ and $\mathbf{x}_-$ are stable nodes, and $\mathbf{y}_+$ and $\mathbf{y}_-$ are unstable nodes.*

Proof. Let (A, B) be the fundamental matrices of the system. Without loss of generality, we can assume that the matrix A is in real Jordan normal form, see Proposition 3.10.2.

Statement (a) is a consequence of Theorem 3.11.1(a).

(b) From Theorem 3.11.1(b) it follows that $\partial\mathbb{D}$ is formed by singular points. Since the matrix A is diagonal, the parameters (t, d) belong to $\mathcal{VB}_1|_{D,T} \cup \mathcal{VB}_2|_{D,T}$ and therefore $t < 0$, see Figure 5.2. The statement follows from Theorems 3.11.5(a) and 3.11.9(a).

(c) Since $t^2 - 4d = 0$ and the matrix A is not diagonal, it follows from Theorem 3.11.1(c) that there exist exactly two singular points at infinity, $\mathbf{x}_+$ and $\mathbf{x}_-$. Moreover, $\mathbf{x}_+ = -\mathbf{x}_-$. For simplicity we denote by $\mathbf{x}_+$ the singular point in $\partial\mathbb{D}_+$ and by $\mathbf{x}_-$ the singular point in $\partial\mathbb{D}_-$. Statements (c.1) and (c.2) follow from Theorems 3.11.6(a), (b) and 3.11.10(a).

(d) By Theorem 3.11.1(d), there exist exactly four singular points at infinity. We denote by $\mathbf{x}_+$ and $\mathbf{y}_+$ the singular points in $\partial\mathbb{D}_+$, and by $\mathbf{x}_-$ and $\mathbf{y}_-$ the singular points in $\partial\mathbb{D}_-$.

Suppose that $d \geq 0$ and $t > 0$. Then $(t, d) \notin W|_{D,T}$, see Figure 5.1. Therefore, $k_1 \neq 0$ and $k_2 \neq 0$, see Lemma 4.7.5(a). From Theorem 3.11.7(a) and (e) it follows that $\mathbf{x}_+$ and $\mathbf{x}_-$ are stable nodes, and from Theorem 3.11.8(a) and (e.3) it follows that $\mathbf{y}_+$ and $\mathbf{y}_-$ are saddle points. Moreover, the unstable manifolds of $\mathbf{y}_+$ and $\mathbf{y}_-$ are contained in $\partial\mathbb{D}$. This proves statement (d.1).

Suppose that $d \geq 0$ and $t < 0$. Then from Theorem 3.11.7(b), (d.3) and Theorem 3.11.11(b) it follows that $\mathbf{x}_+$ and $\mathbf{x}_-$ are saddle points with the stable manifold in $\partial\mathbb{D}$. From Theorem 3.11.8(b), (d) and Theorem 3.11.12(b) it follows that $\mathbf{y}_+$ and $\mathbf{y}_-$ are unstable nodes.

Suppose now that $d < 0$. Then $(t, d) \notin W_1^*|_{D,T} \cup W_2^*|_{D,T}$, see Figure 5.1. Hence $k_1 \neq 0$. Therefore, $\mathbf{x}_+$ and $\mathbf{x}_-$ are stable nodes, see Theorem 3.11.7(c), and $\mathbf{y}_+$ and $\mathbf{y}_-$ are unstable nodes, see Theorem 3.11.8(c) and Theorem 3.11.12(b). □

In Figure 5.2 we summarize the local phase portrait at infinity proved in Proposition 5.2.2. We recall that only when (t, d) are in $\mathcal{VB}_1|_{D,T}$ or in $\mathcal{VB}_2|_{D,T}$ the behaviour at infinity is not uniquely determined.

5.2.4 Periodic orbits

Next we study the existence of Jordan curves formed by solutions. In general these curves are formed by separatrices, and so they play an essential role in the description of the phase portrait. In the next result we distinguish in the parameter space the fundamental systems having this kind of invariant curves and the regions in the phase space where these curves appear.

Proposition 5.2.3. *Consider a fundamental system with parameters $D > 0$ and $T < 0$.*

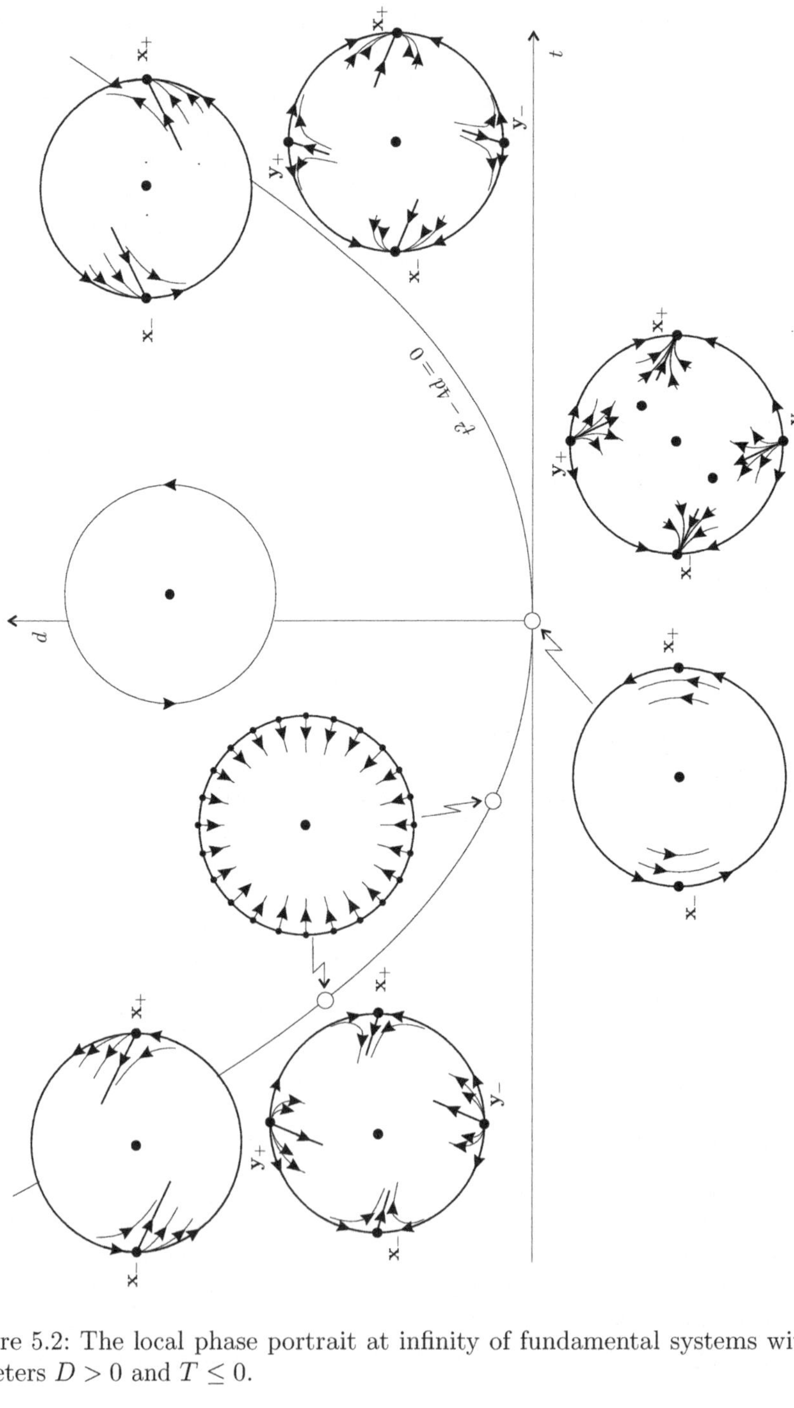

Figure 5.2: The local phase portrait at infinity of fundamental systems with parameters $D > 0$ and $T \leq 0$.

(a) *If $t \leq 0$, then the system has no Jordan curves formed by solutions.*

(b) *Suppose that $t > 0$. If Γ is a Jordan curve formed by solutions, then $\Gamma \cap (S_+ \cup S_-) \neq \varnothing$ and $\Gamma \cap S_0 \neq \varnothing$.*

Proof. (a) Suppose that $t < 0$. In this case the statement is a corollary of Theorem 3.12.2(a). Suppose now that $t = 0$ and let Γ be a Jordan curve formed by solutions. Thus, either $\Gamma \subset S_+ \cup L_+$ or $\Gamma \subset S_- \cup L_-$, see Theorem 3.12.2(b.2). Suppose that $\Gamma \subset S_+ \cup L_+$, otherwise we can apply similar arguments. Since in $S_+ \cup L_+$ and $S_- \cup L_-$ the system is linear, we conclude that in these half-planes the system is in the center case. This implies that the parameter $d > 0$, the invariant curve Γ is a periodic orbit and there exists a singular point $\mathbf{e}$ in $\Sigma_\Gamma \subset S_+$. From the symmetry of the vector field with respect to the origin it is easy to conclude the existence of three singular points, $\mathbf{e}$, $-\mathbf{e}$ and the origin. By Proposition 5.2.1(a), the existence of these singular points contradicts the fact that $d > 0$. Therefore, if $t = 0$, then there are no Jordan curves formed by solutions.

Statement (b) is a corollary of Theorem 3.12.2(b.1). □

According to Proposition 5.2.3, the existence of Jordan curves formed by solutions in fundamental systems with $D > 0$ and $T < 0$ is possible only under the assumption of $t > 0$. Moreover, these curves can only be periodic orbits, homoclinic cycles, or heteroclinic cycles intersecting the straight lines L_+ and L_-, see Lemma 3.12.1 and Theorem 3.12.2(b.1). As it is shown in Subsection 4.6, in this case we can define the return map π in a neighbourhood of the invariant curve Γ. This return map can be expressed depending on the fundamental parameter d, see Theorem4.6.1(b) and (c). Thus, when $d \neq 0$ it follows that $\pi = (\pi^A_{++} \circ \pi^B_{+-})^2$, and when $d = 0$, $\pi = (\widetilde{\pi}^A_{++} \circ \pi^B_{+-})^2$. In these expressions, π^A_{++}, $\widetilde{\pi}^A_{++}$ and π^B_{+-} are the Poincaré maps defined by the flow of the corresponding linear system and associated to the symmetric straight lines L_+ and L_-.

As it is shown in Subsection 4.6, the fixed points of the return map π are the zeros of the Lamerey map $g_A = \pi^{-A}_{++} - \pi^B_{+-}$ when $d \neq 0$ and the Lamerey map $g_A = \widetilde{\pi}^{-A}_{++} - \pi^B_{+-}$ when $d = 0$, see expressions (4.46) and (4.47), respectively.

In the next result we summarize all the information proved in Section 4.4 about the Poincaré map π^B_{+-} when $D > 0$ and $T < 0$.

Lemma 5.2.4. *Given a proper fundamental system with fundamental matrices (A, B) and parameters $D > 0$ and $T < 0$, the Poincaré map π^B_{+-} has the following properties:*

(a) *Let $0 > \Lambda_1 > \Lambda_2$ be the eigenvalues of the matrix B. There exists a value $a_0 > |\Lambda_1|^{-1}$ such that $\pi^B_{+-} : [a_0, +\infty) \to [0, +\infty)$, $\pi^B_{+-}(a_0) = 0$, $\lim_{a \nearrow +\infty} \pi^B_{+-}(a) = +\infty$, and $\pi^B_{+-}(a) < a$.*

(b) *If $a \in (a_0, +\infty)$, then $(\pi^B_{+-})'(a) > 1$ and $\lim_{a \searrow a_0} (\pi^B_{+-})'(a) = +\infty$.*

(c) *If $a \in (a_0, +\infty)$, then $(\pi^B_{+-})''(a) < 0$.*

(d) *The straight line* $b = a + 2T/D$ *is an asymptote of the graph of the map* π_{+-}^{B} *as* $a \nearrow +\infty$.

(e) *The implicit expression of the map* π_{+-}^{B} *depends on the sign of* $T^2 - 4D$. *Specifically,*

(e.1) *if* $T^2 - 4D > 0$, *then* π_{+-}^{B} *is implicitly defined by*

$$\left(\frac{\pi_{+-}^{B}(a)\left(T - \sqrt{T^2-4D}\right) - 2}{a\left(T - \sqrt{T^2-4D}\right) + 2}\right)^{\frac{T+\sqrt{T^2-4D}}{T-\sqrt{T^2-4D}}} = \frac{\pi_{+-}^{B}(a)\left(T + \sqrt{T^2-4D}\right) - 2}{a\left(T + \sqrt{T^2-4D}\right) + 2},$$

(e.2) *if* $T^2 - 4D = 0$, *then* π_{+-}^{B} *is implicitly defined by*

$$\frac{\pi_{+-}^{B}(a)\,T - 2}{aT + 2} = \exp\left(\frac{\left(\pi_{+-}^{B}(a) + a\right)T}{\left(\pi_{+-}^{B}(a) - a\right)T - 2 + 2a\pi_{+-}^{B}(a)\,D}\right),$$

(e.3) *if* $T^2 - 4D < 0$, *then* π_{+-}^{B} *is implicitly defined by*

$$\frac{1 - T\pi_{+-}^{B}(a) + D\left(\pi_{+-}^{B}(a)\right)^2}{1 + Ta + Da^2} = \exp\left(\frac{2T}{\sqrt{4D-T^2}}\arctan\left(\frac{\left(\pi_{+-}^{B}(a) + a\right)\sqrt{4D-T^2}}{\left(\pi_{+-}^{B}(a) - a\right)T - 2 + 2a\pi_{+-}^{B}(a)\,D}\right)\right).$$

Proof. The lemma is a consequence of Corollaries 4.4.4, 4.4.9 and 4.4.14. □

In the next results we summarize all the information about the Poincaré maps π_{++}^{-A} and $\widetilde{\pi}_{++}^{-A}$ obtained in Sections 4.4 and 4.5.2 in the case when the fundamental parameters are $D > 0$, $T < 0$, $d \geq 0$ and $t > 0$.

Lemma 5.2.5. *Consider a fundamental system with fundamental matrices* (A, B) *and parameters* $D > 0$, $T < 0$, $d \geq 0$ *and* $t > 0$. *If* $d > 0$, *then the Poincaré map* π_{++}^{A} *is well defined and its inverse map* π_{++}^{-A} *has the following properties:*

(a) *There exists a value* $r \leq +\infty$ *such that* $\pi_{++}^{-A} : [0, +\infty) \to [0, r)$ *and* $\pi_{++}^{-A}(a) < a$.

(b) *If* $a \in (0, +\infty)$, *then* $0 < (\pi_{++}^{-A})'(a) < 1$ *and* $\lim_{a \searrow 0} (\pi_{++}^{-A})'(a) = 1$.

(c) *If* $t^2 - 4d \geq 0$, *then there exists* $b_0 > 0$ *such that* $b = b_0$ *is a horizontal asymptote of the graph of* $\pi_{++}^{-A}(a)$ *when* a *tends to* $+\infty$. *If* $t^2 - 4d < 0$, *then the straight line* $b = e^{-\gamma\pi}a + t(1 + e^{-\gamma\pi})/d$, *with* $\gamma = t/\sqrt{4d - t^2}$, *is an asymptote of the graph of* $\pi_{++}^{-A}(a)$ *when* a *tends to* $+\infty$.

If $d = 0$, *then the Poincaré map* $\widetilde{\pi}_{++}^{A}$ *is well defined and its inverse map* $\widetilde{\pi}_{++}^{-A}$ *has the following properties:*

(d) *There exists a value* $r \leq +\infty$ *such that* $\widetilde{\pi}^{-A}_{++} : [0, +\infty) \to [0, r)$ *and* $\widetilde{\pi}^{-A}_{++}(a) < a$.

(e) *If* $a \in (0, +\infty)$, *then* $0 < (\widetilde{\pi}^{-A}_{++})'(a) < 1$ *and* $\lim_{a \searrow 0} (\widetilde{\pi}^{-A}_{++})'(a) = 1$.

(f) *There exists* $b_0 > 0$ *such that* $b = b_0$ *is a horizontal asymptote of the graph of* $\widetilde{\pi}^{-A}_{++}(a)$ *when* a *tends to* $+\infty$.

Proof. Since $t > 0$ and $d \geq 0$, it is easy to check that $(D, T, d, t) \notin W|_{D,T}$, see Figure 5.1. By Theorem 4.7.6, the system is a proper fundamental system and the Poincaré maps π^{-A}_{++} and $\widetilde{\pi}^{-A}_{++}$ are well defined depending on whether $d > 0$ or $d = 0$, respectively. The properties of the maps follow from Corollaries 4.4.2, 4.4.7, 4.4.12 and 4.5.10. □

Using the Lamerey map g_A, in the next proposition we prove the existence of a unique limit cycle for any fundamental system with parameters $D > 0$, $T < 0$, $d \geq 0$ and $t > 0$.

Proposition 5.2.6. *Any fundamental system with parameters* $D > 0$, $T < 0$, $d \geq 0$ *and* $t > 0$ *has exactly one limit cycle* Γ. *Moreover,* Γ *is a hyperbolic asymptotically unstable limit cycle which arises via a Hopf bifurcation at infinity when* $t = 0$, *and it satisfies that* $\mathbf{0} \in \Sigma_\Gamma$.

Under the assumption that the system is given in its proper normal form, if t *is suficiently small, then the radius of* Γ *is approximately equal to*

$$R_\Gamma = \sqrt{1 + \left(\frac{Dt - Td}{d} \frac{1 + e^{-\gamma\pi}}{1 - e^{-\gamma\pi}} \right)^2}, \quad \textit{where } \gamma = \frac{t}{\sqrt{4d - t^2}}.$$

Proof. Since $t > 0$ and $d \geq 0$, we have that $(D, T, d, t) \notin W|_{D,T}$ and the system is a proper fundamental system, see Theorem 4.7.6. Therefore, the Poincaré maps π^B_{+-} and π^A_{++} or $\widetilde{\pi}^A_{++}$ are also well defined depending on whether $d > 0$ or $d = 0$, respectively. Consequently, the Lamerey map $g_A = \pi^{-A}_{++} - \pi^B_{+-}$ or $g_A = \widetilde{\pi}^{-A}_{++} - \pi^B_{+-}$ is well defined, depending on whether $d > 0$ or $d = 0$, respectively. We prove the existence of a unique limit cycle by showing that g_A has exactly one zero.

The maps $\pi^B_{+-}, \pi^{-A}_{++}$ and $\widetilde{\pi}^{-A}_{++}$ are described in Lemmas 5.2.4 and 5.2.5, respectively. From this we obtain that g_A is defined in $[a_0, +\infty)$, $g_A(a_0) > 0$ and $\lim_{a \nearrow +\infty} g_A(a) = -\infty$. Thus, g_A has a zero in $[a_0, +\infty)$. Moreover, since $g'_A(a) < 0$ in $[a_0, +\infty)$, the Lamerey map g_A has a unique zero a^* in $[a_0, +\infty)$. Therefore, the system has exactly one limit cycle Γ. In Figure 5.3 we draw the different Lamerey diagrams depending on the sign of $t^2 - 4d$.

To study the local phase portrait of the limit cycle Γ we use the return map $\pi = (\pi^A_{++} \circ \pi^B_{+-})^2$. The case $\pi = (\widetilde{\pi}^A_{++} \circ \pi^B_{+-})^2$ follows by similar arguments. By Lemma 5.2.4(b), $(\pi^B_{+-})'(a) > 1$, and by Lemma 5.2.5(c), $(\pi^{-A}_{++})'(a) < 1$. Hence,

$$\left.\frac{d\pi}{da}\right|_{a^*} = \left(\left.\frac{d\pi^A_{++}}{da}\right|_{\pi^B_{+-}(a^*)} \left.\frac{d\pi^B_{+-}}{da}\right|_{a^*} \right)^2 > 1.$$

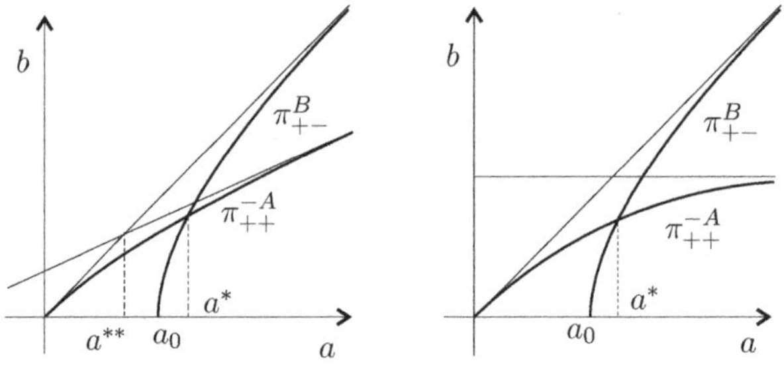

(a) $t^2 - 4d < 0$ (b) $t^2 - 4d \geq 0$

Figure 5.3: Lamerey diagram for parameters $D > 0$, $T < 0$, $d > 0$ and $t > 0$.

Therefore, Γ is an asymptotically unstable hyperbolic limit cycle, see Proposition 4.6.2. Moreover, since any limit cycle has a singular point in its interior and the origin is the unique singular point of the system, see Proposition 5.2.1(b), it follows that $\mathbf{0} \in \Sigma_\Gamma$.

Let a^{**} be the abscissa of the intersection point between the asymptotes of the graphs of π_{++}^{-A} and π_{+-}^{B}, see Figure 5.3(a). Note that if t is sufficiently small we always have that $t^2 - 4d < 0$. Since

$$a^{**} + 2\frac{T}{D} = e^{-\gamma\pi} a^{**} + \frac{t}{d}\left(1 + e^{-\gamma\pi}\right),$$

with $\gamma = t/\sqrt{4d - t^2}$, one has that

$$a^{**} = \frac{Dt\left(e^{\gamma\pi} + 1\right) - 2Tde^{\gamma\pi}}{Dd\left(e^{\gamma\pi} - 1\right)},$$

which is an aproximation of the value of a^*.

Suppose that the system is given in the proper normal form, i.e.,

$$\dot{\mathbf{x}} = \begin{pmatrix} 0 & -d \\ 1 & t \end{pmatrix} \mathbf{x} + \varphi(x_2) \begin{pmatrix} d - D \\ T - t \end{pmatrix}.$$

The contact point $\mathbf{p}$ of the flow of the system with the straight line L_+ satisfies that $\mathbf{p} = (-T, 1)^T$ and $\dot{\mathbf{p}} = (-D, 0)^T$. Moreover, since a^{**} is an approximation of a^*, Γ intersects L_+ close to the points $\mathbf{q}_1 = \mathbf{p} + a^{**}\dot{\mathbf{p}}$ and $\mathbf{q}_0 = \mathbf{p} - \pi_{++}^{-A}(a^{**})\dot{\mathbf{p}}$. Therefore, $R_\Gamma = \|\mathbf{q}_1\|$ is an approximation of the radius of Γ, i.e.,

$$R_\Gamma = \sqrt{1 + (T + a^{**}D)^2} = \sqrt{1 + \left(\frac{Dt - Td}{d}\frac{\left(1 + e^{-\gamma\pi}\right)}{\left(1 - e^{-\gamma\pi}\right)}\right)^2}.$$

Finally, since $\lim_{t \searrow 0} \gamma = 0$, it follows that $\lim_{t \searrow 0} R_\Gamma = +\infty$, which proves that Γ arises from a Hopf bifurcation at infinity when $t = 0$. □

Under the same hypothesis of Proposition 5.2.6 in [42, Theorem 1] the authors give the following approximation for the radius of the limit cycle Γ:

$$J_\Gamma = (Dt - 2t - T)\frac{1 + e^{\gamma\pi}}{1 - e^{-\gamma\pi}}.$$

It is easy to check that

$$\lim_{t \searrow 0} \frac{R_\Gamma}{J_\Gamma} = 1,$$

which proves that the two approximations coincide to first order.

We will define in the quadrant $\mathcal{C}_4 = \{(D, T, t, d) \in \mathbb{R}^4 : t > 0,\ d < 0\}$ a 3-dimensional topological manifold $\mathcal{H}_e\mathcal{L}$ in such a way that if $\mathcal{H}_e\mathcal{L}|_{D,T}$ is the differentiable curve along which $\mathcal{H}_e\mathcal{L}$ intersects the plane $\Pi_{D,T}$, then any fundamental system with parameters contained in $\mathcal{H}_e\mathcal{L}|_{D,T}$ has a heteroclinic cycle connecting the singular points $\mathbf{e}_+$ and $\mathbf{e}_-$.

To define $\mathcal{H}_e\mathcal{L}$ we first search for necessary conditions on the fundamental parameters (D, T, d, t) in order to have a heteroclinic cycle for any fundamental system with these parameters. Since these conditions are obtained by using the Poincaré maps, that is, by assuming that the system is proper, in order to obtain a complete characterization of the fundamental systems having a heteroclinic cycle we need to prove that systems with parameters on $\mathcal{H}_e\mathcal{L}$ are proper. We note that Llibre and Sotomayor [44] obtained only the necessary conditions on the parameters assuming that they are also sufficient.

Since the point $W_1|_{D,T} \cap W_2|_{D,T}$ is not contained in $\mathcal{C}_4$, for any $(t, d) \in \mathcal{C}_4$, there exists a proper fundamental system having these parameters. In the next result we prove that two different proper fundamental systems having these parameters also have the same Poincaré maps.

Proposition 5.2.7. *Consider two different proper fundamental systems having the same fundamental parameters $D > 0$, $T < 0$, $d < 0$, $t > 0$ and with fundamental matrices (A, B) and (A^*, B^*), respectively. Then $\pi^A_{++} = \pi^{A^*}_{++}$ and $\pi^B_{+-} = \pi^{B^*}_{+-}$.*

Proof. Suppose that $T^2 - 4D \neq 0$. It is easy to conclude that B and B^* are equivalent matrices, i.e., there exists a regular matrix M such that $B^* = MBM^{-1}$. Since the Poincaré maps are invariant under linear changes of coordinates, we get $\pi^B_{+-} = \pi^{B^*}_{+-}$. By applying similar arguments to the matrices A and A^* we obtain that $\pi^A_{++} = \pi^{A^*}_{++}$.

Suppose now that $T^2 - 4D = 0$. Since $D > 0$ and both systems are proper, we obtain that the real Jordan normal forms of the matrices B and B^* are not diagonal, see Lemma 4.7.1(c). Moreover, B and B^* are equivalent matrices. Consequently, $\pi^B_{+-} = \pi^{B^*}_{+-}$. By applying similar arguments to the matrices A and A^* we obtain that $\pi^A_{++} = \pi^{A^*}_{++}$. □

Therefore, there exists a unique Poincaré map π^B_{+-} associated to any parameters $D > 0$ and $T < 0$. Similarly, there exists a unique Poincaré map π^A_{++} associated to any pair $(t, d) \in \mathcal{C}_4$.

Now we define the manifold $\mathcal{H}_e\mathcal{L}$. Consider the following set in $\mathbb{R}^4$:

$$S = \left\{(D,T,k,t) : D > 0,\ T < 0 \text{ and } 0 < t < k \le t + a_0^{-1}\right\},$$

where a_0 is the lower boundary of the domain of definition of π_{+-}^B, see Lemma 5.2.4(a). Since π_{+-}^B depends on D and T, it follows that the value a_0 also depends on D and T. To simplify the notation, in the following computations we assume implicitly this dependence. We also consider that the values of the parameters D and T are fixed. We now obtain the expression of $\mathcal{H}_e\mathcal{L}|_{D,T}$. The expression of $\mathcal{H}_e\mathcal{L}$ follows by varying the parameters D and T.

Take $(k,t) \in S$; that is, $0 < t < k$ and $a_0 \le (k-t)^{-1}$. Hence, the function

$$h(k,t) := \pi_{+-}^B\left((k-t)^{-1}\right) - k^{-1} \tag{5.3}$$

is well defined and differentiable in S, see Lemma 5.2.4(a), and satisfies

$$\frac{\partial h}{\partial k} = -\frac{d\pi_{+-}^B}{da}\frac{1}{(t-k)^2} + \frac{1}{k^2} < 0 \ \text{ and } \ \frac{\partial h}{\partial t} = \frac{d\pi_{+-}^B}{da}\frac{1}{(t-k)^2} > 0. \tag{5.4}$$

Note that if $(k,t) \in S$, then $k^2 > (t-k)^2$ and $(\pi_{+-}^B)'(a) > 1$, see Lemma 5.2.4(b).

Set $k_0 > 0$ and take $t \in \left[k_0 - a_0^{-1}, k_0\right)$. Thus, $(k_0,t) \in S$. When t varies in $[k_0 - a_0^{-1}, k_0)$, the value $(k-t)^{-1}$ covers the interval $[a_0, +\infty)$ and the map $\pi_{+-}^B((k-t)^{-1})$ takes any value in $[0,+\infty)$, see Lemma 5.2.4(a). Therefore, given $k_0 > 0$, there exists a unique t_0 such that $\pi_{+-}^B((k_0-t_0)^{-1}) = k_0^{-1}$, which is equivalent to $h(k_0,t_0) = 0$.

By the Implicit Function Theorem, there exists a differentiable function $k(t)$ defined on $(0,+\infty)$ such that

$$h(k(t),t) = \pi_{+-}^B\left((k(t)-t)^{-1}\right) - k(t)^{-1} = 0, \tag{5.5}$$

and

$$\frac{dk}{dt} = -\frac{\partial h/\partial t}{\partial h/\partial k} > 1,$$

see (5.4). Define in $(0,+\infty)$ the function $f \in C^1((0,+\infty))$ by

$$f(t) := k(t)(t - k(t)) < 0. \tag{5.6}$$

It is easy to check that $f'(t) = k'(t)(t-k(t)) + k(t)(1-k'(t)) < 0$. Hence, the graph of f splits $\mathcal{C}_4$ in the following subsets:

$$\begin{aligned}
\mathcal{C}_4^1|_{D,T} &:= \{(t,d) \in \mathcal{C}_4 : t > 0 \ \text{ and } \ d < f(t)\},\\
\mathcal{H}_e\mathcal{L}|_{D,T} &:= \{(t,d) \in \mathcal{C}_4 : t > 0 \ \text{ and } \ d = f(t)\},\\
\mathcal{C}_4^2|_{D,T} &:= \{(t,d) \in \mathcal{C}_4 : t > 0 \ \text{ and } \ d > f(t)\}.
\end{aligned} \tag{5.7}$$

Next we obtain an expression of the differentiable curve $\mathcal{H}_e\mathcal{L}|_{D,T}$ depending on the parameters D and T.

Proposition 5.2.8. *Consider $D > 0$ and $T < 0$.*

(a) *If $T^2 - 4D > 0$, then $\mathcal{H}_e\mathcal{L}|_{D,T}$ is implicitly defined by*

$$\left(\frac{\left(T-\sqrt{T^2-4D}\right)\left(\sqrt{t^2-4d}-t\right)+4d}{\left(T-\sqrt{T^2-4D}\right)\left(\sqrt{t^2-4d}+t\right)-4d}\right)^{\frac{T+\sqrt{T^2-4D}}{T-\sqrt{T^2-4D}}} = \frac{\left(T+\sqrt{T^2-4D}\right)\left(\sqrt{t^2-4d}-t\right)+4d}{\left(T+\sqrt{T^2-4D}\right)\left(\sqrt{t^2-4d}+t\right)-4d}.$$

(b) *If $T^2 - 4D = 0$, then $\mathcal{H}_e\mathcal{L}|_{D,T}$ is implicitly defined by*

$$\frac{T\left(\sqrt{t^2-4d}-t\right)+4d}{T\left(\sqrt{t^2-4d}+t\right)-4d} = \exp\left(\frac{2T\sqrt{t^2-4d}}{T^2-2Tt+4d}\right).$$

(c) *If $T^2 - 4D < 0$, then $\mathcal{H}_e\mathcal{L}|_{D,T}$ is implicitly defined by*

$$\frac{4d^2+\left(t-\sqrt{t^2-4d}\right)\left(D\left(t-\sqrt{t^2-4d}\right)-2Td\right)}{4d^2+\left(t+\sqrt{t^2-4d}\right)\left(D\left(t+\sqrt{t^2-4d}\right)-2Td\right)}$$
$$= \exp\left(\frac{2T}{\sqrt{4D-T^2}}\arctan\left(\frac{\sqrt{t^2-4d}\sqrt{4D-T^2}}{2(D+d)-Tt}\right)\right).$$

Proof. Given $(t,d) \in \mathcal{H}_e\mathcal{L}|_{D,T}$, i.e., $t > 0$, $d < 0$ and $d = k(t)(t-k(t))$, see (5.7) and (5.6), we have

$$k(t) = \frac{t+\sqrt{t^2-4d}}{2} \quad \text{and} \quad t-k(t) = \frac{t-\sqrt{t^2-4d}}{2}.$$

Since $k(t)$ and t satisfy (5.5), we conclude that

$$\pi_{+-}^B\left(\frac{2}{\sqrt{t^2-4d}-t}\right) = \frac{2}{\sqrt{t^2-4d}+t}$$

is an implicit expression of the curve $\mathcal{H}_e\mathcal{L}|_{D,T}$. The proposition follows from Lemma 5.2.4(e). □

In Proposition 5.2.12 we will prove that any proper fundamental system with parameters $(t,d) \in \mathcal{C}_4^1|_{D,T}$ has no Jordan curves formed by solutions. When $(t,d) \in \mathcal{H}_e\mathcal{L}|_{D,T}$, the system has a heteroclinic cycle connecting the singular points $\mathbf{e}_+$ and $\mathbf{e}_-$; and when $(t,d) \in \mathcal{C}_4^2|_{D,T}$, the system has exactly one limit cycle. First we need some technical lemmas.

Lemma 5.2.9. *Let π_{++}^A be the Poincaré map defined by the flow of a proper fundamental system with parameters $d < 0$ and $t > 0$, and let $\lambda_1 > 0 > \lambda_2$ be the eigenvalues of the matrix A. Then:*

(a) $\pi_{++}^{-A} : [0, |\lambda_2|^{-1}) \to [0, \lambda_1^{-1})$, $\pi_{++}^{-A}(0) = 0$ *and* $\lim_{a \nearrow |\lambda_2|^{-1}} \pi_{++}^{-A}(a) = \lambda_1^{-1}$.

(b) *If* $a \in [0, |\lambda_2|^{-1})$, *then* $\pi_{++}^{-A}(a) < a$.

(c) *If* $a \in (0, |\lambda_2|^{-1})$, *then* $0 < (\pi_{++}^{-A})'(a) < 1$ *and* $\lim_{a \nearrow |\lambda_2|^{-1}} (\pi_{++}^{-A})'(a) = 0$.

(d) *If* $a \in (0, |\lambda_2|^{-1})$, *then* $(\pi_{++}^{-A})''(a) < 0$.

(e) *The straight line* $b = \lambda_1^{-1}$ *is a horizontal asymptote of the graph of* π_{++}^{-A} *when* a *tends to* $|\lambda_2|^{-1}$.

Proof. The lemma follows from Corollary 4.4.16. □

Similarly to the case $d > 0$ and $t > 0$, the existence of a limit cycle will be obtained as a consequence of the existence of an isolated zero of the Lamerey map. In the next lemma we obtain the existence of a heteroclinic cycle from this map.

Lemma 5.2.10. *Consider a proper fundamental system with parameters* $D > 0$, $T < 0$, $d < 0$ *and* $t > 0$. *Let* $\lambda_1 > 0 > \lambda_2$ *be the eigenvalues of the matrix* A *and let* a_0 *be the lower boundary of the domain of definition of* π_{+-}^{B}.

(a) *If* $|\lambda_2|^{-1} \leq a_0$, *then the system has no Jordan curves formed by solutions.*

(b) *Suppose that* $a_0 < |\lambda_2|^{-1}$ *and consider the Lamerey map*

$$g(a) := \begin{cases} \pi_{++}^{-A}(a) - \pi_{+-}^{B}(a), & \text{if } a \in \left[a_0, |\lambda_2|^{-1}\right), \\ \lambda_1^{-1} - \pi_{+-}^{B}\left(|\lambda_2|^{-1}\right), & \text{if } a = |\lambda_2|^{-1}. \end{cases}$$

(b.1) g *is continuous in* $[a_0, |\lambda_2|^{-1}]$ *and* $g \in C^\infty((a_0, |\lambda_2|^{-1}))$.

(b.2) g *has a zero* $a^* \in [a_0, |\lambda_2|^{-1})$ *if and only if the system has a limit cycle* Γ. *In that case* Γ *is an asymptotically unstable hyperbolic limit cycle which intersects* L_+ *at the points of coordinates* a^* *and* $\pi_{+-}^{B}(a^*)$. *Also, the origin* $\mathbf{0}$ *is contained in* Σ_Γ.

(b.3) g *has a zero at* $a^* = |\lambda_2|^{-1}$ *if and only if the system has a heteroclinic cycle* Δ *connecting the singular points* $\mathbf{e}_+$ *and* $\mathbf{e}_-$. *Moreover, the cycle* Δ *intersects* L_+ *at the points of coordinates* $|\lambda_2|^{-1}$ *and* λ_1^{-1}.

Proof. (a) Suppose that there exists a Jordan curve Γ formed by solutions of the system. Then Γ intersects the straight lines L_+ and L_-, see Proposition 5.2.3(b). Let a^* be the coordinate of the intersection point. By the continuous dependence of the solutions of a linear system on the initial conditions, the return map π is defined in a neighbourhood of a^*. Since $D > 0$, it follows that $\pi = (\pi_{++}^{A} \circ \pi_{+-}^{B})^2$, see Theorem 4.6.1(b).

Lemmas 5.2.9(a) and 5.2.4(a.1) yield $\pi_{++}^{A}([0, \lambda_1^{-1})) = [0, |\lambda_2|^{-1})$ and $\pi_{+-}^{B}([a_0, +\infty)) = [0, +\infty)$. Therefore, π is defined if $a_0 < |\lambda_2|^{-1}$, which contradicts the hypothesis. From this we conclude that the system has no Jordan curves formed by solutions.

Statement (b.1) follows by noting that the Poincaré map π_{++}^{-A} is analytic and that $\lim_{a\nearrow|\lambda_2|^{-1}} \pi_{++}^{-A}(a) = \lambda_1^{-1}$.

(b.2) Let $a^* \in [a_0, |\lambda_2|^{-1})$ be a zero of g, i.e., $\pi_{+-}^{B}(a^*) = \pi_{++}^{-A}(a^*)$. Hence, $\pi_{++}^{A} \circ \pi_{+-}^{B}(a^*) = a^*$ which implies that a^* is a fixed point of π. Therefore, there exists a periodic orbit Γ intersecting the straight line L_+ at the points with coordinates a^* and $\pi_{++}^{-A}(a^*) = \pi_{+-}^{B}(a^*)$. Finally, since

$$\pi'(a^*) = \left[\left(\pi_{++}^{A}\right)'\left(\pi_{+-}^{B}(a^*)\right)\left(\pi_{+-}^{B}\right)'(a^*)\right]^2 > 1,$$

see Lemmas 5.2.4(b) and 5.2.9(c), the periodic orbit Γ is an asymptotically unstable hyperbolic limit cycle, see Theorem 2.7.5. The other implication follows by reversing the above argument.

(b.3) Let $\mathbf{p}$ be the contact point of the flow with L_+. Note that $\mathbf{p}$ does exist because the Poincaré maps are defined. Since the singular point $\mathbf{e}_+$ is a saddle, the stable and the unstable manifolds of $\mathbf{e}_+$ intersect L_+ at the points $\mathbf{q}_+^s = \mathbf{p} - \lambda_1^{-1}\dot{\mathbf{p}}$ and $\mathbf{q}_+^u = \mathbf{p} + |\lambda_2|^{-1}\dot{\mathbf{p}}$, respectively, see Corollary 4.4.17. Similarly the stable and the unstable manifolds of $\mathbf{e}_-$ intersect L_- at the points $\mathbf{q}_-^s = -\mathbf{p}+\lambda_1^{-1}\dot{\mathbf{p}}$ and $\mathbf{q}_-^u = -\mathbf{p} - |\lambda_2|^{-1}\dot{\mathbf{p}}$, see Figure 5.4.

Let γ_+^{s-} be the stable separatrix of $\mathbf{e}_+$ through the point $\mathbf{q}_+^s$ and let γ_+^{u-} be the unstable separatrix of $\mathbf{e}_+$ through the point $\mathbf{q}_+^u$. Similarly, let γ_-^{s+} and γ_-^{u+} be the stable and unstable manifolds of $\mathbf{e}_-$ through the points $\mathbf{q}_-^s$ and $\mathbf{q}_-^u$, respectively.

Suppose that $|\lambda_2|^{-1}$ is a zero of the Lamerey map g. Thus $\pi_{+-}^{B}(|\lambda_2|^{-1}) = \lambda_1^{-1}$ and therefore the orbit γ_+^{u-} intersects L_- at the point of coordinate λ_1^{-1}, i.e., $\mathbf{q}_-^s$. Analogously, the orbit γ_-^{u+} intersects L_+ at the point of coordinate λ_1^{-1}, i.e., $\mathbf{q}_+^s$. Therefore, $\gamma_+^{u-} = \gamma_-^{s+}$, $\gamma_-^{u+} = \gamma_+^{s-}$ and the singular points $\mathbf{e}_+$ and $\mathbf{e}_-$ form a heteroclinic cycle Δ, see Figure 5.4.

Conversely, if Δ is a heteroclinic cycle connecting the singular points $\mathbf{e}_+$ and $\mathbf{e}_-$, then $\pi_{+-}^{B}(\||\lambda_2|^{-1}) = \lambda_1^{-1}$, see Figure 5.4. □

Lemma 5.2.11. *Consider a proper fundamental system with fundamental parameters $D > 0$, $T < 0$, $d < 0$ and $t > 0$, and let the eigenvalues of the matrix A be $\lambda_1 > 0 > \lambda_2$.*

(a) *The parameters (t, d) belong to $C_4^1|_{D,T}$ if and only if $\pi_{+-}^{B}(|\lambda_2|^{-1}) < \lambda_1^{-1}$ or $|\lambda_2|^{-1}$ does not lie in the domain of π_{+-}^{B}.*

(b) *The parameters (t, d) belong to $\mathcal{H}_e\mathcal{L}_{D,T}$ if and only if $\pi_{+-}^{B}(|\lambda_2|^{-1}) = \lambda_1$.*

(c) *The parameters (t, d) belong to $C_4^2|_{D,T}$ if and only if $\pi_{+-}^{B}(|\lambda_2|^{-1}) > \lambda_1$.*

Proof. Consider the function $f(t) = k(t)(t - k(t))$ defined in (5.6) where $k(t)$ is implicitly defined in (5.5). Since $t - k(t) < 0$ and $k(t) > 0$, it follows that

$$k(t) = \frac{t + \sqrt{t^2 - 4f(t)}}{2}.$$

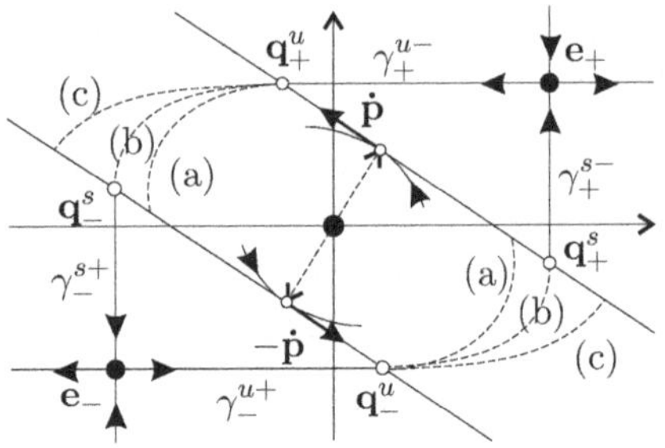

Figure 5.4: Different possibilities for the stable and the unstable manifolds of the singular points $\mathbf{e}_+$ and $\mathbf{e}_-$. In the case (b) these orbits coincide, forming a heteroclinic cycle Δ connecting these singular points.

On the other hand, since $d = \lambda_1(t - \lambda_1)$, one has $k(t) < \lambda_1$ if and only if $d < f(t)$.

(a) Suppose that $(t, d) \in \mathcal{C}_4^1|_{D,T}$. In this case $t > 0$ and $d < f(t)$, which implies that $k(t) < \lambda_1$. Since h is a decreasing function with respect to the first coordinate, see (5.4), it follows that $h(\lambda_1, t) < h(k(t), t) = 0$. Therefore, if π_{+-}^B is defined at $|\lambda_2|^{-1}$, then $\pi_{+-}^B(|\lambda_2|^{-1}) < \lambda_1^{-1}$.

Conversely, if $|\lambda_2|^{-1}$ belongs to the domain of π_{+-}^B and $\pi_{+-}^B(|\lambda_2|^{-1}) < \lambda_1^{-1}$, then $h(\lambda_1, t) < h(k(t), t) = 0$. Therefore, $k(t) < \lambda_1$ and $d < f(t)$, which proves that $(t, d) \in \mathcal{C}_4^1|_{D,T}$. Suppose now that $|\lambda_2|^{-1}$ does not lie in the domain of π_{+-}^B. From Lemma 5.2.4(a) it follows that $0 < |\lambda_2|^{-1} < a_0$, i.e., $-a_0^{-1} > -|\lambda_2| = \lambda_2$, which implies that $t < \lambda_1 - a_0^{-1}$. Since $(k(t), t) \in S$, where S is the domain of h, see (5.3), it follows that $t \geq k(t) - a_0^{-1}$ and consequently $\lambda_1 > k(t)$, $d < f(t)$ and $(t, d) \in \mathcal{C}_4^1|_{D,T}$.

(b) Suppose that $(t, d) \in \mathcal{H}_e\mathcal{L}|_{D,T}$, which implies that $t > 0$ and $d = f(t)$. From this we obtain that $k(t) = \lambda_1$ and $k(t) - t = |\lambda_2|$. Substituting these values in (5.5) it follows that $\pi_{+-}^B(|\lambda_2|^{-1}) = \lambda_1^{-1}$. The converse statement follows in the same way.

(c) Suppose that $(t, d) \in \mathcal{C}_4^2|_{D,T}$, which implies that $t > 0$ and $d > f(t)$. From this we obtain that $\lambda_1 < k(t)$. Since h is a decreasing function with respect to the first coordinate, we get that $h(\lambda_1, t) > h(k(t), t) = 0$, or, equivalently, $\pi_{+-}^B((\lambda_1 - t)^{-1}) > \lambda_1^{-1}$. Therefore, $\pi_{+-}^B(|\lambda_2|^{-1}) > \lambda_1^{-1}$. □

Proposition 5.2.12. *Consider a proper fundamental system with parameters $D > 0$, $T < 0$, $d < 0$ and $t > 0$, so the eigenvalues of the matrix A are real and satisfy that $\lambda_1 > 0 > \lambda_2$.*

(a) *If $(t, d) \in \mathcal{C}_4^1|_{D,T}$, then the system has no Jordan curves formed by solutions.*

(b) *If $(t, d) \in \mathcal{H}_e\mathcal{L}|_{D,T}$, then the system has exactly one Jordan curve Δ formed by solutions. Moreover, Δ is a heteroclinic cycle connecting the singular* *points* $\mathbf{e}_+$ *and* $\mathbf{e}_-$.

(c) *If* $(t,d) \in \mathcal{C}_4^2\big|_{D,T}$, *then the system has exactly one Jordan curve* Γ *formed by solutions. Moreover,* Γ *is a asymptotic unstable hyperbolic limit cycle such that the origin* $\mathbf{0}$ *is contained in* Σ_Γ.

Proof. Note that the Lamerey map $g(a)$ (see Lemma 5.2.10) is not only defined when $(t,d) \in \mathcal{C}_4^1\big|_{D,T}$, see Lemma 5.2.11(a). In this case $|\lambda_2|^{-1} \leq a_0$. Otherwise, when $g(a)$ is defined, it follows that

$$g'(a) = \left.\frac{d\pi_{++}^{-A}}{da}\right|_a - \left.\frac{d\pi_{+-}^{B}}{da}\right|_a < 0$$

in $(a_0, |\lambda_2|^{-1})$, see Lemmas 5.2.4 and 5.2.9(c).

(a) Suppose that $(t,d) \in \mathcal{C}_4^1\big|_{D,T}$ and $|\lambda_2|^{-1} \leq a_0$. Hence, by Lemma 5.2.10(a), there are no Jordan curves formed by solutions.

Suppose now that $|\lambda_2|^{-1} > a_0$. In this case $g(a)$ is defined and satisfies that $g'(a) < 0$ and $g(|\lambda_2|^{-1}) > 0$, see Lemma 5.2.11(a). Therefore, the map g is positive in its domain, i.e., the system has no Jordan curves formed by solutions.

(b) Suppose that $(t,d) \in \mathcal{H}_e\mathcal{L}|_{D,T}$. By Lemma 5.2.11(b), $g(|\lambda_2|^{-1}) = 0$. Hence, the system has a heteroclinic cycle connectingo the singular points $\mathbf{e}_+$ and $\mathbf{e}_-$. Moreover, since $g'(a) < 0$, the function g has no zeros left in its domain $[a_0, |\lambda_2|^{-1})$. This implies that there is no other Jordan curve formed by solutions.

(c) Suppose that $(t,d) \in \mathcal{C}_4^2\big|_{D,T}$. By Lemma 5.2.11(c), $g(|\lambda_2|^{-1}) < 0$. Moreover, from Lemmas 5.2.4 and 5.2.9(a) it follows that $g(a_0) = \pi_{++}^{-A}(a_0) - \pi_{+-}^{B}(a_0) = \pi_{++}^{-A}(a_0) > 0$. Therefore, since $g' < 0$ the function g has a unique zero a^* in its domain $[a_0, |\lambda_2|^{-1})$, which implies that there exists a unique Jordan curve formed by solutions and it is a limit cycle, see Lemma 5.2.10(b.2). □

In the next result we study the non-proper fundamental systems in $\mathcal{C}_4$.

Proposition 5.2.13. *Every non-proper fundamental system with parameters* $D > 0$, $T < 0$ *and* $(t,d) \in \mathcal{C}_4$ *has no Jordan curves formed by solutions.*

Proof. Let Γ be a Jordan curve formed by solutions, i.e., Γ is either a periodic orbit, or a heteroclinic cycle, or a homoclinic cycle, see Lemma 3.12.1. By Proposition 5.2.3(b), Γ intersects the straight lines L_+ and L_-. Thus, we can define a Poincaré map from L_+ to L_+ in a neighbourhood of Γ, which contradicts the fact that the system is non-proper. □

Suppose that $(t^*, d^*) \in W|_{D,T} \cap \mathcal{H}_e\mathcal{L}|_{D,T}$, and hence there exist proper and non-proper fundamental systems with the same parameters (t^*, d^*), i.e., for the same parameters (t^*, d^*) we have different phase portraits. In the next result we prove that this situation is not possible.

Proposition 5.2.14. *Suppose that* $D > 0$ *and* $T < 0$. *Then* $W|_{D,T} \cap \mathcal{H}_e\mathcal{L}|_{D,T} = \varnothing$.

Proof. Suppose that $T^2 - 4D < 0$. In this case $W_1|_{D,T} = W_2|_{D,T} = \varnothing$, and the statement is straightforward.

Suppose that $T^2 - 4D \geq 0$. In this case the set $\mathcal{H}_e\mathcal{L}|_{D,T}$ is the graph of the function $f(t) = k(t)(t - k(t))$ with $t > 0$, see (5.7) and (5.6). We recall that $0 < t < k(t) \leq t + a_0^{-1} < t - \Lambda_1$, where a_0 is the lower bound of the interval of definition of the Poincaré map π^B_{+-}, which depends on the parameters $D > 0$ and $T < 0$.

Since $\lim_{t\searrow 0} f(t) = \lim_{t\searrow 0} -k(t)^2 > \lim_{t\searrow 0} -(t - \Lambda_1)^2 = -\Lambda_1^2$ and the straight line $W_1|_{D,T}$ intersects the line $t = 0$ at the point $(0, -\Lambda_1^2)$, see Figure 5.1, it is sufficient to prove that $\mathcal{H}_e\mathcal{L}|_{D,T}$ and $W_1|_{D,T}$ do not intersect. Suppose that there exists a point $(t^*, d^*) \in W_1|_{D,T} \cap \mathcal{H}_e\mathcal{L}|_{D,T}$. Hence, since $(t^*, d^*) \in \mathcal{H}_e\mathcal{L}|_{D,T}$, we have that $d^* = f(t^*) = k(t^*)(t^* - k(t^*))$, with $k(t^*) > 0$ and $t - k(t^*) < 0$. Moreover, since $(t^*, d^*) \in W_1|_{D,T}$, we have that $d^* = \Lambda_1(t^* - \Lambda_1)$ with $\Lambda_1 < 0$ and $t^* - \Lambda_1 > 0$. Therefore, $k(t^*) = t^* - \Lambda_1$, in contradiction with $k(t^*) < t^* - \Lambda_1$. □

5.2.5 Phase portraits

Let

$$\dot{\mathbf{x}} = \mathbf{f}_{\mathbb{D}}(\mathbf{x}) \tag{5.8}$$

be the Poincaré compactification of a fundamental system with fundamental parameters (D, T, d, t), where $D > 0$ and $T < 0$. In this section we describe the different phase portraits in the Poincaré disc $\mathbb{D}$ of the compactified system (5.8) when the parameters (t, d) vary in the plane $\Pi_{D,T}$.

We note that the orbits in $\partial\mathbb{D}$ are separatrices of the system (5.8). Moreover, if $\partial\mathbb{D}$ contains no singular points, then the system has a periodic orbit on $\partial\mathbb{D}$. We denote by ∞ this periodic orbit and by Σ_∞ the interior of the Poincaré disc.

Proposition 5.2.15. *Consider the Poincaré compactification of a fundamental system with parameters $D > 0$, $T < 0$ and $(t, d) \in \mathcal{H}_\infty \cup \mathcal{C}_2^1$. Then:*

(a) *The separatrices of the system are:*

(a.1) *the hyperbolic singular point at the origin $\mathbf{0}$, which is asymptotically stable, and*

(a.2) *the limit cycle ∞ at infinity.*

(b) *The canonical region is $\Sigma_\infty \setminus \mathbf{0} = W^u(\infty) \cap W^s(\mathbf{0})$.*

(c) *The phase portrait of the Poincaré compactification is topologically equivalent to its correspondent in Figure 5.6.*

Proof. (a) From Propositions 5.2.1(a) and 5.2.2(a) we obtain that the origin is the unique singular point in the Poincaré disc $\mathbb{D}$ and it is an asymptotically stable hyperbolic singular point. From Proposition 5.2.3(a) it follows that ∞ is the unique Jordan curve formed by solutions. Hence, ∞ is a limit cycle and so $\mathbf{0}$ and ∞ are the separatrices of the system.

(b) Let γ be an orbit in $\Sigma_\infty \setminus \mathbf{0}$. Since $\mathbb{D}$ is a compact manifold, the α- and ω-limit sets of γ, i.e., $\alpha(\gamma)$ and $\omega(\gamma)$, are contained in $\mathbb{D}$. By applying the Poincaré–Bendixson Theorem we conclude that $\alpha(\gamma) = \infty$ and $\omega(\gamma) = \mathbf{0}$. Consequently, $\Sigma_\infty \setminus \mathbf{0} = W^u(\infty) \cap W^s(\mathbf{0})$. Notice that this also proves that there are not additional separatrices in the system.

Statement (c) follows from Theorem 2.6.9. □

Proposition 5.2.16. *Consider the Poincaré compactification of a fundamental system with parameters $D > 0$, $T < 0$ and $(t, d) \in \mathcal{C}_1^2$. Then:*

(a) *The separatrices of the system are:*

(a.1) *the hyperbolic singular point at the origin* $\mathbf{0}$, *which is asymptotically stable, and*

(a.2) *the limit cycle ∞ at infinity and a hyperbolic limit cycle $\Gamma \subset \Sigma_\infty$, which is asymptotically unstable.*

(b) *The canonical regions are $\Sigma_\infty \setminus \mathrm{Cl}(\Sigma_\Gamma) = W^u(\Gamma) \cap W^s(\infty)$ and $\Sigma_\Gamma \setminus \mathbf{0} = W^u(\Gamma) \cap W^s(\mathbf{0})$.*

(c) *The phase portrait of the Poincaré compactification is topologically equivalent to its correspondent in Figure* 5.6.

Proof. (a) Note that the existence of those separatrices is a consequence of Propositions 5.2.1(a), 5.2.2(a) and 5.2.6.

(b) Let γ be an orbit in $\Sigma_\Gamma \setminus \mathbf{0}$. Since $\mathrm{Cl}(\Sigma_\Gamma)$ is an invariant compact set, the α- and ω-limit sets of γ are contained in $\mathrm{Cl}(\Sigma_\Gamma)$. The origin is the unique singular point of the system and there are no Jordan curves formed by solutions contained in Σ_Γ, see Propositions 5.2.1(a) and 5.2.6. Hence, the Poincaré–Bendixson Theorem implies that $\alpha(\gamma)$ and $\omega(\gamma)$ are the origin and the limit cycle Γ, respectively. Therefore, $\Sigma_\Gamma \setminus \mathbf{0} = W^u(\Gamma) \cap W^s(\mathbf{0})$.

Similar arguments can be applied to the invariant compact set $\mathbb{D} \setminus \Sigma_\Gamma = \mathrm{Cl}(\Sigma_\infty \setminus \mathrm{Cl}(\Sigma_\Gamma))$. Hence, we obtain that $\Sigma_\infty \setminus \mathrm{Cl}(\Sigma_\Gamma) = W^u(\Gamma) \cap W^s(\infty)$. From this we also conclude that the system has not additional separatrices.

Statement (c) follows from Theorem 2.6.9. □

Proposition 5.2.17. *Consider the Poincaré compactification of a fundamental system with parameters $D > 0$, $T < 0$, $(t, d) \in \mathcal{SN}_\infty$ and $t > 0$. Then:*

(a) *The separatrices of the system are:*

(a.1) *the hyperbolic singular point at the origin* $\mathbf{0}$, *which is asymptotically stable, and the two saddle-nodes at infinity,* $\mathbf{x}_+$ *and* $\mathbf{x}_-$;

(a.2) *the hyperbolic limit cycle Γ, which is asymptotically unstable;*

(a.3) *the center manifolds of* $\mathbf{x}_+$ *and* $\mathbf{x}_-$, *which are contained in $\partial\mathbb{D}$, and the stable manifolds of* $\mathbf{x}_+$ *and* $\mathbf{x}_-$.

(b) *The canonical regions are* $\Sigma_\infty^{0+} = W^u(\Gamma) \cap W^s(\mathbf{x}_+)$, $\Sigma_\infty^{0-} = W^u(\Gamma) \cap W^s(\mathbf{x}_-)$, *and* $\Sigma_\Gamma \setminus \mathbf{0} = W^u(\Gamma) \cap W^s(\mathbf{0})$.

(c) *The phase portrait of the Poincaré compactification is topologically equivalent to its correspondent in Figure* 5.6.

Proof. (a) The existence of the listed separatrices is a consequence of Propositions 5.2.1(a), 5.2.2(c) and 5.2.6.

(b) For the behaviour of the flow in $\Sigma_\Gamma \setminus \mathbf{0}$ see the proof of Proposition 5.2.16(b).

Since the system has no Jordan curves formed by solutions in $\Sigma_\infty \setminus \mathrm{Cl}(\Sigma_\Gamma)$, see Proposition 5.2.6, and the unique singular points in $\mathbb{D} \setminus \Sigma_\Gamma$ are $\mathbf{x}_+$ and $\mathbf{x}_-$, we conclude that there are no Jordan curves formed by solutions in $\mathbb{D} \setminus \Sigma_\Gamma$. By applying the Poincaré–Bendixson Theorem to the invariant set $\mathbb{D} \setminus \Sigma_\Gamma$ it follows that $\Sigma_\infty \setminus \mathrm{Cl}(\Sigma_\Gamma) \subset W^u(\Gamma)$.

Let γ_+^s be the stable separatrix of $\mathbf{x}_+$ contained in Σ_∞, i.e., $\omega(\gamma_+^s) = \mathbf{x}_+$. Since $\gamma_+^s \subset \Sigma_\infty \setminus \mathrm{Cl}(\Sigma_\Gamma)$, we have $\alpha(\gamma_+^s) = \Gamma$. Analogously, if γ_-^s is the stable separatrix of $\mathbf{x}_-$, then $\omega(\gamma_-) = \mathbf{x}_-$ and $\alpha(\gamma_-) = \Gamma$. Therefore, γ_+^s and γ_-^s split $\Sigma_\infty \setminus \mathrm{Cl}(\Sigma_\Gamma)$ into two open connected and invariant regions Σ_∞^{0+} and Σ_∞^{0-}. Let Σ_∞^{0+} be the region containing the parabolic sector of $\mathbf{x}_+$ and let Σ_∞^{0-} be the region containing the hyperbolic sector of $\mathbf{x}_+$. By the symmetry of the flow with respect to the origin Σ_∞^{0-} contains the parabolic sector of $\mathbf{x}_-$ and Σ_∞^{0+} contains the hyperbolic sector of $\mathbf{x}_-$. By Proposition 5.2.2(b), the parabolic sector of $\mathbf{x}_+$ and $\mathbf{x}_-$ is stable. Hence, we conclude that $\Sigma_\infty^{0+} \subset W^s(\mathbf{x}_+)$ and $\Sigma_\infty^{0-} \subset W^s(\mathbf{x}_-)$. From this we also obtain that the system has not additional separatrices.

Statement (c) follows from Theorem 2.6.9. □

Proposition 5.2.18. *Consider the Poincaré compactification of a fundamental system with parameters* $D > 0$, $T < 0$, $(t, d) \in \mathcal{C}_1^1 \cup \mathcal{N}$, *and* $t > 0$. *Then:*

(a) *The separatrices of the system are:*

(a.1) *the hyperbolic singular point at the origin* $\mathbf{0}$, *which is asymptotically stable; the asymptotically stable nodes* $\mathbf{x}_+$, $\mathbf{x}_- \in \partial\mathbb{D}$, *and the saddle points* $\mathbf{y}_+$, $\mathbf{y}_- \in \partial\mathbb{D}$;

(a.2) *the hyperbolic limit cycle* Γ, *which is asymptotically unstable;*

(a.3) *the stable manifolds of* $\mathbf{y}_+$ *and* $\mathbf{y}_-$, *and the unstable manifolds of* $\mathbf{y}_+$ *and* $\mathbf{y}_-$, *which are contained in* $\partial\mathbb{D}$.

(b) *The canonical regions are* $\Sigma_\infty^{0+} = W^u(\Gamma) \cap W^s(\mathbf{x}_+)$, $\Sigma_\infty^{0-} = W^u(\Gamma) \cap W^s(\mathbf{x}_-)$, *and* $\Sigma_\Gamma \setminus \mathbf{0} = W^s(\mathbf{0}) \cap W^u(\Gamma)$.

(c) *The phase portrait of the Poincaré compactification is topologically equivalent to its correspondent in Figure* 5.6.

Proof. (a) The existence of the listed separatrices is a consequence of Propositions 5.2.1(a), 5.2.2(d) and 5.2.6.

(b) For the behaviour of the flow in $\Sigma_\Gamma \setminus \mathbf{0}$ see the proof of Proposition 5.2.16(b).

By Proposition 5.2.6, there are no Jordan curves formed by solutions in $\Sigma_\infty \setminus \mathrm{Cl}(\Sigma_\Gamma)$. Moreover, the unique singular points in $\mathbb{D} \setminus \Sigma_\Gamma$ are the nodes $\mathbf{x}_+$ and $\mathbf{x}_-$ and the saddles $\mathbf{y}_+$ and $\mathbf{y}_-$. Thus, any separatrix cycle with singular points in $\partial\mathbb{D}$ contains either $\mathbf{x}_+$ or $\mathbf{x}_-$, which contradicts the fact that these points are nodes. Hence there are no separatrix cycles in $\mathbb{D} \setminus \Sigma_\Gamma$. By the Poincaré–Bendixson Theorem, the α- and ω-limit sets of any orbit in $\mathbb{D} \setminus \Sigma_\Gamma$ are the singular points $\mathbf{x}_+, \mathbf{x}_-, \mathbf{y}_+, \mathbf{y}_-$ and the limit cycle Γ which is asymptotically unstable. Therefore, $\Sigma_\infty \setminus \mathrm{Cl}(\Sigma_\Gamma) \subset W^u(\Gamma)$ and the stable separatrix γ_+^s of the saddle $\mathbf{y}_+$, i.e., $\omega(\gamma_+^s) = \mathbf{y}_+$, satisfies that $\alpha(\gamma_+^s) = \Gamma$. Analogously, if γ_-^s is the stable separatrix of $\mathbf{y}_-$, then $\alpha(\gamma_-^s) = \Gamma$ and $\omega(\gamma_-^s) = \mathbf{y}_-$. Thus, γ_+^s and γ_-^s split $\Sigma_\infty \setminus \mathrm{Cl}(\Sigma_\Gamma)$ into two open, connected and invariant regions, denoted by Σ_∞^{0+} and Σ_∞^{0-}. Let Σ_∞^{0+} be the region containing the point $\mathbf{x}_+$ in its closure, and let Σ_∞^{0-} be the region containing the point $\mathbf{x}_-$ in its clousure.

By the Poincaré–Bendixson Theorem, the α- and ω-limit sets in $\mathrm{Cl}(\Sigma_\infty^{0+})$ are the saddle points $\mathbf{y}_+$ and $\mathbf{y}_-$, the asymtotically stable node $\mathbf{x}_+$ and the asymptotically unstable limit cycle. Therefore, $\Sigma_\infty^{0+} \subset W^u(\Gamma) \cap W^s(\mathbf{x}_+)$. By using the symmetry of the flow with respect to the origin we conclude that $\Sigma_\infty^{0-} \subset W^u(\Gamma) \cap W^s(\mathbf{x}_-)$. This also proves that there are no additional separatrices in $\mathbb{D}$.

Statement (c) follows from Theorem 2.6.9. □

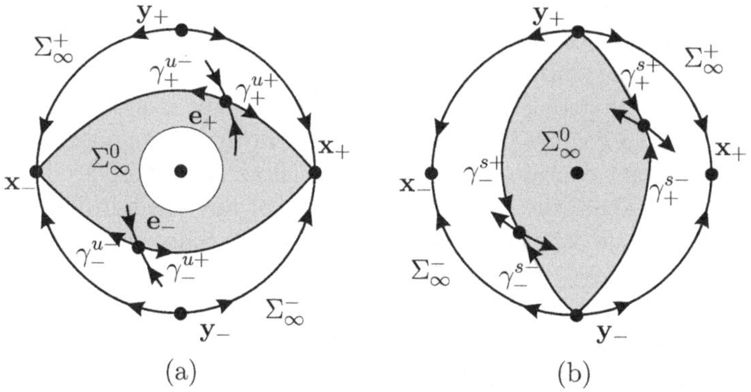

Figure 5.5: Regions Σ_∞^+, Σ_∞^0 and Σ_∞^- when $d < 0$.

Proposition 5.2.19. *Consider the Poincaré compactification of a fundamental system with parameters $D > 0$, $T < 0$, and $(t, d) \in \mathcal{C}_4^2\big|_{D,T}$. Then:*

(a) *The separatrices of the system are:*

(a.1) *the asymptotically stable singular point at the origin* $\mathbf{0}$*, the saddle points* $\mathbf{e}_+$*,* $\mathbf{e}_-$*, the asymptotically stable nodes at infinity,* $\mathbf{x}_+$ *and* $\mathbf{x}_-$*, and the asymptotically unstable nodes at infinity,* $\mathbf{y}_+$ *and* $\mathbf{y}_-$*;*

(a.2) *the asymptotically unstable hyperbolic limit cycle* Γ*;*

(a.3) *the stable and the unstable separatrices of* $\mathbf{e}_+$ *and* $\mathbf{e}_-$*, and the orbits contained in* $\partial\mathbb{D}$*.*

(b) *The canonical regions are:* $\Sigma_\infty^{++} \subset W^u(\mathbf{y}_+) \cap W^s(\mathbf{x}_+)$, $\Sigma_\infty^{+-} \subset W^u(\mathbf{y}_+) \cap W^s(\mathbf{x}_-)$, $\Sigma_\infty^{0-} = W^u(\Gamma) \cap W^s(\mathbf{x}_-)$, $\Sigma_\infty^{--} \subset W^u(\mathbf{y}_-) \cap W^s(\mathbf{x}_-)$, $\Sigma_\infty^{-+} \subset W^u(\mathbf{y}_-) \cap W^s(\mathbf{x}_+)$, $\Sigma_\infty^{0+} = W^u(\Gamma) \cap W^s(\mathbf{x}_+)$, *and* $\Sigma_\Gamma \setminus \mathbf{0} = W^u(\Gamma) \cap W^s(\mathbf{0})$.

(c) *The phase portrait of the Poincaré compactification is topologically equivalent to its correspondent in Figure* 5.6.

Proof. (a) Note that the orbits contained in $\partial\mathbb{D}$ are separatrices. The existence of the others separatrices is a consequence of Propositions 5.2.1(b), 5.2.2(d) and 5.2.12(c).

(b) For the behaviour of the flow in the region $\Sigma_\Gamma \setminus \mathbf{0}$ see the proof of Proposition 5.2.16(b).

Since $\mathbf{x}_+$, $\mathbf{x}_-$, $\mathbf{y}_+$ and $\mathbf{y}_-$ are nodes, it is easy to conclude that there are no separatrix cycles with singular points in $\partial\mathbb{D}$. Hence, there exists exactly one Jordan curve formed by solutions and it is contained in Σ_∞, see Proposition 5.2.12(c). Therefore, the α- and ω-limit sets of any orbit in $\mathbb{D} \setminus \Sigma_\Gamma$ are the singular points $\mathbf{e}_+$, $\mathbf{e}_-$, $\mathbf{x}_+$, $\mathbf{x}_-$, $\mathbf{y}_+$, $\mathbf{y}_-$, and the limit cycle Γ.

Let γ_+^{u+} and γ_+^{u-} be the unstable separatrices of the saddle point $\mathbf{e}_+$. Since $\mathbf{y}_+$, $\mathbf{y}_-$, and Γ are asymptotically unstable, it follows that $\omega(\gamma_+^{u+})$, $\omega(\gamma_+^{u-}) \in \{\mathbf{x}_+, \mathbf{x}_-\}$. Suppose that both separatrices have the same ω-limit set, for instance $\mathbf{x}_+$. Then $W^u(\mathbf{e}_+) \cup \mathbf{x}_+$ splits $\mathbb{D}$ into two open and invariant regions. Let M_+ be the region which does not contain $\mathbf{e}_-$. It is easy to check that there are no separatrix cycles in $\mathrm{Cl}(M_+)$. Thus, the α- and ω-limit sets of any orbit in $\mathrm{Cl}(M_+)$ are the singular points $\mathbf{e}_+$ and $\mathbf{x}_+$. Let γ be an orbit in M_+. Since $\mathbf{x}_+$ is asymptotically stable it follows that $\alpha(\gamma) = \mathbf{e}_+$. This implies that γ is an unstable separatrix of $\mathbf{e}_+$, which contradicts that $\gamma \subset M_+$. Therefore, $\omega(\gamma_+^{u+}) \neq \omega(\gamma_+^{u-})$.

Without loss of generality we assume that $\omega(\gamma_+^{u+}) = \mathbf{x}_+$ and $\omega(\gamma_+^{u-}) = \mathbf{x}_-$. By the symmetry of the flow with respect to the origin, $\omega(\gamma_-^{u+}) = \mathbf{x}_+$ and $\omega(\gamma_-^{u-}) = \mathbf{x}_-$. Then $W^u(\mathbf{e}_+)$ and $W^u(\mathbf{e}_-)$ split $\Sigma_\infty \setminus \mathrm{Cl}(\Sigma_\Gamma)$ into three open, connected and invariant regions, denoted by Σ_∞^+, Σ_∞^0, and Σ_∞^-, see Figure 5.5(a). We study the behaviour of the flow in each of these regions.

The α- and ω-limit sets of any orbit in $\mathrm{Cl}(\Sigma_\infty^+)$ are the singular points $\mathbf{x}_+$, $\mathbf{y}_+$, $\mathbf{x}_-$, and $\mathbf{e}_+$. Let γ be an orbit in Σ_∞^+. Since $\mathbf{x}_+$ and $\mathbf{x}_-$ are asymtotically stable nodes and $\mathbf{e}_+$ has the unstable separatrices contained in $\partial\Sigma_\infty^+$, it is clear that $\alpha(\gamma) = \mathbf{y}_+$. Therefore, $\Sigma_\infty^+ \subset W^u(\mathbf{y}_+)$. By the symmetry of the flow with respect to the origin, we conclude that $\Sigma_\infty^- \subset W^u(\mathbf{y}_-)$.

On the other hand, the α- and ω-limit sets in $\mathrm{Cl}(\Sigma^0_\infty)$ are the nodes $\mathbf{x}_+$ and $\mathbf{x}_-$, the saddle points $\mathbf{e}_+$ and $\mathbf{e}_-$, and the asymptotically unstable limit cyle Γ. From this we obtain that $\Sigma^0_\infty \subset W^u(\Gamma)$.

Since $W^u(\mathbf{e}_+)$ is a common boundary for the regions Σ^+_∞ and Σ^0_∞, a stable separatrix of $\mathbf{e}_+$ denoted by γ^{s+}_+ is contained in Σ^+_∞, and the other, denoted by γ^{s-}_+, is contained in Σ^0_∞, see Figure 5.5(a). By the symmetry of the flow with respect to the origin we obtain that $\gamma^{s-}_- \subset \Sigma^-_\infty$ and $\gamma^{s+}_- \subset \Sigma^0_\infty$. Arguments similar to those used in the study of Σ^+_∞, Σ^0_∞ and Σ^-_∞ show that $\alpha(\gamma^{s+}_+) = \mathbf{y}_+$, $\alpha(\gamma^{s-}_+) = \Gamma$, $\alpha(\gamma^{s+}_-) = \Gamma$, and $\alpha(\gamma^{s-}_-) = \mathbf{y}_-$. Hence, $W^s(\mathbf{e}_+)$ and $W^s(\mathbf{e}_-)$ split Σ^+_∞, Σ^0_∞, and Σ^-_∞ in the open, connected and invariant regions Σ^{++}_∞, Σ^{+-}_∞, Σ^{0+}_∞, Σ^{0-}_∞, Σ^{--}_∞, and Σ^{-+}_∞, see Figure 5.5(a).

Let Σ^{++}_∞ be the region bounded by the orbits γ^{s+}_+ and γ^{u+}_+. From $\Sigma^{++}_\infty \subset \Sigma^+_\infty$ we conclude that $\Sigma^{++}_\infty \subset W^u(\mathbf{y}_+)$. Moreover, since the α- and ω-limit sets in $\mathrm{Cl}(\Sigma^{++}_\infty)$ are the saddle $\mathbf{e}_+$, the asymptotically stable node $\mathbf{x}_+$, and the asymptotically unstable node $\mathbf{y}_+$, we obtain that $\Sigma^{++}_\infty \subset W^s(\mathbf{x}_+)$. Therefore, $\Sigma^{++}_\infty \subset W^u(\mathbf{y}_+) \cap W^s(\mathbf{x}_+)$. The remainder inclusions follow in a similar way.

Statement (c) follows from Theorem 2.6.9. □

Proposition 5.2.20. *Consider the Poincaré compactification of a fundamental system with parameters $D > 0$, $T < 0$, and $(t, d) \in \mathcal{H}_e\mathcal{L}|_{D,T}$. Then:*

(a) *The separatrices of the system are:*

(a.1) *the asymptotically stable point at the origin $\mathbf{0}$, the saddle points $\mathbf{e}_+$, $\mathbf{e}_-$, the asymptotically stable nodes $\mathbf{x}_+$, $\mathbf{x}_- \in \partial\mathbb{D}$, and the asymptotically unstable nodes $\mathbf{y}_+$, $\mathbf{y}_- \in \partial\mathbb{D}$;*

(a.2) *the stable and the unstable separatrices of $\mathbf{e}_+$ and $\mathbf{e}_-$, which form a heteroclinic cycle Δ, and the orbits which are contained in $\partial\mathbb{D}$.*

(b) *The canonical regions are: $\Sigma^{++}_\infty \subset W^u(\mathbf{y}_+) \cap W^s(\mathbf{x}_+)$, $\Sigma^{+-}_\infty \subset W^u(\mathbf{y}_+) \cap W^s(\mathbf{x}_-)$, $\Sigma^{--}_\infty \subset W^u(\mathbf{y}_-) \cap W^s(\mathbf{x}_-)$, $\Sigma^{-+}_\infty \subset W^u(\mathbf{y}_-) \cap W^s(\mathbf{x}_+)$, and $\Sigma_\Delta \backslash \mathbf{0} = W^u(\Delta) \cap W^s(\mathbf{0})$.*

(c) *The phase portrait of the Poincaré compactification is topologically equivalent to its correspondent in Figure 5.6.*

Proof. (a) Note that the orbits contained in $\partial\mathbb{D}$ are separatrices. The existence of the remaining separatrices is a consequence of Propositions 5.2.1(b), 5.2.2(d) and 5.2.12(b).

(b) For the behaviour of the flow in the region $\Sigma_\Delta \setminus \mathbf{0}$, see the proof of Proposition 5.2.17(c).

Since $\mathbf{x}_+$, $\mathbf{x}_-$, $\mathbf{y}_+$, and $\mathbf{y}_-$ are nodes, there are no separatrix cycles with singular points contained in $\partial\mathbb{D}$. Moreover, the unique Jordan curve formed by solutions contained in $\mathbb{D}$ is the heteroclinic cycle Δ, see Proposition 5.2.12(b). Thus, by the Poincaré–Bendixson Theorem, the unique α- and ω-limit sets in $\mathbb{D} \setminus \Sigma_\Delta$ are the singular points $\mathbf{e}_+$, $\mathbf{e}_-$, $\mathbf{x}_+$, $\mathbf{x}_-$, $\mathbf{y}_+$ and $\mathbf{y}_-$, and the heteroclinic cycle Δ.

Since $\mathbf{e}_+$ and $\mathbf{e}_-$ are contained in the cycle Δ and both points have a separatrix contained in $\Sigma_\infty \setminus \mathrm{Cl}(\Sigma_\Delta)$, it is easy to check that Δ cannot be the α- or ω-limit set of any orbit in $\Sigma_\infty \setminus \mathrm{Cl}(\Sigma_\Delta)$.

Let γ_+^{s+} and γ_-^{s-} be the separatrices of the saddle points $\mathbf{e}_+$ and $\mathbf{e}_-$ which are not contained in the cycle Δ. Hence, $\gamma_+^{s+} \cup \gamma_-^{s-} \subset \Sigma_\infty \setminus \mathrm{Cl}(\Sigma_\Delta)$. We conclude that the points $\mathbf{e}_+$, $\mathbf{e}_-$, $\mathbf{x}_+$, and $\mathbf{x}_-$ cannot be the α-limit sets of γ_+^{s+} and γ_-^{s-}. Therefore, $\alpha(\gamma_+^{s+}) = \mathbf{y}_+$ and $\alpha(\gamma_-^{s-}) = \mathbf{y}_-$.

In a similar way, if γ_+^{u+} and γ_-^{u-} are the unstable separatrices of the singular points $\mathbf{e}_+$ and $\mathbf{e}_-$ which are not contained in the cycle Δ, then $\omega(\gamma_+^{u+}) = \mathbf{x}_+$ and $\omega(\gamma_-^{u-}) = \mathbf{x}_-$. Thus, the separatrices γ_+^{s+}, γ_+^{u+}, γ_-^{s-}, and γ_-^{u-} split $\Sigma_\infty \setminus \mathrm{Cl}(\Sigma_\Delta)$ into four open, connected and invariant regions denoted by Σ_∞^{++}, Σ_∞^{+-}, Σ_∞^{--}, and Σ_∞^{-+}.

Let Σ_∞^{jk} be the region which contains the points $\mathbf{y}_j$ and $\mathbf{x}_k$ on the boundary, with $j, k \in \{+, -\}$. The behaviour of the flow in Σ_∞^{jk} follows by noting that $\mathbf{y}_j$ is asymptotically unstable and $\mathbf{x}_k$ is asymptotically stable.

Statement (c) follows from Theorem 2.6.9. □

Proposition 5.2.21. *Consider the Poincaré compactification of a fundamental system with parameters $D > 0$, $T < 0$, and (t, d) contained in the parameter region $\mathcal{C}_4^1\big|_{D,T} \cup \mathcal{C}_3 \cup \{(0, d) \in \Pi_{D,T} : d < 0\}$. Then:*

(a) *The separatrices of the system are:*

(a.1) *the asymptotically stable point at the origin $\mathbf{0}$, the saddle points $\mathbf{e}_+$, $\mathbf{e}_-$, the asymptotically stable points $\mathbf{x}_+$, $\mathbf{x}_- \in \partial\mathbb{D}$, and the asymptotically unstable nodes $\mathbf{y}_+$, $\mathbf{y}_- \in \partial\mathbb{D}$;*

(a.2) *the separatrices of the saddles $\mathbf{e}_+$ and $\mathbf{e}_-$, and the orbits which are contained in $\partial\mathbb{D}$.*

(b) *The canonical regions are: $\Sigma_\infty^{++} \subset W^u(\mathbf{y}_+) \cap W^s(\mathbf{x}_+)$, $\Sigma_\infty^{+0} = W^u(\mathbf{y}_+) \cap W^s(\mathbf{0})$, $\Sigma_\infty^{+-} \subset W^u(\mathbf{y}_+) \cap W^s(\mathbf{x}_-)$, $\Sigma_\infty^{--} \subset W^u(\mathbf{y}_-) \cap W^s(\mathbf{x}_-)$, $\Sigma_\infty^{-0} = W^u(\mathbf{y}_-) \cap W^s(\mathbf{0})$, and $\Sigma_\infty^{-+} \subset W^u(\mathbf{y}_-) \cap W^s(\mathbf{x}_+)$.*

(c) *The phase portrait of the Poincaré compactification is topologically equivalent to its correspondent in Figure* 5.6.

Proof. (a) Orbits in $\partial\mathbb{D}$ are separatrices. The existence of the remaining separatrices is a consequence of Propositions 5.2.1(b), 5.2.2(d) and 5.2.3(a) when $t \leq 0$, and a consequence of Propositions 5.2.12(a) and 5.2.13 when $t > 0$.

(b) Note that there are no Jordan curves formed by solutions in Σ_∞, see Proposition 5.2.3(a) when $t \leq 0$ and Propositions 5.2.12(a) and 5.2.13 when $t > 0$. Moreover, since $\mathbf{x}_+$, $\mathbf{x}_-$, $\mathbf{y}_+$, and $\mathbf{y}_-$ are nodes, there are no separatrix cycles with singular points contained in $\partial\mathbb{D}$. Hence, from the Poincaré–Bendixson Theorem it follows that the α- or ω-limit sets of any orbit in $\mathbb{D}$ are the singular points in the statement (a.1).

Let γ_+^{s+} and γ_+^{s-} be the stable separatrices of the saddle point $\mathbf{e}_+$. Since the singular points $\mathbf{0}$, $\mathbf{x}_+$, and $\mathbf{x}_-$ are asymptotically stable, $\alpha(\gamma_+^{s+}) = \mathbf{y}_+$ and $\alpha(\gamma_+^{s-}) = \mathbf{y}_-$. From the symmetry of the flow with respect to the origin we get $\alpha(\gamma_-^{s+}) = \mathbf{y}_+$ and $\alpha(\gamma_-^{s-}) = \mathbf{y}_-$. Thus, the stable manifolds of $\mathbf{e}_+$ and $\mathbf{e}_-$ split Σ_∞ into three open, connected and invariant regions denoted by Σ_∞^+, Σ_∞^0, and Σ_∞^-. Let Σ_∞^+ be the region containing the point $\mathbf{e}_+$, Σ_∞^- the region containing the point $\mathbf{e}_-$, and Σ_∞^0 the region containing the origin $\mathbf{0}$.

The α- and ω-limit sets in Σ_∞^+ are the nodes $\mathbf{y}_+$ and $\mathbf{y}_-$, the saddle point $\mathbf{e}_+$, and the asymptotically stable node $\mathbf{x}_+$. Therefore, $\Sigma_\infty^+ \subset W^s(\mathbf{x}_+)$. In a similar way it follows that $\Sigma_\infty^0 \subset W^s(\mathbf{0})$ and $\Sigma_\infty^- \subset W^s(\mathbf{x}_-)$.

Since $W^s(\mathbf{e}_+)$ is the common boundary between the regions Σ_∞^+ and Σ_∞^0, one of the unstable separatrices of $\mathbf{e}_+$, denoted by γ_+^{u+}, is contained in Σ_∞^+ and the other, denoted by γ_+^{u-}, is contained in Σ_∞^0. From the symmetry of the flow with respect to the origin we have $\gamma_-^{u-} \subset \Sigma_\infty^-$ and $\gamma_-^{u+} \subset \Sigma_\infty^0$. Hence, the curve defined by $W^u(\mathbf{e}_+) \cup \mathbf{0} \cup W^u(\mathbf{e}_-)$ splits the regions Σ_∞^+, Σ_∞^0, and Σ_∞^- into the six open, connected, and invariant regions Σ_∞^{++}, Σ_∞^{+0}, Σ_∞^{+-}, Σ_∞^{--}, Σ_∞^{-0}, and Σ_∞^{-+}. Let Σ_∞^{jk} be the region containing the points $\mathbf{y}_j$ and $\mathbf{x}_k$ in its boundary, with $j, k \in \{+, -\}$. The statement follows by noting that $\mathbf{y}_j$ is asymptotically unstable and that $\mathbf{x}_k$ is asymptotically stable.

Statement (c) follows from Theorem 2.6.9. □

Proposition 5.2.22. *Consider the Poincaré compactification of a fundamental system with parameters $D > 0$, $T < 0$, $(t, d) \in \mathcal{C}_2^2 \cup \mathcal{N}$, and $t < 0$. Then:*

(a) *The separatrices of the system are:*

  (a.1) *the asymptotically stable point at the origin $\mathbf{0}$, the asymptotically unstable nodes $\mathbf{y}_+, \mathbf{y}_- \in \partial\mathbb{D}$, the saddle points $\mathbf{x}_+, \mathbf{x}_- \in \partial\mathbb{D}$ which have their stable manifold contained in $\partial\mathbb{D}$, and*

  (a.2) *the stable and unstable separatrices of $\mathbf{x}_+$ and $\mathbf{x}_-$.*

(b) *The canonical regions are: $\Sigma_\infty^{+0} = W^u(\mathbf{y}_+) \cap W^s(\mathbf{0})$ and $\Sigma_\infty^{-0} = W^u(\mathbf{y}_-) \cap W^s(\mathbf{0})$.*

(c) *The phase portrait of the Poincaré compactification is topologically equivalent to its correspondent in Figure* 5.6.

Proof. (a) The existence of the listed separatrices is a consequence of Propositions 5.2.1(a) and 5.2.2(d).

(b) From Proposition 5.2.3(a) it follows that there are no Jordan curves formed by solutions contained in Σ_∞. Moreover, there are no separatrix cycles with singular points in $\partial\mathbb{D}$. Otherwise, the singular point $\mathbf{y}_+$ or $\mathbf{y}_-$ would belong to the cycle, which is not possible. By the Poincaré–Bendixson Theorem, the α- and ω-limit sets of any orbit in $\mathbb{D}$ are: the asymptotically stable point at the origin, the asymptotically unstable nodes $\mathbf{y}_+$, $\mathbf{y}_-$, and the saddle points $\mathbf{x}_+$ and $\mathbf{x}_-$. Therefore, $\Sigma_\infty \subset W^s(\mathbf{0})$. We recall that the stable separatrices of $\mathbf{x}_+$ and $\mathbf{x}_-$ are contained in $\partial\mathbb{D}$.

Let γ_+^u be the unstable separatrix of $\mathbf{x}_+$, i.e., $\alpha(\gamma_+^u) = \mathbf{x}_+$. It is easy to conclude that $\omega(\gamma_+^u) = \mathbf{0}$. By the symmetry of the flow with respect to the origin, if γ_-^u is the unstable separatrix of $\mathbf{x}_-$, then $\alpha(\gamma_-^u) = \mathbf{x}_-$ and $\omega(\gamma_-^u) = \mathbf{0}$. Thus, $\gamma_+^u \cup \mathbf{0} \cup \gamma_-^u$ splits Σ_∞ into the two open, connected, and invariant regions Σ_∞^{+0} and Σ_∞^{-0}. Let Σ_∞^{+0} be the region with $\mathbf{y}_+$ on the boundary. The α- and ω-limit sets in $\mathrm{Cl}(\Sigma_\infty^{+0})$ are the origin, the saddle points $\mathbf{x}_+$, $\mathbf{x}_-$, and the asymptotically unstable node $\mathbf{y}_+$. Therefore, $\Sigma_\infty^{+0} \subset W^u(\mathbf{y}_+)$ and $\Sigma_\infty^{+0} = W^u(\mathbf{y}_+) \cap W^s(\mathbf{0})$. From the symmetry of the flow with respect to the origin we obtain that $\Sigma_\infty^{-0} = W^u(\mathbf{y}_-) \cap W^s(\mathbf{0})$.

Statement (c) follows from the Theorem 2.6.9. □

Proposition 5.2.23. *Consider the Poincaré compactification of a fundamental system with parameters $D > 0$, $T < 0$, $(t, d) \in \mathcal{SN}_\infty \setminus \{\mathcal{VB}_1|_{D,T} \cup \mathcal{VB}_2|_{D,T}\}$, and $t < 0$. Then:*

(a) *The separatrices of the system are:*

(a.1) *the asymptotically stable point at the origin $\mathbf{0}$, and the saddle-nodes at infinity $\mathbf{x}_+$ and $\mathbf{x}_-$;*

(a.2) *the center manifold of $\mathbf{x}_+$ and $\mathbf{x}_-$, which is contained in $\partial\mathbb{D}$, and the unstable hyperbolic manifold of $\mathbf{x}_+$ and $\mathbf{x}_-$.*

(b) *The canonical regions are: $\Sigma_\infty^{+0} \subset W^u(\mathbf{x}_+) \cap W^s(\mathbf{0})$ and $\Sigma_\infty^{-0} \subset W^u(\mathbf{x}_-) \cap W^s(\mathbf{0})$.*

(c) *The phase portrait of the Poincaré compactification is topologically equivalent to its correspondent in Figure* 5.6.

Proof. (a) The existence of these separatrices is a consequence of Propositions 5.2.1(a) and 5.2.2(c).

(b) By using arguments similar to those in the proof of Proposition 5.2.22, we conclude that there are no limit cycles and separatrix cycles in $\mathbb{D}$. Thus, the possible α- and ω-limit sets in $\mathbb{D}$ are the asymptotically stable node $\mathbf{0}$ and the saddle-nodes $\mathbf{x}_+$ and $\mathbf{x}_-$.

Let γ_+^u be the unstable separatrix of $\mathbf{x}_+$, i.e., $\alpha(\gamma_+^u) = \mathbf{x}_+$. Then $\omega(\gamma_+^u) = \mathbf{0}$. By the symmetry of the flow with respect to the origin, if γ_-^u is the unstable separatrix of $\mathbf{x}_-$, then $\alpha(\gamma_-^u) = \mathbf{x}_-$ and $\omega(\gamma_-^u) = \mathbf{0}$. Thus, the curve $\gamma_+^u \cup \mathbf{0} \cup \gamma_-^u$ splits Σ_∞ into the two open, connected, and invariant regions Σ_∞^{+0} and Σ_∞^{-0}. Let Σ_∞^{+0} be the region containing the parabolic sector of $\mathbf{x}_+$, and let Σ_∞^{-0} be the region containing the hyperbolic sector of $\mathbf{x}_+$. Since the possible α- and ω-limit sets in $\mathrm{Cl}(\Sigma_\infty^{+0})$ are $\mathbf{0}$, $\mathbf{x}_+$, and $\mathbf{x}_-$, we obtain that $\Sigma_\infty^{+0} \subset W^u(\mathbf{x}_+) \cap W^s(\mathbf{0})$. By using again the symmetry of the flow with respect to the origin, it follows that $\Sigma_\infty^{-0} \subset W^u(\mathbf{x}_-) \cap W^s(\mathbf{0})$.

Statement (c) follows from Theorem 2.6.9. □

In the next result we deal with the fundamental systems that have non-isolated singular points. In this case we cannot apply Theorem 2.6.9. Thus in order to characterize the topological equivalence classes of these fundamental systems we build explicit homeomorphism which establish the equivalence relation.

Proposition 5.2.24. *Consider the Poincaré compactification of a fundamental system with fundamental matrices (A, B) and parameters $D > 0$, $T < 0$, and $(t, d) \in \mathcal{VB}_1|_{D,T} \cup \mathcal{VB}_2|_{D,T}$.*

(a) *If the real Jordan normal form of the matrix A is non-diagonal, then the phase portrait of the Poincaré compactification is the one described in Proposition* 5.2.23.

(b) *Suppose that the real Jordan normal form of the matrix A is diagonal.*

  (b.1) *The boundary of the Poincaré disc $\mathbb{D}$ is an unstable normally hyperbolic manifold. The origin is the unique singular point in Σ_∞ and it is global asymptotically stable, i.e., $\Sigma_\infty \setminus \mathbf{0} \subset W^s(\mathbf{0})$.*

  (b.2) *The phase portrait of the Poincaré compactification is topologically equivalent to its correspondent in Figure* 5.6.

Proof. Statement (a) follows similarly as in Proposition 5.2.23.

(b.1) The local phase portrait of $\partial\mathbb{D}$ and $\mathbf{0}$ is a consequence of Propositions 5.2.1(a) and 5.2.2(b). Moreover, since there are no Jordan curves formed by solutions in $\mathbb{D}$, see Proposition 5.2.3(a), the statement follows from the Poincaré–Bendixson Theorem.

(b.2) From statement (b.1) it follows that any orbit of the system tends to the origin when s tends to ∞. Moreover, since the matrix A is diagonal, so is the matrix B. Therefore, the origin is a diagonal node and the orbits in a neigbourhood are contained in straight lines. Hence, if we consider a sufficiently small circle c centered at the origin, then c intersects every orbit of the system at exactly one point.

Let c' be a circle centered at the origin and contained in the corresponding phase portrait in Figure 5.6. It is easy to check that c' intersects evey orbit of the system at exactly one point. Let $\mathbf{h}$ be a homeomorphism from c to c'. Using $\mathbf{h}$, it is easy to define a homeomorphism between the flows of the two systems which preserves the orientation. Hence, the two phase portraits are topologically equivalent. □

Proposition 5.2.25. *Consider the Poincaré compactification of a fundamental system with parameters $D > 0$, $T < 0$, and $(t, d) \in \mathcal{O}$. Then:*

(a) *The separatrices of the system are:*

  (a.1) *the asymptotically stable singular point at the origin $\mathbf{0}$, the singular points at infinity $\mathbf{x}_+$ and $\mathbf{x}_-$, and*

  (a.2) *the central manifold of $\mathbf{x}_+$ and $\mathbf{x}_-$ which is contained in $\partial\mathbb{D}$, forming a heteroclinic cycle at infinity Δ.*

(b) *The canonical region is $\Sigma_\infty \setminus \mathbf{0} = W^u(\Delta) \cap W^s(\mathbf{0})$.*

(c) *The phase portrait of the Poincaré compactification is topologically equivalent to its correspondent in Figure* 5.6.

Proof. (a) The existence of the separatrices is a consequence of Propositions 5.2.1(a), 5.2.2, and 5.2.3(a).

(b) Since $\mathbf{x}_+$ and $\mathbf{x}_-$ are saddle-nodes with the central manifold contained in $\partial\mathbb{D}$, it is easy to conclude the existence of the heteroclinic cycle Δ. Moreover, there are no Jordan curves formed by solutions in Σ_∞, see Proposition 5.2.3(a). Hence, since the origin is asymptotically stable, we conclude that $\Sigma_\infty \subset W^s(\mathbf{0})$. Consequently, $\Sigma_\infty \setminus \mathbf{0} \subset W^u(\Delta)$.

Statement (c) follows from Theorem 2.6.9. □

5.2.6 The bifurcation set

In this subsection we describe the bifurcations that take place in the phase portrait of the Poincaré compactification of a fundamental system when we vary the parameters (t, d) in $\Pi_{D,T}$, where $D > 0$ and $T < 0$.

Take $(t, d) \in \mathcal{H}_\infty$ and vary the parameters (t, d) clockwise. When the parameters are on $\mathcal{H}_\infty$, the phase portrait is formed by the origin, a globally asymptotically stable fixesd point, and a limit cycle at infinity, i.e., contained in $\partial\mathbb{D}$. Just after crossing the manifold $\mathcal{H}_\infty$, an asymptotically unstable hyperbolic limit cycle Γ emerges from the limit cycle at $\partial\mathbb{D}$. Hence, the parameters in the manifold $\mathcal{H}_\infty$ correspond to a supercritical Hopf bifurcation at infinity.

When we take parameters (t, d) in the manifold $\mathcal{SN}_\infty$ with $t > 0$, then saddle-nodes $\mathbf{x}_+$ and $\mathbf{x}_-$ appear at infinity $\partial\mathbb{D}$. Just after crossing this manifold $\mathcal{SN}_\infty$ both saddle-nodes splits into a saddle $\mathbf{y}_\pm$ and a node $\mathbf{x}_\pm$. Thus, the manifold $\mathcal{SN}_\infty$ coincides with a supercritical saddle-node bifurcation at infinity.

When we cross the straight line $\mathcal{N}$, every saddle point at $\partial\mathbb{D}$ splits into an asymptotically unstable node $\mathbf{y}_\pm$ contained in $\partial\mathbb{D}$ and a saddle point $\mathbf{e}_\pm$ contained in Σ_∞. Since the flow in the Poincaré disc $\mathbb{D}$ is the projection of the Poincaré compactified flow on the Poincaré sphere, it is easy to see that on the sphere every saddle point on the equator splits into three singular points: one asymptotically unstable node on the equator, and two saddle points, each of them contained in one hemisphere. Therefore, the straight half-line $\mathcal{N}$ with $t > 0$ coincides with a supercritical pitchfork bifurcation at infinity.

On $\mathcal{H}_e\mathcal{L}|_{D,T}$ the saddle points $\mathbf{e}_+$ and $\mathbf{e}_-$ which emerge from the pitchfork bifurcation at infinity, collide with the limit cycle Γ which emerges from the Hopf bifurcation at infinity. Thus, the manifold $\mathcal{H}_e\mathcal{L}|_{D,T}$ coincides with a heteroclinic cycle bifurcation.

When $(t, d) \in \mathcal{N}$ and $t < 0$, the saddle points $\mathbf{e}_\pm$ go to infinity and collide with the nodes $\mathbf{x}_\pm$ in $\partial\mathbb{D}$, giving rise to a saddle point. Thus, the straight half-line $\mathcal{N}$ with $t < 0$ coincides with a subcritical pitchfork bifurcation at infinity.

On the manifold $\mathcal{SN}_\infty$ with $t < 0$ the saddle points at infinity $\mathbf{x}_\pm$ and the nodes at infinity $\mathbf{y}_\pm$ coalesce into two saddle-node points at infinity $\mathbf{x}_\pm$. These saddle-node points vanish when we leave $\mathcal{SN}_\infty$. Therefore, the manifold $\mathcal{SN}_\infty$ coincides with a subcritical saddle-node bifurcation at infinity. Note that on this manifold there exist two special bifurcation points, $\mathcal{VB}_1|_{D,T}$ and $\mathcal{VB}_2|_{D,T}$. Taking parameters at these points, there exist fundamental systems having two saddle-nodes at infinity $\mathbf{x}_\pm$, and there exist fundamental systems for which the infinity is formed by singular points.

One of the most complex bifurcations when $D > 0$ and $T < 0$ takes place at the origin and it is a codimension 2 bifurcation. At this point the different bifurcation manifolds $\mathcal{H}_\infty$, $\mathcal{SN}_\infty$, $\mathcal{N}$, and $\mathcal{H}_e\mathcal{L}|_{D,T}$ coincide.

5.3 The case $D > 0$ and $T = 0$

In this section we present the phase portrait of the Poincaré compactification of fundamental systems

$$\dot{\mathbf{x}} = A\mathbf{x} + \varphi\left(\mathbf{k}^T\mathbf{x}\right)\mathbf{b},$$

with fundamental matrices (A, B) and parameters (D, T, d, t), $D > 0$ and $T = 0$. Since we consider the case $T = 0$, we denote the plane $\Pi_{D,T}$ by $\Pi_{D,0}$. In the next result we find the regions in the parameter space $\Pi_{D,0}$ where the fundamental systems are proper.

Lemma 5.3.1. *Assume that $D > 0$ and $T = 0$.*

(a) *The sets $W_1|_{D,T}$ and $W_2|_{D,T}$ are empty. Hence, any fundamental system with $(t, d) \in \Pi_{D,0}$ is proper.*

(b) *Consider a fundamental system with fundamental matrices (A, B) and parameters $D > 0$, $T = 0$, and $(t, d) \in \Pi_{D,0}$. Suppose that the matrix A is in real Jordan normal form. If $t^2 - 4d = 0$, then A is non-diagonal and $k_1 \neq 0$. If $t^2 - 4d > 0$, then $k_1 k_2 \neq 0$.*

Proof. (a) From (4.50) it follows that $W_1|_{D,T}$ and $W_2|_{D,T}$ are empty sets. The statement follows from Lemma 4.7.3(a) and Theorem 4.7.6(a).

(b) Suppose that $t^2 - 4d = 0$. From Lemma 4.7.2 we obtain that only when $(D, T, d, t) \in F_0^A$, the real Jordan normal form of the matrix A is diagonal. Since for these parameters $F_0^A|_{D,0} = \varnothing$, see Lemma 4.7.3(d), we conclude that the matrix A is non-diagonal. From Proposition 4.7.1(b) it follows that $k_1 \neq 0$.

Suppose that $t^2 - 4d > 0$. In this case the proof follows from Lemma 4.7.1(d). □

5.3.1 Singular points

In the next result we summarize the information about the finite singular points of the fundamental systems with parameters $D > 0$ and $T = 0$ obtained in Theorem 3.9.3(a).

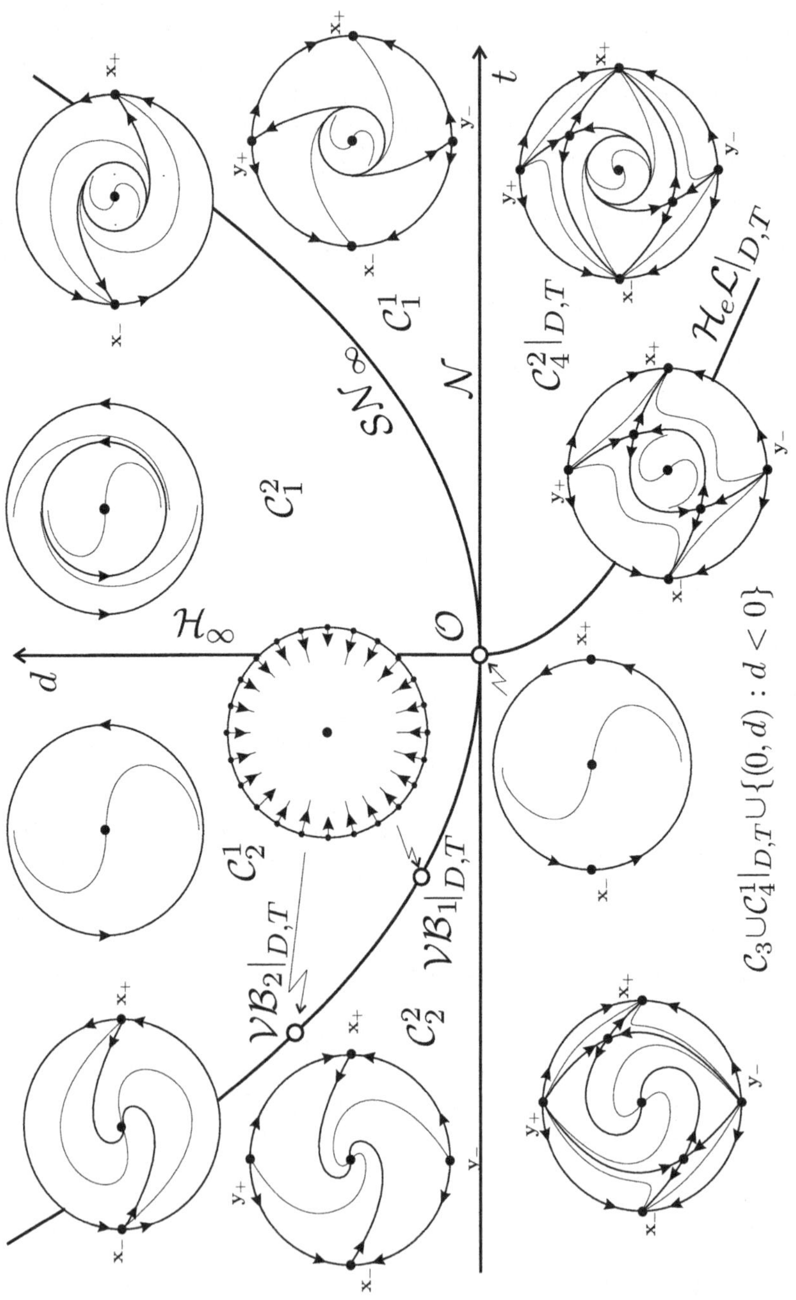

Figure 5.6: Phase portraits and the bifurcation set of fundamental systems with parameters $D > 0$ and $T < 0$.

Proposition 5.3.2. *Consider a fundamental system with fundamental parameters $D > 0$ and $T = 0$.*

(a) *If $d \geq 0$, then the origin is the unique singular point and it is a center.*

(b) *If $d < 0$, then there exist exactly three singular points: the center at the origin and two saddle points $\mathbf{e}_+ \in S_+$ and $\mathbf{e}_- \in S_-$.*

5.3.2 Behaviour at infinity

The different behaviours at infinity when $D > 0$ and $T = 0$ are summarized in Proposition 5.2.2 and represented in Figure 5.2. We recall that in this case the real Jordan normal form of the matrix A is always non-diagonal, see Lemma 5.3.1(b). Therefore, when $D > 0$ and $T = 0$, the manifold $\partial\mathbb{D}$ at infinity is not a normally hyperbolic manifold, see Proposition 5.2.2(b).

5.3.3 Annular region of periodic orbits

Since any fundamental system with parameters $D > 0$ and $T = 0$ is a proper fundamental system, see Lemma 5.3.1(a), there exists a contact point $\mathbf{p}_+$ of the flow with the straight line L_+. Let Γ_M be the orbit through $\mathbf{p}_+$. By Proposition 4.2.7(a), Γ_M is locally contained in $L_- \cup S_0 \cup L_+$. Since the singular point at the origin is a center, every orbit, except the origin, which is contained in the strip $L_- \cup S_0 \cup L_+$, is a periodic orbit. From the symmetry of the flow with respect to the origin it is easy to conclude that $\mathbf{p}_- = -\mathbf{p}_+$ is the contact point of the flow with L_- and, moreover, Γ_M is a periodic orbit tangent to L_+ and L_- at the contact points $\mathbf{p}_+$ and $\mathbf{p}_-$.

Let Σ_{Γ_M} be the region bounded by Γ_M. Thus Σ_{Γ_M} is contained in $L_- \cup S_0 \cup L_+$, which implies that $\mathrm{Cl}(\Sigma_{\Gamma_M}) \setminus \mathbf{0}$ is an annular region formed by periodic orbits, see Figure 5.7. In the next result we prove that these annular regions contain all Jordan curves formed by solutions when $t \neq 0$.

Proposition 5.3.3. *Consider a fundamental system with parameters $D > 0$, $T = 0$ and $t \neq 0$. Any Jordan curve formed by solutions is a periodic orbit and is contained in $\mathrm{Cl}(\Sigma_{\Gamma_M}) \setminus \mathbf{0}$.*

Proof. Let Γ be a Jordan curve formed by solutions. From Theorem 3.12.2(d) it follows that $\Gamma \subset \Gamma_+ \cup S_0 \cup \Gamma_-$, i.e., Γ is a periodic orbit. Since Γ_M is tangent to the straight lines L_+ and L_-, we conclude that $\Gamma \subset \mathrm{Cl}(\Sigma_{\Gamma_M}) \setminus \mathbf{0}$. □

Note that for $t \neq 0$ the periodic orbit Γ_M is maximal in the set of all periodic orbits of the system. Moreover, the interior region bounded by Γ_M contains all Jordan curves formed by solutions. Therefore, the phase portrait of these systems is determined by the stability of Γ_M. In fact in the next result we prove that Γ_M is either the α- or the ω-limit set of any orbit in $\mathbb{R}^2 \setminus \Sigma_{\Gamma_M}$ when $d \geq 0$.

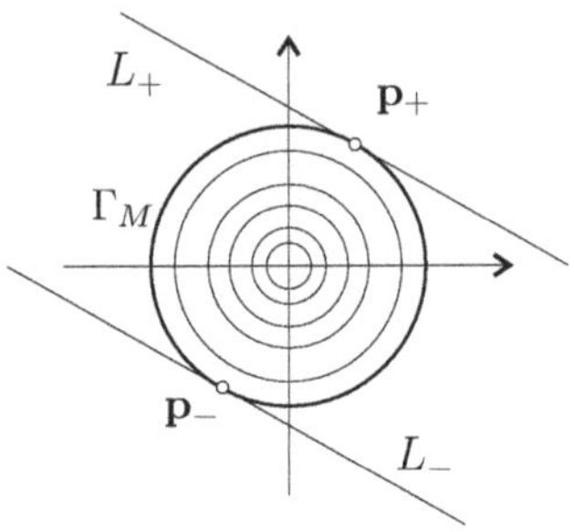

Figure 5.7: The periodic orbit Γ_M and the annular region $\mathrm{Cl}(\Sigma_{\Gamma_M}) \setminus \mathbf{0}$ formed by periodic orbits.

Proposition 5.3.4. *Consider a fundamental system with parameters $D > 0$, $T = 0$, $t \neq 0$, and $d \geq 0$.*

(a) *If $t > 0$, then Γ_M is an outside asymptotically unstable limit cycle. Moreover, $\mathbb{R}^2 \setminus \Sigma_{\Gamma_M} = W^u(\Gamma_M)$.*

(b) *If $t < 0$, then Γ_M is an outside asymptotically stable limit cycle. Moreover, $\mathbb{R}^2 \setminus \Sigma_{\Gamma_M} = W^s(\Gamma_M)$.*

Proof. (a) By Propositions 3.13.3 and 3.13.5(d), every orbit of the system has a bounded α-limit. Since the origin is the unique singular point and every Jordan curve formed by solutions is contained in $\mathrm{Cl}(\Sigma_{\Gamma_M})$, see Proposition 5.3.3, the α-limit set of any orbit in $\mathbb{R}^2 \setminus \Sigma_{\Gamma_M}$ is the limit cycle Γ_M, which proves the statement.

Statement (b) follows from Proposition 3.7.1 by reversing the direction of the time. □

For $d \neq 0$, in order to study the local phase portrait of the periodic orbit Γ_M we use the return map π. In the next lemma we summarize the information about the Poincaré maps π^B_{+-} and π^A_{++}.

Lemma 5.3.5. *Consider a fundamental system with $D > 0$, $T = 0$ and $d \neq 0$.*

(a) *The Poincaré map π^B_{+-} is defined and coincides with the identity map in $[0, +\infty)$.*

(b) *The Poincaré map π^A_{++} is defined.*

(b.1) *If $t > 0$ and $d > 0$, then there exists a value $a_0 > 0$ (where a_0 can be equal to $+\infty$) such that $\pi^A_{++} : [0, a_0) \to [0, +\infty)$, $\pi^A_{++}(0) = 0$, $\pi^A_{++}(a) > a$, $\lim_{a \nearrow a_0} \pi^A_{++}(a) = +\infty$, and $(\pi^A_{++})'(a) > 1$ in the interval $(0, a_0)$.*

(b.2) *If $t = 0$, then the map π^A_{++} is equal to the identity in the interval $[0, a_0)$ where $a_0 = +\infty$ if $d > 0$ or $a_0 = |d|^{-\frac{1}{2}}$ if $d < 0$.*

(b.3) *If $t > 0$ and $d < 0$, then there exist two values $a_0 > 0$ and $b_0 > 0$ such that $\pi_{++}^A : [0, a_0) \to [0, b_0)$, $\pi_{++}^A(0) = 0$, $\pi_{++}^A(a) > a$, $\lim_{a \nearrow a_0} \pi_{++}^A(a) = b_0$, and $(\pi_{++}^A)'(a) > 1$ in $(0, a_0)$.*

(c) *The return map π satisfies $\pi = \pi_{++}^A \circ \pi_{++}^A$.*

Proof. By Lemma 5.3.1(a), the system is proper. Therefore, the Poincaré maps π_{++}^A and π_{+-}^B are defined.

Statement (a) is a consequence of Proposition 4.4.13(b).

Statement (b.1) is a consequence of Propositions 4.4.1, 4.4.6 and 4.4.11. Statement (b.2) follows from Proposition 4.4.11(b) when $d > 0$ and from Proposition 4.4.15(a) when $d < 0$. Statement (b.3) follows from Proposition 4.4.15(b).

From Theorem 4.6.1(b) it follows that the return map π is defined and satisfies that $\pi = (\pi_{++}^A \circ \pi_{+-}^B)^2$, where π_{+-}^B is the identity map. From this we conclude the proof of statement (c). □

In the following result we describe the local phase portrait of the periodic orbit Γ_M in the case $d < 0$.

Proposition 5.3.6. *Consider a fundamental system with parameters $D > 0$, $T = 0$, $t \neq 0$, and $d < 0$.*

(a) *If $t > 0$, then Γ_M is an outside asymptotically unstable limit cycle.*

(b) *If $t < 0$, then Γ_M is an outside asymptotically stable limit cycle.*

Proof. (a) From Lemmas 5.3.5(c) and (b.3) it follows that the unique fixed point of the return map π it located at $a = 0$. Moreover,

$$\pi'(a) > 1 \text{ if } a \in (0, +\infty). \tag{5.9}$$

Since Γ_M intersects L_+ at the contact point $\mathbf{p}_+$ with coordinate $a = 0$, the statement follows from Proposition 4.6.2(c).

Statement (b) follows from statement (a) by reversing the sign of time, see Proposition 3.7.1. □

In Proposition 5.3.3 we have proved that for $t \neq 0$, the annular region bounded by Γ_M and the singular point at the origin contains all the Jordan curves formed by solutions, and that these Jordan curves are periodic orbits. We also have studied the local behaviour of the flow in a neighbourhood of the periodic orbit Γ_M. Now we deal with the case $t = 0$. In the next result we show that for $t = 0$ and $d \geq 0$ the annular region of periodic orbits also exists and it is not bounded. In fact this annular region covers the phase plane, i.e., we have a global center.

Proposition 5.3.7. *Consider a fundamental system with parameters $D > 0$, $T = 0$, $d \geq 0$, and $t = 0$. Any orbit in $\mathbb{R}^2 \setminus \mathbf{0}$ is a periodic orbit.*

Proof. We split the proof depending on $d > 0$ or $d = 0$. Suppose that $d > 0$ and let γ be an orbit in $\mathbb{R}^2 \setminus \mathbf{0}$. Since $d > 0$ and $t = 0$, the system has a virtual center in the half-planes $L_+ \cup S_+$ and $L_- \cup S_-$. Moreover, since $D > 0$ and $T = 0$, the system has a center in S_0. Thus, either $\gamma \subset S_0$, or γ intersects L_+ and L_-. In the first case, we obtain that γ is a periodic orbit. In the second case, since the return map is the identity, see Lemmas 5.3.5(b.2) and (c), it follows that γ is a periodic orbit.

Suppose that $d = 0$. Let $\dot{\mathbf{x}} = A\mathbf{x} + \varphi(\mathbf{k}^T\mathbf{x})\mathbf{b}$ be a fundamental system. Without loss of generality we can assume that the matrix A is in real Jordan normal form, i.e.,

$$A = \begin{pmatrix} 0 & 1 \\ 0 & 0 \end{pmatrix}.$$

From (3.10) it follows that $D = -k_1b_2 > 0$, which is equivalent to $-k_1/b_2 > 0$.

Consider the auxiliary function $f(u) = \int_0^u \varphi(r)dr$, where φ is the piecewise linear characteristic function of the system. It is clear that $f(u)$ satisfies $f(u) = f(-u)$, is non-negative, is strictly increasing in $(0, +\infty)$, and has a global minimum at $u = 0$. Hence, for any $h > 0$ there exists exactly one value $u_h > 0$ such that $h - f(u) > 0$ if and only if $u \in (-u_h, u_h)$ and $f(-u_h) = f(u_h) = h$.

Consider the following function defined on the whole phase plane $\mathbb{R}^2$:

$$V(\mathbf{x}) = \frac{1}{2}\left(-\frac{k_1}{b_2}\right)x_2^2 + f\left(\mathbf{k}^T\mathbf{x}\right) \geq 0,$$

where $\mathbf{x} = (x_1, x_2)^T$. Since the components of the vector field are $\dot{x}_1 = x_2 + \varphi(\mathbf{k}^T\mathbf{x})b_1$ and $\dot{x}_2 = \varphi(\mathbf{k}^T\mathbf{x})b_2$, it is easy to check that V is a first integral of the system, that is,

$$\frac{dV}{ds} = (k_1b_1 + k_2b_2)\,\varphi^2\left(\mathbf{k}^T\mathbf{x}\right) = T\varphi^2\left(\mathbf{k}^T\mathbf{x}\right) = 0.$$

Therefore, the sets

$$I_h = \left\{\mathbf{x} \in \mathbb{R}^2 : V(\mathbf{x}) = h\right\} = \left\{\mathbf{x} \in \mathbb{R}^2 : x_2 = \pm\sqrt{2\left(-\frac{b_2}{k_1}\right)\left(h - f\left(\mathbf{k}^T\mathbf{x}\right)\right)}\right\},$$

with $h \geq 0$, are formed by solutions of the system. Moreover, $\mathbf{x} \in I_h$ if and only if $\mathbf{k}^T\mathbf{x} \in [-u_h, u_h]$. We conclude that I_h is a Jordan curve formed by solutions when $h > 0$ and that $I_0 = \mathbf{0}$. Since the unique singular point is the origin, it follows that I_h is a periodic orbit when $h > 0$. □

5.3.4 Heteroclinic cycles

We deal now with the case $d < 0$ and $t = 0$. As we shall see, in this case the annular region of periodic orbits is bounded by a heteroclinic cycle Δ to the saddle points $\mathbf{e}_+$ and $\mathbf{e}_-$.

Proposition 5.3.8. *Consider a fundamental system with parameters $D > 0$, $T = 0$, $d < 0$, and $t = 0$. The system has a heteroclinic cycle Δ connecting the singular points $\mathbf{e}_+$ and $\mathbf{e}_-$. Moreover, the region $\mathrm{Cl}(\Sigma_\Delta)$ contains any Jordan curve formed by solutions, and the annular region $\Sigma_\Delta \setminus \mathbf{0}$ is formed by all the periodic orbits of the system.*

Proof. Let (A, B) be the fundamental matrices of the system, and let $\lambda_1 > 0 > \lambda_2$ be the eigenvalues of A.

Since $t = 0$, we have that $\lambda_1 = |\lambda_2| = |d|^{\frac{1}{2}}$. Hence, a stable separatrix γ_+^{s-} and an unstable separatrix γ_+^{u-} of the point $\mathbf{e}_+$ intersect the straight line L_+ at the points of coordinates $a = \lambda_1^{-1}$ and $b = |\lambda_2|^{-1}$, respectively, see Corollary 4.4.17.

By the symmetry of the flow with respect to the origin, a stable separatrix γ_-^{s+} and an unstable separatrix γ_-^{u+} of $\mathbf{e}_-$ intersect L_- at the points of coordinates $a = \lambda_1^{-1}$ and $b = |\lambda_2|^{-1}$, respectively.

By Lemma 5.3.5(a), we have $\pi_{+-}^B(|\lambda_2|^{-1}) = |\lambda_2|^{-1} = \lambda_1^{-1}$. Hence, the unstable separatrix γ_+^{u-} of $\mathbf{e}_+$ coincides with the stable separatrix γ_-^{s+} of $\mathbf{e}_-$, and γ_+^{s-} coincides with γ_-^{u+}. This forms a heteroclinic cycle Δ to the singular points $\mathbf{e}_+$ and $\mathbf{e}_-$.

Note that the Poincaré map π_{++}^A is the identity in $[0, |\lambda_2|^{-1})$, see Lemma 5.3.5(b.2). Hence, the return map $\pi = \pi_{++}^A \circ \pi_{++}^A$ is the identity in $[0, |\lambda_2|^{-1})$. Therefore, the annular region $\Sigma_\Delta \setminus \mathbf{0}$ is formed by periodic orbits. □

5.3.5 Phase portraits

We recall that if the compactified flow of a fundamental system has no singular points contained in the manifold at infinity $\partial\mathbb{D}$, then we denote by ∞ the periodic orbit contained in $\partial\mathbb{D}$.

Define on the plane $\Pi_{D,0}$ the half-line

$$\mathcal{H}_e\mathcal{L}|_{D,0} := \{(t, d) : t = 0 \text{ and } d < 0\}.$$

By Proposition 5.2.8(c), $\mathcal{H}_e\mathcal{L}|_{D,0}$ is the limit of the differentiable curve $\mathcal{H}_e\mathcal{L}|_{D,T}$ when T tends to 0.

Proposition 5.3.9. *Consider the Poincaré compactification of a fundamental system with parameters $D > 0$, $T = 0$, and $(t, d) \in \mathcal{H}_\infty$. Then:*

(a) *The separatrices of the system are:*

  (a.1) *the singular point at the origin $\mathbf{0}$, which is a center, and*

  (a.2) *the periodic orbit at infinity ∞.*

(b) *The canonical region is $\Sigma_\infty \setminus \mathbf{0}$ and is formed by periodic orbits.*

(c) *The phase portrait of the Poincaré compactification is topologically equivalent to its correspondent in Figure 5.8.*

Proof. Statements (a) and (b) are a consequence of Propositions 5.3.2(a), 5.2.2(a) and 5.3.7.

Statement (c) follows from Theorem 2.6.9. □

Proposition 5.3.10. *Consider the Poincaré compactification of a fundamental system with parameters $D > 0$, $T = 0$, and $(t, d) \in \mathcal{C}_1^2$. Then:*

(a) *The separatrices of the system are:*

(a.1) *the singular point at the origin* $\mathbf{0}$ *which is a center;*

(a.2) *the limit cycle at infinity* ∞*, and the outside asymptotically unstable limit cycle* Γ_M *such that* $\mathbf{0} \in \Sigma_{\Gamma_M}$.

(b) *The canonical regions are:* $\Sigma_\infty \setminus \mathrm{Cl}(\Sigma_{\Gamma_M}) = W^u(\Gamma_M) \cap W^s(\infty)$*, and* $\Sigma_{\Gamma_M} \setminus \mathbf{0}$ *formed by periodic orbits.*

(c) *The phase portrait of the Poincaré compactification is topologically equivalent to its correspondent in Figure* 5.8.

Proof. (a) The existence of these separatrices is a consequence of Propositions 5.3.2(a), 5.2.2(a), 5.3.3(a) and 5.3.4(a).

(b) By Proposition 5.3.3, any Jordan curve formed by solutions is contained in $\mathrm{Cl}(\Sigma_{\Gamma_M}) \setminus \mathbf{0}$ and is a periodic orbit. Thus, from the Poincaré–Bendixson Theorem it follows that the α- and ω-limit sets of the orbits in $\Sigma_\infty \setminus \mathrm{Cl}(\Sigma_{\Gamma_M})$ are the outside asymptotically unstable limit cycle Γ_M and the limit cycle at infinity ∞. Hence, $\Sigma_\infty \setminus \mathrm{Cl}(\Sigma_{\Gamma_M}) = W^u(\Gamma_M) \cap W^s(\infty)$.

The behaviour of the flow in $\Sigma_{\Gamma_M} \setminus \mathbf{0}$ is a consequence of Proposition 5.3.3. Statement (c) follows from Theorem 2.6.9. □

Proposition 5.3.11. *Consider the Poincaré compactification of a fundamental system with parameters $D > 0$, $T = 0$, $(t, d) \in \mathcal{SN}_\infty$, and $t > 0$. Then:*

(a) *The separatrices of the system are:*

(a.1) *the singular point at the origin* $\mathbf{0}$ *which is a center, the saddle-node points at infinity* $\mathbf{x}_+$, $\mathbf{x}_-$*, which have their stable hyperbolic manifold contained in* Σ_∞*, and their central manifold contained in* $\partial\mathbb{D}$;

(a.2) *the outside asymptotically unstable limit cycle* Γ_M *which satisfies that* $\mathbf{0} \in \Sigma_{\Gamma_M}$;

(a.3) *the separatrices of the singular points* $\mathbf{x}_+$, $\mathbf{x}_-$.

(b) *The canonical regions are: the annular region* $\Sigma_{\Gamma_M} \setminus \mathbf{0}$ *formed by periodic orbits,* $\Sigma_\infty^{0+} \subset W^u(\Gamma_M) \cap W^s(\mathbf{x}_+)$*, and* $\Sigma_\infty^{0-} \subset W^u(\Gamma_M) \cap W^s(\mathbf{x}_-)$.

(c) *The phase portrait of the Poincaré compactification is topologically equivalent to its correspondent in Figure* 5.8.

Proof. (a) The existence of these separatrices is a consequence of Propositions 5.3.2(a), 5.2.2(c), 5.3.3(a) and 5.3.4(a).

(b) The behaviour of the flow in the annular region $\Sigma_{\Gamma_M} \setminus \mathbf{0}$ is a consequence of Proposition 5.3.3(a).

By Proposition 5.3.4(a), $\Sigma_\infty \setminus \mathrm{Cl}(\Sigma_{\Gamma_M}) \subset W^u(\Gamma_M)$. Since the stable separatrices γ_+^s and γ_-^s of the singular points at infinity $\mathbf{x}_+$ and $\mathbf{x}_-$ are contained in Σ_∞, it follows that $\alpha(\gamma_+^s) = \alpha(\gamma_-^s) = \Gamma_M$. Hence, γ_+^s and γ_-^s split $\Sigma_\infty \setminus \mathrm{Cl}(\Sigma_{\Gamma_M})$ into two open, connected, and invariant regions denoted by Σ_∞^{0+} and Σ_∞^{0-}. It is easy to conclude that $\Sigma_\infty^{0+} \subset W^s(\mathbf{x}_+)$ and $\Sigma_\infty^{0-} \subset W^s(\mathbf{x}_-)$, for more details see the proof of Proposition 5.2.17(b).

Statement (c) follows from Theorem 2.6.9. □

Proposition 5.3.12. *Consider the Poincaré compactification of a fundamental system with parameters $D > 0$, $T = 0$, $(t, d) \in \mathcal{C}_1^1 \cup \mathcal{N}$, and $t > 0$. Then:*

(a) *The separatrices of the system are:*

  (a.1) *the singular point at the origin $\mathbf{0}$ which is a center, the asymptotically stable nodes at infinity $\mathbf{x}_+$, $\mathbf{x}_-$, and the saddle points at infinity $\mathbf{y}_+$, $\mathbf{y}_-$, which have their unstable manifold contained in $\partial\mathbb{D}$;*

  (a.2) *the outside asymptotically unstable limit cycle Γ_M which satisfies that $\mathbf{0} \in \Sigma_{\Gamma_M}$;*

  (a.3) *the separatrices of the saddle points $\mathbf{y}_+$ and $\mathbf{y}_-$.*

(b) *The canonical regions are: the annular region $\Sigma_{\Gamma_M} \setminus \mathbf{0}$ formed by periodic orbits, $\Sigma_\infty^{0+} = W^u(\Gamma_M) \cap W^s(\mathbf{x}_+)$, and $\Sigma_\infty^{0-} = W^u(\Gamma_M) \cap W^s(\mathbf{x}_-)$.*

(c) *The phase portrait of the Poincaré compactification is topologically equivalent to its correspondent in Figure* 5.8.

Proof. (a) The existence of these separatrices is a consequence of Propositions 5.3.2(a), 5.2.2(d), 5.3.3 and 5.3.4(a).

(b) By Proposition 5.3.4(a), $\Sigma_\infty \setminus \mathrm{Cl}(\Sigma_{\Gamma_M}) \subset W^u(\Gamma_M)$. Then the stable separatrices γ_+^s and γ_-^s of $\mathbf{y}_+$ and $\mathbf{y}_-$ satisfy that $\alpha(\gamma_+^s) = \alpha(\gamma_-^s) = \Gamma_M$. Hence, the orbits γ_+^s and γ_-^s split $\Sigma_\infty \setminus \mathrm{Cl}(\Sigma_{\Gamma_M})$ into two open, connected, and invariant regions denoted by Σ_∞^{0+} and Σ_∞^{0-}. Let Σ_∞^{0+} be the region containing the point $\mathbf{x}_+$ in its boundary. By the Poincaré–Bendixson Theorem, the α- and ω-limit sets contained in $\mathrm{Cl}(\Sigma_\infty^{0+})$ are: the asymptotically stable node $\mathbf{x}_+$, the saddle points $\mathbf{y}_+$, $\mathbf{y}_-$, and the limit cycle Γ_M. Therefore, $\Sigma_\infty^{0+} \subset W^u(\Gamma_M) \cap W^s(\mathbf{x}_+)$. Moreover, if $\gamma \subset W^u(\Gamma_M)$, then either $\gamma \subset \Sigma_\infty^{0+}$, or $\gamma \subset \Sigma_\infty^{0-}$, or γ is one of the separatrices γ_+^s and γ_-^s.

The behaviour of the flow in $\Sigma_{\Gamma_M} \setminus \mathbf{0}$ is a consequence of Proposition 5.3.3(a). Statement (c) follows from Theorem 2.6.9. □

Proposition 5.3.13. *Consider the Poincaré compactification of a fundamental system with parameters $D > 0$, $T = 0$ and $(t, d) \in \mathcal{C}_4$. Then:*

(a) *The separatrices of the system are:*

(a.1) *the singular point at the origin* $\mathbf{0}$ *which is a center, the saddle points* $\mathbf{e}_+$, $\mathbf{e}_-$, *the asymptotically stable nodes at infinity* $\mathbf{x}_+$, $\mathbf{x}_- \in \partial\mathbb{D}$, *and the asymptotically unstable nodes at infinity* $\mathbf{y}_+$, $\mathbf{y}_- \in \partial\mathbb{D}$;

(a.2) *the outside asymptotically unstable limit cycle* Γ_M, *which satisfies that* $\mathbf{0} \in \Sigma_{\Gamma_M}$;

(a.3) *the separatrices of the singular points* $\mathbf{e}_+$ *and* $\mathbf{e}_-$.

(b) *The canonical regions are: the annular region* $\Sigma_{\Gamma_M} \setminus \mathbf{0}$ *formed by periodic orbits,* $\Sigma_\infty^{++} \subset W^u(\mathbf{y}_+) \cap W^s(\mathbf{x}_+)$, $\Sigma_\infty^{+-} \subset W^u(\mathbf{y}_+) \cap W^s(\mathbf{x}_-)$, $\Sigma_\infty^{0+} = W^u(\Gamma_M) \cap W^s(\mathbf{x}_+)$, $\Sigma_\infty^{0-} = W^u(\Gamma_M) \cap W^s(\mathbf{x}_-)$, $\Sigma_\infty^{-+} \subset W^u(\mathbf{y}_-) \cap W^s(\mathbf{x}_+)$, *and* $\Sigma_\infty^{--} \subset W^u(\mathbf{y}_-) \cap W^s(\mathbf{x}_-)$.

(c) *The phase portrait of the Poincaré compactification is topologically equivalent to its correspondent in Figure* 5.8.

Proof. (a) The existence of these separatrices is a consequence of Propositions 5.3.2(a), 5.2.2(d), 5.3.3 and 5.3.6(a).

(b) By Proposition 5.3.3, the system has no Jordan curves formed by solutions. Since $\mathbf{x}_+$, $\mathbf{x}_-$, $\mathbf{y}_+$, and $\mathbf{y}_-$ are nodes, it is easy to conclude that there are no separatrix cycles having their singular points contained in $\partial\mathbb{D}$. Thus, by the Poincaré–Bendixson Theorem, the α- and ω-limit sets in $\mathbb{D} \setminus \Sigma_{\Gamma_M}$ are the singular points at infinity, $\mathbf{x}_+$, $\mathbf{x}_-$, $\mathbf{y}_+$, and $\mathbf{y}_-$, and the limit cycle Γ_M. The statement follows by using arguments similar to those in the proof of Proposition 5.2.19(b). The behaviour of the flow in $\Sigma_{\Gamma_M} \setminus \mathbf{0}$ is a consequence of Proposition 5.3.3(a).

Statement (c) follows from Theorem 2.6.9. □

Proposition 5.3.14. *Consider the Poincaré compactification of a fundamental system with parameters* $D > 0$, $T = 0$ *and* $(t, d) \in \mathcal{H}_e\mathcal{L}|_{D,0}$. *Then:*

(a) *The separatrices of the system are:*

(a.1) *the singular point at the origin* $\mathbf{0}$ *which is a center, the saddle points* $\mathbf{e}_+$, $\mathbf{e}_-$, *the asymptotically stable nodes at infinity* $\mathbf{x}_+$, $\mathbf{x}_-$, *and the asymptotically unstable nodes at infinity* $\mathbf{y}_+$, $\mathbf{y}_-$;

(a.2) *the separatrices of the saddle points* $\mathbf{e}_+$ *and* $\mathbf{e}_-$, *which form a heteroclinic cycle* Δ *such that* $\mathbf{0} \in \Sigma_\Delta$, *and the separatrices of the singular points at infinity which are contained in* $\partial\mathbb{D}$.

(b) *The canonical regions are: the annular region* $\Sigma_\Delta \setminus \mathbf{0}$ *formed by periodic orbits,* $\Sigma_\infty^{++} \subset W^u(\mathbf{y}_+) \cap W^s(\mathbf{x}_+)$, $\Sigma_\infty^{+-} \subset W^u(\mathbf{y}_+) \cap W^s(\mathbf{x}_-)$, $\Sigma_\infty^{--} \subset W^u(\mathbf{y}_-) \cap W^s(\mathbf{x}_-)$, *and* $\Sigma_\infty^{-+} \subset W^u(\mathbf{y}_-) \cap W^s(\mathbf{x}_+)$.

(c) *The phase portrait of the Poincaré compactification is topologically equivalent to its correspondent in Figure* 5.8.

Proof. (a) The existence of these separatrices is a consequence of Propositions 5.3.2(b), 5.2.2(d) and 5.3.8.

(b) The behaviour of the flow in $\Sigma_\Delta \setminus \mathbf{0}$ is a consequence of Proposition 5.3.8. The behaviour of the flow in $\Sigma_\infty \setminus \mathrm{Cl}(\Sigma_\Delta)$ follows by using arguments similar to those used in the proof of Proposition 5.2.20(d).

Statement (c) follows from Theorem 2.6.9. □

Proposition 5.3.15. *Consider the Poincaré compactification of a fundamental system with parameters $D > 0$, $T = 0$, and $(t, d) \in \mathcal{C}_3$. Then:*

(a) *The separatrices of the system are:*

(a.1) *the singular point at the origin* $\mathbf{0}$ *which is a center, the saddle points* $\mathbf{e}_+$, $\mathbf{e}_-$, *the asymptotically stable nodes at infinity* $\mathbf{x}_+$, $\mathbf{x}_-$, *and the asymptotically unstable nodes at infinity* $\mathbf{y}_+$, $\mathbf{y}_-$*;*

(a.2) *the outside asymptotically stable limit cycle* Γ_M, *which satisfies* $\mathbf{0} \in \Sigma_{\Gamma_M}$*;*

(a.3) *the separatrices of the saddle points* $\mathbf{e}_+$, $\mathbf{e}_-$, *and the separatrices of the singular points at infinity which are contained in* $\partial\mathbb{D}$.

(b) *The canonical regions are: the annular region* $\Sigma_{\Gamma_M} \setminus \mathbf{0}$ *is formed by periodic orbits,* $\Sigma_\infty^{++} \subset W^u(\mathbf{y}_+) \cap W^s(\mathbf{x}_+)$, $\Sigma_\infty^{+0} = W^u(\mathbf{y}_+) \cap W^s(\Gamma_M)$, $\Sigma_\infty^{+-} \subset W^u(\mathbf{y}_+) \cap W^s(\mathbf{x}_-)$, $\Sigma_\infty^{--} \subset W^u(\mathbf{y}_-) \cap W^s(\mathbf{x}_-)$, $\Sigma_\infty^{-0} = W^u(\mathbf{y}_-) \cap W^s(\Gamma_M)$, *and* $\Sigma_\infty^{-+} \subset W^u(\mathbf{y}_-) \cap W^s(\mathbf{x}_+)$.

(c) *The phase portrait of the Poincaré compactification is topologically equivalent to its correspondent in Figure* 5.8.

Proof. (a) The existence of these separatrices is a consequence of Propositions 5.3.2(b), 5.2.2(d), 5.3.3 and 5.3.6(b).

(b) The behaviour of the flow in $\Sigma_{\Gamma_M} \setminus \mathbf{0}$ follows from Proposition 5.3.3(a). The behaviour of the flow in $\Sigma_\infty \setminus \mathrm{Cl}(\Sigma_{\Gamma_M})$ follows by using similar arguments to those in the proof of Proposition 5.2.21(c). We note that the role played there by the singular point at the origin is played here by the limit cycle Γ_M.

Statement (c) follows from Theorem 2.6.9. □

Proposition 5.3.16. *Consider the Poincaré compactification of a fundamental system with parameters $D > 0$, $T = 0$, $(t, d) \in \mathcal{C}_2^2 \cup \mathcal{N}$, and $t < 0$. Then:*

(a) *The separatrices of the system are:*

(a.1) *the singular point at the origin* $\mathbf{0}$, *which is a center, the saddle points at infinity* $\mathbf{x}_+$, $\mathbf{x}_-$, *which have the stable manifold contained in* $\partial\mathbb{D}$, *and the asymptotically unstable nodes at infinity* $\mathbf{y}_+$, $\mathbf{y}_-$*;*

(a.2) *the outside asymptotically stable limit cycle* Γ_M, *which satisfies that* $\mathbf{0} \in \Sigma_{\Gamma_M}$*;*

(a.3) *the separatrices of the saddle points* $\mathbf{x}_+$ *and* $\mathbf{x}_-$.

(b) *The canonical regions are: the annular region* $\Sigma_{\Gamma_M} \setminus \mathbf{0}$ *formed by periodic orbits,* $\Sigma_\infty^{+0} = W^u(\mathbf{y}_+) \cap W^s(\Gamma_M)$*, and* $\Sigma_\infty^{-0} = W^u(\mathbf{y}_-) \cap W^s(\Gamma_M)$.

(c) *The phase portrait of the Poincaré compactification is topologically equivalent to its correspondent in Figure* 5.8.

Proof. (a) The existence of these separatrices is a consequence of Propositions 5.3.2(a), 5.2.2(c), 5.3.3 and 5.3.4(b).

(c) The behaviour of the flow in $\Sigma_{\Gamma_M} \setminus \mathbf{0}$ is a consequence of Proposition 5.3.3(a). The behaviour of the flow in $\Sigma_\infty \setminus \mathrm{Cl}(\Sigma_{\Gamma_M})$ follows by using arguments similar to those in the proof of Proposition 5.2.22(c). We note that the role played there by the singular point at the origin is played here by the limit cycle Γ_M.

Statement (c) follows from Theorem 2.6.9. □

Proposition 5.3.17. *Consider the Poincaré compactification of a fundamental system with parameters* $D > 0$, $T = 0$, $(t, d) \in \mathcal{SN}_\infty$*, and* $t < 0$*. Then:*

(a) *The separatrices of the system are:*

(a.1) *the singular point at the origin* $\mathbf{0}$*, which is a center, the saddle-nodes at infinity* $\mathbf{x}_+, \mathbf{x}_- \in \partial\mathbb{D}$*, which have their unstable hyperbolic manifold contained in* Σ_∞*, and their center manifold contained in* $\partial\mathbb{D}$*;*

(a.2) *the outside asymptotically stable periodic orbit* Γ_M*, which satisfies that* $\mathbf{0} \in \Sigma_{\Gamma_M}$*;*

(a.3) *the separatrices of the saddle-nodes at infinity* $\mathbf{x}_+$ *and* $\mathbf{x}_-$.

(b) *The canonical regions are: the annular region* $\Sigma_{\Gamma_M} \setminus \mathbf{0}$ *formed by periodic orbits,* $\Sigma_\infty^{+0} = W^u(\mathbf{x}_+) \cap W^s(\Gamma_M)$*, and* $\Sigma_\infty^{-0} = W^u(\mathbf{x}_-) \cap W^s(\Gamma_M)$.

(c) *The phase portrait of the Poincaré map is topologically equivalent to its correspondent in Figure* 5.8.

Proof. (a) The existence of these separatrices is a consequence of Propositions 5.3.2(a), 5.2.2(c), 5.3.3 and 5.3.4(b).

(c) The behaviour of the flow in $\Sigma_{\Gamma_M} \setminus \mathbf{0}$ is a consequence of Proposition 5.3.3(a). The behaviour of the flow in $\Sigma_\infty \setminus \mathrm{Cl}(\Sigma_{\Gamma_M})$ follows by using similar arguments to those used in the proof of Proposition 5.2.23(c). We note that the role played there by the singular point $\mathbf{0}$ is played here by the periodic orbit Γ_M.

Statement (c) follows from Theorem 2.6.9. □

Proposition 5.3.18. *Consider the Poincaré compactification of a fundamental system with parameters* $D > 0$, $T = 0$*, and* $(t, d) \in \mathcal{C}_2^1$*. Then:*

(a) *The separatrices of the system are:*

(a.1) *the singular point at the origin* $\mathbf{0}$*, which is a center;*

(a.2) *the limit cycle at infinity* ∞*, and the outside asymptotically stable periodic orbit* Γ_M*, which satisfies that* $\mathbf{0} \in \Sigma_{\Gamma_M}$.

(b) *The canonical regions are: the annular region* $\Sigma_{\Gamma_M} \setminus \mathbf{0}$ *formed by periodic orbits, and* $\Sigma_\infty \setminus \mathrm{Cl}(\Sigma_{\Gamma_M}) = W^u(\infty) \cap W^s(\Gamma_M)$.

(c) *The phase portrait of the Poincaré compactification is topologically equivalent to its correspondent in Figure* 5.8.

Proof. (a) The existence of these separatrices is a consequence of Propositions 5.3.2(a), 5.2.2(a), 5.3.3 and 5.3.4(b).

(b) The behaviour of the flow in $\Sigma_{\Gamma_M} \setminus \mathbf{0}$ and $\Sigma_\infty \setminus \mathrm{Cl}(\Sigma_{\Gamma_M})$ follows from Propositions 5.3.3(a) and 5.3.3, respectively.

Statement (c) follows from Theorem 2.6.9. □

Proposition 5.3.19. *Consider the Poincaré compactification of a fundamental system with parameters* $D > 0$, $T = 0$, *and* $(t, d) \in \mathcal{O}$. *Then:*

(a) *The separatrices of the system are:*

(a.1) *the singular point at the origin* $\mathbf{0}$, *which is a center, and the singular points at infinity* $\mathbf{x}_+$, $\mathbf{x}_-$, *which satisfy that a neighbourhood of each of them is a hyperbolic sector with the separatrices contained in* $\partial\mathbb{D}$;

(a.2) *the separatrices of the singular points at infinity* $\mathbf{x}_+$ *and* $\mathbf{x}_-$.

(b) *The unique canonical region is* $\Sigma_\infty \setminus \mathbf{0}$ *formed by periodic orbits.*

(c) *The phase portrait of the Poincaré compactification is topologically equivalent to its correspondent in Figure* 5.8.

Proof. (a) The existence of these separatrices is a consequence of Proposition 5.3.2(a) and 5.2.2(c).

(b) The behaviour of the flow in $\Sigma_\infty \setminus \mathbf{0}$ follows from Proposition 5.3.7.

Statement (c) follows from Theorem 2.6.9. □

5.3.6 The bifurcation set

In Subsection 5.3.5 we have described the phase portrait of the Poincaré compactification of a fundamental system with parameters $D > 0$ and $T = 0$ as depending on the parameters t and d. We have summarized all descriptions in Figure 5.8. From this and from Subsection 5.2.6 we conclude that in this case the bifurcation set is formed by the half-lines $\mathcal{H}_\infty$ and $\mathcal{H}_e\mathcal{L}|_{D,0}$, the differentiable curve $\mathcal{SN}_\infty$, and the straight line $\mathcal{N}$. The phase portrait of the compactified flow exhibits a Hopf bifurcation at infinity on the straight line $\mathcal{H}_\infty$, a saddle-node bifurcation at infinity on the curve $\mathcal{SN}_\infty$, a pitchfork bifurcation at infinity on the straight line $\mathcal{N}$, and a homoclinic bifurcation on the curve $\mathcal{H}_e\mathcal{L}|_{D,0}$.

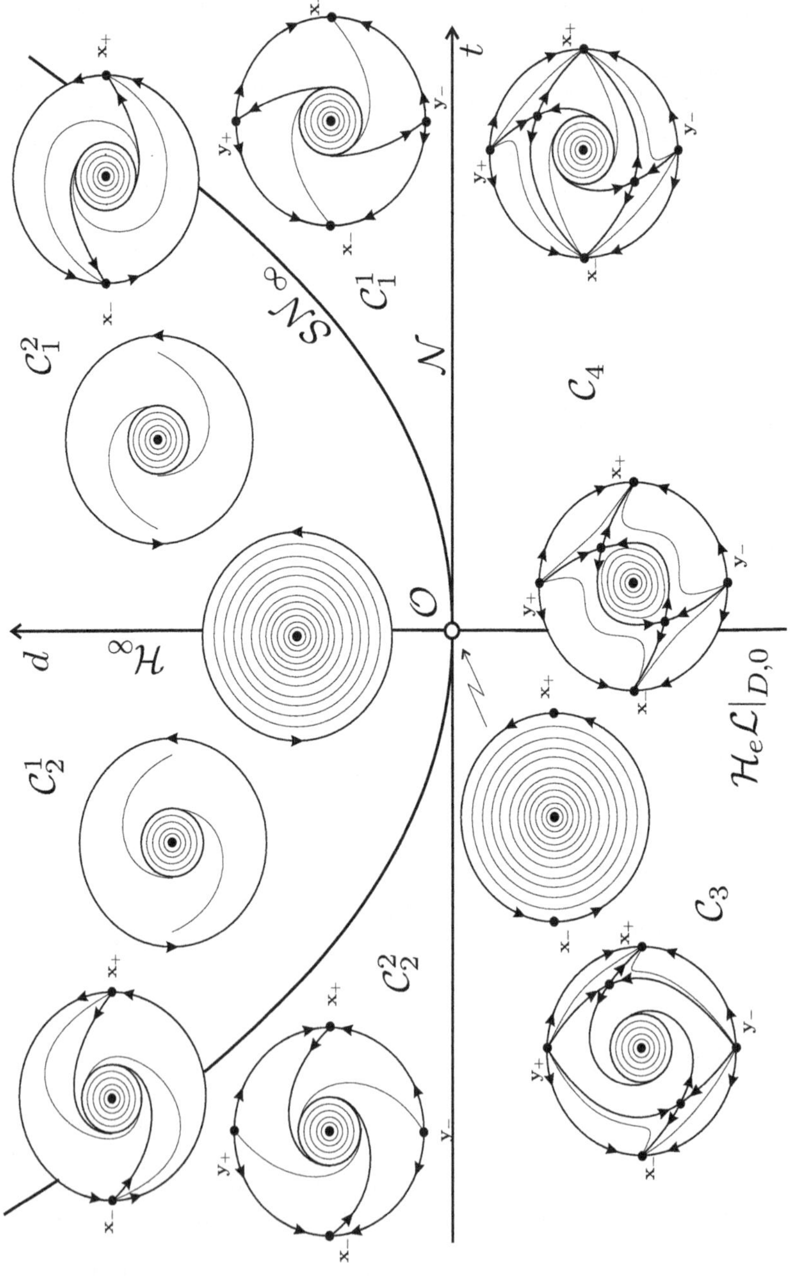

Figure 5.8: Phase portraits and the bifurcation set of the Poincaré compactification of fundamental systems with parameters $D > 0$ and $T = 0$.

5.4 The case $D > 0$ and $T > 0$

Consider a fundamental system $\dot{\mathbf{x}}(s) = \mathbf{f}(\mathbf{x}(s))$ with parameters (D, T, d, t), $D > 0$, and $T > 0$, where $\mathbf{f}(\mathbf{x}) = A\mathbf{x} + \varphi(\mathbf{k}^T\mathbf{x})\mathbf{b}$. Let $\dot{\mathbf{x}} = \mathbf{f}_{\mathbb{D}}(\mathbf{x})$ be its Poincaré compactification. The change of the time variable $\tau = -s$ transforms the system in the new fundamental system $\mathbf{x}'(\tau) = A^*\mathbf{x}(\tau) + \varphi(\mathbf{k}^T\mathbf{x}(\tau))\mathbf{b}^*$, with fundamental parameters (D, T^*, d, t^*), $T^* = -T < 0$, and $t^* = -t$, see Proposition 3.7.1, where $'$ denotes the derivative with respect to τ, $A^* = -A$, and $\mathbf{b}^* = -\mathbf{b}$. The Poincaré compactification of the new system satisfies $\mathbf{x}' = -\mathbf{f}_{\mathbb{D}}(\mathbf{x}(-\tau))$, see Proposition 3.10.1(b), and has the same orbits as $\dot{\mathbf{x}}(s) = \mathbf{f}_{\mathbb{D}}(\mathbf{x}(s))$, but with the opposite orientation. Therefore we will describe the phase portraits of the compactified system $\dot{\mathbf{x}}(s) = \mathbf{f}_{\mathbb{D}}(\mathbf{x}(s))$ by using the phase portraits of the compactified systems $\mathbf{x}' = -\mathbf{f}_{\mathbb{D}}(\mathbf{x}(-\tau))$ described in Subsection 5.2.5. We will also describe the bifurcation set of the fundamental systems with parameters $D > 0$ and $T > 0$ by using the bifurcation set in the case $D > 0$ and $T^* < 0$. In order to do that we define the following regions of parameters in the plane $\Pi_{D,T}$, with $D > 0$ and $T > 0$:

$$\mathcal{C}_3^1\big|_{D,T} := \left\{(t, d) : (-t, d) \in \mathcal{C}_4^2\big|_{D,T^*}\right\},$$
$$\mathcal{C}_3^2\big|_{D,T} := \left\{(t, d) : (-t, d) \in \mathcal{C}_4^1\big|_{D,T^*}\right\},$$

the curve

$$\mathcal{H}_e\mathcal{L}\big|_{D,T} := \left\{(t, d) : (-t, d) \in \mathcal{H}_e\mathcal{L}\big|_{D,T^*}\right\},$$

and the points

$$\mathcal{VB}_1\big|_{D,T} := \left\{(t, d) : (-t, d) \in \mathcal{VB}_1\big|_{D,T^*}\right\},$$
$$\mathcal{VB}_2\big|_{D,T} := \left\{(t, d) : (-t, d) \in \mathcal{VB}_2\big|_{D,T^*}\right\}.$$

Note that these sets are the transformed versions of the sets defined in (5.1), (5.2) and (5.7).

Proposition 5.4.1. *Consider the Poincaré compactification of a fundamental system with parameters $D > 0$ and $T > 0$, and fundamental matrices (A, B).*

(a) *If $(t, d) \in \mathcal{H}_\infty \cup \mathcal{C}_1^2$, then the phase portrait of the Poincaré compactification is topologically equivalent to its correspondent in Figure* 5.9.

(b) *If either $(t, d) \in \mathcal{SN}_\infty \setminus \{\mathcal{VB}_1|_{D,T}, \mathcal{VB}_2|_{D,T}\}$ with $t > 0$, or the parameters $(t, d) \in \{\mathcal{VB}_1|_{D,T}, \mathcal{VB}_2|_{D,T}\}$, and the real Jordan normal form of the matrix A is non-diagonal, then the phase portrait of the Poincaré compactification is topologically equivalent to its correspondent in Figure* 5.9.

(c) *If $(t, d) \in \{\mathcal{VB}_1|_{D,T}, \mathcal{VB}_2|_{D,T}\}$ and the real Jordan normal form of the matrix A is diagonal, then the phase portrait of the Poincaré compactification is topologically equivalent to its correspondent in Figure* 5.9.

(d) *If* $(t, d) \in \mathcal{C}_1^1 \cup \mathcal{N}$ *and* $t > 0$, *then the phase portrait of the Poincaré compactification is topologically equivalent to its correspondent in Figure* 5.9.

(e) *If* $(t, d) \in \mathcal{C}_3^2\big|_{D,T} \cup \mathcal{C}_4 \cup \{(0, d) \in \Pi_{D,T} : d < 0\}$, *then the phase portrait of the Poincaré compactification is topologically equivalent to its correspondent in Figure* 5.9.

(f) *If* $(t, d) \in \mathcal{H}_e\mathcal{L}\big|_{D,T}$, *then the phase portrait of the Poincaré compactification is topologically equivalent to its correspondent in Figure* 5.9.

(g) *If* $(t, d) \in \mathcal{C}_3^1\big|_{D,T}$, *then the phase portrait of the Poincaré compactification is topologically equivalent to its correspondent in Figure* 5.9.

(h) *If* $(t, d) \in \mathcal{C}_2^2 \cup \mathcal{N}$ *and* $t < 0$, *then the phase portrait of the Poincaré compactification is topologically equivalent to its correspondent in Figure* 5.9.

(i) *If* $(t, d) \in \mathcal{SN}_\infty$ *and* $t < 0$, *then the phase portrait of the Poincaré compactification is topologically equivalent to its correspondent in Figure* 5.9.

(j) *If* $(t, d) \in \mathcal{C}_2^1$, *then the phase portrait of the Poincaré compactification is topologically equivalent to its correspondent in Figure* 5.9.

(k) *If* $(t, d) \in \mathcal{O}$, *then the phase portrait of the Poincaré compactification is topologically equivalent to its correspondent in Figure* 5.9.

Proof. All statements follow by reversing the time and applying the correspondig proposition of Subsection 5.2.5. □

5.4.1 The bifurcation set

By Theorem 5.4.1 and Figure 5.9, the half-line $\mathcal{H}_\infty$, the differentiable curve $\mathcal{SN}_\infty$, the straight line $\mathcal{N}$, and the differentiable curve $\mathcal{H}_e\mathcal{L}\big|_{D,T}$ correspond to a Hopf bifurcation at infinity, a saddle-node bifurcation at infinity, a pitchfork bifurcation at infinity, and a homoclinic bifurcation, respectively. For more details, see Subsection 5.2.6.

5.5 The case $D < 0$ and $T < 0$

In this section we describe the phase portraits of the Poincaré compactification of fundamental systems

$$\dot{\mathbf{x}} = A\mathbf{x} + \varphi\left(\mathbf{k}^T\mathbf{x}\right)\mathbf{b},$$

with parameters (D, T, d, t), $D < 0$, $T < 0$, and $(t, d) \in \Pi_{D,T}$. The dynamical richness of this case is due to the potential co-existence of different kinds of Jordan curves formed by solutions.

We start by searching in the parameter plane $\Pi_{D,T}$ for the regions where the system is proper.

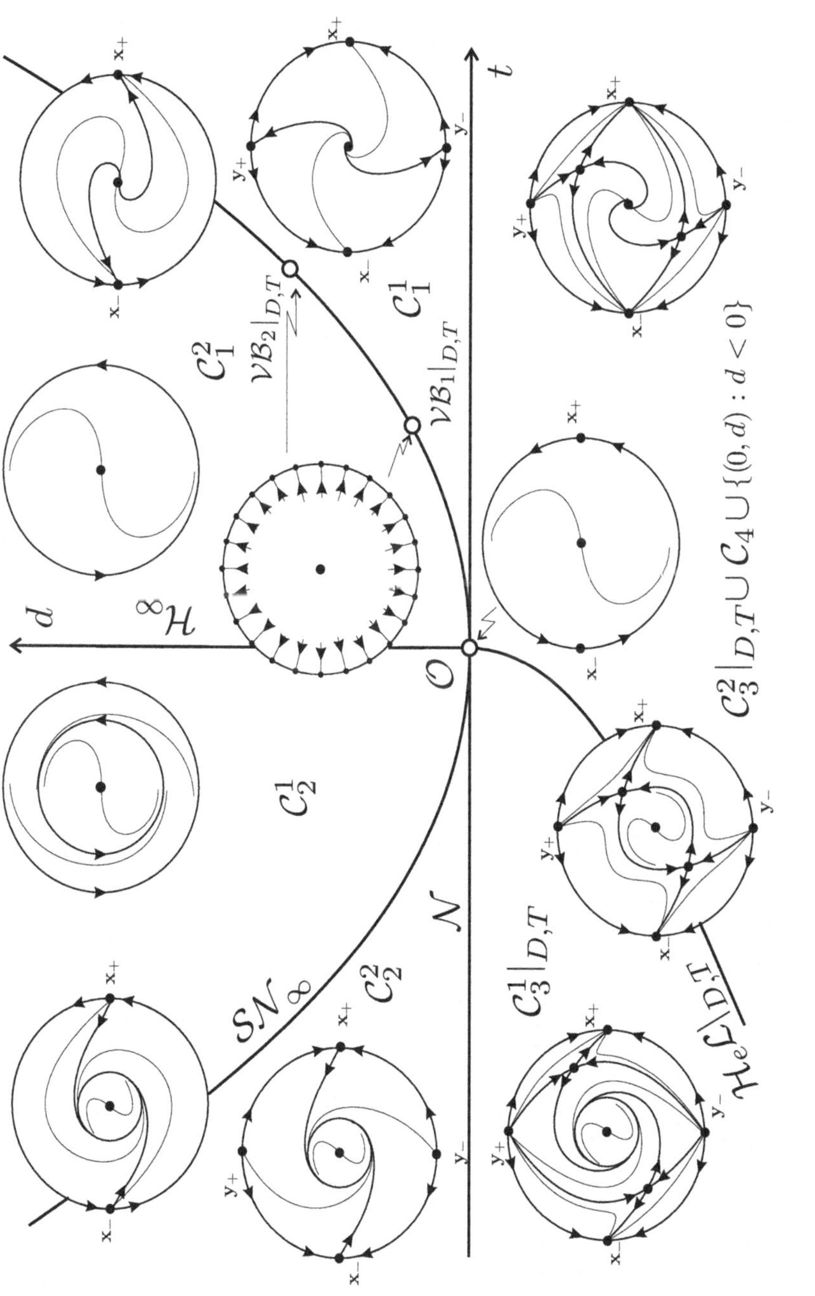

Figure 5.9: Phase portraits and the bifurcation set of the Poincaré compactification of fundamental systems with parameters $D > 0$ and $T > 0$.

5.5.1 Proper fundamental systems

Assume that $D < 0$ and $T < 0$. We recall that fundamental systems with parameters $(t, d) \notin W|_{D,T}$ are proper, see Theorem 4.7.6. Note that systems with parameters (t, d) contained in the set $W_1|_{D,T} \cap W_2|_{D,T}$ (that is, $t = T$ and $d = D$) are not proper. In Figure 5.10 we represent the straight lines $W_1|_{D,T}$ and $W_2|_{D,T}$ for the parameters $D < 0$ and $T < 0$. Finally, observe that for $(t, d) \in W|_{D,T} = W_1|_{D,T} \cup W_2|_{D,T}$ proper and non-proper fundamental systems may co-exist.

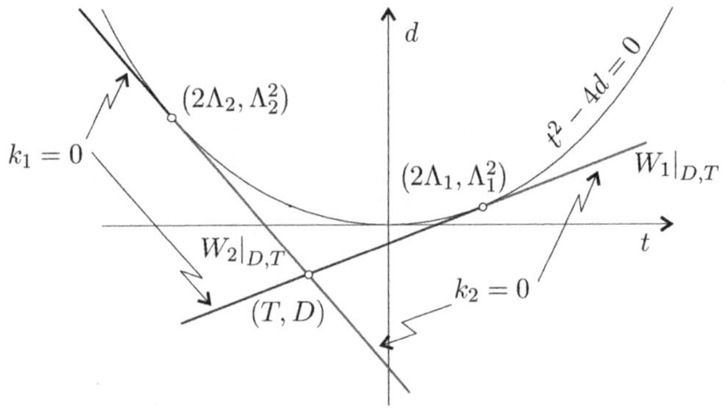

Figure 5.10: Straight lines $W_1|_{D,T}$ and $W_2|_{D,T}$ for the parameters $D < 0$ and $T < 0$.

By Lemmas 4.7.2(c) and (d), when $t^2 - 4d = 0$ and the parameters (t, d) do not belong to $\mathcal{VB}_1|_{D,T} \cup \mathcal{VB}_2|_{D,T}$, the real Jordan normal form of the fundamental matrix A is non-diagonal. Moreover, when $(t, d) \in \mathcal{VB}_1|_{D,T} \cup \mathcal{VB}_2|_{D,T}$, the fundamental parameters do not determine whether the real Jordan normal form of the matrix A is diagonal or not.

5.5.2 Singular points

The following result is a consequence of Theorem 3.9.3. Note that it does not depend on the parameter T.

Proposition 5.5.1. *Consider a fundamental system with parameter $D < 0$.*

(a) *If $d > 0$, then the system has exactly three singular points: a hyperbolic saddle at the origin $\mathbf{0}$, and the points $\mathbf{e}_+ \in S_+$ and $\mathbf{e}_- \in S_-$. If $t = 0$, then $\mathbf{e}_+$ and $\mathbf{e}_-$ are centers. Otherwise, they are hyperbolic and asymptotically stable (when $t < 0$) or unstable (when $t > 0$).*

(b) *If $d \leq 0$, then the system has exactly one hyperbolic saddle at the origin $\mathbf{0}$.*

5.5.3 Behaviour at infinity

The following proposition does not depend on the parameter T.

Proposition 5.5.2. *Consider a fundamental system with parameter $D < 0$ and fundamental matrices (A, B). Suppose that the matrix A is in real Jordan form.*

(a) *If $t^2 - 4d < 0$, then the system has a periodic orbit at infinity ∞.*

(b) *If $t^2 - 4d = 0$ and A is diagonal, then any point on $\partial\mathbb{D} \setminus \{\pm\mathbf{k}^\perp/\|\mathbf{k}\|\}$ is a stable normally hyperbolic singular point when $t > 0$, and a unstable normally hyperbolic singular point when $t < 0$. The local phase portrait of $\pm\mathbf{k}^\perp/\|\mathbf{k}\|$ is topologically equivalent to its correspondent in Figure 5.11(a) when $t > 0$, and in Figure 5.11(b) when $t < 0$.*

(c) *If $t^2 - 4d = 0$ and A is non-diagonal, then there exist exactly two singular points at infinity, $\mathbf{x}_+ \in \partial\mathbb{D}_+$ and $\mathbf{x}_- \in \partial\mathbb{D}_-$. If $t > 0$, then $\mathbf{x}_+$ and $\mathbf{x}_-$ are saddle-nodes with stable hyperbolic manifolds. Moreover, the center manifolds are contained in $\partial\mathbb{D}$. If $t = 0$, then a neighbourhood of the singular points $\mathbf{x}_+$ and $\mathbf{x}_-$ is an elliptic sector. If $t < 0$, then $\mathbf{x}_+$ and $\mathbf{x}_-$ are saddle-nodes with unstable hyperbolic manifolds. The center manifolds are contained in $\partial\mathbb{D}$.*

(d) *If $t^2 - 4d > 0$, then the system has four singular points at infinity, $\mathbf{x}_+$, $\mathbf{y}_+ \in \partial\mathbb{D}_+$, $\mathbf{x}_-$, $\mathbf{y}_- \in \partial\mathbb{D}_-$. If $d \leq 0$, then $\mathbf{x}_+$ and $\mathbf{x}_-$ are asymptotically stable nodes, and $\mathbf{y}_+$ and $\mathbf{y}_-$ are asymptotically unstable nodes. If $d > 0$, $t > 0$ and $k_2 \neq 0$, then $\mathbf{x}_+$ and $\mathbf{x}_-$ are asymptotically stable nodes, and $\mathbf{y}_+$ and $\mathbf{y}_-$ are saddle points with the unstable manifolds contained in $\partial\mathbb{D}$. If $d > 0$, $t > 0$ and $k_2 = 0$, then $\mathbf{x}_+$ and $\mathbf{x}_-$ are asymptotically stable nodes, and the local phase portraits of $\mathbf{y}_+$ and $\mathbf{y}_-$ are topologically equivalent to their correspondents in Figure* 5.11(c). *If $d > 0$, $t < 0$ and $k_1 \neq 0$, then $\mathbf{x}_+$ and $\mathbf{x}_-$ are saddle points with the stable manifolds contained in $\partial\mathbb{D}$, and $\mathbf{y}_+$ and $\mathbf{y}_-$ are asymptotically unstable nodes. If $d > 0$, $t < 0$ and $k_1 = 0$, then the local phase portraits of $\mathbf{x}_+$ and $\mathbf{x}_-$ are topologically equivalent to their correspondents in Figure* 5.11(d), *and $\mathbf{y}_+$ and $\mathbf{y}_-$ are asymptotically unstable nodes.*

Proof. Statement (a) is a consequence of Theorem 3.11.1(a).

(b) Since $t^2 - 4d = 0$ and the matrix A is diagonal, the manifold at infinity $\partial\mathbb{D}$ is formed by singular points, see Theorem 3.11.1(c). Moreover, $(t, d) \in F_0^A|_{D,T} = \mathcal{VB}_1|_{D,T} \cup \mathcal{VB}_2|_{D,T}$, see Lemma 4.7.2(c), which implies that either $t = 2\Lambda_1 > 0$ or $t = 2\Lambda_2 < 0$. The local phase portrait of $\pm\mathbf{k}^\perp/\|\mathbf{k}\|$ follows from Theorem 3.11.9(e). The local phase portraits of the remaining points are as described in Theorem 3.11.5(a).

(c) By Theorem 3.11.1(b), the system has two singular points at infinity. We denote by $\mathbf{x}_+$ the singular point contained in $\partial\mathbb{D}_+$ and by $\mathbf{x}_-$ the singular point contained in $\partial\mathbb{D}_-$. The statement is a consequence of Theorems 3.11.6(a), (b) and 3.11.10(e).

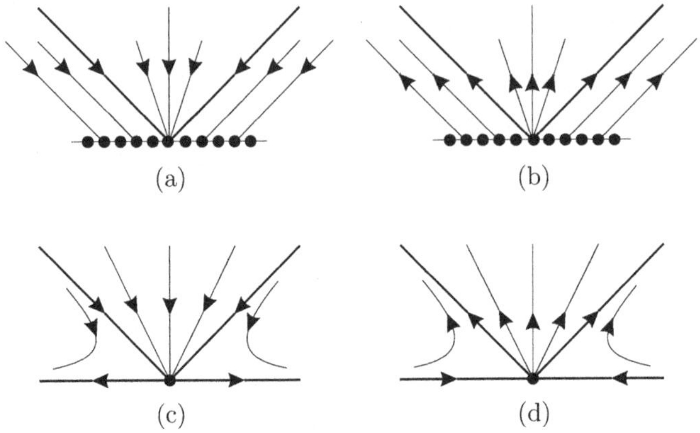

Figure 5.11: Singular points at infinity when $D < 0$, $t^2 - 4d = 0$ (a) and (b), or $t^2 - 4d > 0$ (c) and (d).

(d) By Theorem 3.11.1(b), the system has four singular points at infinity. We denote by $\mathbf{x}_+$, $\mathbf{y}_+$ the singular points contained in $\partial\mathbb{D}_+$ and by $\mathbf{x}_-$, $\mathbf{y}_-$ the singular points contained in $\partial\mathbb{D}_-$.

When $d \leq 0$, the local phase portraits of $\mathbf{x}_+$ and $\mathbf{x}_-$ are as described in Theorems 3.11.7(c) and 3.11.11(f.1). The local phase portraits of $\mathbf{y}_+$ and $\mathbf{y}_-$ are as described in Theorems 3.11.8(c) and 3.11.12(f.1).

When $d > 0$ and $t > 0$, the local phase portraits of $\mathbf{x}_+$ and $\mathbf{x}_-$ are as described in Theorems 3.11.7(a) and 3.11.11(f.1). If $k_2 \neq 0$, then the local phase portraits of $\mathbf{y}_+$ and $\mathbf{y}_-$ are as described in Theorem 3.11.8(a). If $k_2 = 0$, then the local phase portraits of $\mathbf{y}_+$ and $\mathbf{y}_-$ are as described in from Theorem 3.11.12(f.2).

When $d > 0$ and $t < 0$, the local phase portraits of $\mathbf{y}_+$ and $\mathbf{y}_-$ are as described in Theorems 3.11.8(b) and 3.11.12(f.1). If $k_1 \neq 0$, then the local phase portraits of $\mathbf{x}_+$ and $\mathbf{x}_-$ are as described in Theorem 3.11.7(b). If $k_1 = 0$, then the local phase portraits of $\mathbf{x}_+$ and $\mathbf{x}_-$ are as described in Theorem 3.11.11(f.2). □

Following Proposition 5.5.2, in Figure 5.12 we display the behaviour of the flow in a neighbourhood of $\partial\mathbb{D}$. Note that when $t^2 - 4d < 0$, there exists a periodic orbit at infinity. At this time we cannot decide the stability of this periodic orbit. Moreover, it is not clear the behaviour at infinity when $(t, d) \in \mathcal{VB}_1|_{D,T}$ and $(t, d) \in \mathcal{VB}_2|_{D,T}$.

5.5.4 Periodic orbits

First we locate the parameter region in the plane $\Pi_{D,T}$ which contains all the fundamental systems having Jordan curves formed by solutions.

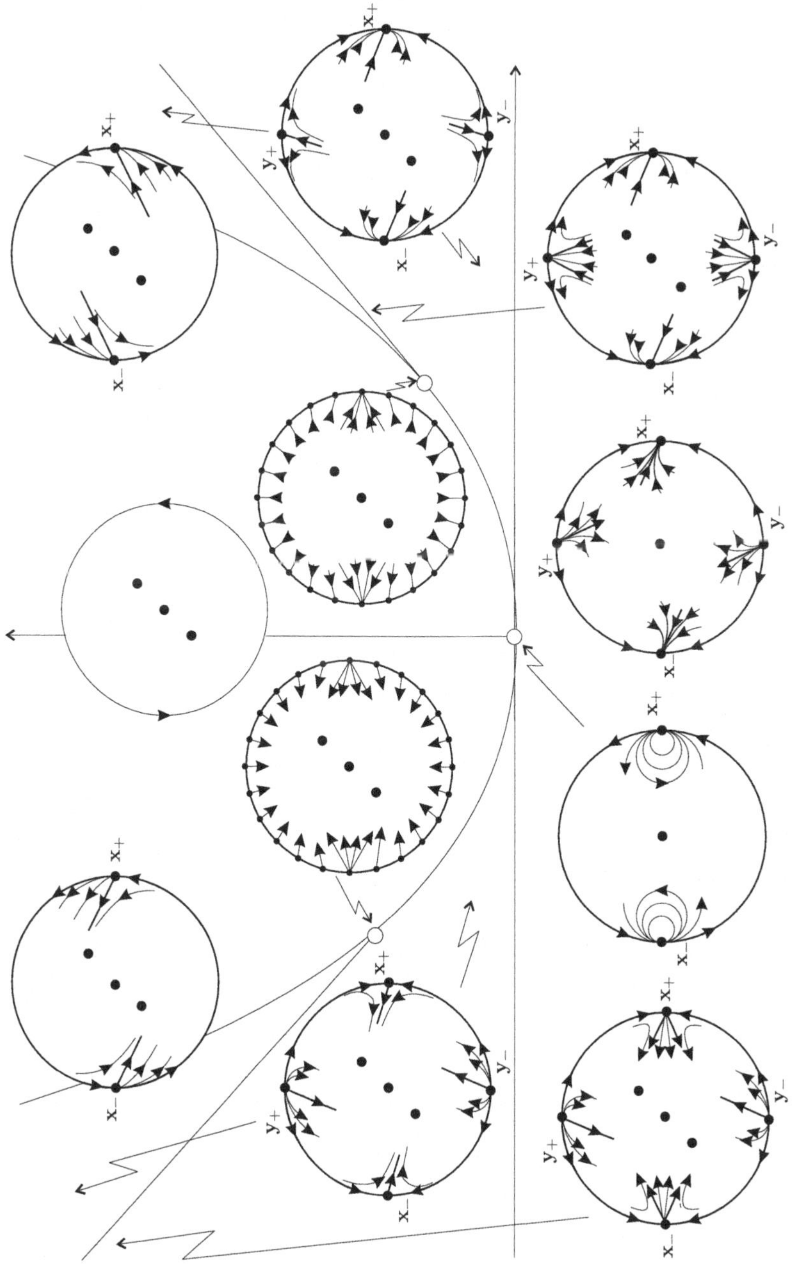

Figure 5.12: Behaviour at infinity when $D < 0$.

Proposition 5.5.3. *Consider a fundamental system with parameters $D < 0$ and $T < 0$.*

(a) *If $t < 0$ or $t = 0$ and $d \le 0$, then the system has no Jordan curves formed by solutions.*

(b) *If $t = 0$ and $d > 0$, then there exist two periodic orbits $\Phi_+ \subset S_+ \cup L_+$ and $\Phi_- \subset S_- \cup L_-$ such that the annular regions $\mathrm{Cl}(\Sigma_{\Phi_+}) \setminus \mathbf{e}_+$ and $\mathrm{Cl}(\Sigma_{\Phi_-}) \setminus \mathbf{e}_-$ are foliated by periodic orbits. Moreover, any Jordan curve formed by solutions is contained in one of these annular regions.*

(c) *If $t > 0$ and Γ is a Jordan curve formed by solutions, then $\Gamma \cap S_0 \neq \varnothing$ and $\Gamma \cap (S_+ \cup S_-) \neq \varnothing$.*

Proof. (a) Suppose that $t < 0$. Under this assumption the statement follows from Theorem 3.12.2(a). Suppose that $t = 0$ and $d \le 0$, and let Γ be a Jordan curve formed by solutions. By Theorem 3.12.2(c), either $\Gamma \subset S_+ \cup L_+$, or $\Gamma \subset S_- \cup L_-$. This contradicts the fact that the systems in the half-planes $S_+ \cup L_+$ and $S_- \cup L_-$ are linear systems with determinant $d \le 0$.

(b) Suppose that $t = 0$ and $d > 0$. Then $(t, d) \notin W|_{D,T}$, which implies that the system is proper. Let $\mathbf{p}_+$ be the contact point of the flow with L_+ and let Φ_+ be the orbit through $\mathbf{p}_+$. By Proposition 4.2.10, the orbit Φ_+ is contained in $S_+ \cup L_+$. Therefore, Φ_+ is a periodic orbit with $\mathbf{e}_+ \in \Sigma_{\Phi_+}$. Since $\mathbf{e}_+$ is a center, the region Σ_{Φ_+} is formed by periodic orbits. By the symmetry of the flow with respect to the origin, there exists a periodic orbit Φ_- through the contact point $\mathbf{p}_- = -\mathbf{p}_+$. Moreover, $\Phi_- \subset S_- \cup L_-$, $\mathbf{e}_- \in \Sigma_{\Phi_-}$, and $\mathrm{Cl}(\Sigma_{\Phi_-}) \setminus \mathbf{e}_-$ is an annular region foliated by periodic orbits, see Figure 5.13. Since any Jordan curve formed by solutions Γ is contained in either $S_+ \cup L_+$ or $S_- \cup L_-$, we conclude that either $\Gamma \subset \mathrm{Cl}(\Sigma_{\Phi_+}) \setminus \mathbf{e}_+$ or $\Gamma \subset \mathrm{Cl}(\Sigma_{\Phi_-}) \setminus \mathbf{e}_-$.

Statement (c) follows from Theorem 3.12.2(b). □

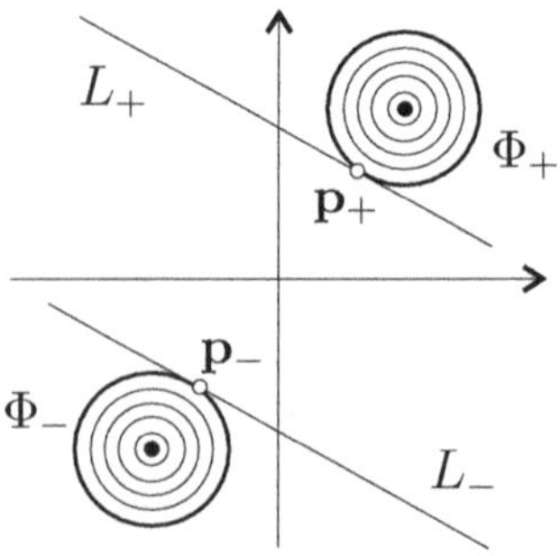

Figure 5.13: Periodic orbits Φ_+ and Φ_-, and the annular regions of periodic orbits when $D < 0$, $T < 0$, $d > 0$, and $t = 0$.

From Proposition 5.5.3 we obtain that only fundamental systems with parameters $t > 0$, or $t = 0$ and $d > 0$, can exhibit Jordan curves Γ formed by solutions. Moreover, when $t > 0$ the Jordan curve Γ intersects L_+ and/or L_-. In this case there exists a contact point of the flow with the straight lines L_+ and L_-, and hence the return map π is defined, see Theorem 4.6.1(a).

In the next lemma we summarize the information about the return map π and the Poincaré maps $\widetilde{\pi}^A_{++}$, π^B_{++} and π^B_{+-}.

Lemma 5.5.4. *Consider a fundamental system with fundamental matrices (A, B) and parameters $D < 0$ and $T \leq 0$. Let $\Lambda_1 > 0 > \Lambda_2$ be the eigenvalues of the matrix B.*

(a) *The return map π is defined. Its behaviour is not trivial if and only if $t^2 - 4d < 0$. In this case*

$$\pi(a) = \begin{cases} \left(\widetilde{\pi}^A_{++} \circ \pi^B_{++}\right)(a), & \text{if } a \in [0, \Lambda_1^{-1}), \\ \left(\widetilde{\pi}^A_{++} \circ \pi^B_{+-}\right)^2(a), & \text{if } a \in \left(\Lambda_1^{-1}, +\infty\right). \end{cases}$$

(b) *If $T = 0$ and the Poincaré maps π^B_{++} and π^B_{+-} are defined, then $\pi^B_{++} : [0, \Lambda_1^{-1}) \to [0, |\Lambda_2|^{-1})$, $\pi^B_{+-} : (\Lambda_1^{-1}, +\infty) \to (|\Lambda_2|^{-1}, +\infty)$, and they are identity maps.*

(c) *If $T < 0$ and the Poincaré map π^B_{++} is defined, then:*

(c.1) $\pi^B_{++} : [0, \Lambda_1^{-1}) \to [0, |\Lambda_2|^{-1})$, $\pi^B_{++}(0) = 0$, $\pi^B_{++}(a) < a$ *in* $(0, \Lambda_1^{-1})$, *and* $\lim_{a \nearrow \Lambda_1^{-1}} \pi^B_{++}(a) = |\Lambda_2|^{-1}$;

(c.2) *if* $a \in (0, \Lambda_1^{-1})$, *then* $0 < (\pi^B_{++})'(a) < 1$, $\lim_{a \searrow 0} (\pi^B_{++})'(a) = 1$, $\lim_{a \nearrow \Lambda_1^{-1}} (\pi^B_{++})'(a) = 0$, *and* $(\pi^B_{++})''(a) < 0$;

(c.3) *the straight line $b = |\Lambda_2|^{-1}$ is a horizontal asymptote of the graph of π^B_{++}.*

(d) *If $T < 0$ and the Poincaré map π^B_{+-} is defined, then:*

(d.1) $\pi^B_{+-} : (\Lambda_1^{-1}, +\infty) \to (|\Lambda_2|^{-1}, +\infty)$, $\pi^B_{+-}(a) < a$, $\lim_{a \searrow \Lambda_1^{-1}} \pi^B_{+-}(a) = |\Lambda_2|^{-1}$, *and* $\lim_{a \nearrow +\infty} \pi^B_{+-}(a) = +\infty$;

(d.2) *if* $a \in (\Lambda_1^{-1}, +\infty)$, *then* $(\pi^B_{+-})'(a) < 1$, $\lim_{a \searrow \Lambda_1^{-1}} d\pi^B_{+-}(a)/da = 0$, $\lim_{a \nearrow +\infty} (\pi^B_{+-})'(a) = 1$, *and* $(\pi^B_{+-})''(a) > 0$;

(d.3) *the straight line $b = a + 2T/D$ is an asymptote of the graph of π^B_{+-}. Moreover, $\pi^B_{+-}(a) > a + 2T/D$ for every $a \in (\Lambda_1^{-1}, +\infty)$.*

(e) *If $t = 0$ and the Poincaré map $\widetilde{\pi}^A_{++}$ is defined, then $\widetilde{\pi}^A_{++}$ coincides with the identity in the interval $[0, +\infty)$.*

(f) *If* $t > 0$ *and the Poincaré map* $\widetilde{\pi}^A_{++}$ *is defined, then:*

(f.1) *there exists a value* $b^* > 0$ *such that* $\widetilde{\pi}^A_{++} : [0, +\infty) \to [b^*, +\infty)$, $\widetilde{\pi}^A_{++}(a) > a$ *in* $(0, +\infty)$, *and* $\widetilde{\pi}^A_{++}(0) = b^*$;

(f.2) $(\widetilde{\pi}^A_{++})'(a) > 0$, $\lim\limits_{a \searrow 0} (\widetilde{\pi}^A_{++})'(a) = 0$, *and* $(\widetilde{\pi}^A_{++})''(a) > 0$;

(f.3) *the straight line* $b = e^{\gamma\pi}a + t(1 + e^{\gamma\pi})/d$ *is an asymptote of the graph of* $\widetilde{\pi}^A_{++}$, *where* $\gamma = t/\sqrt{4d - t^2}$. *Moreover,* $\widetilde{\pi}^A_{++}(a) > e^{\gamma\pi}a + t(1 + e^{\gamma\pi})/d$ *in* $[0, +\infty)$.

Proof. (a) Suppose that $t^2 - 4d < 0$. Thus $(t, d) \notin W|_{D,T}$, see Lemma 4.7.3(a), and the system is proper. From this we conclude that the return map π is defined. The expression of π is a consequence of Theorem 4.6.1(d). Thus, if π is defined and is non-trivial, then from Theorem 4.6.1(d) it follows that $t^2 - 4d < 0$.

The remaining statements except (d.3) and (f.3) follow from Corollary 4.4.16, 4.4.19 and Propositions 4.5.7, 4.4.15 and 4.4.18.

(d.3) Define the auxiliary function $\psi(a) = \pi^B_{+-}(a) - a - 2T/D$ in the interval $(\Lambda_1^{-1}, +\infty)$. From statement (d.2) it follows that $\psi'(a) < 0$, i.e., $\psi(a)$ is strictly decreasing in $(\Lambda_1^{-1}, +\infty)$. On the other hand, since $b = a + 2T/D$ is the asymptote of the graph of π^B_{+-}, we get that $\lim\limits_{a \nearrow +\infty} \psi(a) = 0$. Therefore, $\psi(a) > 0$ in $(\Lambda_1^{-1}, +\infty)$.

(f.3) Define $\psi(a) = \widetilde{\pi}^A_{++}(a) - e^{\gamma\pi}a - t(1 + e^{\gamma\pi})/d$ in $[0, +\infty)$. Since $\psi'(a) = (\widetilde{\pi}^A_{++})'(a) - e^{\gamma\pi}$ and $\psi''(a) = (\widetilde{\pi}^A_{++})''(a)$, statement (f.2) implies that $\psi''(a) > 0$, i.e., $\psi'(a)$ is strictly increasing. On the other hand, since the straight line $b = e^{\gamma\pi}a + t(1+e^{\gamma\pi})/d$ is an asymptote of the graph of $\widetilde{\pi}^A_{++}$, we get that $\lim\limits_{a \nearrow +\infty} \psi'(a) = 0$ and $\lim\limits_{a \nearrow +\infty} \psi(a) = 0$. Therefore, $\psi'(a) < 0$ and $\psi(a) > 0$ in $[0, +\infty)$. □

The return map π is defined only when $t^2 - 4d < 0$, see Lemma 5.5.4(a). Therefore, $\{(t, d) \in \Pi_{D,T} : t^2 - 4d \leq 0 \text{ and } t \geq 0\}$ is the parameter region in the plane $\Pi_{D,T}$ which contains the fundamental systems with Jordan curves formed by solutions. From this we obtain the following proposition.

Proposition 5.5.5. *Fundamental systems with parameters* $D < 0$, $T < 0$, $t > 0$, *and* $t^2 - 4d \geq 0$ *have no Jordan curves formed by solutions.*

In Proposition 5.5.3(b) we have proved that when $t = 0$ and $d > 0$, there exist two periodic orbits Φ_+ and Φ_- such that the annular regions $\mathrm{Cl}(\Sigma_{\Phi_+}) \setminus \{\mathbf{e}_+\}$ and $\mathrm{Cl}(\Sigma_{\Phi_-}) \setminus \{\mathbf{e}_-\}$ contain all of the Jordan curves formed by solutions. These Jordan curves are periodic orbits. Using the return map π described in Lemma 5.5.4 we study the local phase portraits of Φ_+ and Φ_-.

Proposition 5.5.6. *Consider a fundamental system with parameters* $D < 0$, $T < 0$, $d > 0$, *and* $t = 0$. *Then the periodic orbits* Φ_+ *and* Φ_- *are non-hyperbolic and they are outside asymptotically stable.*

Proof. Since $t^2 - 4d < 0$, the Poincaré maps $\widetilde{\pi}^A_{++}$, π^B_{++} and π^B_{+-} and the return map π are defined. Moreover, $\pi = \widetilde{\pi}^A_{++} \circ \pi^B_{++}$ in $[0, \Lambda_1^{-1})$, $\pi(0) = 0$, and $0 < \pi'(a) < 1$, see Lemma 5.5.4. Since Φ_+ and Φ_- intersect L_+ and L_-, respectively, at points with coordinate $a = 0$, the result follows from Proposition 4.6.2. □

Lamerey map

As we have seen, only fundamental systems with parameters $t \geq 0$ and $t^2 - 4d < 0$ can exhibit Jordan curves formed by solutions. Moreover, when $t = 0$, the regions $\mathrm{Cl}(\Sigma_{\Phi_+})$ and $\mathrm{Cl}(\Sigma_{\Phi_-})$ contain all Jordan curves formed by solutions, and the periodic orbits Φ_+ and Φ_- are outside asymptotically stable.

Now we study the Jordan curves formed by solutions when the fundamental parameters lie in $\mathcal{C}_1^2 = \{(t, d) : t > 0 \text{ and } t^2 - 4d < 0\}$. Note that in the following result we also include the case $T = 0$, which we consider in the next section.

Lemma 5.5.7. *Consider a fundamental system with parameters $D < 0$, $T \leq 0$, and $(t, d) \in \mathcal{C}_1^2$.*

(a) *The Lamerey map*

$$g(a) = \begin{cases} \widetilde{\pi}^A_{++}(a) - \pi^{-B}_{++}(a), & \text{if } a \in \left[0, |\Lambda_2|^{-1}\right), \\ \widetilde{\pi}^A_{++}(a) - \Lambda_1^{-1}, & \text{if } a = |\Lambda_2|^{-1}, \\ \widetilde{\pi}^A_{++}(a) - \pi^{-B}_{+-}(a), & \text{if } a \in \left(|\Lambda_2|^{-1}, +\infty\right), \end{cases}$$

is defined and continuous in $[0, +\infty)$, and it satisfies that $g \in C^1((0, +\infty))$ and $g \in C^\omega((0, \infty) \setminus |\Lambda_2|^{-1})$.

(b) *The map g has a zero at a value $a^* \neq |\Lambda_2|^{-1}$ if and only if the system has a periodic orbit Γ. In this case, the periodic orbit Γ intersects the straight line L_+ at the point of coordinate $\widetilde{\pi}^A_{++}(a^*)$. Moreover, if $a^* < |\Lambda_2|^{-1}$, then $\mathbf{e}_+ \in \Sigma_\Gamma$ and $\mathbf{0} \notin \Sigma_\Gamma$; and if $a^* > |\Lambda_2|^{-1}$, then $\{\mathbf{0}, \mathbf{e}_+, \mathbf{e}_-\} \subset \Sigma_\Gamma$.*

(b.1) *If $g'(a^*) > 0$, then Γ is an asymptotically unstable hyperbolic limit cycle.*

(b.2) *If $g'(a^*) = 0$ and there exists $\varepsilon > 0$ such that $g' < 0$ in $(a^* - \varepsilon, a^*)$ and $g' > 0$ in $(a^*, a^* + \varepsilon)$, then Γ is a non-hyperbolic limit cycle which is inside asymptotically stable and outside asymptotically unstable.*

(b.3) *If $g'(a^*) = 0$ and there exists $\varepsilon > 0$ such that $g' > 0$ in $(a^* - \varepsilon, a^*)$ and $g' < 0$ in $(a^*, a^* + \varepsilon)$, then Γ is a non-hyperbolic limit cycle which is inside asymptotically unstable and outside asymptotically stable.*

(b.4) *If $g'(a^*) < 0$, then Γ is an asymptotically stable hyperbolic limit cycle.*

(c) *The map g has a zero at the value $|\Lambda_2|^{-1}$ if and only if the system has two homoclinic cycles Δ_+ and Δ_- to a common singular point at the origin $\mathbf{0}$. Moreover, $\Delta = \Delta_+ \cup \Delta_-$ is a double homoclinic cycle.*

Proof. (a) Suppose that $(t,d)\in\mathcal{C}_1^2$. By Lemma 5.5.4(a), the maps $\widetilde{\pi}_{++}^A$, π_{++}^B and π_{+-}^B are defined. Since $\widetilde{\pi}_{++}^A$, π_{++}^B and π_{+-}^B are analytic functions, the Lamerey map g is analytic in $(0,-\Lambda_2^{-1})\cup(-\Lambda_2^{-1},+\infty)$. From Lemma 5.5.4(c.1)(d.1)(c.2) and (d.2) it follows that g and g' are continuous at $|\Lambda_2|^{-1}$.

(b) Consider the Lamerey map g_B defined in (4.48). It is easy to check that g is the extension of g_B to the value $|\Lambda_2|^{-1}$. Thus, $a^*\neq|\Lambda_2|^{-1}$ is a zero of g if and only if a^* is a zero of g_B. In this case, $\widetilde{\pi}_{++}^A(a^*)$ is a fixed point of the return map π. Therefore, there exists a periodic orbit Γ intersecting L_+ at $\widetilde{\pi}_{++}^A(a^*)$. Moreover, when $a^*<|\Lambda_2|^{-1}$, the return map in a neigbourhood of a^* is given by $\widetilde{\pi}_{++}^A\circ\pi_{++}^B$. Hence, $\mathbf{e}_+\in\Sigma_\Gamma$ (see the definition of $\widetilde{\pi}_{++}^A$) and Σ_Γ does not contain the origin (see the definition of π_{++}^B). Analogously, when $a^*>|\Lambda_2|^{-1}$, the singular points $\mathbf{e}_+$, $\mathbf{e}_-$ and $\mathbf{0}$ are contained in Σ_Γ.

(b.1) Suppose that $a^*\in[0,|\Lambda_2|^{-1})$, $g(a^*)=0$, and $g'(a^*)>0$. Hence, $\widetilde{\pi}_{++}^A(a^*)=\pi_{++}^{-B}(a^*)$ and $(\widetilde{\pi}_{++}^A)'(a^*)>(\pi_{++}^{-B})'(a^*)$. From Lemma 5.5.4(a) it follows that

$$\begin{aligned}\left.\frac{d\pi}{da}\right|_{\widetilde{\pi}_{++}^A(a^*)}&=\left.\frac{d\widetilde{\pi}_{++}^A}{da}\right|_{\pi_{++}^B\left(\widetilde{\pi}_{++}^A(a^*)\right)}\left.\frac{d\pi_{++}^B}{da}\right|_{\widetilde{\pi}_{++}^A(a^*)}\\&>\left.\frac{d\pi_{++}^{-B}}{da}\right|_{\pi_{++}^B\left(\widetilde{\pi}_{++}^A(a^*)\right)}\left.\frac{d\pi_{++}^B}{da}\right|_{\widetilde{\pi}_{++}^A(a^*)}=1.\end{aligned}$$

Therefore the limit cycle Γ is asymptoticaly unstable.

The statement follows by using similar arguments when we suppose that $a^*\in(|\Lambda_2|^{-1},+\infty)$.

Statements (b.2), (b.3) and (b.4) follows in a similar way.

(c) Suppose that $|\Lambda_2|^{-1}$ is a zero of the Lamerey map g. Then $\widetilde{\pi}_{++}^A(|\Lambda_2|^{-1})=\Lambda_1^{-1}$. Since the stable and the unstable separatrices of the saddle at the origin intersect L_+ at the points with coordinates Λ_1^{-1} and $|\Lambda_2|^{-1}$, respectively, see Corollary 4.4.17, we conclude that the two separatrices connect and they form a homoclinic cycle. □

We now introduce the auxiliary functions

$$y(a)=e^{\gamma\pi}a+\frac{t}{d}\left(1+e^{\gamma\pi}\right),$$

where

$$\gamma=\frac{t}{\sqrt{4d-t^2}},$$

and

$$h(a)=\begin{cases}y(a)-\pi_{++}^{-B}(a), & \text{if } a\in\left[0,|\Lambda_2|^{-1}\right),\\ y(a)-\Lambda_1^{-1}, & \text{if } a=|\Lambda_2|^{-1},\\ y(a)-\pi_{+-}^{-B}(a), & \text{if } a\in\left(|\Lambda_2|^{-1},+\infty\right).\end{cases}\tag{5.10}$$

Note that $y(a)$ is the asymptote of the graph of the Poincaré map $\widetilde{\pi}^A_{++}$, see Lemma 5.5.4 (f.3). Thus h is an approximation of the Lamerey map g. In the following result we present some properties of the function h.

Lemma 5.5.8. *Suppose that $D > 0$ and $T < 0$, and let $g(a)$ and $h(a)$ be the functions defined in Lemma* 5.5.7(a) *and in* (5.10). *Then $g(a) > h(a)$ in $(0, +\infty)$, $\lim_{a \nearrow +\infty} g(a) - h(a) = 0$, $g'(a) < h'(a)$ in $[0, +\infty)$, and $\lim_{a \nearrow +\infty} g'(a) - h'(a) = 0$.*

Proof. From Lemma 5.5.4(f.3) we obtain that $\widetilde{\pi}^A_{++}(a) > y(a)$, which implies that $g(a) > h(a)$. Moreover, since $y(a)$ is an asymptote of the graph of the map $\widetilde{\pi}^A_{++}(a)$, it follows that $\lim_{a \nearrow +\infty} g(a) - h(a) = 0$ and $\lim_{a \nearrow +\infty} g'(a) - h'(a) = 0$.

On the other hand, since $(\widetilde{\pi}^A_{++})''(a) > 0$ (see Lemma 5.5.4(f.3)), the function $(\widetilde{\pi}^A_{++})'(a)$ is strictly increasing in $(0, +\infty)$. From this we conclude that

$$\frac{d\widetilde{\pi}^A_{++}}{da} < \lim_{a \nearrow +\infty} \frac{d\widetilde{\pi}^A_{++}}{da} = \lim_{a \nearrow +\infty} y'(a) = e^{\gamma\pi}.$$

Since $y(a)$ is a straight line, it follows that $(\widetilde{\pi}^A_{++})'(a) < y'(a)$ in $(0, +\infty)$, and therefore $g'(a) < h'(a)$ in $(0, +\infty)$. □

Lemma 5.5.9. *Suppose that $D < 0$ and $T < 0$, and let g be the map defined in Lemma* 5.5.7(a).

(a) *If a_1 is a zero of g in $[0, |\Lambda_2|^{-1}]$, then $g' < 0$ in $[a_1, |\Lambda_2|^{-1}]$. Therefore, g has at most one zero in $[0, |\Lambda_2|^{-1}]$.*

(b) *The map g' is strictly increasing in $(|\Lambda_2|^{-1}, +\infty)$ and there is a unique value $a^* \in (|\Lambda_2|^{-1}, +\infty)$ such that $g'(a^*) = 0$. Moreover, a^* is a minimum of g in $(|\Lambda_2|^{-1}, +\infty)$. Finally, $\lim_{a \nearrow +\infty} g(a) = +\infty$.*

(c) *If a_1 is a zero of g in $[0, |\Lambda_2|^{-1}]$, then there exists exactly one zero a_2 of g in $(|\Lambda_2|^{-1}, +\infty)$ and $g'(a_2) > 0$.*

(d) *If the map g has no zeros in $[0, |\Lambda_2|^{-1}]$, then either g has two zeros, $a_1 < a_2$, in $(|\Lambda_2|^{-1}, +\infty)$ which satisfy $g'(a_1) < 0 < g'(a_2)$, or g has exactly one zero in $(|\Lambda_2|^{-1}, +\infty)$ with multiplicity greater than* 1, *or g has no zeros in $(|\Lambda_2|^{-1}, +\infty)$.*

Proof. (a) Suppose that a_1 is a zero of g in the interval $[0, |\Lambda_2|^{-1}]$, and assume that there exists $\xi \in [a_1, |\Lambda_2|^{-1}]$ such that $g'(\xi) \geq 0$. Then $h'(\xi) > 0$, see Lemma 5.5.8. Since $h''(a) = -(\pi^{-B}_{++})''(a)$, it is easy to check that $h''(a) < 0$ in $(0, |\Lambda_2|^{-1})$ and $h''(|\Lambda_2|^{-1}) = 0$, see Lemma 5.5.4(c). Thus $h' > 0$ in $(0, \xi]$. Therefore, since $h(0) = t(1 + e^{\gamma\pi})/d > 0$, we obtain that $h > 0$ in $[0, \xi]$, which contradicts the fact that $h(a_1) < g(a_1) = 0$. Then we conclude that $g'(a) < 0$ in $[a_1, |\Lambda_2|^{-1}]$.

(b) From Lemmas 5.5.4(d.2) and (f.2) it follows that

$$g''(a) = \left.\frac{d^2\widetilde{\pi}^A_{++}}{da^2}\right|_a - \left.\frac{d^2\pi^{-B}_{+-}}{da^2}\right|_a > 0$$

in $(|\Lambda_2|^{-1}, +\infty)$. Thus, $g'(a)$ is strictly increasing in $(|\Lambda_2|^{-1}, +\infty)$.

On the other hand, from Lemma 5.5.4, we conclude that $\lim_{a \nearrow |\Lambda_2|^{-1}} g'(a) = -\infty$ and $\lim_{a \nearrow +\infty} g'(a) = e^{\gamma\pi} - 1 > 0$. Therefore, there exists a unique value a^* in the interval $(|\Lambda_2|^{-1}, +\infty)$ such that $g'(a^*) = 0$. Moreover, g takes its minimum value at a^*.

Since the straight line $b = a + 2T/D$ is an asymptote of the graph of the map π^B_{+-}, see Lemma 5.5.4(d.3), we obtain that

$$\lim_{a \nearrow +\infty} h(a) = \lim_{a \nearrow +\infty} \left[(e^{\gamma\pi} - 1)\, a + \left(\frac{t}{d} \left(1 + e^{\gamma\pi}\right) + 2\frac{T}{D} \right) \right] = +\infty.$$

From Lemma 5.5.8 we conclude that $\lim_{a \nearrow +\infty} g(a) = +\infty$.

(c) Suppose that a_1 is a zero of the map g in the interval $[0, |\Lambda_2|^{-1}]$. By statement (a), a_1 is the unique zero in $[0, |\Lambda_2|^{-1}]$, and $g' < 0$ on $[0, |\Lambda_2|^{-1}]$. Thus $g(|\Lambda_2|^{-1}) < 0$. By statement (b), g is strictly increasing in the interval $(|\Lambda_2|^{-1}, +\infty)$ and $\lim_{a \nearrow +\infty} g(a) = +\infty$. Hence, there exists a unique zero a_2 of g in $(|\Lambda_2|^{-1}, +\infty)$.

(d) Since $g(0) > 0$ and there are no zeros of g in the interval $(0, |\Lambda_2|^{-1}]$, it follows that $g > 0$ in $(0, |\Lambda_2|^{-1}]$. By statement (b), there are three possibilities: (i) the value of g at the minimum is negative, which implies that g has two zeros, $a_1 \in (|\Lambda_2|^{-1}, a^*)$ and $a_2 \in (a^*, +\infty)$; (ii) the value of g at the minimum is zero, which implies that a^* has multiplicity greater that 1; (iii) the value of g at the minimum is positive, which implies that $g > 0$. □

In Lemma 5.5.9 we have summarized the possible qualitative behaviours of the graph of the map g when the parameters (t, d) are in the region $\mathcal{C}^2_1$. Each of these possible graphs are drawn in Figure 5.14.

Now we will prove that the five possibilities depicted in Figure 5.14 can be obtained by varying the parameters (t, d) in $\mathcal{C}^2_1$. Specifically, we exhibit two differentiable curves $\mathcal{H}_o\mathcal{L}|_{D,T}$ and $\mathcal{NH}_{lc}|_{D,T}$ in $\mathcal{C}^2_1$ such that, if we take the parameters in $\mathcal{H}_o\mathcal{L}|_{D,T}$, then the graph of the corresponding map g is the one sketched in Figure 5.14(b); and if we take the parameters in $\mathcal{NH}_{lc}|_{D,T}$, then the graph of the corresponding map g is sketched in Figure 5.14(d). We will also prove that these curves split $\mathcal{C}^2_1$ into three regions, and in each of these regions the graph of the corresponding map g is equivalent to one of the graphs shown in Figure 5.14(a), (c) and (e), respectively.

For a fixed value $d_0 > 0$ we define the family of maps $g(a,t) = g(a)$ and $\widetilde{\pi}^A_{++}(a,t) = \widetilde{\pi}^A_{++}(a)$, where $0 < t < 2\sqrt{d_0}$. Note that the Poincaré maps π^{-B}_{++} and π^{-B}_{+-} and the eigenvalues of the matrix B, denoted by Λ_1 and Λ_2, depend only on the parameters T and D and they are fixed along this section.

Lemma 5.5.10. *Suppose that $D < 0$, $T < 0$, and $d_0 > 0$ and consider the family of maps $g(a,t)$ with $0 < t < 2\sqrt{d_0}$.*

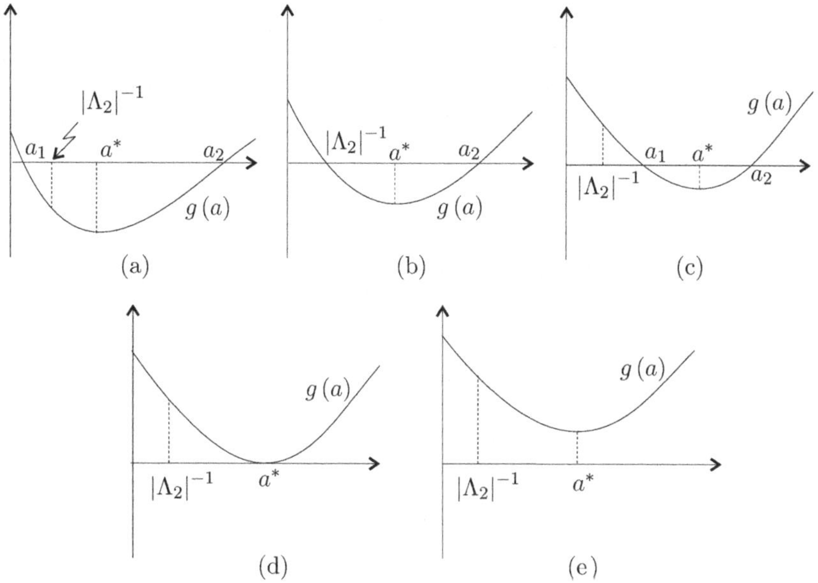

Figure 5.14: The qualitative behaviour of the graph of the map g when (t, d) varies in the parameter region $\mathcal{C}_1^2$.

(a) *Given* $a \in (0, +\infty)$, $\lim_{t \searrow 0} g(a,t) < 0$.

(b) *There exists a value* $t(d_0)$ *in* $(0, 2\sqrt{d_0})$ *such that* $g(a, t(d_0)) > 0$ *in* $[0, +\infty)$. *Moreover,* $\lim_{t \nearrow 2\sqrt{d_0}} g(a,t) = +\infty$ *uniformly in* a.

(c) *If* $a \in [0, +\infty)$ *and* $t \in (0, 2\sqrt{d_0})$, *then* $\partial g/\partial t|_{(a,t)} > 0$.

(d) *Given* $a \in (0, +\infty)$, *there exists a value* $t(a, d_0)$ *in* $(0, 2\sqrt{d_0})$ *with* $g(a, t(a, d_0)) = 0$.

Proof. (a) Since the Poincaré maps depend differentiably on the parameters, see Lemma 4.3.5, it follows that the Poincaré map $\widetilde{\pi}_{++}^{A}(a,t)$ and the map g are differentiable in t. Thus $\lim_{t \searrow 0} \widetilde{\pi}_{++}^{A}(a,t) = \widetilde{\pi}_{++}^{A}(a,0)$ for any $a \in [0, +\infty)$. On the other hand, since the map $\widetilde{\pi}_{++}^{A}(a,0)$ is the identity, see Lemma 5.5.4(e), it follows that $\lim_{t \searrow 0} \widetilde{\pi}_{++}^{A}(a,t) = a$ for any $a \in [0, +\infty)$. Therefore,

$$\lim_{t \searrow 0} g(a,t) = \begin{cases} a - \pi_{++}^{-B}(a), & \text{if } a \in \left(0, |\Lambda_2|^{-1}\right), \\ |\Lambda_2|^{-1} - \Lambda_1^{-1}, & \text{if } a = |\Lambda_2|^{-1}, \\ a - \pi_{+-}^{-B}(a), & \text{if } a \in \left(|\Lambda_2|^{-1}, +\infty\right). \end{cases}$$

The statement follows by noting that $\pi_{++}^{-B}(a) > a$ and $\pi_{+-}^{-B}(a) > a$, see Lemmas 5.5.4(c.1) and (d.1), and that $|\Lambda_2|^{-1} - \Lambda_1^{-1} = -T/D < 0$.

(b) Suppose that $a \in [0, |\Lambda_2|^{-1})$. Hence $g(a,t) = \widetilde{\pi}_{++}^{A}(a,t) - \pi_{++}^{-B}(a)$. The map π_{++}^{-B} satisfies that $0 \le \pi_{++}^{-B}(a) < \Lambda_1^{-1}$ and the map $\widetilde{\pi}_{++}^{A}$ satisfies that

$$\widetilde{\pi}_{++}^{A}(a,t) > e^{\gamma\pi}a + \frac{t}{d_0}\left(1 + e^{\gamma\pi}\right) > \frac{t}{d_0}\left(1 + e^{\gamma\pi}\right),$$

where $\gamma = t/\sqrt{4d_0 - t^2}$, see Lemma 5.5.4. It follows that $g(a,t) > t(1+e^{\gamma\pi})/d_0 - \Lambda_1^{-1}$.

Note that this inequality also holds when $a = |\Lambda_2|^{-1}$. In this case $g(a,t) = \widetilde{\pi}_{++}^{A}(a,t) - \Lambda_1^{-1}$ and the map $\widetilde{\pi}_{++}^{A}$ satisfies the same inequality as before, see Lemma 5.5.4.

Suppose now that $a \in (|\Lambda_2|^{-1}, +\infty)$. Then $g(a,t) = \widetilde{\pi}_{++}^{A}(a,t) - \pi_{+-}^{-B}(a)$. By Lemma 5.5.4, it follows that $\Lambda_1^{-1} < \pi_{+-}^{-B}(a) < a - 2T/D$. Thus $g(a,t) > e^{\gamma\pi}a + t(1+e^{\gamma\pi})/d_0 - a - 2T/D$. Therefore, we can write that

$$g(a,t) > \frac{t}{d_0}\left(1 + e^{\gamma\pi}\right) - \max\left\{\Lambda_1^{-1}, 2\frac{T}{D}\right\}.$$

Taking the value $t(d_0)$ in such a way that $t(d_0)/d_0 > \max\{\Lambda_1^{-1}, 2\frac{T}{D}\}$, we obtain that $g(a,t(d_0)) > 0$ in $[0,+\infty)$. Moreover, we conclude that $\lim_{t \nearrow 2\sqrt{d_0}} g(a,t) = +\infty$ uniformly in the variable a.

(c) For any given $a_0 \in [0,+\infty)$ we consider the map $b(t) := \widetilde{\pi}_{++}^{A}(a_0,t)$ defined in the interval $(0, 2\sqrt{d_0})$. From the definition of $g(a,t)$ it follows that

$$\left.\frac{\partial g}{\partial t}\right|_{(a_0,t)} = \frac{db}{dt}.$$

Thus we need to prove that $db/dt > 0$ in $(0, 2\sqrt{d_0})$.

From Proposition 4.5.7(a.4) and Corollary 4.5.8(e) it follows that $1 - bt + b^2 d_0 = (1 + a_0 t + a_0^2 d_0)e^{2\gamma(t)\tau(b,t)}$, where $b = b(t)$, $\gamma(t) = \frac{t}{\sqrt{4d_0 - t^2}}$, and $\tau(b,t) = \arctan\left(\frac{(a_0+b)\sqrt{4d_0-t^2}}{(b-a_0)t-2+2a_0bd}\right)$. Thus, the map b is implicitly defined by

$$\ln\left(\frac{1 - bt + b^2 d_0}{1 + a_0 t + a_0^2 d_0}\right) = 2\gamma(t)\,\tau(b,t). \tag{5.11}$$

Consider the auxiliary maps

$$\psi_1(x,y) := \ln\left(\frac{1 - xy + x^2 d_0}{1 + a_0 y + a_0^2 d_0}\right) \quad \text{and} \quad \psi_2(x,y) := 2\gamma(y)\,\tau(x,y).$$

Using this maps we can rewrite expression (5.11) by $\psi_1(b,t) - \psi_2(b,t) = 0$. Differentiating this expression with respect to t we obtain

$$0 = \frac{d}{dt}\left(\psi_1(b,t) - \psi_2(b,t)\right)$$
$$= \left(\left.\frac{\partial \psi_1}{\partial x}\right|_{(b,t)} \frac{db}{dt} + \left.\frac{\partial \psi_1}{\partial y}\right|_{(b,t)} \right) - \left(\left.\frac{\partial \psi_2}{\partial x}\right|_{(b,t)} \frac{db}{dt} + \left.\frac{\partial \psi_2}{\partial y}\right|_{(b,t)} \right),$$

which is equivalent to

$$\left(\left.\frac{\partial \psi_1}{\partial x}\right|_{(b,t)} - \left.\frac{\partial \psi_2}{\partial x}\right|_{(b,t)} \right) \frac{db}{dt} = \left(\left.\frac{\partial \psi_2}{\partial y}\right|_{(b,t)} - \left.\frac{\partial \psi_1}{\partial y}\right|_{(b,t)} \right). \tag{5.12}$$

Since

$$\left.\frac{\partial \psi_1}{\partial x}\right|_{(b,t)} = \frac{2bd_0 - t}{1 - bt + b^2 d_0},$$
$$\left.\frac{\partial \psi_1}{\partial y}\right|_{(b,t)} = -\frac{(b + a_0)(1 + a_0 b d_0)}{(1 - bt + b^2 d_0)(1 + a_0 t + a_0^2 d_0)},$$
$$\left.\frac{\partial \psi_2}{\partial x}\right|_{(b,t)} = -\frac{t}{1 - bt + b^2 d_0},$$

and

$$\left.\frac{\partial \psi_2}{\partial y}\right|_{(b,t)} = \frac{2}{4d_0 - t^2}\left(\frac{4d_0 \tau}{\sqrt{4d_0 - t^2}}\right)$$
$$- \frac{2}{4d_0 - t^2}\left(\frac{(b + a_0)\left[(a_0 b d_0 - 1)t^2 + (b - a_0)2d_0 t\right]}{2(1 - bt + b^2 d_0)(1 + a_0 t + a_0^2 d_0)}\right),$$

relation (5.12) can be recast as

$$\left(\frac{2bd_0 - t}{1 - bt + b^2 d_0} + \frac{t}{1 - bt + b^2 d_0}\right)\frac{db}{dt}$$
$$= \frac{2}{4d_0 - t^2}\left(\frac{4d_0 \tau}{\sqrt{4d_0 - t^2}} - \frac{(b + a_0)\left[(a_0 b d_0 - 1)t^2 + (b - a_0)2d_0 t\right]}{2(1 - bt + b^2 d_0)(1 + a_0 t + a_0^2 d_0)}\right)$$
$$+ \frac{(b + a_0)(1 + a_0 b d_0)}{(1 - bt + b^2 d_0)(1 + a_0 t + a_0^2 d_0)}.$$

Operating in both sides of the equality we obtain that

$$\frac{2bd_0\left(4d_0 - t^2\right)}{1 - bt + b^2 d_0}\frac{db}{dt} = \frac{8d_0 \tau}{\sqrt{4d_0 - t^2}}$$
$$+ \frac{2d_0(b + a_0)(2 + 2a_0 b d_0 + a_0 t)}{(1 - bt + b^2 d_0)(1 + a_0 t + a_0^2 d_0)}$$
$$- \frac{2d_0(b + a_0)\left(bt + a_0 b t^2\right)}{(1 - bt + b^2 d_0)(1 + a_0 t + a_0^2 d_0)}.$$

Isolating db/dt in this equality we get

$$\frac{db}{dt} = \frac{1 - bt + b^2 d_0}{b\,(4d_0 - t^2)} \left(\frac{4\tau}{\sqrt{4d_0 - t^2}} \right) + \frac{1}{b\,(4d_0 - t^2)} \left(\frac{(b + a_0)\left(2 + 2a_0bd_0 + a_0t - bt - a_0bt^2\right)}{(1 + a_0t + a_0^2 d_0)} \right),$$

which can be simplified to

$$\frac{db}{dt} = \frac{4\tau\left(1 - b\,t + b^2 d_0\right)\left(1 + a_0t + a_0^2 d_0\right)}{b\,(4d_0 - t^2)\sqrt{4d_0 - t^2}\,(1 + a_0t + a_0^2 d_0)} + \frac{\sqrt{4d_0 - t^2}\,(b + a_0)\left(2 + 2a_0bd_0 + a_0t - bt - a_0bt^2\right)}{b\,(4d_0 - t^2)\sqrt{4d_0 - t^2}\,(1 + a_0t + a_0^2 d_0)}.$$

Defining now the map

$$f(b,t) := 4\tau\left(1 - bt + b^2 d_0\right)\left(1 + a_0t + a_0^2 d_0\right) + \sqrt{4d_0 - t^2}\,(b + a_0)\left(2 + 2a_0bd_0 + a_0t - bt - a_0bt^2\right),$$

the expression of db/dt can be written as

$$\frac{db}{dt} = \frac{f(b,t)}{b\,(4d_0 - t^2)\sqrt{4d_0 - t^2}\,(1 + a_0t + a_0^2 d_0)}.$$

We recall that $b = b(t) = \widetilde{\pi}^A_{++}(a_0, t)$. Since $b \geq 0$, $4d_0 - t^2 > 0$, and $a_0 \geq 0$, it follows that

$$b\left(4d_0 - t^2\right)\sqrt{4d_0 - t^2}\left(1 + a_0t + a_0^2 d_0\right) \geq 0.$$

Hence, the sign of db/dt is the sign of the function $f(b,t)$, so let us compute the sign of $f(b,t)$ when t varies in $(0, 2\sqrt{d_0})$.

In the proof of Proposition 4.5.7 we showed that $\tau(b,t) \in (\pi, \tau^*]$, where $\tau^* < 2\pi$. Hence, since $b \geq 0$ and $a_0 \geq 0$, it follows that $1 - b\,t + b^2 d_0 > 0$ and $1 + a_0t + a_0^2 d_0 > 0$, which implies that $4\tau(1 - b\,t + b^2 d_0)(1 + a_0t + a_0^2 d_0) > 4\pi(1 - b\,t + b^2 d_0)(1 + a_0t + a_0^2 d_0)$. Therefore, if we consider the auxiliary map

$$p(b,t) := 4\pi\left(1 - bt + b^2 d_0\right)\left(1 + a_0t + a_0^2 d_0\right) + \sqrt{4d_0 - t^2}\,(b + a_0)\left(2 + 2a_0bd_0 + a_0t - bt - a_0bt^2\right),$$

then it is easy to conclude that $f(b,t) > p(b,t)$.

Since

$$p(b,0) = 4\pi\left(1 + b^2 d_0\right)\left(1 + a_0^2 d_0\right) + 2\sqrt{d_0}\,(b + a_0)\,(2 + 2a_0bd_0),$$

we have that $p(b,0) > 0$. Note that if we prove that $p(b,t) \neq 0$ for $t \in (0, 2\sqrt{d_0})$, then $p(b,t) > 0$ for $t \in (0, 2\sqrt{d_0})$, which implies that $f(b,t) > 0$ for $t \in (0, 2\sqrt{d_0})$.

Since the signs of the functions f and db/dt coincide, we conclude that $db/dt > 0$ for $t \in (0, 2\sqrt{d_0})$, which proves the statement.

Writing $p(b,t)$ as a polynomial in the variable b, we have that $p(b,t) = b^2 n_2(t) + b n_1(t) + n_0(t)$, where

$$n_2(t) := 4\pi d_0 \left(1 + a_0 t + a_0^2 d_0\right) + \sqrt{4d_0 - t^2}\left(2a_0 d_0 - a_0 t^2 - t\right),$$
$$n_1(t) := \sqrt{4d_0 - t^2}\left(2 + a_0^2\left(2d_0 - t^2\right)\right) - 4\pi t\left(1 + a_0 t + a_0^2 d_0\right),$$

and

$$n_0(t) := 4\pi\left(1 + a_0 t + a_0^2 d_0\right) + \sqrt{4d_0 - t^2}\left(2a_0 + a_0^2 t\right).$$

We are going to prove that the discriminant $\triangle(t) := n_1(t)^2 - 4n_2(t)n_0(t)$ of the polynomial $p(b,t)$ is negative. Therefore, $p(b,t)$ has no real zeros.

Inserting the expressions of the functions $n_k(t)$ in $\triangle(t)$ it is easy to check that

$$\triangle(t) = -\left(4d_0 - t^2\right)\left(c_4(t)\,a_0^4 + 4c_3(t)\,a_0^3 + 8c_2(t)\,a_0^2 + 8c_1(t)\,a_0 + c_0(t)\right),$$

where

$$4d_0 - t^2 > 0,$$

$$c_0(t) := 4(2\pi - 1)(2\pi + 1) > 0,$$
$$c_1(t) := \left(4\pi^2 - 1\right)t + 2\pi\sqrt{4d_0 - t^2} > 0,$$
$$c_2(t) := \left(4\pi^2 + 1\right)d_0 + \left(2\pi^2 - 1\right)t^2 + 3\pi t\sqrt{4d_0 - t^2} > 0,$$
$$c_3(t) := \left(8\pi^2 d_0 - t^2\right)t + 4d_0\pi\sqrt{4d_0 - t^2} + 2d_0 t + 2t^2\pi\sqrt{4d_0 - t^2} > 0$$

and

$$c_4(t) := \left(16\pi^2 - 4\right)d_0^2 + 8d_0\pi\, t\sqrt{4d_0 - t^2} + t^2\left(4d_0 - t^2\right) > 0.$$

Therefore, $\triangle(t) < 0$ in $(0, 2\sqrt{d_0})$ and statement (c) follows.

(d) By statements (a) and (b), if we take $a_0 \in (0, +\infty)$, then there exist values t_1 and t_2 in $(0, 2\sqrt{d_0})$ such that $g(a_0, t_1) < 0$ and $g(a_0, t_2) > 0$. Hence, there exists a value t_3 in (t_1, t_2) such that $g(a_0, t_3) = 0$. The uniqueness of this value can be obtained from statement (c). □

The curve $\mathcal{H}_o\mathcal{L}|_{D,T}$ of homoclinic cycles

Given $d > 0$ and $a \in (0, +\infty)$, there exists a value $t = t(a,d)$ in $(0, 2\sqrt{d})$ such that $g(a, t(a,d)) = 0$, see Lemma 5.5.10(d). In particular, when $a = |\Lambda_2|^{-1}$, we can define a function w_1 of d by

$$w_1(d) := t\left(|\Lambda_2|^{-1}, d\right) \tag{5.13}$$

such that $g(|\Lambda_2|^{-1}, w_1(d)) = 0$. It is easy to check that for any fixed $d > 0$ the qualitative behaviour of the graph of $g(a, w_1(d))$ is the one shown in Figure 5.14(b), see Lemma 5.5.9(a) and (b). This implies that the system has two homoclinic cycles to a common singular point at the origin, see Lemma 5.5.7(c).

Define $\mathcal{H}_o\mathcal{L}|_{D,T} := \{(w_1(d), d) : d > 0\}$. Since $g(|\Lambda_2|^{-1}, w_1(d)) = 0$, we obtain that $\widetilde{\pi}^A_{++}(|\Lambda_2|^{-1}) = \Lambda_1^{-1}$. Hence, by Proposition 4.5.7(a.4), an implicit expression of the curve $\mathcal{H}_o\mathcal{L}|_{D,T}$ is given by

$$\exp\left(\frac{2t}{\sqrt{4d-t^2}}\arctan\left(\frac{\sqrt{T^2-4D}\sqrt{4d-t^2}}{2(D+d)-Tt}\right)\right) = \frac{4D^2+\left(T-\sqrt{T^2-4D}\right)\left[d\left(T-\sqrt{T^2-4D}\right)-2Dt\right]}{4D^2+\left(T+\sqrt{T^2-4D}\right)\left[d\left(T+\sqrt{T^2-4D}\right)-2Dt\right]}.$$

We note that the curve $\mathcal{H}_o\mathcal{L}|_{D,T}$ tends to the straight line $\mathcal{H}_\infty = \{t = 0, d > 0\}$ as T tends to 0.

The curve $\mathcal{NH}_{lc}|_{D,T}$ of non-hyperbolic limit cycles

We now define a function $w_2(d)$ in the interval $(0, +\infty)$ in such a way that any fundamental system with parameters $(t = w_2(d), d)$ has a non-hyperbolic limit cycle which is inside asymptotically stable and outside asymptotically unstable.

For any value $d > 0$ we consider the function $g(a, t) = \widetilde{\pi}^A_{++}(a, t) - \pi^{-B}_{+-}(a)$ defined in the domain $(|\Lambda_2|^{-1}, \infty) \times (0, 2\sqrt{d})$. For any $t \in (0, 2\sqrt{d})$ there exist a unique value $a^*(t) \in (|\Lambda_2|^{-1}, \infty)$ such that $\partial g/\partial a|_{(a^*(t),t)} = 0$, see Lemma 5.5.9(b). Moreover, $g(a, t)$ takes its minimum value at $a^*(t)$. From

$$\left.\frac{\partial^2 g}{\partial a^2}\right|_{(a,t)} = \left.\frac{d^2\widetilde{\pi}^A_{++}}{da^2}\right|_a - \left.\frac{d^2\pi^{-B}_{+-}}{da^2}\right|_a > 0,$$

see Lemmas 5.5.4(f.2) and (d.2), and the Implicit Function Theorem we obtain that $a^*(t)$ is a differentiable function.

Consider the auxiliary function $\mu(t) := g(a^*(t), t)$ defined in the interval $(0, 2\sqrt{d})$. Since g and $a^*(t)$ are differentiable functions, it follows that the function $\mu(t)$ is differentiable. Moreover,

$$\mu'(t) = \left.\frac{\partial g}{\partial a}\right|_{(a^*(t),t)} \left.\frac{da^*}{dt}\right|_t + \left.\frac{\partial g}{\partial t}\right|_{(a^*(t),t)} = \left.\frac{\partial g}{\partial t}\right|_{(a^*(t),t)} > 0,$$

see Lemma 5.5.10(c). Then $\mu(t)$ is strictly increasing in $(0, 2\sqrt{d})$.

Since $g(|\Lambda_2|^{-1}, w_1(d)) = 0$ and $a^*(w_1(d))$ is the minimum of the function $g(a, w_1(d))$ in the interval $(|\Lambda_2|^{-1}, +\infty)$, it follows that $\mu(w_1(d)) < 0$. On the other hand, $g(a, t(d)) > 0$ in the interval $(|\Lambda_2|^{-1}, +\infty)$, see Lemma 5.5.10(b). Therefore, $\mu(t(d)) > 0$. We conclude that there exists a unique value $w_2(d) = (w_1(d), t(d))$ such that $\mu(w_2(d)) = 0$. Equivalently, $g(a^*(w_2(d)), w_2(d)) = 0$, where $a^*(w_2(d))$ is the solution of the equation $\partial g/\partial a|_{(a,w_2(d))} = 0$.

Thus, given $d > 0$ and considering the values $a = a^*(w_2(d))$ and $t = w_2(d)$, there exists a unique solution of the system

$$g(a,t) = 0, \quad \left.\frac{\partial g}{\partial a}\right|_{(a,t)} = 0. \tag{5.14}$$

Note that the qualitative behaviour of the graph of $g(a, w_2(d))$ is the one shown in Figure 5.14(d), see Lemma 5.5.9.

Consider the differentiable curve $\mathcal{NH}_{lc}|_{D,T} := \{(w_2(d), d) : d > 0\}$. Since $g(a^*(w_2(d)), w_2(d)) = 0$, any fundamental system with parameters $(t, d) \in \mathcal{NH}_{lc}|_{D,T}$ has a periodic orbit, see Lemmas 5.5.7(b.2) and (b.3). Moreover, since

$$\frac{\partial^2 g}{\partial a^2} > 0 \text{ and } \left.\frac{\partial g}{\partial a}\right|_{(a^*(w_2(d)),w_2(d))} = 0,$$

this periodic orbit is inside asymptotically stable and outside asymptotically unstable, see Lemmas 5.5.7(b.2) and (b.3).

Equations (5.14) can be expressed in terms of the Poincaré maps as

$$\pi_{+-}^{-B}(a^*(w_2(d))) = \widetilde{\pi}_{++}^{A}(a^*(w_2(d))),$$

where $a^*(w_2(d)) \in (|\Lambda_2|^{-1}, +\infty)$ is the solution of $d\pi_{+-}^{-B}/da = d\widetilde{\pi}_{++}^{A}/da$, see Lemma 5.5.9(b). Note that for $T = 0$ the Poincaré map π_{+-}^{B} is the identity. Thus, we conclude that the curve $\mathcal{NH}_{lc}|_{D,0}$ tends to the straight line $\mathcal{H}_\infty = \{t = 0 \text{ and } d > 0\}$ as T tends to 0.

Behaviour of the Lamerey map

Consider in $\mathcal{C}_1^2$ the following regions bounded by the curves $\mathcal{H}_o\mathcal{L}|_{D,T}$ and $\mathcal{NH}_{lc}|_{D,T}$:

$$\begin{aligned} \mathcal{C}_1^{21}\big|_{D,T} &= \left\{(t,d) \in \mathcal{C}_1^2 : w_2(d) < t < 2\sqrt{d}\right\}, \\ \mathcal{C}_1^{22}\big|_{D,T} &= \left\{(t,d) \in \mathcal{C}_1^2 : w_1(d) < t < w_2(d)\right\}, \\ \mathcal{C}_1^{23}\big|_{D,T} &= \left\{(t,d) \in \mathcal{C}_1^2 : 0 < t < w_1(d)\right\}, \end{aligned} \tag{5.15}$$

see Figure 5.15.

Next we prove that given parameters (t, d) contained in either the region $\mathcal{C}_1^{21}\big|_{D,T}$, or in the region $\mathcal{C}_1^{22}\big|_{D,T}$, or in the region $\mathcal{C}_1^{23}\big|_{D,T}$, the qualitative behaviour of the graph of $g(a)$ is represented in Figure 5.14(e), (c) or (a), respectively.

Lemma 5.5.11. *Suppose that $(t, d) \in \mathcal{C}_1^2$ and let $g(a)$ be the Lamerey map as defined in Lemma 5.5.7(a).*

(a) *If $(t, d) \in \mathcal{C}_1^{23}\big|_{D,T}$, then the map $g(a)$ has exactly two zeros: a_1 in the interval $(0, |\Lambda_2|^{-1})$ and a_2 in the interval $(|\Lambda_2|^{-1}, +\infty)$. Furthermore, $g'(a_1) < 0$ and $g'(a_2) > 0$. The qualitative behaviour of the graph of $g(a)$ is shown in Figure 5.14(a).*

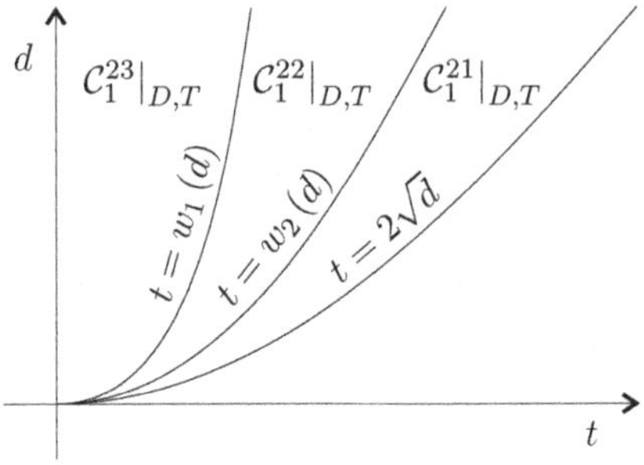

Figure 5.15: Partition of the parameter region $\mathcal{C}_1^2$ by the curves $w_1(d)$ and $w_2(d)$.

(b) *If* $(t,d) \in \mathcal{H}_0\mathcal{L}|_{D,T}$, *then the map* $g(a)$ *has exactly two zeros:* $a_1 = |\Lambda_2|^{-1}$ *and* $a_2 \in (|\Lambda_2|^{-1}, +\infty)$. *Furthermore,* $g'(a_1) < 0$ *and* $g'(a_2) > 0$. *The qualitative behaviour of the graph of* $g(a)$ *is shown in Figure* 5.14(b).

(c) *If* $(t,d) \in \mathcal{C}_1^{22}|_{D,T}$, *then the map* $g(a)$ *has exactly two zeros* a_1 *and* a_2, *and these zeros lie in* $(|\Lambda_2|^{-1}, +\infty)$. *Furthermore,* $g'(a_1) < 0$ *and* $g'(a_2) > 0$. *The qualitative behaviour of the graph of* $g(a)$ *is shown in Figure* 5.14(c).

(d) *If* $(t,d) \in \mathcal{NH}_{lc}|_{D,T}$, *then the map* $g(a)$ *has exactly one zero* a^* *and this zero lies in* $(|\Lambda_2|^{-1}, +\infty)$. *Furthermore,* $g'(a^*) = 0$, $g' < 0$ *in* $(|\Lambda_2|^{-1}, a^*)$, *and* $g' > 0$ *in* $(a^*, +\infty)$. *The qualitative behaviour of the graph of* $g(a)$ *is shown in Figure* 5.14(d).

(e) *If* $(t,d) \in \mathcal{C}_1^{21}|_{D,T}$, *then* $g(a) > 0$ *in* $[0, +\infty)$. *The qualitative behaviour of the graph of* $g(a)$ *is shown in Figure* 5.14(e).

Proof. (a) Suppose that $(t,d) \in \mathcal{C}_1^{23}|_{D,T}$. Hence, $t < w_1(d)$, which implies that $g(a) = g(a,t) < g(a, w_1(d))$. We recall that $\partial g/\partial t > 0$, see Lemma 5.5.10(c). Thus $g(|\Lambda_2|^{-1}) < g(|\Lambda_2|^{-1}, w_1(d)) = 0$. On the other hand, since $g(0) = g(0,t) = \widetilde{\pi}_{++}^{A}(0,t) > 0$, we conclude that there exists a value $a_1 \in (0, |\Lambda_2|^{-1})$ such that $g(a_1) = 0$. The statement follows from Lemmas 5.5.9(a) and (c).

(b) Suppose that $(t,d) \in \mathcal{H}_0\mathcal{L}|_{D,T}$. Hence, $t = w_1(d)$, which implies that $g(|\Lambda_2|^{-1}, w_1(d)) = 0$. Thus, the map $g(a) = g(a, w_1(d))$ has a zero at $a_1 = |\Lambda_2|^{-1}$. The statement follows from Lemmas 5.5.9(a) and (c).

(c) Suppose that $(t,d) \in \mathcal{C}_1^{22}|_{D,T}$. Hence, $w_1(d) < t < w_2(d)$, which implies that

$$0 = g\left(|\Lambda_2|^{-1}, w_1(d)\right) < g\left(|\Lambda_2|^{-1}\right) = g\left(|\Lambda_2|^{-1}, t\right) < g\left(|\Lambda_2|^{-1}, w_2(d)\right).$$

We recall that $\partial g/\partial t > 0$.

Suppose that $g(a)$ has a zero a_1 in $[0, |\Lambda_2|^{-1})$. By Lemma 5.5.9(a), the map $g(a)$ is strictly decreasing in $[a_1, |\Lambda_2|^{-1})$. Hence, $g(|\Lambda_2|^{-1}) < g(a_1) = 0$, which contradicts the above inequality. Therefore, $g(a)$ has no zeros in $[0, |\Lambda_2|^{-1})$.

Consider the map $\mu(t) = g(a^*(t), t)$. Since $t < w_2(d)$, it follows that $\mu'(t) = \partial g/\partial t|_{(a^*(t),t)} > 0$, see Lemma 5.5.10(c). Therefore, $g(a^*(t)) = g(a^*(t), t) < g(a^*(w_2(d)), w_2(d)) = 0$. By Lemma 5.5.9(b), $\lim_{a \nearrow +\infty} g(a) = +\infty$.

Summarizing all the above information, we obtain that $g(|\Lambda_2|^{-1}) > 0$, $g(a^*(t)) < 0$, and $\lim_{a \nearrow +\infty} g(a) = +\infty$. Hence, the map $g(a)$ has exactly two zeros a_1 and a_2 in the interval $(|\Lambda_2|^{-1}, +\infty)$. The sign of g' at the points a_1 and a_2 follows from Lemma 5.5.9(d).

(d) Suppose that $(t, d) \in \mathcal{NH}_{lc}|_{D,T}$. Thus $t = w_2(d)$, which implies that $a^*(t)$ is the unique zero of $g(a)$ and $g'(a)$.

(e) Suppose that $(t, d) \in \mathcal{C}_1^{21}|_{D,T}$. Thus $w_2(d) < t$ and $g(a, w_2(d)) < g(a)$, see Lemma 5.5.10(c). Since $a^*(w_2(d))$ is the minimum of $g(a, w_2(d))$ in $(|\Lambda_2|^{-1}, +\infty)$ and $g(a^*(w_2(d)), w_2(d)) = 0$, it follows that $g(a) > 0$ when $a > |\Lambda_2|^{-1}$. From Lemma 5.5.9(c) it follows that $g(a)$ has no zeros in the interval $[0, |\Lambda_2|^{-1}]$, which proves the statement. □

Now we present a result about the number of periodic orbits and homoclinic cycles for fundamental systems with parameters $(t, d) \in \mathcal{C}_1^2$.

Theorem 5.5.12. *Consider a fundamental system with parameters $D < 0$, $T < 0$, and $(t, d) \in \mathcal{C}_1^2$.*

(a) *If $(t, d) \in \mathcal{C}_1^{23}|_{D,T}$, then the system has exactly three Jordan curves formed by solutions, Φ_+, Φ_- and Γ. Moreover, Φ_+ and Φ_- are asymptotically stable hyperbolic limit cycles such that $\mathbf{e}_+ \in \Sigma_{\Phi_+}$, $\mathbf{e}_- \in \Sigma_{\Phi_-}$, and Γ is an asymptotically unstable hyperbolic limit cycle such that $\Phi_+ \cup \Phi_- \cup \mathbf{0} \in \Sigma_\Gamma$.*

(b) *If $(t, d) \in \mathcal{H}_o\mathcal{L}|_{D,T}$, then the system has exactly three Jordan curves formed by solutions, Δ_+, Δ_- and Γ. Moreover, Δ_+ and Δ_- are homoclinic cycles to a common singular point at the origin $\mathbf{0}$, and Γ is an asymptotically unstable hyperbolic limit cycle such that $\Delta_+ \cup \Delta_- \subset \Sigma_\Gamma$.*

(c) *If $(t, d) \in \mathcal{C}_1^{22}|_{D,T}$, then the system has exactly two Jordan curves formed by solutions, Φ and Γ. Moreover, Φ is an asymptotically stable hyperbolic limit cycle, and Γ is an symptotically unstable hyperbolic limit cycle such that $\{\mathbf{e}_+, \mathbf{e}_-, \mathbf{0}\} \subset \Sigma_\Phi \subset \Sigma_\Gamma$.*

(d) *If $(t, d) \in \mathcal{NH}_{lc}|_{D,T}$, then there exists exactly one Jordan curve formed by solutions, Γ. Moreover, Γ is a non-hyperbolic limit cycle which is inside asymptotically stable and outside asymptotically unstable and such that $\{\mathbf{e}_+, \mathbf{e}_-, \mathbf{0}\} \subset \Sigma_\Gamma$.*

(e) *If $(t, d) \in \mathcal{C}_1^{21}|_{D,T}$, then the system has no Jordan curves formed by solutions.*

Proof. (a) Suppose that $(t, d) \in \mathcal{C}_1^{23}|_{D,T}$. Hence, the map $g(a)$ has exactly two zeros $a_1 \in (0, |\Lambda_2|^{-1})$ and $a_2 \in (|\Lambda_2|^{-1}, +\infty)$, see Lemma 5.5.11(a). Each of these zeros is associated to a periodic orbit which intersects the straight line L_+. Let Φ_+

be the periodic orbit which intersects L_+ at the point of coordinate $\widetilde{\pi}_{++}^A(a_1)$ and let Γ be the periodic orbit which intersects L_+ at the point of coordinate $\widetilde{\pi}_{++}^A(a_2)$.

Since $a_1 < |\Lambda_2|^{-1}$, it follows that the origin is not contained in the region Σ_{Φ_+} and the singular point $\mathbf{e}_+$ is contained in the region Σ_{Φ_+}, see Lemma 5.5.7(b). By the symmetry of the flow with respect to the origin, there exists a periodic orbit Φ_- such that $\mathbf{e}_- \in \Sigma_{\Phi_-}$. Since $a_2 > |\Lambda_2|^{-1} > a_1$, the origin is contained in Σ_Γ and the periodic orbits Φ_+ and Φ_- are contained in Σ_Γ.

From Lemma 5.5.11(a) we obtain that $g'(a_1) < 0$ and $g'(a_2) > 0$. The statement now follows from Lemmas 5.5.7(b.4) and (b.1).

(b) Suppose that $(t,d) \in \mathcal{H}_o\mathcal{L}|_{D,T}$. Then the zeros of the map $g(a)$ satisfy that $a_1 = |\Lambda_2|^{-1}$ and $a_2 \in (|\Lambda_2|^{-1}, +\infty)$, see Lemma 5.5.11(b). By Lemma 5.5.7(c), and associated to the zero a_1, there exists a double homoclinic cycle Δ to a common singular point at the origin which is formed by two homoclinic cycles Δ_+ and Δ_-. Associated to the zero a_2 there exists a periodic orbit Γ surrounding the origin, see Lemma 5.5.7(b). Moreover, since $a_2 > a_1$ it follows that $\Sigma_\Delta \subset \Sigma_\Gamma$.

The return map π in a neighbourhood of the double homoclinic cycle Δ satisfies that $d\pi/da < 1$. Note that in Lemma 5.5.11(b) we proved that $g'(a_1) < 0$. Thus, Δ_+ and Δ_- are inside asymptotically stable. Therefore, the double cycle Δ is outside asymptotically stable. By Lemma 5.5.11(b), $g'(a_2) > 0$, which implies that Γ is an asymptotically unstable hyperbolic limit cycle.

Statements (c), (d) and (e) follow similarly by using Lemmas 5.5.11 and 5.5.7. □

From Theorem 5.5.12(c) we conclude that a fundamental system with parameters $D < 0$, $T < 0$, and $(t,d) \in C_1^{22}|_{D,T}$ has two hyperbolic limit cycles, Φ and Γ. One of these limit cycles is asymptotically stable and the other is asymptotically unstable. Moreover, they satisfy that $\Phi \subset \Sigma_\Gamma$. This solves in the negative a conjecture appearing in the work of Chua and Lum [47, p. 4].

5.5.5 Phase portraits

In this subsection we describe the different phase portraits in the Poincaré disc $\mathbb{D}$ of the compactified system of a fundamental system with parameters $D < 0$, $T < 0$, and (t,d) varying in $\Pi_{D,T}$.

Define in $\Pi_{D,T}$ the following half-lines:

$$\begin{aligned} W_1|_{D,T}^* &:= W_1|_{D,T} \cap \{(t,d) : t > 2\Lambda_1\}, \\ W_2|_{D,T}^* &:= W_2|_{D,T} \cap \{(t,d) : t < 2\Lambda_2\}. \end{aligned} \tag{5.16}$$

Proposition 5.5.13. *Consider the Poincaré compactification of a fundamental system with parameters $D < 0$, $T < 0$, and $(t,d) \in \mathcal{H}_\infty$. Then:*

(a) *The separatrices of the system are:*

(a.1) *the saddle point at the origin $\mathbf{0}$ and the singular points $\mathbf{e}_+$ and $\mathbf{e}_-$, which are of center type;*

(a.2) *the limit cycle at infinity* ∞ *and the outside asymptotically stable periodic orbits* Φ_+ *and* Φ_- *such that* $\mathbf{e}_+ \in \Sigma_{\Phi_+}$ *and* $\mathbf{e}_- \in \Sigma_{\Phi_-}$*;*

(a.3) *the stable and the unstable manifolds of the saddle point at the origin.*

(b) *The canonical regions are: the annular regions* $\Sigma_{\Phi_+} \setminus \mathbf{e}_+$ *and* $\Sigma_{\Phi_-} \setminus \mathbf{e}_-$ *foliated by periodic orbits,* $\Sigma_\infty^+ = W^u(\infty) \cap W^s(\Phi_+)$*, and* $\Sigma_\infty^- = W^u(\infty) \cap W^s(\Phi_-)$*.*

(c) *The phase portrait of the Poincaré compactification is topologically equivalent to its correspondent in Figure* 5.18.

Proof. (a) The existence of the listed separatrices is a consequence of Propositions 5.5.1(a), 5.5.2(a), 5.5.3(b) and 5.5.6.

(b) The behaviour of the flow in the annular regions $\Sigma_{\Phi_+} \setminus \mathbf{e}_+$ and $\Sigma_{\Phi_-} \setminus \mathbf{e}_-$ follows from Proposition 5.5.3(b).

Since there are no Jordan curves formed by solutions contained in the region $\Sigma_\infty \setminus \mathrm{Cl}(\Sigma_{\Phi_+} \cup \Sigma_{\Phi_-})$, see Proposition 5.5.3(b), the separatrices of the origin do not connect in a homoclinic cycle. Moreover, the periodic orbits Φ_+ and Φ_- are outside asymptotically stable, see Proposition 5.5.6. Let γ^{s+} and γ^{s-} be the stable separatrices of the saddle point at the origin and let γ^{u+} and γ^{u-} be the unstable separatrices of the singular point at the origin. From the Poincaré–Bendixson Theorem we conclude that $\alpha(\gamma^{s+}) = \alpha(\gamma^{s-}) = \infty$, $\omega(\gamma^{u+}) = \Phi_+$, and $\omega(\gamma^{u-}) = \Phi_-$. Therefore, the stable manifold of the origin splits $\Sigma_\infty \setminus \mathrm{Cl}(\Sigma_{\Phi_+} \cup \Sigma_{\Phi_-} \cup W^u(\mathbf{0}))$ into the two open, connected, and invariant regions Σ_∞^+ and Σ_∞^-.

Let Σ_∞^+ be the region containing the orbit Φ_+ in its boundary, and let Σ_∞^- be the region containing the orbit Φ_- in its boundary. We conclude that $\Sigma_\infty^+ = W^u(\infty) \cap W^s(\Phi_+)$ and $\Sigma_\infty^- = W^u(\infty) \cap W^s(\Phi_-)$, which implies that no other separatrices than those in statement (a) appear in the system.

Statement (c) follows from Theorem 2.6.9. □

Proposition 5.5.14. *Consider the Poincaré compactification of a fundamental system with parameters* $D < 0$*,* $T < 0$*, and* $(t, d) \in \mathcal{C}_1^{23}\big|_{D,T}$*. Then:*

(a) *The separatrices of the system are:*

(a.1) *the saddle point at the origin* $\mathbf{0}$ *and the asymptotically unstable hyperbolic foci* $\mathbf{e}_+$ *and* $\mathbf{e}_-$*;*

(a.2) *the limit cycle at infinity* ∞*, the asymptotically unstable hyperbolic limit cycle* Γ *and the asymptotically stable hyperbolic limit cycles* Φ_+ *and* Φ_-*, which satisfy that* $\mathbf{e}_+ \in \Sigma_{\Phi_+}$*,* $\mathbf{e}_- \in \Sigma_{\Phi_-}$*,* $\Phi_+ \cup \Phi_- \cup \mathbf{0} \subset \Sigma_\Gamma$*;*

(a.3) *the stable and the unstable manifolds of the saddle point at the origin.*

(b) *The canonical regions are:* $\Sigma_\infty \setminus \mathrm{Cl}(\Sigma_\Gamma) = W^u(\Gamma) \cap W^s(\infty)$*,* $\Sigma_\Gamma^+ = W^u(\Gamma) \cap W^s(\Phi_+)$*,* $\Sigma_\Gamma^- = W^u(\Gamma) \cap W^s(\Phi_-)$*,* $\Sigma_{\Phi_+} \setminus \mathbf{e}_+ = W^u(\mathbf{e}_+) \cap W^s(\Phi_+)$*, and* $\Sigma_{\Phi_-} \setminus \mathbf{e}_- = W^u(\mathbf{e}_-) \cap W^s(\Phi_-)$*.*

(c) *The phase portrait of the Poincaré compactification is topologically equivalent to its correspondent in Figure* 5.18.

Proof. (a) The existence of the listed separatrices is a consequence of Propositions 5.5.1(a), 5.5.2(a) and Theorem 5.5.12(a).

(b) From Theorem 5.5.12(a) and the Poincaré–Bendixson Theorem it follows that the α- and the ω-limit sets of the orbits contained in $\mathbb{D} \setminus \Sigma_\Gamma$ are the periodic orbit at infinity ∞ and the asymptotically unstable limit cycle Γ. Consequently, $\Sigma_\infty \setminus \mathrm{Cl}(\Sigma_\Gamma) = W^u(\Gamma) \cap W^s(\infty)$. In a similar way, the α- and the ω-limit sets in $\mathrm{Cl}(\Sigma_{\Phi_+})$ are the unstable focus $\mathbf{e}_+$ and the asymptotically stable limit cycle Φ_+. Therefore, $\Sigma_{\Phi_+} \setminus \mathbf{e}_+ = W^u(\mathbf{e}_+) \cap W^s(\Phi_+)$. By the symmetry of the flow with respect to the origin, it follows that $\Sigma_{\Phi_-} \setminus \mathbf{e}_- = W^u(\mathbf{e}_-) \cap W^s(\Phi_-)$.

The stable manifold of the origin splits $\Sigma_\Gamma \setminus \mathrm{Cl}(\Sigma_{\Phi_+} \cup \Sigma_{\Phi_-} \cup W^u(\mathbf{0}))$ into the two open, connected and invariant regions Σ_Γ^+ and Σ_Γ^-. For more details, see the proof of Proposition 5.5.13(b). The behaviour of the flow in Σ_Γ^+ and Σ_Γ^- is easily checked.

Statement (c) follows from Theorem 2.6.9. □

Proposition 5.5.15. *Consider the Poincaré compactification of a fundamental system with parameters $D < 0$, $T < 0$, and $(t, d) \in \mathcal{H}_o\mathcal{L}|_{D,T}$. Then:*

(a) *The separatrices of the system are:*

(a.1) *the saddle point at the origin* $\mathbf{0}$ *and the asymptotically unstable hyperbolic foci* $\mathbf{e}_+$ *and* $\mathbf{e}_-$;

(a.2) *the limit cycle at infinity* ∞ *and the asymptotically unstable hyperbolic limit cycle* Γ*;*

(a.3) *the separatrices of the saddle at the origin, which form two homoclinic cycles* Δ_+ *and* Δ_- *to a common singular point at the origin. Moreover,* $\mathbf{e}_+ \in \Sigma_{\Delta_+}$, $\mathbf{e}_- \in \Sigma_{\Delta_-}$, *and* $\Delta = \Delta_+ \cup \Delta_-$ *is a double homoclinic cycle contained in* Σ_Γ.

(b) *The canonical regions are:* $\Sigma_\infty \setminus \mathrm{Cl}(\Sigma_\Gamma) = W^u(\Gamma) \cap W^s(\infty)$, $\Sigma_\Gamma \setminus \mathrm{Cl}(\Sigma_\Delta) = W^u(\Gamma) \cap W^s(\Delta)$, $\Sigma_{\Delta_+} \setminus \mathbf{e}_+ = W^u(\mathbf{e}_+) \cap W^s(\Delta_+)$, *and* $\Sigma_{\Delta_-} \setminus \mathbf{e}_- = W^u(\mathbf{e}_-) \cap W^s(\Delta_-)$.

(c) *The phase portrait of the Poincaré compactification is topologically equivalent to its correspondent in Figure* 5.18.

Proof. (a) The existence of the listed separatrices is a consequence of Propositions 5.5.1(a), 5.5.2(a) and Theorem 5.5.12(b).

(b) The α- and the ω-limit sets in the region $\mathbb{D} \setminus \Sigma_\Gamma$ are the limit cycles ∞ and Γ. From this we conclude that $\Sigma_\infty \setminus \mathrm{Cl}(\Sigma_\Gamma) = W^u(\Gamma) \cap W^s(\infty)$.

Define $\Sigma_\Delta = \Sigma_{\Delta_+} \cup \Sigma_{\Delta_-}$. The α- and the ω-limit sets contained in the region $\mathrm{Cl}(\Sigma_\Gamma) \setminus \Sigma_\Delta$ are the asymptotically unstable limit cycle Γ and the double homoclinic cycle Δ. From this we conclude that $\Sigma_\Gamma \setminus \mathrm{Cl}(\Sigma_\Delta) = W^u(\Gamma) \cap W^s(\Delta)$.

The α- and the ω-limit sets contained in the region $\mathrm{Cl}(\Sigma_{\Delta_+})$ are the homoclinic cycle Δ_+ and the asymptotically unstable singular point $\mathbf{e}_+$. Thus $\Sigma_{\Delta_+} \setminus$

$\mathbf{e}_+ = W^u(\mathbf{e}_+) \cap W^s(\Delta_+)$. The behaviour in the region $\Sigma_{\Delta_-} \setminus \mathbf{e}_-$ follows by the symmetry of the flow with respect to the origin.

Statement (c) is a consequence of Theorem 2.6.9. □

Proposition 5.5.16. *Consider the Poincaré compactification of a fundamental system with parameters $D < 0$, $T < 0$, and $(t, d) \in \mathcal{C}_1^{22}|_{D,T}$. Then:*

(a) *The separatrices of the system are:*

(a.1) *the saddle point at the origin $\mathbf{0}$ and the asymptotically unstable foci $\mathbf{e}_+$ and $\mathbf{e}_-$;*

(a.2) *the limit cycle at infinity ∞, the asymptotically unstable hyperbolic limit cycle Γ, and the asymptotically stable hyperbolic limit cycle Φ, which satisfies that $\{\mathbf{0}, \mathbf{e}_+, \mathbf{e}_-\} \subset \Sigma_\Phi \subset \Sigma_\Gamma$;*

(a.3) *the separatrices of the saddle at the origin.*

(b) *The canonical regions are: $\Sigma_\infty \setminus \mathrm{Cl}(\Sigma_\Gamma) = W^u(\Gamma) \cap W^s(\infty)$, $\Sigma_\Gamma \setminus \mathrm{Cl}(\Sigma_\Phi) = W^u(\Gamma) \cap W^s(\Phi)$, $\Sigma_\Phi^+ = W^u(\mathbf{e}_+) \cap W^s(\Phi)$, and $\Sigma_\Phi^- = W^u(\mathbf{e}_-) \cap W^s(\Phi)$.*

(c) *The phase portrait of the Poincaré compactification is topologically equivalent to its correspondent in Figure* 5.18.

Proof. (a) The existence of the listed separatrices is a consequence of Propositions 5.5.1(a), 5.5.2(a) and Theorem 5.5.12(c).

(b) The α- and the ω-limit sets contained in the region $\Sigma_\infty \setminus \mathrm{Cl}(\Sigma_\Gamma)$ (respectively, in the region $\Sigma_\Gamma \setminus \mathrm{Cl}(\Sigma_\Phi)$) are the limit cycles ∞ and Γ (respectively, the limit cycles Γ and Φ). From this we conclude the behaviour of the flow in the region $\Sigma_\infty \setminus \mathrm{Cl}(\Sigma_\Gamma)$ (respectively, in the region $\Sigma_\Gamma \setminus \mathrm{Cl}(\Sigma_\Phi)$).

Since there are no Jordan curves formed by solutions in Σ_Φ, see Theorem 5.5.12(c), the separatrices of the saddle at the origin are not contained in any homoclinic cycle. On the other hand, the singular points $\mathbf{e}_+$ and $\mathbf{e}_-$ are asymptotically unstable. Thus, if γ^{s+} and γ^{s-} are the stable separatrices of the origin and γ^{u+} and γ^{u-} are the unstable separatrices of the origin, then, by the Poincaré–Bendixson Theorem, we have $\alpha(\gamma^{s+}) = \mathbf{e}_+, \alpha(\gamma^{s-}) = \mathbf{e}_-$ and $\omega(\gamma^{u+}) = \omega(\gamma^{u-}) = \Phi$). Hence, the separatrices γ^{u+} and γ^{u-} split the region $\Sigma_\Phi \setminus \mathrm{Cl}(W^s(\mathbf{0}))$ into the two open, connected, and invariant regions Σ_Φ^+ and Σ_Φ^-.

Let Σ_Φ^+ be the region containig $\mathbf{e}_+$ in its boundary. The α- and the ω-limit sets in $\mathrm{Cl}(\Sigma_\Phi^+)$ are the singular points $\mathbf{0}$ and $\mathbf{e}_+$, and the asymptotically stable limit cycle Φ. Therefore, $\Sigma_\Phi^+ = W^u(\mathbf{e}_+) \cap W^s(\Phi)$. By the symmetry of the flow with respect to the origin, we conclude that $\Sigma_\Phi^- = W^u(\mathbf{e}_-) \cap W^s(\Phi)$.

Statement (c) follows from Theorem 2.6.9. □

Proposition 5.5.17. *Consider the Poincaré compactification of a fundamental system with parameters $D < 0$, $T < 0$, and $(t, d) \in \mathcal{NH}_{lc}|_{D,T}$. Then:*

(a) *The separatrices of the system are:*

(a.1) *the saddle point at the origin* $\mathbf{0}$ *and the asymptotically unstable foci* $\mathbf{e}_+$ *and* $\mathbf{e}_-$;

(a.2) *the limit cycle at infinity* ∞, *and the non-hyperbolic limit cycle* $\Gamma = \Phi$, *which is inside asymptotically stable and outside asymptotically unstable and satisfies that* $\{\mathbf{0}, \mathbf{e}_+, \mathbf{e}_-\} \subset \Sigma_\Gamma$;

(a.3) *the separatrices of the saddle at the origin.*

(b) *The canonical regions are:* $\Sigma_\infty \setminus \mathrm{Cl}(\Sigma_\Gamma) = W^u(\Gamma) \cap W^s(\infty)$, $\Sigma_\Gamma^+ = W^u(\mathbf{e}_+) \cap W^s(\Gamma)$, *and* $\Sigma_\Gamma^- = W^u(\mathbf{e}_-) \cap W^s(\Gamma)$.

(c) *The phase portrait of the Poincaré compactification is topologically equivalent to its correspondent in Figure* 5.18.

Proof. (a) The existence of the listed separatrices is a consequence of Propositions 5.5.1(a), 5.5.2(a) and Theorem 5.5.12(d).

(b) The behaviour of the flow in the regions $\Sigma_\infty \setminus \mathrm{Cl}(\Sigma_\Gamma)$ and Σ_Γ follows by applying arguments similar to those in the proof of Propositions 5.5.16(b) and 5.5.16(b) respectively.

Statement (c) is a consequence of Theorem 2.6.9. □

Proposition 5.5.18. *Consider the Poincaré compactification of a fundamental system with parameters* $D < 0$, $T < 0$, *and* $(t, d) \in \mathcal{C}_1^{21}\big|_{D,T}$. *Then:*

(a) *The separatrices of the system are:*

(a.1) *the saddle point at the origin* $\mathbf{0}$ *and the asymptotically unstable foci* $\mathbf{e}_+$, $\mathbf{e}_-$;

(a.2) *the limit cycle at infinity* ∞;

(a.3) *the separatrices of the saddle at the origin.*

(b) *The canonical regions are:* $\Sigma_\infty^+ = W^u(\mathbf{e}_+) \cap W^s(\infty)$ *and* $\Sigma_\infty^- = W^u(\mathbf{e}_-) \cap W^s(\infty)$.

(c) *The phase portrait of the Poincaré compactification is topologically equivalent to its correspondent in Figure* 5.18.

Proof. (a) The existence of the listed separatrices is a consequence of Propositions 5.5.1(a), 5.5.2(a) and Theorem 5.5.12(e).

Statement (b) follows by using arguments similar to those in the proof of Proposition 5.5.16(b).

Statement (c) is a consequence of Theorem 2.6.9. □

Now we are going to describe the phase portrait of the compactified flow of the fundamental systems with parameters $t^2 - 4d \geq 0$. To do this, we need to control the characteristic directions of the singular points at infinity and the location of the singular points $\mathbf{e}_+$ and $\mathbf{e}_-$. In Section 3.11 we described this under the assumption that the fundamental matrix A is given in its real Jordan normal form. We recall that this assumption was not a restriction, see Proposition 3.10.2.

Proposition 5.5.19. *Consider the Poincaré compactification of a fundamental system with parameters $D < 0$, $T < 0$, $(t, d) \in \mathcal{SN}_\infty \setminus \mathcal{VB}_1|_{D,T}$, and $t > 0$. Then:*

(a) *The separatrices of the system are:*

(a.1) *the saddle point at the origin $\mathbf{0}$, the asymptotically unstable nodes $\mathbf{e}_+$, $\mathbf{e}_-$, and the saddle-nodes at infinity $\mathbf{x}_+$, $\mathbf{x}_-$, which have their central manifold contained in $\partial\mathbb{D}$ and their stable hyperbolic manifolds contained in Σ_∞;*

(a.2) *the separatrices of the saddle at origin, and the separatrices of the saddle-nodes $\mathbf{x}_+$ and $\mathbf{x}_-$.*

(b) *The canonical regions are: $\Sigma_\infty^{++} \subset W^u(\mathbf{e}_+) \cap W^s(\mathbf{x}_+)$, $\Sigma_\infty^{+-} = W^u(\mathbf{e}_+) \cap W^s(\mathbf{x}_-)$, $\Sigma_\infty^{--} \subset W^u(\mathbf{e}_-) \cap W^s(\mathbf{x}_-)$, and $\Sigma_\infty^{-+} = W^u(\mathbf{e}_-) \cap W^s(\mathbf{x}_+)$.*

(c) *The phase portrait of the Poincaré compactification is topologically equivalent to its correspondent in Figure 5.17.*

Proof. Let (A, B) be the fundamental matrices of the system and suppose that the matrix A is in real Jordan normal form. In this case, since $(t, d) \notin \mathcal{VB}_1|_{D,T} \cup \mathcal{VB}_2|_{D,T}$, the matrix A is non-diagonal, see Lemma 4.7.2(a).

(a) The existence of these separatrices is a consequence of Propositions 5.5.1(a) and 5.5.2(c).

(b) Since $(t, d) \notin \mathcal{VB}_1|_{D,T}$, we obtain that $k_1 \neq 0$, see Figure 5.10. From Proposition 3.6.2 we can assume that $k_1 = 1$.

Let γ_+^s and γ_-^s be the stable separatrices of saddle-nodes at infinity $\mathbf{x}_+$ and $\mathbf{x}_-$, which are contained in their hyperbolic manifolds. From Theorem 3.11.6(a) it follows that γ_+^s and γ_-^s are contained in the straight lines $x_2 = -2b_2/t$ and $x_2 = 2b_2/t$, respectively.

Since

$$A = \begin{pmatrix} \lambda & 1 \\ 0 & \lambda \end{pmatrix},$$

the vector $(1, 0)^T$ is an eigenvector of A and

$$\mathbf{e}_+ = -A^{-1}\mathbf{b} = -\frac{4}{t^2}\begin{pmatrix} tb_1/2 - b_2 \\ tb_2/2 \end{pmatrix}.$$

From this we conclude that $\alpha(\gamma_+^s) = \mathbf{e}_+$. By the symmetry of the flow with respect to the origin, $\alpha(\gamma_-^s) = \mathbf{e}_-$, see Figure 5.16.

From Proposition 5.5.5 it follows that there are no Jordan curves formed by solutions contained in Σ_∞. Moreover, there are no separatrix cycles to singular points contained in $\partial\mathbb{D}$. Otherwise, the singular points $\mathbf{e}_+$ or $\mathbf{e}_-$ would belong to this separatrix cycle, which is impossible because they are nodes. Therefore there are no limit cycles and separatrix cycles.

Let γ_0^{u+} and γ_0^{u-} be the unstable separatrices of the origin. The α- and the ω-limit sets contained in $\mathbb{D}$ are the singular points $\mathbf{0}$, $\mathbf{e}_+$, $\mathbf{e}_-$, $\mathbf{x}_+$, and $\mathbf{x}_-$. Since $\mathbf{e}_+$

and $\mathbf{e}_-$ are asymptotically unstable and there are no separatrix cycles, we obtain that $\omega(\gamma_0^{u+}) = \mathbf{x}_+$ and $\omega(\gamma_0^{u-}) = \mathbf{x}_-$. Thus, $W^u(\mathbf{0})$ splits the region Σ_∞ into the two open, connected, and invariant regions Σ_∞^+ and Σ_∞^-.

Let Σ_∞^+ be the region containing $\mathbf{e}_+$, see Figure 5.16, and let γ_0^{s+} be the stable separatrix of the origin in Σ_∞^+. The α- and the ω-limit sets in $\mathrm{Cl}(\Sigma_\infty^+)$ are the singular points $\mathbf{0}$, $\mathbf{e}_+$, $\mathbf{x}_+$, and $\mathbf{x}_-$. Hence $\alpha(\gamma_0^{s+}) = \mathbf{e}_+$ and the invariant curve $\gamma^{s+} \cup \mathbf{e}_+ \cup \gamma_+^s$ splits the region Σ_∞^+ into the two open, connected, and invariant regions Σ_∞^{++} and Σ_∞^{+-}. Let Σ_∞^{++} be the region containing γ^{u+} in its boundary, and let Σ_∞^{+-} be the region containing γ^{u-} in its boundary, see Figure 5.16. Note that the hyperbolic sector of the saddle-node $\mathbf{x}_+$ is contained in Σ_∞^{+-}. In a similar way we define the regions Σ_∞^{--} and Σ_∞^{-+} in Σ_∞^-. Then the hyperbolic sector of $\mathbf{x}_-$ is contained in Σ_∞^{-+}.

The behaviour of the flow in each of this regions Σ_∞^{jk} follows by observing that the α- and the ω-limit sets in $\mathrm{Cl}(\Sigma_\infty^{jk})$ are the singular points $\mathbf{0}$, $\mathbf{e}_j$, and $\mathbf{x}_k$ with $j,k \in \{+,-\}$.

Statement (c) follows from Theorem 2.6.9. □

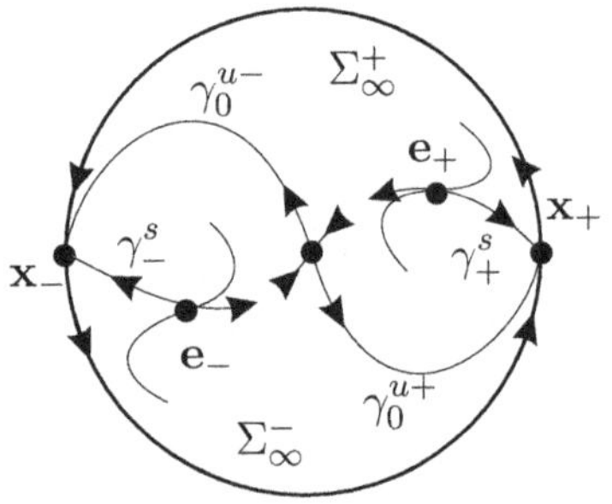

Figure 5.16: Regions Σ_∞^+ and Σ_∞^- defined in the phase space of the Poincaré compactification of a fundamental system with parameters $(t,d) \in \mathcal{SN}_\infty \setminus VB_1|_{D,T}$ and $t > 0$.

In the following result we deal with the special case where the boundary of the Poincaré disc $\partial\mathbb{D}$ is formed by singular points. Under this condition, Theorem 2.6.9 cannot be applied. Hence, in order to classify the phase portraits of this family of fundamental systems we build a homeomorphism between their phase portraits and the corresponding sketches represented in Figure 5.17.

Proposition 5.5.20. *Consider the Poincaré compactification of a fundamental system with parameters $D < 0$, $T < 0$, and $(t,d) \in \mathcal{VB}_1|_{D,T}$, and fundamental matrices (A,B).*

(a) *If the real Jordan normal form of the matrix A is non-diagonal, then the phase portrait of the Poincaré compactification is the one described in Proposition 5.5.19.*

(b) *Suppose that the real Jordan normal form of the matrix A is diagonal. Then:*

(b.1) *Every point in $\partial\mathbb{D}$ is a singular point and $\partial\mathbb{D} \setminus \{\pm\mathbf{k}^\perp/\|\mathbf{k}\|\}$ is a stable normally hyperbolic manifold. There exist exactly three singular points in Σ_∞: the saddle point at the origin $\mathbf{0}$ and the asymptotically unstable degenerated nodes $\mathbf{e}_+$ and $\mathbf{e}_-$.*

(b.2) *The straight lines $\eta_+ = \{\mathbf{e}_+ + r\mathbf{k}^\perp : r \in \mathbb{R}\}$ and $\eta_- = \{\mathbf{e}_- + r\mathbf{k}^\perp : r \in \mathbb{R}\}$ split Σ_∞ into three open and connected regions, Σ_∞^+, Σ_∞^0, and Σ_∞^-, which are invariant under the flow. These regions satisfy that $\Sigma_\infty^+ \subset W^u(\mathbf{e}_+)$, $\Sigma_\infty^- \subset W^u(\mathbf{e}_-)$, and the ω-limit set of every orbit contained in $\Sigma_\infty^+ \cup \Sigma_\infty^+$ is one of the singular points contained in $\partial\mathbb{D} \setminus \{\pm\mathbf{k}^\perp/\|\mathbf{k}\|\}$. The separatrices of the saddle at the origin split Σ_∞^0 into four open, connected, and invariant regions Σ_∞^{++}, Σ_∞^{+-}, Σ_∞^{-+}, and Σ_∞^{--}, such that: $\Sigma_\infty^{++} \subset W^u(\mathbf{e}_+) \cap W^s(\mathbf{k}^\perp/\|\mathbf{k}\|)$, $\Sigma_\infty^{-+} \subset W^u(\mathbf{e}_-) \cap W^s(\mathbf{k}^\perp/\|\mathbf{k}\|)$, $\Sigma_\infty^{--} \subset W^u(\mathbf{e}_-) \cap W^s(-\mathbf{k}^\perp/\|\mathbf{k}\|)$, and $\Sigma_\infty^{+-} \subset W^u(\mathbf{e}_+) \cap W^s(-\mathbf{k}^\perp/\|\mathbf{k}\|)$.*

(b.3) *The phase portrait of the Poincaré compactification is topologically equivalent to its correspondent in Figure* 5.17.

Proof. Let (A, B) be the fundamental matrices of the system and suppose that A is in real Jordan normal form.

(a) Assume that the matrix A is non-diagonal. Since $(t, d) \in \mathcal{VB}_1|_{D,T}$, we have divided the proof into two parts depending on the first coordinate of the vector $\mathbf{k}$.

Suppose that $k_1 \neq 0$. In this case the proof is identical to the proof of Proposition 5.5.19. Suppose that $k_1 = 0$. Then $k_2 \neq 0$ and without lost of generality we can consider $k_2 = -1$. Thus the hyperbolic manifolds of the singular points at infinity $\mathbf{x}_+$ and $\mathbf{x}_-$ are contained in the straight lines $x_2 = -2b_2/t$ and $x_2 = 2b_2/t$, respectively, see Theorem 3.11.10(e). The rest of the proof is identical to the proof of Proposition 5.5.19.

Statement (b.1) is a consequence of Propositions 5.5.1(a) and 5.5.2(b).

(b.2) The straight lines η_+ and η_- are parallel to L_+ and L_- and intersect the boundary of the Poincaré disc $\partial\mathbb{D}$ at the singular points at infinity $\pm\mathbf{k}^\perp/\|\mathbf{k}\|$. Since $\mathbf{e}_+$ and $\mathbf{e}_-$ are diagonal nodes, it follows that η_+ and η_- are invariant under the flow. Thus, the straight lines η_+ and η_- split the region Σ_∞ into the three open, connected, and invariant regions Σ_∞^+, Σ_∞^0, and Σ_∞^-. We denote by Σ_∞^+ and Σ_∞^- the regions which are contained in the half planes S_+ and S_-, respectively. Hence the α- and the ω-limit sets in $\mathrm{Cl}(\Sigma_\infty^+)$ and $\mathrm{Cl}(\Sigma_\infty^-)$ are the asymptotically unstable singular points $\mathbf{e}_+$ and $\mathbf{e}_-$ and the normally hyperbolic manifold at infinity which is asymptotically stable. From this we conclude the behaviour of the flow in Σ_∞^+ and Σ_∞^-.

The α- and the ω-limit sets in the region $\mathrm{Cl}(\Sigma_\infty^0)$ are the saddle point at the origin, the diagonal nodes $\mathbf{e}_+$ and $\mathbf{e}_-$, and the singular points at infinity $\pm\mathbf{k}^\perp/\|\mathbf{k}\| \in \partial\mathbb{D}$. Let γ^{s+} and γ^{s-} be the stable separatrices of the saddle point at the origin, and let γ^{u+} and γ^{u-} be the unstable separatrices of the saddle point at the origin. It is easy to check that $\alpha(\gamma^{s+}) = \mathbf{e}_+$, $\alpha(\gamma^{s-}) = \mathbf{e}_-$, $\omega(\gamma^{u+}) = \mathbf{k}^\perp/\|\mathbf{k}\|$, and $\omega(\gamma^{u-}) = -\mathbf{k}^\perp/\|\mathbf{k}\|$. Thus, the curves $W^s(\mathbf{0})$ and $W^u(\mathbf{0})$ split the region Σ^0_∞ in four open, connected, and invariant regions Σ^{++}_∞, Σ^{-+}_∞, Σ^{--}_∞, and Σ^{+-}_∞. Let Σ^{++}_∞ be the region such that $\{\mathbf{0}, \mathbf{e}_+$, let $\mathbf{k}^\perp/\|\mathbf{k}\|\} \subset \partial\Sigma^{++}_\infty$, Σ^{-+}_∞ be such that $\{\mathbf{0}, \mathbf{e}_-, \mathbf{k}^\perp/\|\mathbf{k}\|\} \subset \partial\Sigma^{-+}_\infty$, Σ^{--}_∞ such that $\{\mathbf{0}, \mathbf{e}_-, -\mathbf{k}^\perp/\|\mathbf{k}\|\} \subset \partial\Sigma^{--}_\infty$, and Σ^{+-}_∞ such that $\{\mathbf{0}, \mathbf{e}_+, -\mathbf{k}^\perp/\|\mathbf{k}\|\} \subset \partial\Sigma^{+-}_\infty$. The behaviour of the flow in each of these regions follow from the stability of the singular points which belong to its boundary.

(b.3) The separatrices in the central region $\mathrm{Cl}(\Sigma^0_\infty)$ are the singular points $\mathbf{0}, \mathbf{e}_+, \mathbf{e}_-$, the singular points at infinity $\pm\mathbf{k}^\perp/\|\mathbf{k}\|$, the orbits on the straight lines η_+ and η_-, and the separatrices of the saddle point at the origin. The canonical regions contained in $\mathrm{Cl}(\Sigma^0_\infty)$ are Σ^{++}_∞, Σ^{-+}_∞, Σ^{--}_∞, and Σ^{+-}_∞. Since in $\mathrm{Cl}(\Sigma^0_\infty)$ there exists a finite number of singular points, Theorem 2.6.9 applies. Thus, the phase portrait in $\mathrm{Cl}(\Sigma^0_\infty)$ is topologically equivalent to the central region of the correspondig phase portrait in Figure 5.17. Let h_0 be the topological equivalence.

Consider a circle centered at the singular point $\mathbf{e}_+$ and contained in Σ^+_∞. Since $\mathbf{e}_+$ is a diagonal node, it is clear that this circle intersects exactly once any orbit contained in $\mathrm{Cl}(\Sigma^+_\infty)$. Hence, we can define an orientation preserving homeomorphism h_+ from the region $\mathrm{Cl}(\Sigma^+_\infty)$ to its correspondent in the picture in Figure 5.17. Similarly, we define an orientation preserving homemorphism h_- from the region $\mathrm{Cl}(\Sigma^-_\infty)$ to its correspondent in the picture in Figure 5.17. Therefore,

$$h = \begin{cases} h_+ & \text{in } \mathrm{Cl}\left(\Sigma^+_\infty\right), \\ h_0 & \text{in } \mathrm{Cl}\left(\Sigma^0_\infty\right), \\ h_- & \text{in } \mathrm{Cl}\left(\Sigma^-_\infty\right), \end{cases}$$

is an orientation preserving homeomorphism from $\mathbb{D}$ to its correspondent picture in Figure 5.17. This proves the statement. □

Proposition 5.5.21. *Consider the Poincaré compactification of a fundamental system with parameters $D < 0$, $T < 0$, and $(t, d) \in \mathcal{C}^1_1 \setminus W_1|^*_{D,T}$. Then:*

(a) *The separatrices of the system are:*

(a.1) *the saddle point at the origin $\mathbf{0}$, the asymptotically unstable nodes $\mathbf{e}_+$ and $\mathbf{e}_-$, the asymptotically stable nodes at infinity $\mathbf{x}_+$ and $\mathbf{x}_-$, and the saddle points at infinity $\mathbf{y}_+$ and $\mathbf{y}_-$, which have their unstable manifold contained in $\partial\mathbb{D}$;*

(a.2) *the separatrices of the saddle point at the origin, and the separatrices of the singular points $\mathbf{y}_+$ and $\mathbf{y}_-$.*

(b) *The canonical regions are: $\Sigma^{++}_\infty = W^u(\mathbf{e}_+) \cap W^s(\mathbf{x}_+)$, $\Sigma^{+-}_\infty = W^u(\mathbf{e}_+) \cap W^s(\mathbf{x}_-)$, $\Sigma^{--}_\infty = W^u(\mathbf{e}_-) \cap W^s(\mathbf{x}_-)$, and $\Sigma^{-+}_\infty = W^u(\mathbf{e}_-) \cap W^s(\mathbf{x}_+)$.*

(c) *The phase portrait of the Poincaré compactification is topologically equivalent to its correspondent in Figure 5.17.*

Proof. Let (A, B) be the fundamental matrices of the system and suppose that the matrix A is in real Jordan normal form. If $(t, d) \notin W_1|^*_{D,T}$, then $t > 2\Lambda_1$, see (5.16) and $k_2 \neq 0$, see Figure 5.10. Without loss of generality we can consider that $k_2 = 1$.

(a) The existence of the listed separatrices is a consequence of Proposition 5.5.1(a) and 5.5.2(d).

(b) Let γ^s_+ and γ^s_- be the unstable separatrices of the saddles $\mathbf{y}_+$ and $\mathbf{y}_-$, respectively. By Theorem 3.11.8(a), the separatrices γ^s_+ and γ^s_- are contained in the straight lines $x = -b_1/\lambda_1$ and $x = b_1/\lambda_1$, respectively, where $\lambda_1 > \lambda_2$ denote the eigenvalues of the matrix A. Since A is in real Jordan normal form, it follows that $(0, 1)^T$ is an eigenvector of A and $\mathbf{e}_+ = -A^{-1}\mathbf{b} = (-b_1/\lambda_1, -b_2/\lambda_2)^T$. Therefore, we conclude that $\alpha(\gamma^s_+) = \mathbf{e}_+$ and $\alpha(\gamma^s_-) = \mathbf{e}_-$.

By Proposition 5.5.5, the system has no Jordan curves formed by solutions contained in the region Σ_∞. Moreover, since $\mathbf{x}_+$, $\mathbf{x}_-$, $\mathbf{e}_+$ and $\mathbf{e}_-$ are nodes, we conclude that there are no separatrix cycles with singular points in $\partial\mathbb{D}$. Thus, the system has no periodic orbits and separatrix cycles, and the α- and the ω-limit sets contained in $\mathbb{D}$ are the seven singular points.

Let γ^{u+} and γ^{u-} be the unstable separatrices of the origin. Since the singular points $\mathbf{e}_+$ and $\mathbf{e}_-$ are asymptotically unstable nodes and the singular points at infinity $\mathbf{y}_+$and $\mathbf{y}_-$ are saddle points, the ω-limit sets of the separatrices γ^{u+} and γ^{u-} are $\mathbf{x}_+$ and $\mathbf{x}_-$. We can suppose that $\omega(\gamma^{u+}) = \mathbf{x}_+$ and $\omega(\gamma^{u-}) = \mathbf{x}_-$. Hence, the curve $W^u(\mathbf{0})$ splits the region Σ_∞ into the two open, connected, and invariant regions Σ^+_∞, Σ^-_∞. Let Σ^+_∞ be the region containing the singular point $\mathbf{e}_+$ and Σ^-_∞ the region containing the singular point $\mathbf{e}_-$.

Let γ^{s+}_+ be the stable separatrix of the origin contained in Σ^+_∞, and γ^{s-} be the stable separatrix of the origin contained in Σ^-_∞. The α- and the ω-limit sets of the system in Σ^+_∞ are the singular points $\mathbf{0}$, $\mathbf{e}_+$, $\mathbf{x}_+$, $\mathbf{y}_+$, and $\mathbf{x}_-$. Therefore, $\alpha(\gamma^{s+}) = \mathbf{e}_+$. Moreover, the curve $W^s(\mathbf{0})$ together with $W^s(\mathbf{y}_+)$ and $W^s(\mathbf{y}_-)$ split the regions Σ^+_∞ and Σ^-_∞ into the four open, connected, and invariant regions Σ^{++}_∞, Σ^{+-}_∞, Σ^{--}_∞, and Σ^{-+}_∞. Let Σ^{jk}_∞ be the region containing the singular points $\mathbf{e}_j$ and $\mathbf{x}_k$ in its boundary, with $j, k \in \{+, -\}$. The statement follows by noting that the singular points $\mathbf{e}_j$ are asymptotically unstable and the singular points $\mathbf{x}_k$ are asymptotically stable. This also proves that there are no additional separatrices.

Statement (c) is a consequence of Theorem 2.6.9. □

Proposition 5.5.22. *Consider the Poincaré compactification of a fundamental system with parameters $D < 0$, $T < 0$, and $(t, d) \in W_1|^*_{D,T}$.*

(a) *If the fundamental system is proper, then the qualitative behaviour of the Poincaré compactification is described in Proposition* 5.5.21.

(b) *Suppose that the fundamental system is not proper. Then:*

(b.1) *The separatrices of the system are:*

(b.1.1) *the saddle point at the origin* $\mathbf{0}$, *the asymptotically unstable nodes* $\mathbf{e}_+$ *and* $\mathbf{e}_-$, *the asymptotically stable nodes at infinity* $\mathbf{x}_+$ *and* $\mathbf{x}_- \in$

$\partial\mathbb{D}$, and the singular points at infinity $\mathbf{y}_+$, $\mathbf{y}_- \in \partial\mathbb{D}$, which have the local phase portrait equivalent to that in Figure 5.11(c)*;*

(b.1.2) *the separatrices of the saddle point at the origin, and the separatrices of the singular points $\mathbf{y}_+$ and $\mathbf{y}_-$.*

(b.2) *The canonical regions are: $\Sigma_\infty^+ = W^u(\mathbf{e}_+) \cap W^s(\mathbf{x}_+)$, $\Sigma_\infty^- = W^u(\mathbf{e}_-) \cap W^s(\mathbf{x}_-)$, $\Sigma_\infty^{++} \subset W^u(\mathbf{e}_+) \cap W^s(\mathbf{y}_+)$, $\Sigma_\infty^{-+} \subset W^u(\mathbf{e}_-) \cap W^s(\mathbf{y}_+)$, $\Sigma_\infty^{--} \subset W^u(\mathbf{e}_-) \cap W^s(\mathbf{y}_-)$, and $\Sigma_\infty^{+-} \subset W^u(\mathbf{e}_+) \cap W^s(\mathbf{y}_-)$.*

(b.3) *The phase portrait of the Poincaré compactification is topologically equivalent to its correspondent in Figure* 5.17.

Proof. Let (A, B) be the fundamental matrices of the system and suppose that the matrix A is in real Jordan normal form.

(a) Suppose that the fundamental system is proper. From Lemma 4.7.1(d) we obtain that $k_2 \neq 0$. The proof is identical to the proof of Proposition 5.5.21.

(b.1) Suppose that the system is not proper. In this case $k_2 = 0$ and we can assume without lost of generality that $k_1 = 1$. Thus the existence of the listed separatrices is a consequence of Propositions 5.5.1(a) and 5.5.2(d).

(b.2) Let $\lambda_1 > \lambda_2 > 0$ be the eigenvalues of the matrix A. Since A is in real Jordan normal form, it follows that the separatrices of $\mathbf{y}_+$ and $\mathbf{y}_-$ are on the straight lines $x = -b_1/\lambda_1$ and $x = b_1/\lambda_1$, see Figure 5.11(c) and Proposition 3.11.12(f.2).

On the other hand, $\mathbf{e}_+ = -A^{-1}\mathbf{b} = (-b_1/\lambda_1, -b_2/\lambda_2)^T$, and the straight lines $\eta_+ = \{\mathbf{e}_+ + r\mathbf{k}^\perp : r \in\}$ and $\eta_- = \{\mathbf{e}_- + r\mathbf{k}^\perp : r \in \mathbb{R}\}$ are contained in the half-planes S_+ and S_-, respectively. Then η_+ and η_- are invariant under the flow and they contain the stable separatrices of the singular points at infinity $\mathbf{y}_+$ and $\mathbf{y}_-$. Therefore, the straight lines η_+ and η_- split the region Σ_∞ into the three open, connected and invariant regions $\Sigma_\infty^+, \Sigma_\infty^0$, and Σ_∞^-. Let Σ_∞^+ be the region containing the singular point $\mathbf{x}_+$ in its boundary, Σ_∞^- the region containing the singular point $\mathbf{x}_-$ in its boundary, and Σ_∞^0 the region limited by the straight lines η_+ and η_-.

By Proposition 5.5.5, the system has no Jordan curves formed by solutions contained in Σ_∞. Moreover, since the singular points $\mathbf{x}_+$, $\mathbf{x}_-$, $\mathbf{e}_+$, and $\mathbf{e}_-$ are nodes, there are no separatrix cycles to singular points in $\partial\mathbb{D}$. We conclude that the system has neither periodic orbits nor separatrix cycles.

Since the separatrices of $\mathbf{y}_+$ and $\mathbf{y}_-$ are contained in η_+ and η_-, respectively, the regions Σ_∞^+ and Σ_∞^- contain the hyperbolic sectors of the singular points $\mathbf{y}_+$ and $\mathbf{y}_-$. Then the α- and the ω-limit sets in $\mathrm{Cl}(\Sigma_\infty^+)$ are the singular points $\mathbf{e}_+$, $\mathbf{y}_+$, $\mathbf{x}_+$, and $\mathbf{y}_-$. Therefore, we conclude that $\Sigma_\infty^+ = W^u(\mathbf{e}_+) \cap W^s(\mathbf{x}_+)$ and $\Sigma_\infty^- = W^u(\mathbf{e}_-) \cap W^s(\mathbf{x}_-)$.

The α- and the ω-limit sets contained in $\mathrm{Cl}(\Sigma_\infty^0)$ are the singular points $\mathbf{0}$, $\mathbf{e}_+$, $\mathbf{e}_-$, $\mathbf{y}_+$, and $\mathbf{y}_-$. Let γ^{s+}, γ^{s-}, γ^{u+}, and γ^{u-} be the stable and the unstable separatrices of the origin, respectively. Thus $\omega(\gamma^{u+}) = \mathbf{y}_+$, $\omega(\gamma^{u-}) = \mathbf{y}_-$, $\alpha(\gamma^{s+}) = \mathbf{e}_+$, and $\alpha(\gamma^{s-}) = \mathbf{e}_-$. Therefore, the curve $W^s(\mathbf{0})$ together with the curve $W^u(\mathbf{0})$ split Σ^0_∞ into the four open, connected, and invariant regions Σ^{++}_∞, Σ^{-+}_∞, Σ^{--}_∞, and Σ^{-+}_∞. Let Σ^{jk}_∞ be the region containing the singular point $\mathbf{e}_j$ inside of it and the singular point at infinity $\mathbf{y}_k$ in its boundary. The behaviour of the flow in Σ^{jk}_∞ follows by noting that $\mathbf{e}_j$ is asymptotically unstable and $\mathbf{y}_k$ is asymptotically stable.

Statement (b.3) is a consequence of Theorem 2.6.9. □

Proposition 5.5.23. *Consider the Poincaré compactification of a fundamental system with parameters $D < 0$, $T < 0$, and $(t, d) \in \{(t, d) : d \leq 0\} \setminus \mathcal{O}$. Then:*

(a) *The separatrices of the system are:*

(a.1) *the saddle point at the origin $\mathbf{0}$, the asymptotically stable nodes at infinity $\mathbf{x}_+$, $\mathbf{x}_- \in \partial\mathbb{D}$, and the asymptotically at infinity unstable nodes $\mathbf{y}_+$, $\mathbf{y}_- \in \partial\mathbb{D}$;*

(a.2) *the separatrices of the origin, and the orbits contained in $\partial\mathbb{D}$.*

(b) *The canonical regions are: $\Sigma^{++}_\infty \subset W^u(\mathbf{y}_+) \cap W^s(\mathbf{x}_+)$, $\Sigma^{+-}_\infty \subset W^u(\mathbf{y}_+) \cap W^s(\mathbf{x}_-)$, $\Sigma^{--}_\infty \subset W^u(\mathbf{y}_-) \cap W^s(\mathbf{x}_-)$, and $\Sigma^{-+}_\infty \subset W^u(\mathbf{y}_-) \cap W^s(\mathbf{x}_+)$.*

(c) *The phase portrait of the Poincaré compactification is topologically equivalent to its correspondent in Figure* 5.17.

Proof. (a) The existence of these separatrices is a consequence of Propositions 5.5.1(b) and 5.5.2(d).

(b) From Proposition 5.5.5 it follows that the system has no Jordan curves formed by solutions contained in Σ_∞. Moreover, since the singular points $\mathbf{x}_+$, $\mathbf{x}_-$, $\mathbf{y}_+$ and $\mathbf{y}_-$ are nodes, we conclude that there are no separatrix cycles to the singular points contained in $\partial\mathbb{D}$. Then the α- and the ω-limit sets in the Poincaré disc $\mathbb{D}$ are the singular points.

Let γ^{s+}, γ^{s-}, γ^{u+}, and γ^{u-} be the stable and the unstable separatrices of the origin, respectively. Hence $\alpha(\gamma^{s+}) = \mathbf{y}_+$, $\alpha(\gamma^{s-}) = \mathbf{y}_-$, $\omega(\gamma^{u+}) = \mathbf{x}_+$, $\omega(\gamma^{u-}) = \mathbf{x}_-$, and the curve $W^s(\mathbf{0})$ together with the curve $W^u(\mathbf{0})$ split the region Σ_∞ into the four open, connected, and invariant regions Σ^{jk}_∞ for $j, k \in \{+, -\}$. Let Σ^{jk}_∞ be the region containing the singular points $\mathbf{y}_j$ and $\mathbf{x}_k$ in its boundary. The behaviour of the flow in Σ^{jk}_∞ follows by noting that the α- and the ω-limit sets containing in $\mathrm{Cl}(\Sigma^{jk}_\infty)$ are the singular points at infinity $\mathbf{y}_j$ and $\mathbf{x}_k$, respectively.

Statement (d) is a consequence of Theorem 2.6.9. □

Proposition 5.5.24. *Consider the Poincaré compactification of a fundamental system with parameters $D < 0$, $T < 0$, and $(t, d) \in \mathcal{C}^2_2 \setminus W_2|^*_{D,T}$. Then:*

(a) *The separatrices of the system are:*

(a.1) *the saddle point at the origin $\mathbf{0}$, the asymptotically stable nodes $\mathbf{e}_+$ and $\mathbf{e}_-$, the saddle points at infinity $\mathbf{x}_+$, $\mathbf{x}_- \in \partial\mathbb{D}$, which have their stable manifold contained in $\partial\mathbb{D}$, and the asymptotically unstable nodes at infinity $\mathbf{y}_+$, $\mathbf{y}_- \in \partial\mathbb{D}$;*

(a.2) *the separatrices of the saddle point at the origin, and the separatrices of the singular points at infinity* $\mathbf{x}_+$ *and* $\mathbf{x}_-$.

(b) *The canonical regions are:* $\Sigma_\infty^{++} = W^u(\mathbf{y}_+) \cap W^s(\mathbf{e}_+)$, $\Sigma_\infty^{+-} = W^u(\mathbf{y}_+) \cap W^s(\mathbf{e}_-)$, $\Sigma_\infty^{--} = W^u(\mathbf{y}_-) \cap W^s(\mathbf{e}_-)$, *and* $\Sigma_\infty^{-+} = W^u(\mathbf{y}_-) \cap W^s(\mathbf{e}_+)$.

(c) *The phase portrait of the Poincaré compactification is topologically equivalent to its correspondent in Figure* 5.17.

Proof. Let (A, B) be the fundamental matrices of the system and suppose that the matrix A is in real Jordan normal form. Since $(t, d) \notin W_2|_{D,T}^*$, we have $t < 2\Lambda_2$. Therefore, the first coordinate of the vector $\mathbf{k}$ satisfies $k_1 \neq 0$ and we can assume without loss of generality that $k_1 = 1$.

(a) The existence of the listed separatrices is a consequence of Propositions 5.5.1(a) and 5.5.2(d).

(b) Let $0 > \lambda_1 > \lambda_2$ be the eigenvalues of the matrix A. Let γ_+^u and γ_-^u the unstable separatrices of $\mathbf{x}_+$ and $\mathbf{x}_-$, respectively. Thus γ_+^u and γ_-^u are contained in the straight lines $x_2 = \pm b_2/\lambda_2$, see Theorem 3.11.7(b).

On the other hand, $\mathbf{e}_+ = -A^{-1}\mathbf{b} = (-b_1/\lambda_1, -b_2/\lambda_2)^T$ and $\mathbf{e}_- = A^{-1}\mathbf{b} = (b_1/\lambda_1, b_2/\lambda_2)^T$. Hence, we conclude that $\omega(\gamma_+^u) = \mathbf{e}_+$ and $\omega(\gamma_-^u) = \mathbf{e}_-$.

By Proposition 5.5.3(a), it follows that the system has no Jordan curves formed by solutions contained in Σ_∞. Thus, since the singular points $\mathbf{y}_+$, $\mathbf{y}_-$, $\mathbf{e}_+$, and $\mathbf{e}_-$ are nodes, we obtain that there are no separatrix cycles to singular points in $\partial\mathbb{D}$. Therefore the α- and the ω-limit sets in $\mathbb{D}$ are the singular points $\mathbf{0}$, $\mathbf{e}_+$, $\mathbf{e}_-$, $\mathbf{x}_+$, $\mathbf{x}_-$, $\mathbf{y}_+$, and $\mathbf{y}_-$.

Let γ^{s+}, γ^{s-}, γ^{u+}, and γ^{u-} be the stable and the unstable separatrices of the origin, respectively. We conclude that $\alpha(\gamma^{s+}) = \mathbf{y}_+$, $\alpha(\gamma^{s-}) = \mathbf{y}_-$, $\omega(\gamma^{u+}) = \mathbf{e}_+$, and $\omega(\gamma^{u-}) = \mathbf{e}_-$. Thus, the separatrices of the origin together with the unstable separatrices of the singular points at infinity $\mathbf{x}_+$ and $\mathbf{x}_-$ split Σ_∞ into the four open, connected, and invariant regions Σ_∞^{jk} for $j, k \in \{+, -\}$. Let Σ_∞^{jk} be the region containing the singular points $\mathbf{y}_j$ and $\mathbf{e}_k$ in its boundary. The statement follows by noting that the singular point $\mathbf{y}_j$ is asymptotically unstable and the singular point $\mathbf{e}_k$ is asymptotically stable.

Statement (d) is a consequence of Theorem 2.6.9. $\square$

Proposition 5.5.25. *Consider the Poincaré compactification of a fundamental system with parameters* $D < 0$, $T < 0$, *and* $(t, d) \in W_2|_{D,T}^*$.

(a) *If the system is proper, then the qualitative behaviour of the Poincaré compactification is described in Proposition* 5.5.24.

(b) *Suppose that the system is not proper. Then:*

(b.1) *The separatrices of the system are:*

(b.1.1) *the saddle point at the origin* $\mathbf{0}$, *the asymptotically stable nodes* $\mathbf{e}_+$, $\mathbf{e}_-$, *the singular points at infinity* $\mathbf{x}_+$, $\mathbf{x}_- \in \partial\mathbb{D}$, *see Figure 5.11(d), and the asymptotically unstable nodes at infinity* $\mathbf{y}_+$, $\mathbf{y}_- \subset \partial\mathbb{D}$;

(b.1.2) *the separatrices of the saddle at the origin, and the separatrices of the singular points* $\mathbf{x}_+$ *and* $\mathbf{x}_-$.

(b.2) *The canonical regions of the system are:* $\Sigma_\infty^+ = W^u(\mathbf{y}_+) \cap W^s(\mathbf{e}_+)$, $\Sigma_\infty^- = W^u(\mathbf{y}_-) \cap W^s(\mathbf{e}_-)$, $\Sigma_\infty^{++} \subset W^u(\mathbf{x}_+) \cap W^s(\mathbf{e}_+)$, $\Sigma_\infty^{-+} \subset W^u(\mathbf{x}_-) \cap W^s(\mathbf{e}_+)$, $\Sigma_\infty^{--} \subset W^u(\mathbf{x}_-) \cap W^s(\mathbf{e}_-)$, *and* $\Sigma_\infty^{+-} \subset W^u(\mathbf{x}_+) \cap W^s(\mathbf{e}_-)$.

(b.3) *The phase portrait of the Poincaré compactification is topologically equivalent to its correspondent in Figure* 5.17.

Proof. Let (A, B) be the fundamental matrices of the system and assume that the matrix A is in real Jordan normal form.

(a) Since the system is proper, $k_1 \neq 0$, see Lemma 4.7.1(d). The proof is identical to the proof of Proposition 5.5.24.

(b.1) Assume that the system is not proper and $(t, d) \in W_2|_{D,T}^*$. Hence $k_2 = 0$ and $k_1 \neq 0$. We take $k_1 = 1$, and then the existence of the listed separatrices is a consequence of Propositions 5.5.1(a) and 5.5.2(d).

(b.2) Since the matrix A is in real Jordan normal form, $(0, 1)^T$ is an eigenvector of A. Then the straight lines $\eta_+ = \{\mathbf{e}_+ + r\mathbf{k}^\perp : r \in \mathbb{R}\}$ and $\eta_- = \{\mathbf{e}_- + r\mathbf{k}^\perp : r \in \mathbb{R}\}$ contain the unstable separatrices of the singular points at infinity $\mathbf{x}_+$ and $\mathbf{x}_-$, see Theorem 3.11.11(f.2). Moreover, these straight lines are invariant under the flow, see the proof of Proposition 5.5.22 for more details. On the other hand, η_+ and η_- are parallel to the straight lines L_+ and L_-, and they are contained in the half-planes S_+ and S_-, respectively. Therefore η_+ and η_- split Σ_∞ into the three open, connected and invariant regions Σ_∞^+, Σ_∞^0, and Σ_∞^-. Let Σ_∞^k be the region containing the singular point at infinity $\mathbf{y}_k$ in its boundary, with $k \in \{+, -\}$; and let Σ_∞^0 be the region containing the origin.

From Proposition 5.5.5 we obtain that the system has no Jordan curves formed by solutions contained in Σ_∞. Moreover, it is easy to conclude that there are no separatrix cycles to singular points contained in $\partial\mathbb{D}$. Therefore the α- and the ω-limit sets in $\mathrm{Cl}(\Sigma_\infty^+)$ are the singular points $\mathbf{e}_+$, $\mathbf{x}_+$, $\mathbf{y}_+$, and $\mathbf{x}_-$. It is easy to check that the hyperbolic sectors of $\mathbf{x}_+$ and $\mathbf{x}_-$ are contained in Σ_∞^+ and Σ_∞^-, respectively. Then $\Sigma_\infty^+ = W^u(\mathbf{y}_+) \cap W^s(\mathbf{e}_+)$ and $\Sigma_\infty^- = W^u(\mathbf{y}_-) \cap W^s(\mathbf{e}_-)$.

Similar arguments to those in the proof of Proposition 5.5.22(b.2) show that the stable and the unstable manifold of the origin split Σ_∞^0 into the four open, connected and invariant regions, Σ_∞^{++}, Σ_∞^{-+}, Σ_∞^{--}, and Σ_∞^{+-}. The statement follows by denoting by Σ_∞^{jk} the region containing the singular points $\mathbf{x}_j$ and $\mathbf{e}_k$ in its boundary, where $j, k \in \{+, -\}$.

Statement (b.3) is a consequence of Theorem 2.6.9. □

Proposition 5.5.26. *Consider the Poincaré compactification of a fundamental system with parameters* $D < 0$, $T < 0$, $(t, d) \in \mathcal{SN}_\infty \setminus \mathcal{VB}_2|_{D,T}$, *and* $t < 0$. *Then:*

(a) *The separatrices of the system are:*

(a.1) *the saddle point at the origin* $\mathbf{0}$, *the degenerated nodes* $\mathbf{e}_+$ *and* $\mathbf{e}_-$, *the saddle-nodes at infinity* $\mathbf{x}_+, \mathbf{x}_- \in \partial\mathbb{D}$, *which have their central manifold* *contained in $\partial\mathbb{D}$, and their unstable hyperbolic manifold contained in Σ_∞;*

(a.2) *the separatrices of the saddle at the origin, and the separatrices of the singular points $\mathbf{x}_+$ and $\mathbf{x}_-$.*

(b) *The canonical regions are: $\Sigma_\infty^{++} = W^u(\mathbf{x}_+) \cap W^s(\mathbf{e}_+)$, $\Sigma_\infty^{+-} = W^u(\mathbf{x}_+) \cap W^s(\mathbf{e}_-)$, $\Sigma_\infty^{--} = W^u(\mathbf{x}_-) \cap W^s(\mathbf{e}_-)$, and $\Sigma_\infty^{-+} = W^u(\mathbf{x}_-) \cap W^s(\mathbf{e}_+)$.*

(c) *The phase portrait of the Poincaré compactification is topologically equivalent to its correspondent in Figure 5.17.*

Proof. Let (A, B) be the fundamental matrices of the system and assume that the matrix A is in real Jordan normal form. Since $(t, d) \notin \mathcal{VB}_2|_{D,T}$, it follows that the matrix A is non-diagonal, see Lemma 4.7.2. Thus we can assume that $k_1 \neq 0$.

(a) The existence of these separatrices is a consequence of Propositions 5.5.1(a) and 5.5.2(c).

(b) Arguments similar to those in Proposition 5.5.19(c) show that: the singular points $\mathbf{e}_+$ and $\mathbf{e}_-$ are the ω-limit sets of the unstable separatrices of the singular points at infinity $\mathbf{x}_+$ and $\mathbf{x}_-$, respectively; the system has no separatrix cycles to singular points contained in $\partial\mathbb{D}$; and the separatrices of the saddle at the origin together with the unstable separatrices of $\mathbf{x}_+$ and $\mathbf{x}_-$ split the region Σ_∞ into the four open, connected, and invariant regions, Σ_∞^{++}, Σ_∞^{+-}, Σ_∞^{--}, and Σ_∞^{-+}. The statement follows immediately if we denote by Σ_∞^{jk} the region which contain the singular points $\mathbf{x}_j$ and $\mathbf{e}_k$ in its boundary, where $j, k \in \{+, -\}$.

Statement (d) is a consequence of Theorem 2.6.9. □

Proposition 5.5.27. *Consider the Poincaré compactification of a fundamental system with parameters $D < 0$, $T < 0$, and $(t, d) \in \mathcal{VB}_2|_{D,T}$ and with fundamental matrices (A, B).*

(a) *If the real Jordan normal form of the matrix A is non-diagonal, then the qualitative behaviour of the Poincaré compactification is described in Proposition 5.5.26.*

(b) *Suppose that the real Jordan normal form of the matrix A is diagonal. Then:*

(b.1) *The boundary $\partial\mathbb{D}$ of the Poincaré disc is formed by singular points in such a way that $\partial\mathbb{D} \setminus \{\pm\mathbf{k}^\perp/\|\mathbf{k}\|\}$ is an unstable normally hyperbolic manifold and the local phase portrait of the singular points $\pm\mathbf{k}^\perp/\|\mathbf{k}\|$ is topologically equivalent to that in Figure 5.11(b). The system has three singular points in Σ_∞: the saddle point at the origin $\mathbf{0}$ and the degenerated diagonal nodes $\mathbf{e}_+$ and $\mathbf{e}_-$.*

(b.2) *The straight lines $\eta_+ = \{\mathbf{e}_+ + r\mathbf{k}^\perp : r \in \mathbb{R}\}$ and $\eta_- = \{\mathbf{e}_- + r\mathbf{k}^\perp : r \in \mathbb{R}\}$ split the interior of the Poincaré disc Σ_∞ into the three open, connected and invariant regions Σ_∞^+, Σ_∞^0, and Σ_∞^-, satisfying that $\Sigma_\infty^+ \subset W^s(\mathbf{e}_+)$, $\Sigma_\infty^- \subset W^s(\mathbf{e}_-)$. Moreover, the α-limit set of every orbit in $\Sigma_\infty^+ \cup \Sigma_\infty^-$*

is contained in $\partial\mathbb{D} \setminus \{\pm\mathbf{k}^\perp/\|\mathbf{k}\|\}$. *The separatrices of the origin split* Σ_∞^0 *into the four open, connected, and invariant regions* Σ_∞^{++}, Σ_∞^{+-}, Σ_∞^{--}, *and* Σ_∞^{-+}, *satisfying that* $\Sigma_\infty^{++} \subset W^u(\mathbf{k}^\perp/\|\mathbf{k}\|) \cap W^s(\mathbf{e}_+)$, $\Sigma_\infty^{+-} \subset W^u(\mathbf{k}^\perp/\|\mathbf{k}\|) \cap W^s(\mathbf{e}_-)$, $\Sigma_\infty^{--} \subset W^u(-\mathbf{k}^\perp/\|\mathbf{k}\|) \cap W^s(\mathbf{e}_-)$, *and* $\Sigma_\infty^{-+} \subset W^u(-\mathbf{k}^\perp/\|\mathbf{k}\|) \cap W^s(\mathbf{e}_+)$.

(b.3) *The phase portrait of the Poincaré compactification is topologically equivalent to its correspondent in Figure* 5.17.

Proof. Without loss of generality we can assume that the matrix A is in real Jordan normal form.

(a) Suppose that $k_1 \neq 0$. Then the statement follows by using arguments similar to those used in the proof of Proposition 5.5.26.

Now suppose that $k_1 = 0$. We assume that $k_2 = 1$. In this case, the unstable separatrices of the singular points at infinity $\mathbf{x}_+$ and $\mathbf{x}_-$ are contained in the straight lines $x_2 = \pm b_2/\lambda$, see Theorem 3.11.10(e). The statement follows by using arguments similar to those in the proof of Proposition 5.5.26.

Statement (b.1) is a consequence of Propositions 5.5.1(a) and 5.5.2(b).

(b.2) From Proposition 5.5.3(a) we conclude that the system has no Jordan curves formed by solutions. As in Proposition 5.5.20(b.2), it can be shown that: the straight lines η_+ and η_- are invariant under the flow and contain the separatrices of the singular points at infinity $\pm\mathbf{k}^\perp/\|\mathbf{k}\|$; the lines η_+ and η_- split the interior of the Poincaré disc Σ_∞ into the three open, connected and invariant regions Σ_∞^+, Σ_∞^0, and Σ_∞^-; and the behaviour of the flow in these regions, as described.

(b.3) The separatrices of the system in $\mathrm{Cl}(\Sigma_\infty^0)$ are the singular points $\mathbf{0}$, $\mathbf{e}_+$, $\mathbf{e}_-$, $\pm\mathbf{k}^\perp/\|\mathbf{k}\|$, the orbits contained in η_+ and η_-, and the separatrices of the origin. Therefore, the canonical regions contained in $\mathrm{Cl}(\Sigma_\infty^0)$ are Σ_∞^{++}, Σ_∞^{+-}, Σ_∞^{--}, and Σ_∞^{-+}. From Theorem 2.6.9 we conclude that there exists an orientation preserving homeomorphism h_0 from $\mathrm{Cl}(\Sigma_\infty^0)$ to the corresponding region in Figure 5.17. Following the proof of Proposition 5.5.20(b.3), we can construct two orientation preserving homeomorphisms h_+ and h_- from Σ_∞^+ and Σ_∞^- to the corresponding regions in Figure 5.17, respectively. The statement follows by noting that

$$h = \begin{cases} h_+ & \text{in } \mathrm{Cl}\left(\Sigma_\infty^+\right), \\ h_0 & \text{in } \mathrm{Cl}\left(\Sigma_\infty^0\right), \\ h_- & \text{in } \mathrm{Cl}\left(\Sigma_\infty^-\right), \end{cases}$$

is a topological equivalence. □

Proposition 5.5.28. *Consider the Poincaré compactification of a fundamental system with parameters* $D < 0$, $T < 0$, *and* $(t, d) \in \mathcal{C}_2^1$. *Then:*

(a) *The separatrices of the system are:*

(a.1) *the saddle point at the origin* $\mathbf{0}$, *and the asymptotically stable foci* $\mathbf{e}_+$, $\mathbf{e}_-$*;*

(a.2) *the limit cycle at infinity* $\infty = \partial\mathbb{D}$;

(a.3) *the separatrices of the saddle point at the origin.*

(b) *The canonical regions are:* $\Sigma_\infty^+ = W^u(\infty) \cap W^s(\mathbf{e}_+)$ *and* $\Sigma_\infty^- = W^u(\infty) \cap W^s(\mathbf{e}_-)$.

(c) *The phase portrait of the Poincaré compactification is topologically equivalent to its correspondent in Figure* 5.17.

Proof. (a) The existence of the listed separatrices is a consequence of Propositions 5.5.1(a) and 5.5.2(a).

(b) The system has no Jordan curves formed by solutions contained in Σ_∞, see Proposition 5.5.3(a). Then the α- and the ω-limit sets in the Poincaré disc $\mathbb{D}$ are the limit cycle ∞ and the singular points $\mathbf{0}$, $\mathbf{e}_+$, and $\mathbf{e}_-$. Let γ^{u+}, γ^{u-}, γ^{s+}, and γ^{s-} be the unstable and the stable separatrices of the origin, respectively. Hence $\omega(\gamma^{u+}) = \mathbf{e}_+$, $\omega(\gamma^{u-}) = \mathbf{e}_-$, and $\alpha(\gamma^{s+}) = \alpha(\gamma^{s-}) = \infty$. Therefore, the separatrices γ^{s+} and γ^{s-} split the region $\Sigma_\infty \setminus \mathrm{Cl}(W^u(\mathbf{0}))$ into the two open, connected, and invariant regions Σ_∞^+ and Σ_∞^-. Let Σ_∞^k be the region containing $\mathbf{e}_k$ in its boundary, where $k \in \{+, -\}$. The behaviour of the flow in Σ_∞^+ and Σ_∞^- follows by observing that the singular points $\mathbf{e}_+$ and $\mathbf{e}_-$ are asymptotically stable and the limit cyle ∞ is asymptotically unstable.

Statement (c) is a consequence of Theorem 2.6.9. □

Proposition 5.5.29. *Consider the Poincaré compactification of a fundamental system with parameters* $D < 0$, $T < 0$, *and* $(t, d) \in \mathcal{O}$. *Then:*

(a) *The separatrices of the system are:*

(a.1) *the saddle point at the origin* $\mathbf{0}$, *and the singular points at infinity* $\mathbf{x}_+$, $\mathbf{x}_- \in \partial\mathbb{D}$, *which have a neighbourhood contained in* $\mathbb{D}$ *formed by an elliptic sector;*

(a.2) *the separatrices of the saddle point at the origin, and the separatrices of the singular points* $\mathbf{x}_+$ *and* $\mathbf{x}_-$, *which form two heteroclinic cycles* Δ_+ *and* Δ_- *to the common singular point at the origin.*

(b) *The canonical regions are:* $\Sigma_{\Delta_+} \subset W^u(\mathbf{x}_+) \cap W^s(\mathbf{x}_+)$, $\Sigma_{\Delta_-} \subset W^u(\mathbf{x}_-) \cap W^s(\mathbf{x}_-)$, $\Sigma_\infty^{+-} \subset W^u(\mathbf{x}_+) \cap W^s(\mathbf{x}_-)$, *and* $\Sigma_\infty^{-+} \subset W^u(\mathbf{x}_-) \cap W^s(\mathbf{x}_+)$.

(c) *The phase portrait of the Poincaré compactification is topologically equivalent to its correspondent in Figure* 5.17.

Proof. Statement (a.1) is a consequence of Propositions 5.5.2(c) and 5.5.1(b).

(a.2) From Proposition 5.5.3(a) we conclude that there are no Jordan curves formed by solutions in the interior of the Poincaré disc Σ_∞. Hence, the separatrices of the origin cannot connect in a separatrix cycle contained in Σ_∞. On the other hand, the origin is the unique singular point in Σ_∞. Let γ^{s+}, γ^{s-}, γ^{u+}, and γ^{u-} be the stable and the unstable separatrices of the origin. Therefore, $\alpha(\gamma_0^{s+}) = \omega(\gamma_0^{u+}) = \mathbf{x}_+$ and $\alpha(\gamma_0^{s-}) = \omega(\gamma_0^{u-}) = \mathbf{x}_-$, which form the heteroclinic cycles Δ_+ and Δ_-.

(b) The behaviour of the flow in Σ_{Δ_+} and Σ_{Δ_-} is straightforward.

Let Σ_∞^{+-} and Σ_∞^{-+} be the connected components of $\Sigma_\infty \setminus \mathrm{Cl}(\Sigma_{\Delta_+} \cup \Sigma_{\Delta_-})$. It is easy to conclude that the boundaries of these regions, $\partial\Sigma_\infty^{+-}$ and $\partial\Sigma_\infty^{-+}$, are not contained in a separatrix cycle. Thus, the α- and the ω-limit sets in $\mathrm{Cl}(\Sigma_\infty^{+-})$ and $\mathrm{Cl}(\Sigma_\infty^{-+})$ are $\mathbf{x}_+$, $\mathbf{x}_+$, and $\mathbf{0}$, which proves the statement.

Statement (c) is a consequence of Theorem 2.6.9. □

5.5.6 The bifurcation set

Using the results proved in Subsection 5.5.5 and the phase portraits sketched in Figures 5.17 and 5.18, in this subsection we describe the bifurcations that take place in the phase portrait of the Poincaré compactification of a fundamental system with parameters $D < 0$ and $T < 0$ and (t, d) varying in the plane $\Pi_{D,T}$.

Take $(t, d) \in \mathcal{C}_2^1$ and vary the parameters clockwise. On the straight line $\mathcal{H}_\infty$ the singular points $\mathbf{e}_+$ and $\mathbf{e}_-$ are bounded centers limited by two periodic orbits. Just after crossing $\mathcal{H}_\infty$ only two periodic orbits persist, becoming two stable limit cycles. Also, another limit cycle appears bifurcating from infinity. Then on the straight line $\mathcal{H}_\infty$ two different bifurcations occur: the first one is a Hopf bifurcation from the singular point at infinity; the other one is a focus-center-limit cycle bifurcation at the singular points $\mathbf{e}_+$ and $\mathbf{e}_-$.

On the curve $\mathcal{H}_o\mathcal{L}|_{D;T}$ the two limit cycles which emerge at the focus-center-limit cycle bifurcation collide with the separatrices of the saddle at the origin forming two homoclinic cycles to the common singular point at the origin. These homoclinic cycles disappear just after crossing the curve $\mathcal{H}_o\mathcal{L}|_{D;T}$ and one unstable limit cycle emerges. Therefore, on the curve $\mathcal{H}_o\mathcal{L}|_{D;T}$ the family of fundamental systems exhibits a homoclinic bifurcation.

On the curve $(t, d) \in \mathcal{NH}_{lc}|_{D,T}$ the limit cycle born at the homoclinic bifurcation collides with the one born in the Hopf bifurcation at infinity. From this collision appears a non-hyperbolic limit cycle which is inside asymptotically stable and outside asymptotically unstable. This limit cycle disappears just after crossing the curve $\mathcal{NH}_{lc}|_{D,T}$. Thus, on the curve $\mathcal{NH}_{lc}|_{D,T}$ the family exhibits a saddle-node bifurcation of limit cycles.

In much the same way as in Subsection 5.2.6, a saddle-node bifurcation of the singular points at infinity occurs on the curve $\mathcal{SN}_\infty$.

The fundamental systems with parameters lying on the straight half-line $W_1|_{D,T}^*$ have two different compactified phase portrait, depending on whether the system is proper or not. If the system is proper, then the phase portrait is topologically equivalent to the phase portrait of systems with parameters in $\mathcal{C}_1^1$. If the system is not proper (an eigenvector of the fundamental matrix A is parallel to the straight line L_+), then the straight lines through the singular points $\mathbf{e}_+$ and $\mathbf{e}_-$ which are parallel to L_+ define a parabolic sector of the saddle points at infinity. Therefore, a parabolic sector bifurcation of the saddle points at infinity occurs on $W_1|_{D,T}^*$.

As in Subsection 5.2.6, a pitchfork bifurcation of the saddles at infinity occurs on the straight line $\mathcal{N}$. In this case (for $t > 0$) it is a subcritical bifurcation.

On the straight half-line $W_2|_{D,T}^*$, a parabolic sector bifurcation of the saddle points at infinity takes place.

5.6 The case $D < 0$ and $T = 0$

In this section we consider the fundamental systems

$$\dot{\mathbf{x}} = A\mathbf{x} + \varphi\left(\mathbf{k}^T\mathbf{x}\right)\mathbf{b},$$

with parameters $D < 0$ and $T = 0$. The dynamical behaviour of these systems is similar to the dynamical behaviour of fundamental systems with parameters $D < 0$ and $T < 0$. Only when $t^2 - 4d < 0$, their phase portraits are different, see Subsection 5.6.1. Therefore, to describe the Poincaré compactification of the fundamental systems with parameters $D < 0$, $T = 0$, and $t^2 - 4d \geq 0$ we refer the reader to Propositions 5.6.3 and 5.6.4, and Figure 5.17.

5.6.1 Proper fundamental systems

According to Subsection 5.5.1, for the non-proper fundamental systems with parameters $D < 0$ and $T = 0$, the parameters (t, d) lie on the straight lines $W_1|_{D,0} = W_1 \cap \Pi_{D,0}$ and $W_2|_{D,0} = W_2 \cap \Pi_{D,0}$, see Figure 5.10. Note that the intersection of these straight lines is a point which belongs to the half-line $t = 0$ and $d < 0$.

From Lemma 4.7.2 we conclude that the fundamental systems with parameters $D < 0$, $T = 0$, and $t^2 - 4d = 0$ and such that the real Jordan normal form of the matrix A is diagonal satisfy that $(t, d) \in \mathcal{VB}_1|_{D,0} \cup \mathcal{VB}_2|_{D,0}$. Here $\mathcal{VB}_1|_{D,0}$ and $\mathcal{VB}_2|_{D,0}$ are the intersection points of the straight lines $W_1|_{D,T}$ and $W_2|_{D,T}$ with the parabola $t^2 - 4d = 0$, respectively.

5.6.2 Finite singular points and singular points at infinity

The existence, number and location of finite and infinite singular points, together with the associated local phase portraits are described in Propositions 5.5.2 and 5.5.1.

5.6.3 Periodic orbits

In the following we prove that only fundamental systems with parameters $t = 0$ and $d > 0$ can exhibit Jordan curves formed by solutions.

Proposition 5.6.1. *Consider a fundamental system with parameters $D < 0$ and $T = 0$.*

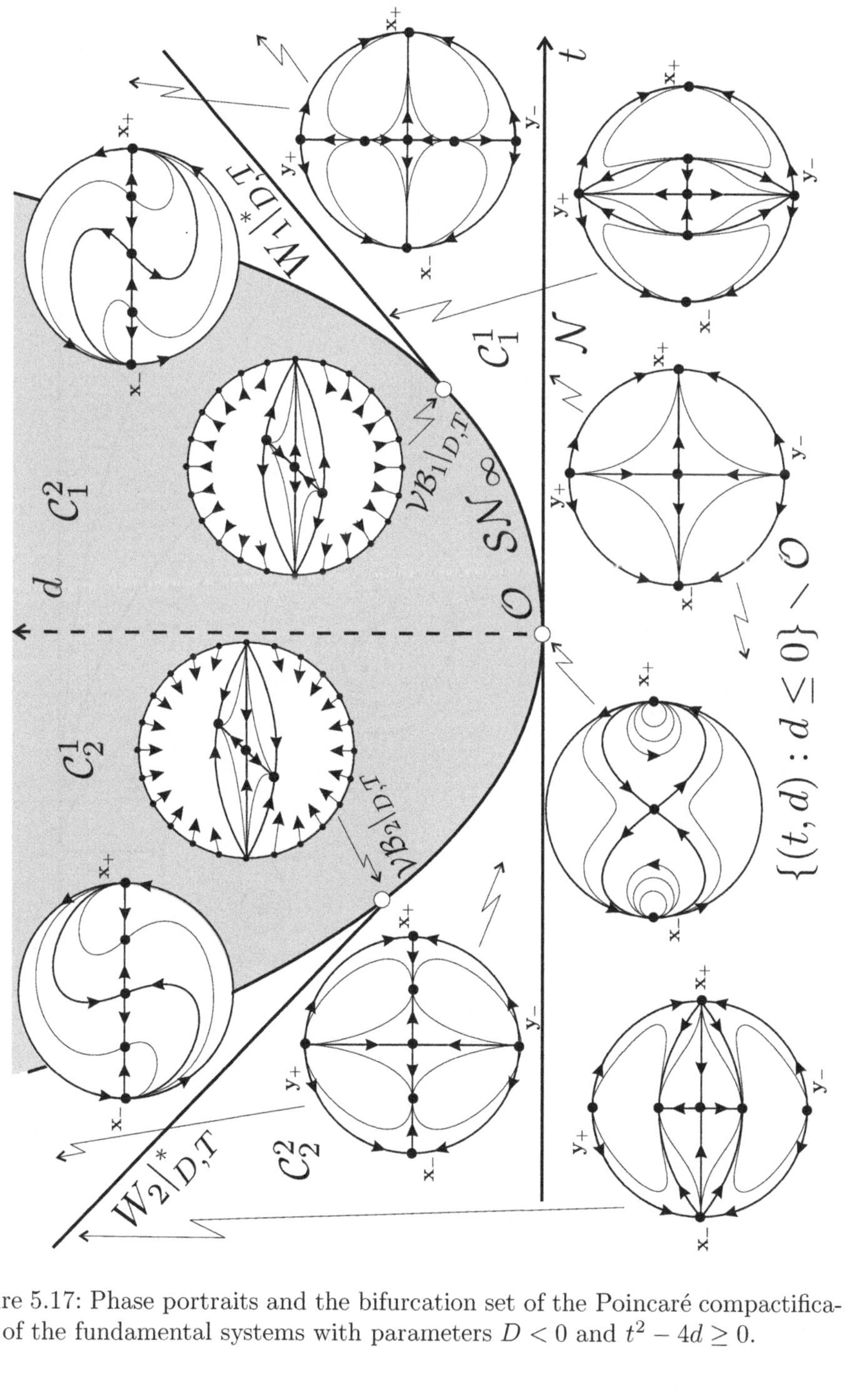

Figure 5.17: Phase portraits and the bifurcation set of the Poincaré compactification of the fundamental systems with parameters $D < 0$ and $t^2 - 4d \geq 0$.

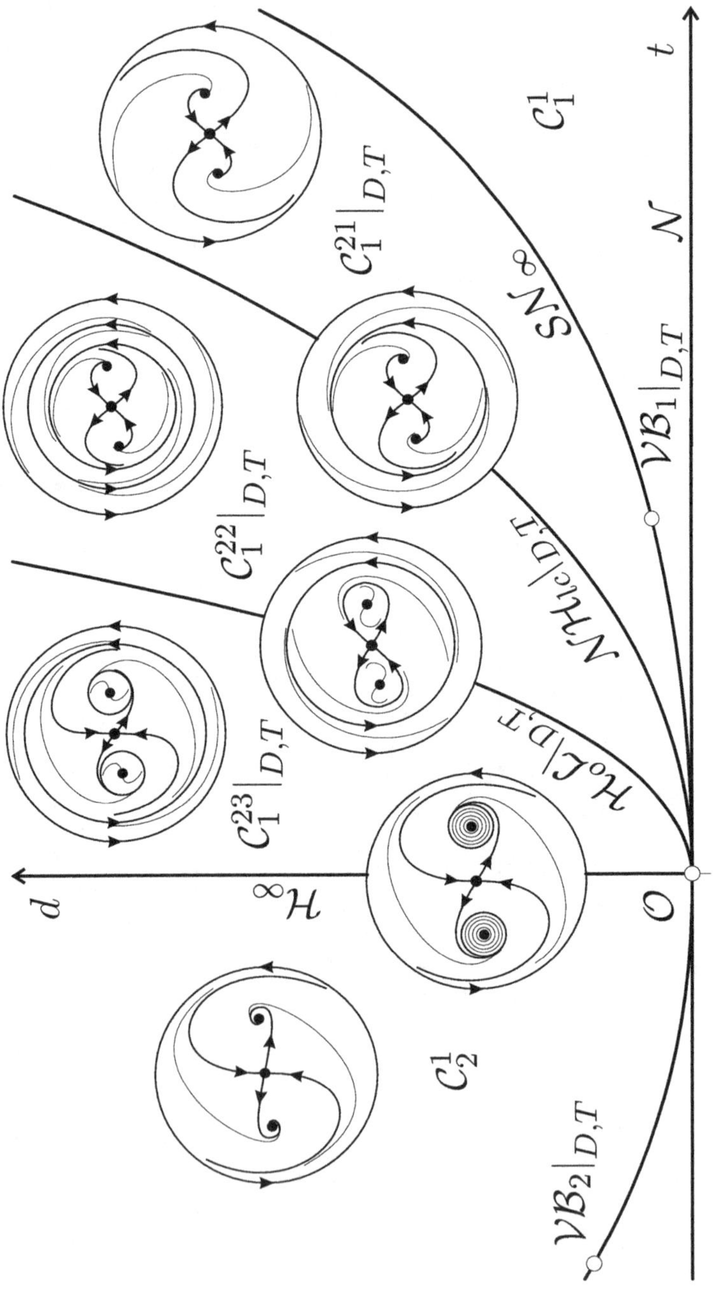

Figure 5.18: Phase portraits and the bifurcation set of the Poincaré compactification of the fundamental systems with parameters $D < 0$, $T < 0$, and $t^2 - 4d < 0$.

(a) *If either* $t \neq 0$ *or* $t = 0$ *and* $d \leq 0$, *then there are no Jordan curves formed by solutions.*

(b) *If* $t = 0$ *and* $d > 0$, *then there exist two periodic orbits* $\Phi_+ \subset L_+ \cup S_+$ *and* $\Phi_- \subset L_- \cup S_-$ *such that the annular regions* $\mathrm{Cl}(\Sigma_{\Phi_+}) \setminus \mathbf{e}_+$ *and* $\mathrm{Cl}(\Sigma_{\Phi_-}) \setminus \mathbf{e}_-$ *are formed by periodic orbits.*

Proof. (a) Suppose that $t \neq 0$ and let Γ be a Jordan curve formed by solutions. From Theorem 3.12.2(d) it follows that $\Gamma \subset L_+ \cup S_0 \cup L_-$. But this is not possible, because the system in $L_+ \cup S_0 \cup L_-$ becomes a linear system with matrix B and $\det(B) = D < 0$.

Suppose now that $t = 0$ and $d \leq 0$. From Lemma 5.5.4(a) it follows that the return map π is not defined. Hence, every Jordan curve formed by solutions is contained in one of the regions S_+, S_0 or S_-. In these regions the system becomes linear with parameters $D < 0$ or $d \leq 0$. This implies that any Jordan curve formed by solutions must be contained in them. Therefore, the systems has no Jordan curves formed by solutions.

(b) The existence of the periodic orbits Φ_+ and Φ_-, and the behaviour of the flow in $\mathrm{Cl}(\Sigma_{\Phi_+}) \setminus \mathbf{e}_+$ and $\mathrm{Cl}(\Sigma_{\Phi_-}) \setminus \mathbf{e}_-$ can be concluded by applying arguments similar to those in the proof of Proposition 5.5.3(b). □

In the next result we show that for a fundamental system with parameters $t = 0$ and $d > 0$ every orbit not contained in $\mathrm{Cl}(\Sigma_{\Phi_+}) \setminus \mathbf{e}_+$ or in $\mathrm{Cl}(\Sigma_{\Phi_-}) \setminus \mathbf{e}_-$ is either a periodic orbit or a homoclinic cycle to the singular point at the origin.

Proposition 5.6.2. *Consider a fundamental system with parameters* $D < 0$, $T = 0$, $t = 0$, *and* $d > 0$. *The stable and the unstable separatrices of the origin form two homoclinic cycles* Δ_+ *and* Δ_- *to the common singular point at the origin. Moreover, every orbit different from* Δ_+, Δ_-, $\mathbf{0}$, $\mathbf{e}_+$, *and* $\mathbf{e}_-$ *is a periodic orbit.*

Proof. Since $T = 0$, $t = 0$, and $t^2 - 4d < 0$, it follows that the Poincaré maps $\widetilde{\pi}^A_{++}$, π^B_{++}, and π^B_{+-} are the identity on their respective domains, see Lemma 5.5.4(b) and (e). Hence the Lamerey map defined in Lemma 5.5.7(a) satisfies that $g\,(a) = 0$ for every $a \neq |\Lambda_2|^{-1}$. Therefore, any orbit γ intersecting either L_+ or L_- at a point of coordinate $a \neq |\Lambda_2|^{-1}$ is a periodic orbit, see Lemma 5.5.7(b).

Let Λ_1 and Λ_2 be the eigenvalues of the matrix B. Since $T = 0$, we have $-\Lambda_1 = \Lambda_2 < 0$. Hence $g(|\Lambda_2|^{-1}) = 0$. From Lemma 5.5.7(c) we conclude that the separatrices at the origin form two homoclinic cycles Δ_+ and Δ_- to the common singular point at the origin.

The periodic orbits Φ_+ and Φ_- defined in Proposition 5.6.1(b) intersect L_+ and L_- at the contact points $\mathbf{p}_+$ and $\mathbf{p}_-$, respectively (see the proof of Proposition 5.5.3 for more details). Since the points $\mathbf{p}_+$ and $\mathbf{p}_-$ have coordinate $a = 0$, it follows that $\Phi_+ \subset \Sigma_{\Delta_+}$ and $\Phi_- \subset \Sigma_{\Delta_-}$. □

5.6.4 Phase portraits

Now we describe the compactified phase portraits of the fundamental systems with parameters $D < 0$ and $T = 0$. We present these results in two separate propositions, depending on the existence of a previously studied phase portrait. In the first of these results we describe a phase portrait which is not topologically equivalent to any of the phase portraits studied before.

Following expression (5.16) we define the half-lines $W_1|^*_{D,0}$ and $W_2|^*_{D,0}$ for the parameter $D < 0$.

Proposition 5.6.3. *Consider the Poincaré compactification of a fundamental system with parameters $D < 0$, $T = 0$, and $(t, d) \in \mathcal{H}_\infty$. Then:*

(a) *The separatrices of the system are:*

(a.1) *the saddle point at the origin $\mathbf{0}$, and the singular points $\mathbf{e}_+$ and $\mathbf{e}_-$, which are centers;*

(a.2) *the limit cycle at infinity ∞;*

(a.3) *the separatrices of the saddle at the origin, which form two homoclinic cycles Δ_+ and Δ_- to the singular point at the origin in such a way that $\mathbf{e}_+ \in \Sigma_{\Delta_+}$ and $\mathbf{e}_- \in \Sigma_{\Delta_-}$.*

(b) *The canonical regions are: $\Sigma_{\Delta_+} \setminus \mathbf{e}_+$, $\Sigma_{\Delta_-} \setminus \mathbf{e}_-$, and $\Sigma_\infty \setminus \mathrm{Cl}(\Sigma_{\Delta_+} \cup \Sigma_{\Delta_-})$. Moreover, every orbit in these regions is periodic.*

(c) *The phase portrait of the Poincaré compactification is topologically equivalent to its correspondent in Figure 5.19.*

Proof. Statements (a.1) and (a.2) follow from Propositions 5.5.1(a) and 5.5.2(a). Statements (a.3) and (b) follow from Proposition 5.6.2. Statement (c) follows from Theorem 2.6.9. □

Proposition 5.6.4. *Consider the Poincaré compactification of a fundamental system with parameters $D < 0$ and $T = 0$ and fundamental matrices (A, B).*

(a) *If $(t, d) \in \mathcal{C}_1^2$, then the phase portrait of the Poincaré compactification is topologically equivalent to its correspondent in Figure 5.19.*

(b) *If either $(t, d) \in \mathcal{SN}_\infty \setminus \mathcal{VB}_1|_{D,0}$ and $t > 0$, or $(t, d) \in \mathcal{VB}_1|_{D,0}$ and the real Jordan form of the matrix A is non-diagonal, then the phase portrait of the Poincaré compactification is topologically equivalent to its correspondent in Figure 5.17.*

(c) *If $(t, d) \in \mathcal{VB}_1|_{D,0}$ and the real Jordan normal form of the matrix A is diagonal, then the phase portrait of the Poincaré compactification is topologically equivalent to its correspondent in Figure 5.17.*

(d) *If either $(t, d) \in \mathcal{C}_1^1 \setminus W_1|^*_{D,0}$ or $(t, d) \in W_1|^*_{D,0}$ and the system is proper, then the phase portrait of the Poincaré compactification is topologically equivalent to its correspondent in Figure 5.17.*

(e) *If* $(t, d) \in W_1|^*_{D,0}$ *and the system is not proper, then the phase portrait of the Poincaré compactification is topologically equivalent to its correspondent in Figure* 5.17.

(f) *If* $t \neq 0$ *and* $d \leq 0$*, then the phase portrait of the Poincaré compactification is topologically equivalent to its correspondent in Figure* 5.17.

(g) *If either* $(t, d) \in \mathcal{C}_2^2 \setminus W_2|^*_{D,0}$ *or* $(t, d) \in W_2|^*_{D,0}$ *and the system is proper, then the phase portrait of the Poincaré compactification is topologically equivalent to its correspondent in Figure* 5.17.

(h) *If* $(t, d) \in W_2|^*_{D,0}$ *and the system is not proper, then the phase portrait of the Poincaré compactification is topologically equivalent to its correspondent in Figure* 5.17.

(i) *If either* $(t, d) \in \mathcal{SN}_\infty \setminus \mathcal{VB}_2|_{D,0}$ *and* $t < 0$*, or* $(t, d) \in \mathcal{VB}_2|_{D,0}$ *and the real Jordan normal form of the matrix* A *is non-diagonal, then the phase portrait of the Poincaré compactification is topologically equivalent to its correspondent in Figure* 5.17.

(j) *If* $(t, d) \in \mathcal{VB}_2|_{D,0}$ *and the real Jordan normal form of the matrix* A *is diagonal, then the phase portrait of the Poincaré compactification is topologically equivalent to its correspondent in Figure* 5.17.

(k) *If* $(t, d) \in \mathcal{C}_2^1$*, then the phase portrait of the Poincaré compactification is topologically equivalent to its correspondent in Figure* 5.19.

Proof. The proof of statement (a) is identical to the proof of the Proposition 5.5.18. The proof of the remainder statements may be obtained from the proof of the corresponding proposition in Subsection 5.5.5. □

5.6.5 The bifurcation set

The bifurcations which take place in the region $t^2 - 4d \geq 0$ of the parameter space $\Pi_{D,0}$ when $D < 0$ are identical to those in the region $t^2 - 4d \geq 0$ of the plane $\Pi_{D,T}$ with $D < 0$ and $T < 0$, see Subsection 5.5.6. We refer the reader to this subsection for a description of these bifurcations.

In the region $t^2 - 4d < 0$ there exists an unique bifurcation curve, namely, the straight line $\mathcal{H}_\infty$. This curve corresponds to a focus-center-limit cycle bifurcation in a neighbourhood of the singular points $\mathbf{e}_+$ and $\mathbf{e}_-$, together with a homoclinic bifurcation.

5.7 The case $D < 0$ and $T > 0$

Following Section 5.4, under a time-reversal transformation, a fundamental system with parameters (D, T, d, t) becomes a fundamental system of parameters (D, T^*, d, t^*), where $T^* = -T$ and $t^* = -t$. Hence, the two systems have identical

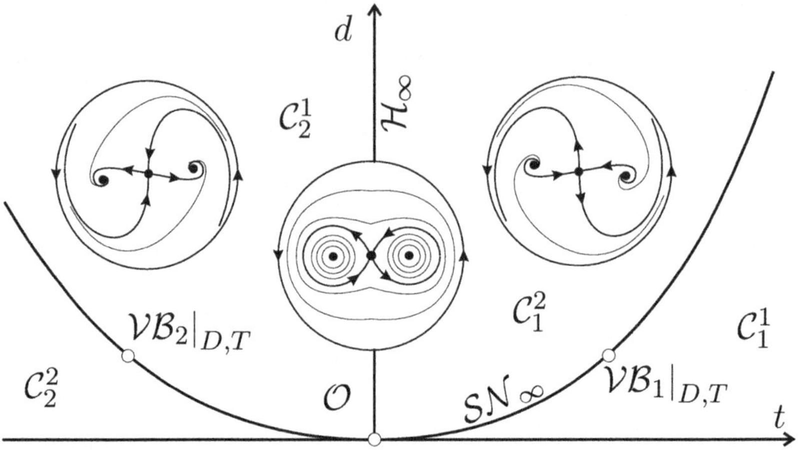

Figure 5.19: Phase portrait and the bifurcation set of the Poincaré compactification of the fundamental systems with parameters $D < 0$, $T = 0$, and $t^2 - 4d < 0$.

orbits but oppositely oriented. Therefore, using the regions $\mathcal{C}_1^{21}|_{D,T^*}$, $\mathcal{C}_1^{22}|_{D,T^*}$, and $\mathcal{C}_1^{23}|_{D,T^*}$, the curves $\mathcal{H}_o\mathcal{L}|_{D,T^*}$ and $\mathcal{NH}_{lc}|_{D,T^*}$, the half-lines $W_1|_{D,T^*}^*$ and $W_2|_{D,T^*}^*$, and the points $\mathcal{VB}_1|_{D,T^*}$ and $\mathcal{VB}_2|_{D,T^*}$ in the plane Π_{D,T^*}, we define in the plane $\Pi_{D,T}$ (where $T = -T^* > 0$) the regions

$$\mathcal{C}_2^{11}|_{D,T} := \left\{(t,d) : (-t,d) \in \mathcal{C}_1^{23}|_{D,T^*}\right\},$$
$$\mathcal{C}_2^{12}|_{D,T} := \left\{(t,d) : (-t,d) \in \mathcal{C}_1^{22}|_{D,T^*}\right\},$$
$$\mathcal{C}_2^{13}|_{D,T} := \left\{(t,d) : (-t,d) \in \mathcal{C}_1^{21}|_{D,T^*}\right\},$$

the curves

$$\mathcal{H}_o\mathcal{L}|_{D,T} := \left\{(t,d) : (-t,d) \in \mathcal{H}_o\mathcal{L}|_{D,T^*}\right\},$$
$$\mathcal{NH}_{lc}|_{D,T} := \left\{(t,d) : (-t,d) \in \mathcal{NH}_{lc}|_{D,T^*}\right\},$$

the half-lines

$$W_1|_{D,T}^* := \left\{(t,d) : (-t,d) \in W_1|_{D,T^*}^*\right\},$$
$$W_2|_{D,T}^* := \left\{(t,d) : (-t,d) \in W_2|_{D,T^*}^*\right\},$$

and the points

$$\mathcal{VB}_1|_{D,T} := \left\{(t,d) : (-t,d) \in \mathcal{VB}_1|_{D,T^*}\right\},$$
$$\mathcal{VB}_2|_{D,T} := \left\{(t,d) : (-t,d) \in \mathcal{VB}_2|_{D,T^*}\right\}.$$

Proposition 5.7.1. *Consider the Poincaré compactification of a fundamental system with parameters $D < 0$ and $T > 0$ and fundamental matrices (A, B).*

(a) *If $(t,d) \in \mathcal{H}_\infty$, then the phase portrait of the Poincaré compactification is topologically equivalent to its correspondent in Figure* 5.20.

(b) *If $(t,d) \in \mathcal{C}_1^2$, then the phase portrait of the Poincaré compactification is topologically equivalent to its correspondent in Figure* 5.20.

(c) *If either $(t,d) \in \mathcal{SN}_\infty \setminus \mathcal{VB}_1|_{D,T}$ and $t > 0$, or $(t,d) \in \mathcal{VB}_1|_{D,T}$ and the real Jordan normal form of the matrix A is non-diagonal, then the phase portrait of the Poincaré compactification is topologically equivalent to its correspondent in Figure* 5.17.

(d) *If $(t,d) \in \mathcal{VB}_1|_{D,T}$ and the real Jordan normal form of the matrix A is diagonal, then the phase portrait of the Poincaré compactification is topologically equivalent to its correspondent in Figure* 5.17.

(e) *If either $(t,d) \in \mathcal{C}_1^1 \setminus W_1|_{D,T}^*$, or $(t,d) \in W_1|_{D,T}^*$ and the system is proper, then the phase portrait of the Poincaré compactification is topologically equivalent to its correspondent in Figure* 5.17.

(f) *If $(t,d) \in W_1|_{D,T}^*$ and the system is not proper, then the phase portrait of the Poincaré compactification is topologically equivalent to its correspondent in Figure* 5.17.

(g) *If $t \neq 0$ and $d \leq 0$, then the phase portrait of the Poincaré compactification is topologically equivalent to its correspondent in Figure* 5.17.

(h) *If either $(t,d) \in \mathcal{C}_2^2 \setminus W_2|_{D,T}^*$, or $(t,d) \in W_2|_{D,T}^*$ and the system is proper, then the phase portrait of the Poincaré compactification is topologically equivalent to its correspondent in Figure* 5.17.

(i) *If $(t,d) \in W_2|_{D,T}^*$ and the system is not proper, then the phase portrait of the Poincaré compactification is topologically equivalent to its correspondent in Figure* 5.17.

(j) *If either $(t,d) \in \mathcal{SN}_\infty \setminus \mathcal{VB}_2|_{D,T}$ and $t < 0$, or $(t,d) \in \mathcal{VB}_2|_{D,T}$ and the real Jordan normal form of the matrix A is non-diagonal, then the phase portrait of the Poincaré compactification is topologically equivalent to its correspondent in Figure* 5.17.

(k) *If $(t,d) \in \mathcal{VB}_2|_{D,T}$ and the real Jordan normal form of the matrix A is diagonal, then the phase portrait of the Poincaré compactification is topologically equivalent to its correspondent in Figure* 5.17.

(l) *If* $(t,d) \in \mathcal{C}_2^{13}\big|_{D,T}$*, then the phase portrait of the Poincaré compactification is topologically equivalent to its correspondent in Figure* 5.20.

(m) *If* $(t,d) \in \mathcal{NH}_{lc}|_{D,T}$*, then the phase portrait of the Poincaré compactification is topologically equivalent to its correspondent in Figure* 5.20.

(n) *If* $(t,d) \in \mathcal{C}_2^{12}\big|_{D,T}$*, then the phase portrait of the Poincaré compactification is topologically equivalent to its correspondent in Figure* 5.20.

(o) *If* $(t,d) \in \mathcal{H}_o\mathcal{L}|_{D,T}$*, then the phase portrait of the Poincaré compactification is topologically equivalent to its correspondent in Figure* 5.20.

(p) *If* $(t,d) \in \mathcal{C}_2^{11}\big|_{D,T}$*, then the phase portrait of the Poincaré compactification is topologically equivalent to its correspondent in Figure* 5.20.

Proof. All the statements follow by applying a time-reversal and by employing the corresponding results of Subsection 5.5.5. □

5.7.1 The bifurcation set

For a description of the bifurcations we refer the reader to Subsection 5.5.6.

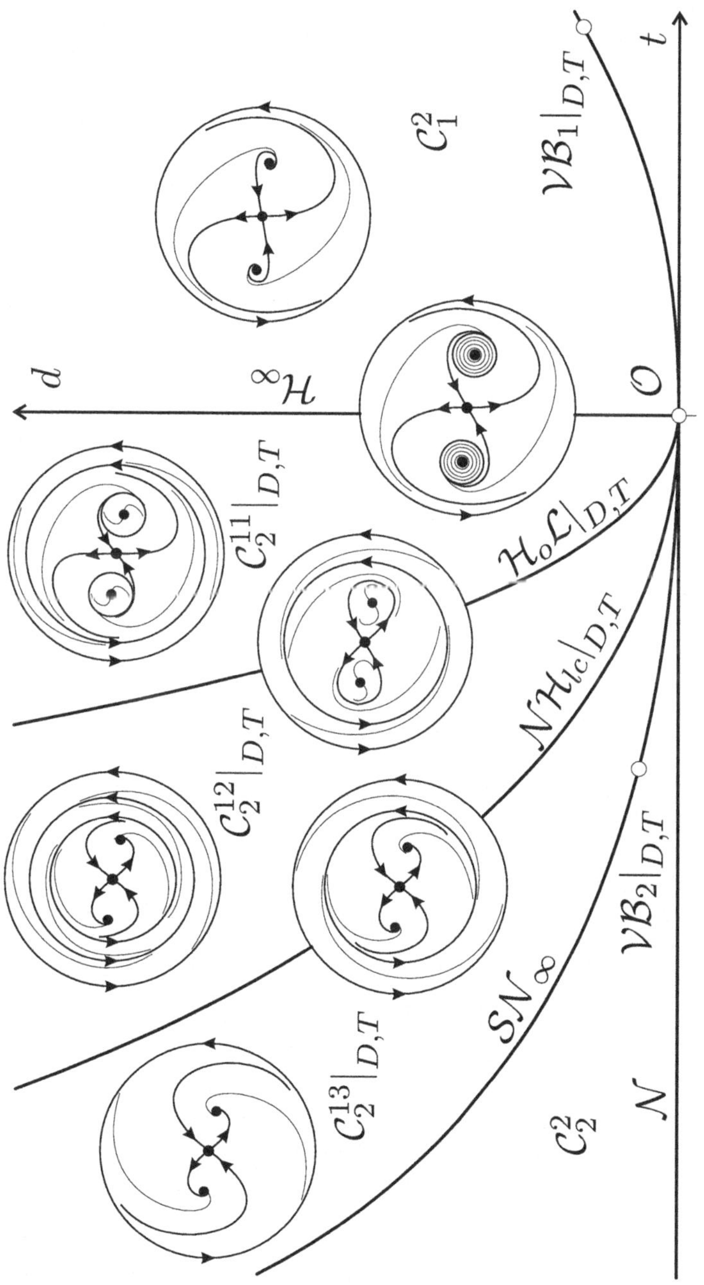

Figure 5.20: Phase portraits and the bifurcation set of the Poincaré compactification of the fundamental systems with parameters $D < 0$, $T > 0$, and $t^2 - 4d < 0$.

Bibliography

[1] J. Alvarez, R. Suárez and J. Alvarez, *Planar linear systems with single saturated feedback.* Systems Control Lett. **20** (1993), 319–326.

[2] A.F. Andreev, *Investigation of the behaviour of the integral curves of a system of two differential equations in the neighborhood of a singular point.* Trans. Amer. Math. Soc. **8** (1958), 183–207.

[3] A.A. Andronov, A.A. Vitt and S.E. Khaikin *Theory of Oscillators.* Dover, New York, 1987.

[4] A.A. Andronov, E.A. Leontovitch, I.I. Gordon and A.G. Maier, *Qualitative Theory of Second-Order Dynamic Systems.* John Wiley & Sons, Ltd. New York, 1973.

[5] A.A. Andronov, E.A. Leontovitch, I.I. Gordon and A.G. Maier, *Theory of Bifurcation of Dynamic Systems on a Plane.* John Wiley & Sons, Ltd. New York, 1973.

[6] T.M. Apostol, *Mathematical Analysis.* Addison-Wesley Publishing Co., Reading-MA, London-Don Mills, 1974.

[7] V.I. Arnold, *Ordinary Differential Equations.* MIT Press, Cambridge, MA, 1973.

[8] V.I. Arnold and Y.S. Ilyashenko, *Dynamical Systems I. Ordinary differential equations.* Encyclopedia of Math. Sciences. Springer-Verlag, Heidelberg, 1988.

[9] D.P. Atherton, *Nonlinear Control Engineering.* Van Nostrand Reinhold Co. Ltd., New York, 1975.

[10] V. Carmona, F. Fernández-Sanchez and A.E. Teruel, *Existence of a reversible T-point heteroclinic cycle in a piecewise linear version of the Michelson system.* SIAM J. Appl. Dyn. Syst., **7** (2008). 1032–1048.

[11] V. Carmona, F. Fernández-Sanchez, E. García-Medina and A.E. Teruel, *Existence of homoclinic connections in continuous piecewise linear systems.* Chaos. **20** (2010).

[12] V. Carmona, F. Fernández-Sanchez, E. García-Medina and A.E. Teruel, *Reversible periodic orbits in a class of 3D continuous piecewise linear systems of differential equations.* Nonlinear Anal. **75** (2012), 5866–5883.

[13] V. Carmona, E. Freire, E. Ponce and F. Torres, *On simplifying and classifiying piecewise-linear systems,* IEEE Trans. Circuits Syst. **49** (2002), 609–620.

[14] C. Chicone, *Ordinary Differential Equations with Applications,* Springer-Verlag, New York, 1999.

[15] L.O. Chua and A.C. Deng, *Canonical piecewise-linear representation.* IEEE Trans. Circuits and Systems **35** (1988), 101–111.

[16] L.O. Chua, C.A. Desoer and E.S. Kuh, *Linear and Nonlinear Circuits.* McGraw-Hill, Singapore, 1987.

[17] S.N. Chow and J.K. Hale, *Methods of Bifurcation Theory.* Springer Verlag, New York, 1982.

[18] E.A. Coddington and N. Levinson, *Theory of Ordinary Differential Equations.* McGraw-Hill, New York, 1955.

[19] M. di Bernardo, C.J. Budd, A.R. Champneys and P. Kowalczyk, *Piecewise-smooth Dynamical Systems: Theory and Applications.* Springer-Verlag, London, 2008.

[20] F. Dumortier, *Singularities of vector fields on the plane.* J. Differential Equations **23** (1977), 53–106.

[21] F. Dumortier, J. Llibre and J.C. Artés, *Qualitative Theory of Planar Differential Systems.* Universitext, Springer, 2006.

[22] A.F. Filippov, *Differential Equations with Discontinuous Righthand Side.* Kluwer Academic Publishers, Dordrecht, 1988.

[23] E. Freire, E. Ponce and J. Ros, *Limit cycle bifurcation from center in symmetric piecewise-linear systems.* Int. J. Bifur. Chaos **9** (1999), 895–907.

[24] E. Freire, E. Ponce and F. Torres, *Hopf-like bifurcations in planar piecewise linear systems.* Publ. Mat. **41** (1997), 135–148.

[25] E. Freire, E. Ponce and F. Torres, *Hopf bifurcations in piecewise linear planar dynamical systems.* Proceedings of Nonlinear Dynamics of Electronic Systems NDES-96 (1996), 129–134.

[26] E. Freire, E. Ponce, F. Rodrigo and F. Torres, *Bifurcation sets of continuous piecewise linear systems with two zones.* Internat. J. Bifur. Chaos **8** (1998), 2073–2097.

[27] F. Giannakopoulos and K. Pliete, *Planar systems of piecewise linear differential equations with a line of discontinuity.* Nonlinearity **14** (2001), 1611–1632.

[28] A. Gray, *Modern Differential Geometry of Curves and Surfaces with Mathematica.* 2nd ed., Boca Raton, FL. CRC Press, 1998.

[29] J. Guckenheimer and P.J. Holmes, *Nonlinear Oscillations, Dynamical Systems, and Bifurcations of Vector Fields.* Springer-Verlag, New York, 1983.

[30] P. Hartman, *Ordinary Differential Equations*, John Wiley & Sons, New York, 1964.

[31] J. Hale and H. Koçak, *Dynamics and Bifurcations*, Springer-Verlag, New York, 1991.

[32] M.A. Henson and D.E. Seborg, *Nonlinear Process Control*, Prentice–Hall, New Jersey, 1997.

[33] M.W. Hirsch and S. Smale, *Differential Equations, Dynamical Systems and Linear Algebra*, Academic Press, New York, 1974.

[34] M.W. Hirsch, C.C. Pugh and M. Shub, *Invariant Manifolds*, Lectures Notes in Math. Springer-Verlag, New York, 1977.

[35] A. Isidori, *Nonlinear Control Systems*, Springer-Verlag, London, 1995.

[36] O. Katsuhiko, *Modern Control Engineering, 5th ed.*, Prentice–Hall, USA, 2009.

[37] M. Komuro, *Normal forms of continuous piecewise linear vector fields and chaotic attractor, Part II: Chaotic attractors.* Japan J. Appl. Math. **5** (1998), 503–549.

[38] J. Milnor, *Topology from the Differentiable Viewpoint.* Based on notes by David W. Weaver. The University Press of Virginia, Charlottesville, VA, 1965.

[39] S. Lefschetz, *Stability of Non-linear Control Systems*, Academic Press, New York, 1965.

[40] S. Lefschetz, *Differential Equations: Geometric Theory*, Dover Publications, New York, 1983.

[41] J. Llibre and E. Ponce, *Global first harmonic bifurcation diagram for odd piecewise lineat control systems*, Dynam. Stability of Systems **11** (1996), 49–88.

[42] J. Llibre and E. Ponce, *Bifurcation of a periodic orbit from infinity in planar piecewise linear vector fields*, Nonlinear Anal. **36** (1999), no. 5, Ser. B: Real World Appl., 623–653.

[43] J. Llibre, E. Ponce and A.E. Teruel, *Horseshoes near homoclinic orbits for piecewise linear differential systems in* $\mathbb{R}^3$. Internat. J. Bifur. Chaos Appl. Sci. Engrg., **17**, (2007), 1171–1184.

[44] J. Llibre and J. Sotomayor, *Phase portraits of planar control systems*, Nonlinear Anal., Theory, Methods and Applications **27** (1996), 1177–1197.

[45] J. Llibre and A.E. Teruel, *Existence of Poincaré maps in piecewise linear differential systems in* $\mathbb{R}^n$, Internat. J. Bifur. Chaos Appl. Sci. Engrg. **14**, (2004) 2843–2851.

[46] R. Lum and L.O. Chua, *Global properties of continuous picewise-linear vector fields. Part I: simplest case in* $\mathbf{R}^2$, Memorandum UCB/ERL/M90/22, University of California at Berkeley, 1990.

[47] R. Lum and L.O. Chua, *Global properties of continuous picewise linear vector fields. Part II: simplest symmetric case in* $\mathbf{R}^2$, Memorandum UCB/ERL/M90/29, University of California at Berkeley, 1990.

[48] L. Markus, *Global structure of ordinary differential equations in the plane.* Trans. Amer. Math Soc. **76** (1954), 127–148.

[49] K.S. Narendra and J.M. Taylor, *Frequency Domain Criteria for Absolute Stability,* Academic Press, New York, 1973.

[50] D.A. Neumann, *Classification of continuous flows on 2-manifolds,* Proc. Amer. Math. Soc. **48** (1975), 73–81.

[51] M.M. Peixoto, *On the classification of flows on 2-manifolds,* in: *Dynamical Systems. Proceedings of a Symposium held at the University of Bahia,* 389–420, Academic Press, New York, 1973.

[52] M.M. Peixoto, *Structural stability on two-dimensional manifolds,* Topology **1** (1962), 101–120.

[53] L. Perko, *Differential Equations and Dynamical Systems,* 3rd ed., Springer-Verlag, New York, 2001.

[54] R. Prohens and A.E. Teruel, *Canard trajectories in 3D piecewise linear systems,* Cont. Disc. Dyn. Syst., **33** (2013), 4595–4611.

[55] F. Rodrigo, *Comportamiento dinámico de osciladores electrónicos del tipo Van der Pol–Duffing,* Tesis Doctoral, Universidad de Sevilla, 1997.

[56] C.T. Sparrow, *Chaos in a three-dimensional single loop feedback system with a piecewise linear feedback function,* J. Math. Anal. Appl., **81** (1981), 275–291.

[57] J. Sotomayor, *Liçoes de Ecuaçoes Diferenciais Ordinárias,* Instituto de Matemática Pura e Aplicada, Rio de Janeiro, 1979.

[58] J. Sotomayor, *Curvas Definidas por Equações Diferenciais no Plano,* Instituto de Matemática Pura e Aplicada, Rio de Janeiro, 1981.

[59] Chai Wah Wu and L.O. Chua, *On the generality of the unfolded Chua's circuit,* Intern. J. Bifur. Chaos **5** (1996), 801–832.

[60] S. Wiggins, *Global Bifurcations and Chaos.* Applied Mathematical Sciences **73**, Springer-Verlag, New York, 1988.

[61] Ye Yan-Qian, *Theory of Limit Cycles,* Translations of Mathematical Monographs **66**, Amer. Math. Soc, Providence, Rhode Island, 1988.

Index

The manufacturer's authorised representative in the EU is Springer Nature Customer Service Centre GmbH, Europaplatz 3, 69115 Heidelberg, Germany. If you have any concerns regarding our products, please contact ProductSafety@springernature.com

Printed and bound by CPI Group (UK) Ltd, Croydon, CR0 4YY
23/03/2026
02076667-0015